CERTIFICATE MA

CERTIFICATE
MATHEMATICS

By

R. W. FOX

Deputy Headmaster,
Fort Luton Secondary Boys' School,
Kent

BOOK III

2nd Edition

DECIMAL AND METRIC

EDWARD ARNOLD

© R. W. FOX 1969
First published 1963
by Edward Arnold (Publishers) Ltd.,
25 Hill Street,
London WIX 8LL

Reprinted 1966
Reprinted 1968
Second edition (Decimal and Metric) 1969
Reprinted 1970, 1972, 1973

Without answers   ISBN: 0 7131 1570 X
With answers   ISBN: 0 7131 1569 6

The full course comprises:

BOOKS I, II, III.

Each issued With and Without Answers.
Book III contains a condensed version of Books I and II with further extension of the work in all branches. The inclusion of a section on Elementary Calculus provides a complete course for either Syllabus 'A' or 'B', or variations of these Syllabuses.

Printed in Great Britain by Fletcher & Son Ltd, Norwich

PREFACE

BOOK III of this series comprises a much abbreviated form of Books I and II together with the necessary additional material to complete an 'O' Level course. The Topic sections employed in Books I and II have been replaced by an introductory course in Calculus, and this, together with the extended work in other sections, allows an effective cover of both types of syllabus (A. and B.).

The content of the various Examination Syllabuses imposes certain difficulties in preparing material to satisfy all needs. It must be left to the individual teacher, therefore, to decide the extent to which he will follow a particular topic, and in consequence some topics may actually be omitted from the course. However, such material (e.g. the sections on Calculus, A.P.s, G.P.s) might well provide a sustaining stimulus, in the post-examination period following 'O' Level, to those considering Additional Mathematics or 'A' Level.

In Book III the work has no longer been arranged in terms, for this would seem to have little value in the fifth year, when the nature of the work, though largely revision, will depend upon the requirements of the students. Nevertheless, the work continues to be arranged under separate subject headings to comply with the author's belief that average pupils prefer it so. The author feels strongly that neither pupil nor teacher should feel a sense of shame in preferring to master the techniques of mathematics as a number of separate skills, to be related as and when the individual becomes capable of so doing, if such a learning process achieves ultimate success and with it a deeper sense of satisfaction.

The author gratefully acknowledges the encouragement of Mr. L. P. F. Miller, Headmaster of Vincent County Secondary School, and to his colleague Miss Xenia V. Carley he offers his sincere thanks for the valuable assistance given in the preparation of this volume.

R. W. Fox

HARMONOGRAPHS

In the field of visual art, it is not uncommon to hear references to 'form' and 'composition' and, similarly, in the field of music, comparable descriptions are made.

From time to time one may hear mathematicians refer to the 'beauty' of a particular symbolic expression or mathematical proof. Such a choice of words may sound strange when used in connection with this subject, but there is a beauty of form to be found and recognized if the interest is deep enough.

Although the inter-relationship of mathematics and the generally accepted art-forms is not always obvious, some have found it an absorbing subject. (The lengths of piano strings certainly obey a mathematical law.)

The harmonograph illustrations separating the various sections of Book III were produced by a machine of the author's design. Surely, they reveal a certain beauty? Yet they depend for their form on the lengths of two interacting pendulums and their relative movement, one to the other. (Harmonic motion.) Many other diagrams have been produced, too delicate and intricate to withstand the rigours of the printing press. Furthermore, since the development of electronic calculating devices, similar figures have been produced by means of analogue computers.

Perhaps the true artist might criticize such computer art as 'too photographic', but the mathematician–artist finds it exciting and satisfying.

> "*As for the difficulty of these Sciences, it must be granted, that the first Aspect of them may seem uncouth and difficult, yet there is no reason why any should be deterr'd hereby, but rather animated with desire to go on as far as others, or (at least) to arrive to a competent Proficiency in them; in order to which I hope this Cursus will, as Ariadne's Thread, lead them through the intricate Labyrinth of these Studies; which that it may do, is the earnest desire of him who is a Lover of the Mathematicks.*"

W. LEYBOURN
Cursus Mathematicus (*1690*)

PREFACE TO SECOND EDITION

DECIMALISATION AND METRICATION

THIS edition of Certificate Mathematics is being published at a time when the United Kingdom is about to change to a decimal currency, and to adopt a new system of weights and measures—the 'International System of Units' (SI).

In the first edition the ratio of the use of imperial units to metric was about four to one, and this ratio has been completely reversed for the new edition. The small number of imperial questions has been retained, as the author feels that there will be many occasions during the changeover period when pupils will need an understanding of the old system.

The revisions to the text in this Second Edition can be summarized as follows:

(i) The Four Rules of Decimals are now dealt with very early in the arithmetic course to facilitate an understanding of the new currency.

(ii) All money questions are now in terms of decimal coinage.

(iii) Most of the 'Weights and Measures' questions have been rationalized either in terms of one unit only, thus avoiding a need for knowledge of tables (e.g. 12 in = 1 ft), or questions have been rewritten in terms of metric units. Additional material has been provided on Metric Weights and Measures.

R.W.F.

ACKNOWLEDGEMENTS

The author wishes to acknowledge the courtesy of the following Examining Bodies in granting permission to use questions set in their examinations. The answers to such questions are the author's.

	Key
The Associated Examining Board for the General Certificate of Education	(A.B.)
University of Cambridge Local Examinations Syndicate	(C.)
University of Durham School Examinations Board	(D.)
Joint Matriculation Board	(J.)
London University Entrance and School Examinations Council	(L.)
Oxford Local Examinations	(O.)
Oxford and Cambridge Schools Examination Board	(O.C.)
Southern Universities' Joint Board for School Examinations	(S.U.)
Welsh Joint Education Committee	(W.)
Scottish Education Department*	(S.)
University of Adelaide	(A.)
University of Melbourne	(M.)
Department of Education, New South Wales†	(N.S.W.)
East Midland Educational Union (Secondary School Certificate)	(E.M.)
Northern Counties Technical Examinations Council (School Certificate)	(N.C.)
Union of Educational Institutions (Secondary School Certificate)	(E.I.)
Union of Lancashire and Cheshire Institutes (Secondary School Certificate)	(U.L.C.I.)
Royal Society of Arts (School Certificate)	(R.S.A.)
The London Chamber of Commerce (Intermediate Stage)	(L.C.C.)
The College of Preceptors (Certificate Examination)	(C.P.)

At convenient stages in the work suitable examination questions have been selected and grouped together, their source being indicated by the key letters at the end of each group.

* *By permission of The Controller, H.M.S.O.*
† *The copyright of these papers is vested in the Crown.*

CONTENTS

PART A: ALGEBRA   1

PART D: TRIGONOMETRY 297

PART E: CALCULUS 333

TEST PAPERS 365

TABLES

Note: The main Imperial units have still been shown for reference purposes.

LENGTH

METRIC

10 millimetres (mm)	= 1 centimetre (cm)
10 centimetres	= 1 decimetre (dm)
10 decimetres	= 1 metre (m)
10 metres	= 1 decametre (dam)
10 decametres	= 1 hectometre (hm)
10 hectometres	= 1 kilometre (km)

BRITISH

12 inches (in)	= 1 foot (ft)
3 feet (ft)	= 1 yard (yd)
22 yards (yd)	= 1 chain (ch)
10 chains	= 1 furlong (fur)
8 furlongs	= 1 mile
220 yards	= 1 furlong
1760 yards	= 1 mile
6 feet	= 1 fathom
6080 feet	= 1 nautical mile

APPROXIMATE EQUIVALENTS

1 inch	= 2·54 cm
1 yard	= 0·9144 m
1 mile	= 1·609 km
1 centimetre	= 0·394 in
1 metre	= 39·37 in
1 kilometre	= 0·6214 ($\frac{5}{8}$) miles

AREA

100 (= 10^2) mm^2	= 1 cm^2
100 (= 10^2) cm^2	= 1 dm^2
etc.	
1 are (a)	= 100 (=10^2) m^2
1 hectare (ha)	= 10 000 (=100^2) m^2

640 acres	= 1 square mile
4840 yd^2	= 1 acre

APPROXIMATE EQUIVALENTS

1 in^2	= 6·45 cm^2
1 cm^2	= 0·155 in^2
1 km^2	= 247·105 acres

VOLUME

$1000\ (=10^3)\ \text{mm}^3 = 1\ \text{cm}^3$
$1000\ (=10^3)\ \text{cm}^3\ = 1\ \text{dm}^3$
etc.

$1728\ (=12^3)\ \text{in}^3 = 1\ \text{ft}^3$
$27\ (=3^3)\ \text{ft}^3\ = 1\ \text{yd}^3$

APPROXIMATE EQUIVALENTS
$1\ \text{in}^3\ = 16 \cdot 39\ \text{cm}^3$
$1\ \text{cm}^3 = 0 \cdot 061\ \text{in}^3$

CAPACITY

METRIC	BRITISH
10 centilitres (cl) = 1 decilitre (dl)	2 pints = 1 quart (qt)
10 decilitres = 1 litre (l)	4 quarts = 1 gallon (gal)
10 litres = 1 decalitre (dal)	
10 decalitres = 1 hectolitre (hl)	
10 hectolitres = 1 kilolitre (kl)	
1 litre = 1 dm³	
= 1000 cm³	

APPROXIMATE EQUIVALENTS
1 gallon = $4 \cdot 546$ litres
= $4 \cdot 546$ dm³
1 litre = $1 \cdot 76$ ($1\frac{3}{4}$) pints
= $61 \cdot 03$ in³

WEIGHT

10 milligrammes (mg) = 1 centigramme (cg)	16 ounces (oz) = 1 pound (lb)	
10 centigrammes = 1 decigramme (dg)	28 pounds = 1 quarter (qr)	
10 decigrammes = 1 gramme (g)	20 hundredweights = 1 ton	
10 grammes = 1 decagramme (dag)	14 pounds = 1 stone	
10 decagrammes = 1 hectogramme (hg)	112 pounds = 1 hundred-weight	
10 hectogrammes = 1 kilogramme (kg)		
1000 kg = 1 tonne (Mg) or metric ton (t)	2240 pounds = 1 ton	

1 cm³ of water (at 4° C.) weighs 1 g
1 litre of water weighs 1 kg

APPROXIMATE EQUIVALENTS
1 pound = $0 \cdot 4536$ kg
1 ton = $1 \cdot 016$ Mg
1 kilogramme = $2 \cdot 205$ ($2\frac{1}{5}$) lb
1 metric ton (tonne) = $0 \cdot 9842$ tons

MONEY

France ⎫ Belgium ⎬ Switzerland ⎭	100 centimes (c.) = 1 franc (fr.)	
U.S.A. ⎫ Canada ⎬	100 cents	= 1 dollar ($)
Germany	100 pfennig	= 1 mark (D.M.)
India	100 Naye Paise	= 1 rupee
Italy	100 centesimi	= 1 lira
U.S.S.R.	100 copeks	= 1 rouble
Spain	100 centimos	= 1 peseta
Malaya ⎫ Hong Kong ⎬	100 cents	= 1 dollar
Pakistan	100 Paisa	= 1 rupee
Argentina	100 centavos	= 1 peso
Brazil	100 centavos	= 1 cruzeiro
South Africa	100 cents	= 1 rand

NOTE:—In East Africa it is customary to express money values in shillings and cents, e.g., 127s., 50c. = £6·375.

MISCELLANEOUS DATA

20 fluid ounces = 1 pint   1 tablespoon = $\frac{1}{2}$ fl oz
1 dessertspoon = $\frac{1}{4}$ fl oz   1 teaspoon = $\frac{1}{8}$ fl oz

Freezing point 32° F. = 0° C   Boiling point 212° F. = 100° C
Cold weather 41° F. = 5° C   Hot weather 86° F. = 30° C
Warm room 68° F. = 20° C   Blood heat 98·4° F. ≏ 36·8° C

Full-size football pitch = 130 yd × 100 yd ≏ 2·7 acre
Full-size rugby pitch = 110 yd × 75 yd ≏ 1·7 acre

60 mile/h = 88 ft/s 1 knot = 1 sea mile/h ≏ 1·7 ft/s
Acceleration due to gravity = 32 ft/s² ≏ 981 cm/s²
Normal atmospheric pressure ≏ 29·9 in or 760 mm of mercury.
≏ 14·7 lb/in² ≏ 1·03 kg/cm²

Velocity of sound in air ≏ 760 mile/h ≏ 1100 ft/s
Velocity of light in vacuum ≏ 186 000 mile/s

Earth's mean radius ≏ 3960 miles ≏ 6380 km
Distance from Sun ≏ 93 × 10⁶ miles ≏ 150 × 10⁶ km
Distance from Moon ≏ 238 000 miles ≏ 381 000 km
Speed of Earth in orbit round Sun ≏ 66 600 mile/h
Speed of Earth spinning on axis ≏ 1038 mile/h

FOUR FIGURE TABLES

SQUARES

	0	1	2	3	4	5	6	7	8	9	1 2 3 4	5 6 7 8 9
10	1000	1020	1040	1061	1082	1103	1124	1145	1166	1188	2 4 6 8	10 13 15 17 19
11	1210	1232	1254	1277	1300	1323	1346	1369	1392	1416	2 5 7 9	11 14 16 18 21
12	1440	1464	1488	1513	1538	1563	1588	1613	1638	1664	2 5 7 10	12 15 17 20 22
13	1690	1716	1742	1769	1796	1823	1850	1877	1904	1932	3 5 8 11	13 16 19 22 24
14	1960	1988	2016	2045	2074	2103	2132	2161	2190	2220	3 6 9 12	14 17 20 23 26
15	2250	2280	2310	2341	2372	2403	2434	2465	2496	2528	3 6 9 12	15 19 22 25 28
16	2560	2592	2624	2657	2690	2723	2756	2789	2822	2856	3 7 10 13	16 20 23 26 30
17	2890	2924	2958	2993	3028	3063	3098	3133	3168	3204	3 7 10 14	17 21 24 28 31
18	3240	3276	3312	3349	3386	3423	3460	3497	3534	3572	4 7 11 15	18 22 26 30 33
19	3610	3648	3686	3725	3764	3803	3842	3881	3920	3960	4 8 12 16	19 23 27 31 35
20	4000	4040	4080	4121	4162	4203	4244	4285	4326	4368	4 8 12 16	20 25 29 33 37
21	4410	4452	4494	4537	4580	4623	4666	4709	4752	4796	4 9 13 17	21 26 30 34 39
22	4840	4884	4928	4973	5018	5063	5108	5153	5198	5244	4 9 13 18	22 27 31 36 40
23	5290	5336	5382	5429	5476	5523	5570	5617	5664	5712	5 9 14 19	23 28 33 38 42
24	5760	5808	5856	5905	5954	6003	6052	6101	6150	6200	5 10 15 20	24 29 34 39 44
25	6250	6300	6350	6401	6452	6503	6554	6605	6656	6708	5 10 15 20	25 31 36 41 46
26	6760	6812	6864	6917	6970	7023	7076	7129	7182	7236	5 11 16 21	26 32 37 42 48
27	7290	7344	7398	7453	7508	7563	7618	7673	7728	7784	5 11 16 22	27 33 38 44 49
28	7840	7896	7952	8009	8066	8123	8180	8237	8294	8352	6 11 17 23	28 34 40 46 51
29	8410	8468	8526	8585	8644	8703	8762	8821	8880	8940	6 12 18 24	30 35 41 47 53
30	9000	9060	9120	9181	9242	9303	9364	9425	9486	9548	6 12 18 24	31 37 43 49 55
31	9610	9672	9734	9797	9860	9923	9986				6 13 19 25	31 38 44 50 56
31							1005	1011	1018		1 1 2 3	3 4 5 5 6
32	1024	1030	1037	1043	1050	1056	1063	1069	1076	1082	1 1 2 3	3 4 5 5 6
33	1089	1096	1102	1109	1116	1122	1129	1136	1142	1149	1 1 2 3	3 4 5 5 6
34	1156	1163	1170	1176	1183	1190	1197	1204	1211	1218	1 1 2 3	3 4 5 6 6
35	1225	1232	1239	1246	1253	1260	1267	1274	1282	1289	1 1 2 3	4 4 5 6 6
36	1296	1303	1310	1318	1325	1332	1340	1347	1354	1362	1 1 2 3	4 4 5 6 7
37	1369	1376	1384	1391	1399	1406	1414	1421	1429	1436	1 2 2 3	4 5 5 6 7
38	1444	1452	1459	1467	1475	1482	1490	1498	1505	1513	1 2 2 3	4 5 5 6 7
39	1521	1529	1537	1544	1552	1560	1568	1576	1584	1592	1 2 2 3	4 5 6 6 7
40	1600	1608	1616	1624	1632	1640	1648	1656	1665	1673	1 2 2 3	4 5 6 6 7
41	1681	1689	1697	1706	1714	1722	1731	1739	1747	1756	1 2 2 3	4 5 6 7 7
42	1764	1772	1781	1789	1798	1806	1815	1823	1832	1840	1 2 3 3	4 5 6 7 8
43	1849	1858	1866	1875	1884	1892	1901	1910	1918	1927	1 2 3 3	4 5 6 7 8
44	1936	1945	1954	1962	1971	1980	1989	1998	2007	2016	1 2 3 4	4 5 6 7 8
45	2025	2034	2043	2052	2061	2070	2079	2088	2098	2107	1 2 3 4	5 5 6 7 8
46	2116	2125	2134	2144	2153	2162	2172	2181	2190	2200	1 2 3 4	5 6 7 7 8
47	2209	2218	2228	2237	2247	2256	2266	2275	2285	2294	1 2 3 4	5 6 7 8 9
48	2304	2314	2323	2333	2343	2352	2362	2372	2381	2391	1 2 3 4	5 6 7 8 9
49	2401	2411	2421	2430	2440	2450	2460	2470	2480	2490	1 2 3 4	5 6 7 8 9
50	2500	2510	2520	2530	2540	2550	2560	2570	2581	2591	1 2 3 4	5 6 7 8 9
51	2601	2611	2621	2632	2642	2652	2663	2673	2683	2694	1 2 3 4	5 6 7 8 9
52	2704	2714	2725	2735	2746	2756	2767	2777	2788	2798	1 2 3 4	5 6 7 8 9
53	2809	2820	2830	2841	2852	2862	2873	2884	2894	2905	1 2 3 4	5 6 7 9 10
54	2916	2927	2938	2948	2959	2970	2981	2992	3003	3014	1 2 3 4	5 7 8 9 10

The position of the decimal point should be found by inspection

SQUARES

	0	1	2	3	4	5	6	7	8	9	1	2	3	4	5	6	7	8	9
55	3025	3036	3047	3058	3069	3080	3091	3102	3114	3125	1	2	3	4	6	7	8	9	10
56	3136	3147	3158	3170	3181	3192	3204	3215	3226	3238	1	2	3	5	6	7	8	9	10
57	3249	3260	3272	3283	3295	3306	3318	3329	3341	3352	1	2	3	5	6	7	8	9	10
58	3364	3376	3387	3399	3411	3422	3434	3446	3457	3469	1	2	4	5	6	7	8	9	11
59	3481	3493	3505	3516	3528	3540	3552	3564	3576	3588	1	2	4	5	6	7	8	10	11
60	3600	3612	3624	3636	3648	3660	3672	3684	3697	3709	1	2	4	5	6	7	8	10	11
61	3721	3733	3745	3758	3770	3782	3795	3807	3819	3832	1	2	4	5	6	7	9	10	11
62	3844	3856	3869	3881	3894	3906	3919	3931	3944	3956	1	3	4	5	6	8	9	10	11
63	3969	3982	3994	4007	4020	4032	4045	4058	4070	4083	1	3	4	5	6	8	9	10	11
64	4096	4109	4122	4134	4147	4160	4173	4186	4199	4212	1	3	4	5	6	8	9	10	12
65	4225	4238	4251	4264	4277	4290	4303	4316	4330	4343	1	3	4	5	7	8	9	10	12
66	4356	4369	4382	4396	4409	4422	4436	4449	4462	4476	1	3	4	5	7	8	9	11	12
67	4489	4502	4516	4529	4543	4556	4570	4583	4597	4610	1	3	4	5	7	8	9	11	12
68	4624	4638	4651	4665	4679	4692	4706	4720	4733	4747	1	3	4	5	7	8	10	11	12
69	4761	4775	4789	4802	4816	4830	4844	4858	4872	4886	1	3	4	6	7	8	10	11	13
70	4900	4914	4928	4942	4956	4970	4984	4998	5013	5027	1	3	4	6	7	8	10	11	13
71	5041	5055	5069	5084	5098	5112	5127	5141	5155	5170	1	3	4	6	7	9	10	11	13
72	5184	5198	5213	5227	5242	5256	5271	5285	5300	5314	1	3	4	6	7	9	10	12	13
73	5329	5344	5358	5373	5388	5402	5417	5432	5446	5461	1	3	4	6	7	9	10	12	13
74	5476	5491	5506	5520	5535	5550	5565	5580	5595	5610	1	3	4	6	7	9	10	12	13
75	5625	5640	5655	5670	5685	5700	5715	5730	5746	5761	2	3	5	6	8	9	11	12	14
76	5776	5791	5806	5822	5837	5852	5868	5883	5898	5914	2	3	5	6	8	9	11	12	14
77	5929	5944	5960	5975	5991	6006	6022	6037	6053	6068	2	3	5	6	8	9	11	12	14
78	6084	6100	6115	6131	6147	6162	6178	6194	6209	6225	2	3	5	6	8	9	11	13	14
79	6241	6257	6273	6288	6304	6320	6336	6352	6368	6384	2	3	5	6	8	10	11	13	14
80	6400	6416	6432	6448	6464	6480	6496	6512	6529	6545	2	3	5	6	8	10	11	13	14
81	6561	6577	6593	6610	6626	6642	6659	6675	6691	6708	2	3	5	7	8	10	11	13	15
82	6724	6740	6757	6773	6790	6806	6823	6839	6856	6872	2	3	5	7	8	10	12	13	15
83	6889	6906	6922	6939	6956	6972	6989	7006	7022	7039	2	3	5	7	8	10	12	13	15
84	7056	7073	7090	7106	7123	7140	7157	7174	7191	7208	2	3	5	7	8	10	12	14	15
85	7225	7242	7259	7276	7293	7310	7327	7344	7362	7379	2	3	5	7	9	10	12	14	15
86	7396	7413	7430	7448	7465	7482	7500	7517	7534	7552	2	3	5	7	9	10	12	14	16
87	7569	7586	7604	7621	7639	7656	7674	7691	7709	7726	2	4	5	7	9	11	12	14	16
88	7744	7762	7779	7797	7815	7832	7850	7868	7885	7903	2	4	5	7	9	11	12	14	16
89	7921	7939	7957	7974	7992	8010	8028	8046	8064	8082	2	4	5	7	9	11	13	14	16
90	8100	8118	8136	8154	8172	8190	8208	8226	8245	8263	2	4	5	7	9	11	13	14	16
91	8281	8299	8317	8336	8354	8372	8391	8409	8427	8446	2	4	5	7	9	11	13	15	16
92	8464	8482	8501	8519	8538	8556	8575	8593	8612	8630	2	4	6	7	9	11	13	15	17
93	8649	8668	8686	8705	8724	8742	8761	8780	8798	8817	2	4	6	7	9	11	13	15	17
94	8836	8855	8874	8892	8911	8930	8949	8968	8987	9006	2	4	6	8	9	11	13	15	17
95	9025	9044	9063	9082	9101	9120	9139	9158	9178	9197	2	4	6	8	10	11	13	15	17
96	9216	9235	9254	9274	9293	9312	9332	9351	9370	9390	2	4	6	8	10	12	14	15	17
97	9409	9428	9448	9467	9487	9506	9526	9545	9565	9584	2	4	6	8	10	12	14	16	18
98	9604	9624	9643	9663	9683	9702	9722	9742	9761	9781	2	4	6	8	10	12	14	16	18
99	9801	9821	9841	9860	9880	9900	9920	9940	9960	9980	2	4	6	8	10	12	14	16	18

The position of the decimal point should be found by inspection

SQUARE ROOTS

	0	1	2	3	4	5	6	7	8	9	1	2	3	4	5	6	7	8	9
10	1000	1005	1010	1015	1020	1025	1030	1034	1039	1044	0	1	1	2	2	3	3	4	4
	3162	3178	3194	3209	3225	3240	3256	3271	3286	3302	2	3	5	6	8	9	11	12	14
11	1049	1054	1058	1063	1068	1072	1077	1082	1086	1091	0	1	1	2	2	3	3	4	4
	3317	3332	3347	3362	3376	3391	3406	3421	3435	3450	1	3	4	6	7	9	10	12	13
12	1095	1100	1105	1109	1114	1118	1122	1127	1131	1136	0	1	1	2	2	3	3	4	4
	3464	3479	3493	3507	3521	3536	3550	3564	3578	3592	1	3	4	6	7	9	10	11	13
13	1140	1145	1149	1153	1158	1162	1166	1170	1175	1179	0	1	1	2	2	3	3	3	4
	3606	3619	3633	3647	3661	3674	3688	3701	3715	3728	1	3	4	5	7	8	10	11	12
14	1183	1187	1192	1196	1200	1204	1208	1212	1217	1221	0	1	1	2	2	3	3	3	4
	3742	3755	3768	3782	3795	3808	3821	3834	3847	3860	1	3	4	5	7	8	9	10	12
15	1225	1229	1233	1237	1241	1245	1249	1253	1257	1261	0	1	1	2	2	2	3	3	4
	3873	3886	3899	3912	3924	3937	3950	3962	3975	3987	1	3	4	5	6	8	9	10	11
16	1265	1269	1273	1277	1281	1285	1288	1292	1296	1300	0	1	1	2	2	2	3	3	4
	4000	4012	4025	4037	4050	4062	4074	4087	4099	4111	1	2	4	5	6	7	9	10	11
17	1304	1308	1311	1315	1319	1323	1327	1330	1334	1338	0	1	1	2	2	2	3	3	3
	4123	4135	4147	4159	4171	4183	4195	4207	4219	4231	1	2	4	5	6	7	8	10	11
18	1342	1345	1349	1353	1356	1360	1364	1367	1371	1375	0	1	1	1	2	2	3	3	3
	4243	4254	4266	4278	4290	4301	4313	4324	4336	4347	1	2	3	5	6	7	8	9	10
19	1378	1382	1386	1389	1393	1396	1400	1404	1407	1411	0	1	1	1	2	2	3	3	3
	4359	4370	4382	4393	4405	4416	4427	4438	4450	4461	1	2	3	5	6	7	8	9	10
20	1414	1418	1421	1425	1428	1432	1435	1439	1442	1446	0	1	1	1	2	2	2	3	3
	4472	4483	4494	4506	4517	4528	4539	4550	4561	4572	1	2	3	4	6	7	8	9	10
21	1449	1453	1456	1459	1463	1466	1470	1473	1476	1480	0	1	1	1	2	2	2	3	3
	4583	4593	4604	4615	4626	4637	4648	4658	4669	4680	1	2	3	4	5	6	7	9	10
22	1483	1487	1490	1493	1497	1500	1503	1507	1510	1513	0	1	1	1	2	2	2	3	3
	4690	4701	4712	4722	4733	4743	4754	4764	4775	4785	1	2	3	4	5	6	7	8	10
23	1517	1520	1523	1526	1530	1533	1536	1539	1543	1546	0	1	1	1	2	2	2	3	3
	4796	4806	4817	4827	4837	4848	4858	4868	4879	4889	1	2	3	4	5	6	7	8	9
24	1549	1552	1556	1559	1562	1565	1568	1572	1575	1578	0	1	1	1	2	2	2	3	3
	4899	4909	4919	4930	4940	4950	4960	4970	4980	4990	1	2	3	4	5	6	7	8	9
25	1581	1584	1587	1591	1594	1597	1600	1603	1606	1609	0	1	1	1	2	2	2	2	3
	5000	5010	5020	5030	5040	5050	5060	5070	5079	5089	1	2	3	4	5	6	7	8	9
26	1612	1616	1619	1622	1625	1628	1631	1634	1637	1640	0	1	1	1	2	2	2	2	3
	5099	5109	5119	5128	5138	5148	5158	5167	5177	5187	1	2	3	4	5	6	7	8	9
27	1643	1646	1649	1652	1655	1658	1661	1664	1667	1670	0	1	1	1	1	2	2	2	3
	5196	5206	5215	5225	5235	5244	5254	5263	5273	5282	1	2	3	4	5	6	7	8	9
28	1673	1676	1679	1682	1685	1688	1691	1694	1697	1700	0	1	1	1	1	2	2	2	3
	5292	5301	5310	5320	5329	5339	5348	5357	5367	5376	1	2	3	4	5	6	7	7	8
29	1703	1706	1709	1712	1715	1718	1720	1723	1726	1729	0	1	1	1	1	2	2	2	3
	5385	5394	5404	5413	5422	5431	5441	5450	5459	5468	1	2	3	4	5	6	6	7	8
30	1732	1735	1738	1741	1744	1746	1749	1752	1755	1758	0	1	1	1	1	2	2	2	3
	5477	5486	5495	5505	5514	5523	5532	5541	5550	5559	1	2	3	4	5	5	6	7	8
31	1761	1764	1766	1769	1772	1775	1778	1780	1783	1786	0	1	1	1	1	2	2	2	3
	5568	5577	5586	5595	5604	5612	5621	5630	5639	5648	1	2	3	4	4	5	6	7	8
32	1789	1792	1794	1797	1800	1803	1806	1808	1811	1814	0	1	1	1	1	2	2	2	3
	5657	5666	5675	5683	5692	5701	5710	5718	5727	5736	1	2	3	4	4	5	6	7	8

The position of the decimal point and the first significant figure should be found by inspection.

SQUARE ROOTS

	0	1	2	3	4	5	6	7	8	9	1	2	3	4	5	6	7	8	9
32	1789	1792	1794	1797	1800	1803	1806	1808	1811	1814	0	1	1	1	1	2	2	2	3
	5657	5666	5675	5683	5692	5701	5710	5718	5727	5736	1	2	3	4	4	5	6	7	8
33	1817	1819	1822	1825	1828	1830	1833	1836	1838	1841	0	1	1	1	1	2	2	2	2
	5745	5753	5762	5771	5779	5788	5797	5805	5814	5822	1	2	3	3	4	5	6	7	8
34	1844	1847	1849	1852	1855	1857	1860	1863	1865	1868	0	1	1	1	1	2	2	2	2
	5831	5840	5848	5857	5865	5874	5882	5891	5899	5908	1	2	3	3	4	5	6	7	8
35	1871	1873	1876	1879	1881	1884	1887	1889	1892	1895	0	1	1	1	1	2	2	2	2
	5916	5925	5933	5941	5950	5958	5967	5975	5983	5992	1	2	3	3	4	5	6	7	8
36	1897	1900	1903	1905	1908	1910	1913	1916	1918	1921	0	1	1	1	1	2	2	2	2
	6000	6008	6017	6025	6033	6042	6050	6058	6066	6075	1	2	2	3	4	5	6	7	7
37	1924	1926	1929	1931	1934	1936	1939	1942	1944	1947	0	1	1	1	1	2	2	2	2
	6083	6091	6099	6107	6116	6124	6132	6140	6148	6156	1	2	2	3	4	5	6	6	7
38	1949	1952	1954	1957	1960	1962	1965	1967	1970	1972	0	1	1	1	1	2	2	2	2
	6164	6173	6181	6189	6197	6205	6213	6221	6229	6237	1	2	2	3	4	5	6	6	7
39	1975	1977	1980	1982	1985	1987	1990	1992	1995	1997	0	1	1	1	1	1	2	2	2
	6245	6253	6261	6269	6277	6285	6293	6301	6309	6317	1	2	2	3	4	5	6	6	7
40	2000	2002	2005	2007	2010	2012	2015	2017	2020	2022	0	0	1	1	1	2	2	2	2
	6325	6332	6340	6348	6356	6364	6372	6380	6387	6395	1	2	2	3	4	5	5	6	7
41	2025	2027	2030	2032	2035	2037	2040	2042	2045	2047	0	0	1	1	1	2	2	2	2
	6403	6411	6419	6427	6434	6442	6450	6458	6465	6473	1	2	2	3	4	5	5	6	7
42	2049	2052	2054	2057	2059	2062	2064	2066	2069	2071	0	0	1	1	1	2	2	2	2
	6481	6488	6496	6504	6512	6519	6527	6535	6542	6550	1	2	2	3	4	5	5	6	7
43	2074	2076	2078	2081	2083	2086	2088	2090	2093	2095	0	0	1	1	1	2	2	2	2
	6557	6565	6573	6580	6588	6595	6603	6611	6618	6626	1	2	2	3	4	5	5	6	7
44	2098	2100	2102	2105	2107	2110	2112	2114	2117	2119	0	0	1	1	1	2	2	2	2
	6633	6641	6648	6656	6663	6671	6678	6686	6693	6701	1	1	2	3	4	4	5	6	7
45	2121	2124	2126	2128	2131	2133	2135	2138	2140	2142	0	0	1	1	1	2	2	2	2
	6708	6716	6723	6731	6738	6745	6753	6760	6768	6775	1	1	2	3	4	4	5	6	7
46	2145	2147	2149	2152	2154	2156	2159	2161	2163	2166	0	0	1	1	1	2	2	2	2
	6782	6790	6797	6804	6812	6819	6826	6834	6841	6848	1	1	2	3	4	4	5	6	7
47	2168	2170	2173	2175	2177	2179	2182	2184	2186	2189	0	0	1	1	1	2	2	2	2
	6856	6863	6870	6877	6885	6892	6899	6907	6914	6921	1	1	2	3	4	4	5	6	6
48	2191	2193	2195	2198	2200	2202	2205	2207	2209	2211	0	0	1	1	1	2	2	2	2
	6928	6935	6943	6950	6957	6964	6971	6979	6986	6993	1	1	2	3	4	4	5	6	6
49	2214	2216	2218	2220	2223	2225	2227	2229	2232	2234	0	0	1	1	1	2	2	2	2
	7000	7007	7014	7021	7029	7036	7043	7050	7057	7064	1	1	2	3	4	4	5	6	6
50	2236	2238	2241	2243	2245	2247	2249	2252	2254	2256	0	0	1	1	1	2	2	2	2
	7071	7078	7085	7092	7099	7106	7113	7120	7127	7134	1	1	2	3	3	4	5	6	6
51	2258	2261	2263	2265	2267	2269	2272	2274	2276	2278	0	0	1	1	1	2	2	2	2
	7141	7148	7155	7162	7169	7176	7183	7190	7197	7204	1	1	2	3	3	4	5	6	6
52	2280	2283	2285	2287	2289	2291	2293	2296	2298	2300	0	0	1	1	1	2	2	2	2
	7211	7218	7225	7232	7239	7246	7253	7259	7266	7273	1	1	2	3	3	4	5	6	6
53	2302	2304	2307	2309	2311	2313	2315	2317	2319	2322	0	0	1	1	1	2	2	2	2
	7280	7287	7294	7301	7308	7314	7321	7328	7335	7342	1	1	2	3	3	4	5	5	6
54	2324	2326	2328	2330	2332	2335	2337	2339	2341	2343	0	0	1	1	1	1	1	2	2
	7348	7355	7362	7369	7376	7382	7389	7396	7403	7409	1	1	2	3	3	4	5	5	6

The position of the decimal point and the first significant figure should be found by inspection.

	0	1	2	3	4	5	6	7	8	9	1 2 3 4	5	6 7 8 9
55	2345	2347	2349	2352	2354	2356	2358	2360	2362	2364	0 0 1 1	1	1 1 2 2
	7416	7423	7430	7436	7443	7450	7457	7463	7470	7477	1 1 2 3	3	4 5 5 6
56	2366	2369	2371	2373	2375	2377	2379	2381	2383	2385	0 0 1 1	1	1 1 2 2
	7483	7490	7497	7503	7510	7517	7523	7530	7537	7543	1 1 2 3	3	4 5 5 6
57	2387	2390	2392	2394	2396	2398	2400	2402	2404	2406	0 0 1 1	1	1 1 2 2
	7550	7556	7563	7570	7576	7583	7589	7596	7603	7609	1 1 2 3	3	4 5 5 6
58	2408	2410	2412	2415	2417	2419	2421	2423	2425	2427	0 0 1 1	1	1 1 2 2
	7616	7622	7629	7635	7642	7649	7655	7662	7668	7675	1 1 2 3	3	4 5 5 6
59	2429	2431	2433	2435	2437	2439	2441	2443	2445	2447	0 0 1 1	1	1 1 2 2
	7681	7688	7694	7701	7707	7714	7720	7727	7733	7740	1 1 2 3	3	4 5 5 6
60	2449	2452	2454	2456	2458	2460	2462	2464	2466	2468	0 0 1 1	1	1 1 2 2
	7746	7752	7759	7765	7772	7778	7785	7791	7797	7804	1 1 2 3	3	4 4 5 6
61	2470	2472	2474	2476	2478	2480	2482	2484	2486	2488	0 0 1 1	1	1 1 2 2
	7810	7817	7823	7829	7836	7842	7849	7855	7861	7868	1 1 2 3	3	4 4 5 6
62	2490	2492	2494	2496	2498	2500	2502	2504	2506	2508	0 0 1 1	1	1 1 2 2
	7874	7880	7887	7893	7899	7906	7912	7918	7925	7931	1 1 2 3	3	4 4 5 6
63	2510	2512	2514	2516	2518	2520	2522	2524	2526	2528	0 0 1 1	1	1 1 2 2
	7937	7944	7950	7956	7962	7969	7975	7981	7987	7994	1 1 2 3	3	4 4 5 6
64	2530	2532	2534	2536	2538	2540	2542	2544	2546	2548	0 0 1 1	1	1 1 2 2
	8000	8006	8012	8019	8025	8031	8037	8044	8050	8056	1 1 2 2	3	4 4 5 6
65	2550	2551	2553	2555	2557	2559	2561	2563	2565	2567	0 0 1 1	1	1 1 2 2
	8062	8068	8075	8081	8087	8093	8099	8106	8112	8118	1 1 2 2	3	4 4 5 6
66	2569	2571	2573	2575	2577	2579	2581	2583	2585	2587	0 0 1 1	1	1 1 2 2
	8124	8130	8136	8142	8149	8155	8161	8167	8173	8179	1 1 2 2	3	4 4 5 5
67	2588	2590	2592	2594	2596	2598	2600	2602	2604	2606	0 0 1 1	1	1 1 2 2
	8185	8191	8198	8204	8210	8216	8222	8228	8234	8240	1 1 2 2	3	4 4 5 5
68	2608	2610	2612	2613	2615	2617	2619	2621	2623	2625	0 0 1 1	1	1 1 2 2
	8246	8252	8258	8264	8270	8276	8283	8289	8295	8301	1 1 2 2	3	4 4 5 5
69	2627	2629	2631	2632	2634	2636	2638	2640	2642	2644	0 0 1 1	1	1 1 2 2
	8307	8313	8319	8325	8331	8337	8343	8349	8355	8361	1 1 2 2	3	4 4 5 5
70	2646	2648	2650	2651	2653	2655	2657	2659	2661	2663	0 0 1 1	1	1 1 2 2
	8367	8373	8379	8385	8390	8396	8402	8408	8414	8420	1 1 2 2	3	4 4 5 5
71	2665	2666	2668	2670	2672	2674	2676	2678	2680	2681	0 0 1 1	1	1 1 1 2
	8426	8432	8438	8444	8450	8456	8462	8468	8473	8479	1 1 2 2	3	4 4 5 5
72	2683	2685	2687	2689	2691	2693	2694	2696	2698	2700	0 0 1 1	1	1 1 1 2
	8485	8491	8497	8503	8509	8515	8521	8526	8532	8538	1 1 2 2	3	3 4 5 5
73	2702	2704	2706	2707	2709	2711	2713	2715	2717	2718	0 0 1 1	1	1 1 1 2
	8544	8550	8556	8562	8567	8573	8579	8585	8591	8597	1 1 2 2	3	3 4 5 5
74	2720	2722	2724	2726	2728	2729	2731	2733	2735	2737	0 0 1 1	1	1 1 1 2
	8602	8608	8614	8620	8626	8631	8637	8643	8649	8654	1 1 2 2	3	3 4 5 5
75	2739	2740	2742	2744	2746	2748	2750	2751	2753	2755	0 0 1 1	1	1 1 1 2
	8660	8666	8672	8678	8683	8689	8695	8701	8706	8712	1 1 2 2	3	3 4 5 5
76	2757	2759	2760	2762	2764	2766	2768	2769	2771	2773	0 0 1 1	1	1 1 1 2
	8718	8724	8729	8735	8741	8746	8752	8758	8764	8769	1 1 2 2	3	3 4 5 5
77	2775	2777	2778	2780	2782	2784	2786	2787	2789	2791	0 0 1 1	1	1 1 1 2
	8775	8781	8786	8792	8798	8803	8809	8815	8820	8826	1 1 2 2	3	3 4 5 5

The position of the decimal point and the first significant figure should be found by inspection.

	0	1	2	3	4	5	6	7	8	9	1 2 3 4	5	6 7 8 9
77	2775	2777	2778	2780	2782	2784	2786	2787	2789	2791	0 0 1 1	1	1 1 1 2
	8775	8781	8786	8792	8798	8803	8809	8815	8820	8826	1 1 2 2	3	3 4 5 5
78	2793	2795	2796	2798	2800	2802	2804	2805	2807	2809	0 0 1 1	1	1 1 1 2
	8832	8837	8843	8849	8854	8860	8866	8871	8877	8883	1 1 2 2	3	3 4 4 5
79	2811	2812	2814	2816	2818	2820	2821	2823	2825	2827	0 0 1 1	1	1 1 1 2
	8888	8894	8899	8905	8911	8916	8922	8927	8933	8939	1 1 2 2	3	3 4 4 5
80	2828	2830	2832	2834	2835	2837	2839	2841	2843	2844	0 0 1 1	1	1 1 1 2
	8944	8950	8955	8961	8967	8972	8978	8983	8989	8994	1 1 2 2	3	3 4 4 5
81	2846	2848	2850	2851	2853	2855	2857	2858	2860	2862	0 0 1 1	1	1 1 1 2
	9000	9006	9011	9017	9022	9028	9033	9039	9044	9050	1 1 2 2	3	3 4 4 5
82	2864	2865	2867	2869	2871	2872	2874	2876	2877	2879	0 0 1 1	1	1 1 1 2
	9055	9061	9066	9072	9077	9083	9088	9094	9099	9105	1 1 2 2	3	3 4 4 5
83	2881	2883	2884	2886	2888	2890	2891	2893	2895	2897	0 0 1 1	1	1 1 1 2
	9110	9116	9121	9127	9132	9138	9143	9149	9154	9160	1 1 2 2	3	3 4 4 5
84	2898	2900	2902	2903	2905	2907	2909	2910	2912	2914	0 0 1 1	1	1 1 1 2
	9165	9171	9176	9182	9187	9192	9198	9203	9209	9214	1 1 2 2	3	3 4 4 5
85	2915	2917	2919	2921	2922	2924	2926	2927	2929	2931	0 0 1 1	1	1 1 1 2
	9220	9225	9230	9236	9241	9247	9252	9257	9263	9268	1 1 2 2	3	3 4 4 5
86	2933	2934	2936	2938	2939	2941	2943	2944	2946	2948	0 0 1 1	1	1 1 1 2
	9274	9279	9284	9290	9295	9301	9306	9311	9317	9322	1 1 2 2	3	3 4 4 5
87	2950	2951	2953	2955	2956	2958	2960	2961	2963	2965	0 0 1 1	1	1 1 1 2
	9327	9333	9338	9343	9349	9354	9359	9365	9370	9375	1 1 2 2	3	3 4 4 5
88	2966	2968	2970	2972	2973	2975	2977	2978	2980	2982	0 0 0 1	1	1 1 1 2
	9381	9386	9391	9397	9402	9407	9413	9418	9423	9429	1 1 2 2	3	3 4 4 5
89	2983	2985	2987	2988	2990	2992	2993	2995	2997	2998	0 0 0 1	1	1 1 1 1
	9434	9439	9445	9450	9455	9460	9466	9471	9476	9482	1 1 2 2	3	3 4 4 5
90	3000	3002	3003	3005	3007	3008	3010	3012	3013	3015	0 0 0 1	1	1 1 1 1
	9487	9492	9497	9503	9508	9513	9518	9524	9529	9534	1 1 2 2	3	3 4 4 5
91	3017	3018	3020	3022	3023	3025	3027	3028	3030	3032	0 0 0 1	1	1 1 1 1
	9539	9545	9550	9555	9560	9566	9571	9576	9581	9586	1 1 2 2	3	3 4 4 5
92	3033	3035	3036	3038	3040	3041	3043	3045	3046	3048	0 0 0 1	1	1 1 1 1
	9592	9597	9602	9607	9612	9618	9623	9628	9633	9638	1 1 2 2	3	3 4 4 5
93	3050	3051	3053	3055	3056	3058	3059	3061	3063	3064	0 0 0 1	1	1 1 1 1
	9644	9649	9654	9659	9664	9670	9675	9680	9685	9690	1 1 2 2	3	3 4 4 5
94	3066	3068	3069	3071	3072	3074	3076	3077	3079	3081	0 0 0 1	1	1 1 1 1
	9695	9701	9706	9711	9716	9721	9726	9731	9737	9742	1 1 2 2	3	3 4 4 5
95	3082	3084	3085	3087	3089	3090	3092	3094	3095	3097	0 0 0 1	1	1 1 1 1
	9747	9752	9757	9762	9767	9772	9778	9783	9788	9793	1 1 2 2	3	3 4 4 5
96	3098	3100	3102	3103	3105	3106	3108	3110	3111	3113	0 0 0 1	1	1 1 1 1
	9798	9803	9808	9813	9818	9823	9829	9834	9839	9844	1 1 2 2	3	3 4 4 5
97	3114	3116	3118	3119	3121	3122	3124	3126	3127	3129	0 0 0 1	1	1 1 1 1
	9849	9854	9859	9864	9869	9874	9879	9884	9889	9894	0 1 1 2	2	3 3 4 4
98	3130	3132	3134	3135	3137	3138	3140	3142	3143	3145	0 0 0 1	1	1 1 1 1
	9899	9905	9910	9915	9920	9925	9930	9935	9940	9945	0 1 1 2	2	3 3 4 4
99	3146	3148	3150	3151	3153	3154	3156	3158	3159	3161	0 0 0 1	1	1 1 1 1
	9950	9955	9960	9965	9970	9975	9980	9985	9990	9995	0 1 1 2	2	3 3 4 4

The position of the decimal point and the first significant figure should be found by inspection.

LOGARITHMS

	0	1	2	3	4	5	6	7	8	9	1	2	3	4	5	6	7	8	9
10	0000	0043	0086	0128	0170	0212	0253	0294	0334	0374	4	8	12	17	21	25	29	33	37
11	0414	0453	0492	0531	0569	0607	0645	0682	0719	0755	4	8	11	15	19	23	26	30	34
12	0792	0828	0864	0899	0934	0969	1004	1038	1072	1106	3	7	10	14	17	21	24	28	31
13	1139	1173	1206	1239	1271	1303	1335	1367	1399	1430	3	6	10	13	16	19	23	26	29
14	1461	1492	1523	1553	1584	1614	1644	1673	1703	1732	3	6	9	12	15	18	21	24	27
15	1761	1790	1818	1847	1875	1903	1931	1959	1987	2014	3	6	8	11	14	17	20	22	25
16	2041	2068	2095	2122	2148	2175	2201	2227	2253	2279	3	5	8	11	13	16	18	21	24
17	2304	2330	2355	2380	2405	2430	2455	2480	2504	2529	2	5	7	10	12	15	17	20	22
18	2553	2577	2601	2625	2648	2672	2695	2718	2742	2765	2	5	7	9	12	14	16	19	21
19	2788	2810	2833	2856	2878	2900	2923	2945	2967	2989	2	4	7	9	11	13	16	18	20
20	3010	3032	3054	3075	3096	3118	3139	3160	3181	3201	2	4	6	8	11	13	15	17	19
21	3222	3243	3263	3284	3304	3324	3345	3365	3385	3404	2	4	6	8	10	12	14	16	18
22	3424	3444	3464	3483	3502	3522	3541	3560	3579	3598	2	4	6	8	10	12	14	15	17
23	3617	3636	3655	3674	3692	3711	3729	3747	3766	3784	2	4	6	7	9	11	13	15	17
24	3802	3820	3838	3856	3874	3892	3909	3927	3945	3962	2	4	5	7	9	11	12	14	16
25	3979	3997	4014	4031	4048	4065	4082	4099	4116	4133	2	3	5	7	9	10	12	14	15
26	4150	4166	4183	4200	4216	4232	4249	4265	4281	4298	2	3	5	7	8	10	11	13	15
27	4314	4330	4346	4362	4378	4393	4409	4425	4440	4456	2	3	5	6	8	9	11	13	14
28	4472	4487	4502	4518	4533	4548	4564	4579	4594	4609	2	3	5	6	8	9	11	12	14
29	4624	4639	4654	4669	4683	4698	4713	4728	4742	4757	1	3	4	6	7	9	10	12	13
30	4771	4786	4800	4814	4829	4843	4857	4871	4886	4900	1	3	4	6	7	9	10	11	13
31	4914	4928	4942	4955	4969	4983	4997	5011	5024	5038	1	3	4	6	7	8	10	11	12
32	5051	5065	5079	5092	5105	5119	5132	5145	5159	5172	1	3	4	5	7	8	9	11	12
33	5185	5198	5211	5224	5237	5250	5263	5276	5289	5302	1	3	4	5	6	8	9	10	12
34	5315	5328	5340	5353	5366	5378	5391	5403	5416	5428	1	3	4	5	6	8	9	10	11
35	5441	5453	5465	5478	5490	5502	5514	5527	5539	5551	1	2	4	5	6	7	9	10	11
36	5563	5575	5587	5599	5611	5623	5635	5647	5658	5670	1	2	4	5	6	7	8	10	11
37	5682	5694	5705	5717	5729	5740	5752	5763	5775	5786	1	2	3	5	6	7	8	9	10
38	5798	5809	5821	5832	5843	5855	5866	5877	5888	5899	1	2	3	5	6	7	8	9	10
39	5911	5922	5933	5944	5955	5966	5977	5988	5999	6010	1	2	3	4	5	7	8	9	10
40	6021	6031	6042	6053	6064	6075	6085	6096	6107	6117	1	2	3	4	5	6	8	9	10
41	6128	6138	6149	6160	6170	6180	6191	6201	6212	6222	1	2	3	4	5	6	7	8	9
42	6232	6243	6253	6263	6274	6284	6294	6304	6314	6325	1	2	3	4	5	6	7	8	9
43	6335	6345	6355	6365	6375	6385	6395	6405	6415	6425	1	2	3	4	5	6	7	8	9
44	6435	6444	6454	6464	6474	6484	6493	6503	6513	6522	1	2	3	4	5	6	7	8	9
45	6532	6542	6551	6561	6571	6580	6590	6599	6609	6618	1	2	3	4	5	6	7	8	9
46	6628	6637	6646	6656	6665	6675	6684	6693	6702	6712	1	2	3	4	5	6	7	7	8
47	6721	6730	6739	6749	6758	6767	6776	6785	6794	6803	1	2	3	4	5	5	6	7	8
48	6812	6821	6830	6839	6848	6857	6866	6875	6884	6893	1	2	3	4	4	5	6	7	8
49	6902	6911	6920	6928	6937	6946	6955	6964	6972	6981	1	2	3	4	4	5	6	7	8
50	6990	6998	7007	7016	7024	7033	7042	7050	7059	7067	1	2	3	3	4	5	6	7	8
51	7076	7084	7093	7101	7110	7118	7126	7135	7143	7152	1	2	3	3	4	5	6	7	8
52	7160	7168	7177	7185	7193	7202	7210	7218	7226	7235	1	2	2	3	4	5	6	7	7
53	7243	7251	7259	7267	7275	7284	7292	7300	7308	7316	1	2	2	3	4	5	6	6	7
54	7324	7332	7340	7348	7356	7364	7372	7380	7388	7396	1	2	2	3	4	5	6	6	7

LOGARITHMS

	0	1	2	3	4	5	6	7	8	9	1 2 3 4	5	6 7 8 9
55	7404	7412	7419	7427	7435	7443	7451	7459	7466	7474	1 2 2 3	4	5 5 6 7
56	7482	7490	7497	7505	7513	7520	7528	7536	7543	7551	1 2 2 3	4	5 5 6 7
57	7559	7566	7574	7582	7589	7597	7604	7612	7619	7627	1 2 2 3	4	5 5 6 7
58	7634	7642	7649	7657	7664	7672	7679	7686	7694	7701	1 1 2 3	4	4 5 6 7
59	7709	7716	7723	7731	7738	7745	7752	7760	7767	7774	1 1 2 3	4	4 5 6 7
60	7782	7789	7796	7803	7810	7818	7825	7832	7839	7846	1 1 2 3	4	4 5 6 6
61	7853	7860	7868	7875	7882	7889	7896	7903	7910	7917	1 1 2 3	4	4 5 6 6
62	7924	7931	7938	7945	7952	7959	7966	7973	7980	7987	1 1 2 3	3	4 5 6 6
63	7993	8000	8007	8014	8021	8028	8035	8041	8048	8055	1 1 2 3	3	4 5 5 6
64	8062	8069	8075	8082	8089	8096	8102	8109	8116	8122	1 1 2 3	3	4 5 5 6
65	8129	8136	8142	8149	8156	8162	8169	8176	8182	8189	1 1 2 3	3	4 5 5 6
66	8195	8202	8209	8215	8222	8228	8235	8241	8248	8254	1 1 2 3	3	4 5 5 6
67	8261	8267	8274	8280	8287	8293	8299	8306	8312	8319	1 1 2 3	3	4 5 5 6
68	8325	8331	8338	8344	8351	8357	8363	8370	8376	8382	1 1 2 3	3	4 4 5 6
69	8388	8395	8401	8407	8414	8420	8426	8432	8439	8445	1 1 2 2	3	4 4 5 6
70	8451	8457	8463	8470	8476	8482	8488	8494	8500	8506	1 1 2 2	3	4 4 5 6
71	8513	8519	8525	8531	8537	8543	8549	8555	8561	8567	1 1 2 2	3	4 4 5 5
72	8573	8579	8585	8591	8597	8603	8609	8615	8621	8627	1 1 2 2	3	4 4 5 5
73	8633	8639	8645	8651	8657	8663	8669	8675	8681	8686	1 1 2 2	3	4 4 5 5
74	8692	8698	8704	8710	8716	8722	8727	8733	8739	8745	1 1 2 2	3	4 4 5 5
75	8751	8756	8762	8768	8774	8779	8785	8791	8797	8802	1 1 2 2	3	3 4 5 5
76	8808	8814	8820	8825	8831	8837	8842	8848	8854	8859	1 1 2 2	3	3 4 5 5
77	8865	8871	8876	8882	8887	8893	8899	8904	8910	8915	1 1 2 2	3	3 4 4 5
78	8921	8927	8932	8938	8943	8949	8954	8960	8965	8971	1 1 2 2	3	3 4 4 5
79	8976	8982	8987	8993	8998	9004	9009	9015	9020	9025	1 1 2 2	3	3 4 4 5
80	9031	9036	9042	9047	9053	9058	9063	9069	9074	9079	1 1 2 2	3	3 4 4 5
81	9085	9090	9096	9101	9106	9112	9117	9122	9128	9133	1 1 2 2	3	3 4 4 5
82	9138	9143	9149	9154	9159	9165	9170	9175	9180	9186	1 1 2 2	3	3 4 4 5
83	9191	9196	9201	9206	9212	9217	9222	9227	9232	9238	1 1 2 2	3	3 4 4 5
84	9243	9248	9253	9258	9263	9269	9274	9279	9284	9289	1 1 2 2	3	3 4 4 5
85	9294	9299	9304	9309	9315	9320	9325	9330	9335	9340	1 1 2 2	3	3 4 4 5
86	9345	9350	9355	9360	9365	9370	9375	9380	9385	9390	1 1 2 2	3	3 4 4 5
87	9395	9400	9405	9410	9415	9420	9425	9430	9435	9440	0 1 1 2	2	3 3 4 4
88	9445	9450	9455	9460	9465	9469	9474	9479	9484	9489	0 1 1 2	2	3 3 4 4
89	9494	9499	9504	9509	9513	9518	9523	9528	9533	9538	0 1 1 2	2	3 3 4 4
90	9542	9547	9552	9557	9562	9566	9571	9576	9581	9586	0 1 1 2	2	3 3 4 4
91	9590	9595	9600	9605	9609	9614	9619	9624	9628	9633	0 1 1 2	2	3 3 4 4
92	9638	9643	9647	9652	9657	9661	9666	9671	9675	9680	0 1 1 2	2	3 3 4 4
93	9685	9689	9694	9699	9703	9708	9713	9717	9722	9727	0 1 1 2	2	3 3 4 4
94	9731	9736	9741	9745	9750	9754	9759	9763	9768	9773	0 1 1 2	2	3 3 4 4
95	9777	9782	9786	9791	9795	9800	9805	9809	9814	9818	0 1 1 2	2	3 3 4 4
96	9823	9827	9832	9836	9841	9845	9850	9854	9859	9863	0 1 1 2	2	3 3 4 4
97	9868	9872	9877	9881	9886	9890	9894	9899	9903	9908	0 1 1 2	2	3 3 4 4
98	9912	9917	9921	9926	9930	9934	9939	9943	9948	9952	0 1 1 2	2	3 3 4 4
99	9956	9961	9965	9969	9974	9978	9983	9987	9991	9996	0 1 1 2	2	3 3 3 4

ANTILOGARITHMS

	0	1	2	3	4	5	6	7	8	9	1 2 3 4	5	6 7 8 9
·00	1000	1002	1005	1007	1009	1012	1014	1016	1019	1021	0 0 1 1	1	1 2 2 2
·01	1023	1026	1028	1030	1033	1035	1038	1040	1042	1045	0 0 1 1	1	1 2 2 2
·02	1047	1050	1052	1054	1057	1059	1062	1064	1067	1069	0 0 1 1	1	1 2 2 2
·03	1072	1074	1076	1079	1081	1084	1086	1089	1091	1094	0 0 1 1	1	1 2 2 2
·04	1096	1099	1102	1104	1107	1109	1112	1114	1117	1119	0 1 1 1	1	2 2 2 2
·05	1122	1125	1127	1130	1132	1135	1138	1140	1143	1146	0 1 1 1	1	2 2 2 2
·06	1148	1151	1153	1156	1159	1161	1164	1167	1169	1172	0 1 1 1	1	2 2 2 2
·07	1175	1178	1180	1183	1186	1189	1191	1194	1197	1199	0 1 1 1	1	2 2 2 2
·08	1202	1205	1208	1211	1213	1216	1219	1222	1225	1227	0 1 1 1	1	2 2 2 3
·09	1230	1233	1236	1239	1242	1245	1247	1250	1253	1256	0 1 1 1	1	2 2 2 3
·10	1259	1262	1265	1268	1271	1274	1276	1279	1282	1285	0 1 1 1	1	2 2 2 3
·11	1288	1291	1294	1297	1300	1303	1306	1309	1312	1315	0 1 1 1	2	2 2 2 3
·12	1318	1321	1324	1327	1330	1334	1337	1340	1343	1346	0 1 1 1	2	2 2 2 3
·13	1349	1352	1355	1358	1361	1365	1368	1371	1374	1377	0 1 1 1	2	2 2 3 3
·14	1380	1384	1387	1390	1393	1396	1400	1403	1406	1409	0 1 1 1	2	2 2 3 3
·15	1413	1416	1419	1422	1426	1429	1432	1435	1439	1442	0 1 1 1	2	2 2 3 3
·16	1445	1449	1452	1455	1459	1462	1466	1469	1472	1476	0 1 1 1	2	2 2 3 3
·17	1479	1483	1486	1489	1493	1496	1500	1503	1507	1510	0 1 1 1	2	2 2 3 3
·18	1514	1517	1521	1524	1528	1531	1535	1538	1542	1545	0 1 1 1	2	2 2 3 3
·19	1549	1552	1556	1560	1563	1567	1570	1574	1578	1581	0 1 1 1	2	2 3 3 3
·20	1585	1589	1592	1596	1600	1603	1607	1611	1614	1618	0 1 1 1	2	2 3 3 3
·21	1622	1626	1629	1633	1637	1641	1644	1648	1652	1656	0 1 1 2	2	2 3 3 3
·22	1660	1663	1667	1671	1675	1679	1683	1687	1690	1694	0 1 1 2	2	2 3 3 3
·23	1698	1702	1706	1710	1714	1718	1722	1726	1730	1734	0 1 1 2	2	2 3 3 4
·24	1738	1742	1746	1750	1754	1758	1762	1766	1770	1774	0 1 1 2	2	2 3 3 4
·25	1778	1782	1786	1791	1795	1799	1803	1807	1811	1816	0 1 1 2	2	2 3 3 4
·26	1820	1824	1828	1832	1837	1841	1845	1849	1854	1858	0 1 1 2	2	3 3 3 4
·27	1862	1866	1871	1875	1879	1884	1888	1892	1897	1901	0 1 1 2	2	3 3 3 4
·28	1905	1910	1914	1919	1923	1928	1932	1936	1941	1945	0 1 1 2	2	3 3 4 4
·29	1950	1954	1959	1963	1968	1972	1977	1982	1986	1991	0 1 1 2	2	3 3 4 4
·30	1995	2000	2004	2009	2014	2018	2023	2028	2032	2037	0 1 1 2	2	3 3 4 4
·31	2042	2046	2051	2056	2061	2065	2070	2075	2080	2084	0 1 1 2	2	3 3 4 4
·32	2089	2094	2099	2104	2109	2113	2118	2123	2128	2133	0 1 1 2	2	3 3 4 4
·33	2138	2143	2148	2153	2158	2163	2168	2173	2178	2183	0 1 1 2	2	3 3 4 4
·34	2188	2193	2198	2203	2208	2213	2218	2223	2228	2234	1 1 2 2	3	3 4 4 5
·35	2239	2244	2249	2254	2259	2265	2270	2275	2280	2286	1 1 2 2	3	3 4 4 5
·36	2291	2296	2301	2307	2312	2317	2323	2328	2333	2339	1 1 2 2	3	3 4 4 5
·37	2344	2350	2355	2360	2366	2371	2377	2382	2388	2393	1 1 2 2	3	3 4 4 5
·38	2399	2404	2410	2415	2421	2427	2432	2438	2443	2449	1 1 2 2	3	3 4 5 5
·39	2455	2460	2466	2472	2477	2483	2489	2495	2500	2506	1 1 2 2	3	3 4 5 5
·40	2512	2518	2523	2529	2535	2541	2547	2553	2559	2564	1 1 2 2	3	4 4 5 5
·41	2570	2576	2582	2588	2594	2600	2606	2612	2618	2624	1 1 2 2	3	4 4 5 5
·42	2630	2636	2642	2649	2655	2661	2667	2673	2679	2685	1 1 2 2	3	4 4 5 6
·43	2692	2698	2704	2710	2716	2723	2729	2735	2742	2748	1 1 2 3	3	4 4 5 6
·44	2754	2761	2767	2773	2780	2786	2793	2799	2805	2812	1 1 2 3	3	4 4 5 6
·45	2818	2825	2831	2838	2844	2851	2858	2864	2871	2877	1 1 2 3	3	4 5 5 6
·46	2884	2891	2897	2904	2911	2917	2924	2931	2938	2944	1 1 2 3	3	4 5 5 6
·47	2951	2958	2965	2972	2979	2985	2992	2999	3006	3013	1 1 2 3	3	4 5 5 6
·48	3020	3027	3034	3041	3048	3055	3062	3069	3076	3083	1 1 2 3	4	4 5 6 6
·49	3090	3097	3105	3112	3119	3126	3133	3141	3148	3155	1 1 2 3	4	4 5 6 6

ANTILOGARITHMS

	0	1	2	3	4	5	6	7	8	9	1 2 3 4	5	6 7 8 9
·50	3162	3170	3177	3184	3192	3199	3206	3214	3221	3228	1 1 2 3	4	4 5 6 7
·51	3236	3243	3251	3258	3266	3273	3281	3289	3296	3304	1 2 2 3	4	5 5 6 7
·52	3311	3319	3327	3334	3342	3350	3357	3365	3373	3381	1 2 2 3	4	5 5 6 7
·53	3388	3396	3404	3412	3420	3428	3436	3443	3451	3459	1 2 2 3	4	5 6 6 7
·54	3467	3475	3483	3491	3499	3508	3516	3524	3532	3540	1 2 2 3	4	5 6 6 7
·55	3548	3556	3565	3573	3581	3589	3597	3606	3614	3622	1 2 2 3	4	5 6 7 7
·56	3631	3639	3648	3656	3664	3673	3681	3690	3698	3707	1 2 3 3	4	5 6 7 8
·57	3715	3724	3733	3741	3750	3758	3767	3776	3784	3793	1 2 3 3	4	5 6 7 8
·58	3802	3811	3819	3828	3837	3846	3855	3864	3873	3882	1 2 3 4	4	5 6 7 8
·59	3890	3899	3908	3917	3926	3936	3945	3954	3963	3972	1 2 3 4	5	5 6 7 8
·60	3981	3990	3999	4009	4018	4027	4036	4046	4055	4064	1 2 3 4	5	6 6 7 8
·61	4074	4083	4093	4102	4111	4121	4130	4140	4150	4159	1 2 3 4	5	6 7 8 9
·62	4169	4178	4188	4198	4207	4217	4227	4236	4246	4256	1 2 3 4	5	6 7 8 9
·63	4266	4276	4285	4295	4305	4315	4325	4335	4345	4355	1 2 3 4	5	6 7 8 9
·64	4365	4375	4385	4395	4406	4416	4426	4436	4446	4457	1 2 3 4	5	6 7 8 9
·65	4467	4477	4487	4498	4508	4519	4529	4539	4550	4560	1 2 3 4	5	6 7 8 9
·66	4571	4581	4592	4603	4613	4624	4634	4645	4656	4667	1 2 3 4	5	6 7 9 10
·67	4677	4688	4699	4710	4721	4732	4742	4753	4764	4775	1 2 3 4	5	7 8 9 10
·68	4786	4797	4808	4819	4831	4842	4853	4864	4875	4887	1 2 3 4	6	7 8 9 10
·69	4898	4909	4920	4932	4943	4955	4966	4977	4989	5000	1 2 3 5	6	7 8 9 10
·70	5012	5023	5035	5047	5058	5070	5082	5093	5105	5117	1 2 4 5	6	7 8 9 11
·71	5129	5140	5152	5164	5176	5188	5200	5212	5224	5236	1 2 4 5	6	7 8 10 11
·72	5248	5260	5272	5284	5297	5309	5321	5333	5346	5358	1 2 4 5	6	7 9 10 11
·73	5370	5383	5395	5408	5420	5433	5445	5458	5470	5483	1 3 4 5	6	8 9 10 11
·74	5495	5508	5521	5534	5546	5559	5572	5585	5598	5610	1 3 4 5	6	8 9 10 12
·75	5623	5636	5649	5662	5675	5689	5702	5715	5728	5741	1 3 4 5	7	8 9 10 12
·76	5754	5768	5781	5794	5808	5821	5834	5848	5861	5875	1 3 4 5	7	8 9 11 12
·77	5888	5902	5916	5929	5943	5957	5970	5984	5998	6012	1 3 4 5	7	8 10 11 12
·78	6026	6039	6053	6067	6081	6095	6109	6124	6138	6152	1 3 4 6	7	8 10 11 13
·79	6166	6180	6194	6209	6223	6237	6252	6266	6281	6295	1 3 4 6	7	9 10 11 13
·80	6310	6324	6339	6353	6368	6383	6397	6412	6427	6442	1 3 4 6	7	9 10 12 13
·81	6457	6471	6486	6501	6516	6531	6546	6561	6577	6592	2 3 5 6	8	9 11 12 14
·82	6607	6622	6637	6653	6668	6683	6699	6714	6730	6745	2 3 5 6	8	9 11 12 14
·83	6761	6776	6792	6808	6823	6839	6855	6871	6887	6902	2 3 5 6	8	9 11 13 14
·84	6918	6934	6950	6966	6982	6998	7015	7031	7047	7063	2 3 5 6	8	10 11 13 15
·85	7079	7096	7112	7129	7145	7161	7178	7194	7211	7228	2 3 5 7	8	10 12 13 15
·86	7244	7261	7278	7295	7311	7328	7345	7362	7379	7396	2 3 5 7	8	10 12 13 15
·87	7413	7430	7447	7464	7482	7499	7516	7534	7551	7568	2 3 5 7	9	10 12 14 16
·88	7586	7603	7621	7638	7656	7674	7691	7709	7727	7745	2 4 5 7	9	11 12 14 16
·89	7762	7780	7798	7816	7834	7852	7870	7889	7907	7925	2 4 5 7	9	11 13 14 16
·90	7943	7962	7980	7998	8017	8035	8054	8072	8091	8110	2 4 6 7	9	11 13 15 17
·91	8128	8147	8166	8185	8204	8222	8241	8260	8279	8299	2 4 6 8	9	11 13 15 17
·92	8318	8337	8356	8375	8395	8414	8433	8453	8472	8492	2 4 6 8	10	12 14 15 17
·93	8511	8531	8551	8570	8590	8610	8630	8650	8670	8690	2 4 6 8	10	12 14 16 18
·94	8710	8730	8750	8770	8790	8810	8831	8851	8872	8892	2 4 6 8	10	12 14 16 18
·95	8913	8933	8954	8974	8995	9016	9036	9057	9078	9099	2 4 6 8	10	12 15 17 19
·96	9120	9141	9162	9183	9204	9226	9247	9268	9290	9311	2 4 6 8	11	13 15 17 19
·97	9333	9354	9376	9397	9419	9441	9462	9484	9506	9528	2 4 7 9	11	13 15 17 20
·98	9550	9572	9594	9616	9638	9661	9683	9705	9727	9750	2 4 7 9	11	13 16 18 20
·99	9772	9795	9817	9840	9863	9886	9908	9931	9954	9977	2 5 7 9	11	14 16 18 20

LOGARITHMS OF SINES

°	0′	6′	12′	18′	24′	30′	36′	42′	48′	54′	1′ 2′ 3′ 4′ 5′
0	Inf. Neg.	3̄·242	3̄·543	3̄·719	3̄·844	3̄·941	2̄·020	2̄·087	2̄·145	2̄·196	
1	2̄·2419	2832	3210	3558	3880	4179	4459	4723	4971	5206	Use
2	2̄·5428	5640	5842	6035	6220	6397	6567	6731	6889	7041	interpolation
3	2̄·7188	7330	7468	7602	7731	7857	7979	8098	8213	8326	
4	2̄·8436	8543	8647	8749	8849	8946	9042	9135	9226	9315	16 32 48 64 80
5	2̄·9403	9489	9573	9655	9736	9816	9894	9970	**0046**	**0120**	13 26 39 52 65
6	1̄·0192	0264	0334	0403	0472	0539	0605	0670	0734	0797	11 22 33 44 55
7	1̄·0859	0920	0981	1040	1099	1157	1214	1271	1326	1381	10 19 29 38 48
8	1̄·1436	1489	1542	1594	1646	1697	1747	1797	1847	1895	8 17 25 34 42
9	1̄·1943	1991	2038	2085	2131	2176	2221	2266	2310	2353	8 15 23 30 38
10	1̄·2397	2439	2482	2524	2565	2606	2647	2687	2727	2767	7 14 20 27 34
11	1̄·2806	2845	2883	2921	2959	2997	3034	3070	3107	3143	6 12 19 25 31
12	1̄·3179	3214	3250	3284	3319	3353	3387	3421	3455	3488	6 11 17 23 28
13	1̄·3521	3554	3586	3618	3650	3682	3713	3745	3775	3806	5 11 16 21 26
14	1̄·3837	3867	3897	3927	3957	3986	4015	4044	4073	4102	5 10 15 20 24
15	1̄·4130	4158	4186	4214	4242	4269	4296	4323	4350	4377	5 9 14 18 23
16	1̄·4403	4430	4456	4482	4508	4533	4559	4584	4609	4634	4 9 13 17 21
17	1̄·4659	4684	4709	4733	4757	4781	4805	4829	4853	4876	4 8 12 16 20
18	1̄·4900	4923	4946	4969	4992	5015	5037	5060	5082	5104	4 8 11 15 19
19	1̄·5126	5148	5170	5192	5213	5235	5256	5278	5299	5320	4 7 11 14 18
20	1̄·5341	5361	5382	5402	5423	5443	5463	5484	5504	5523	3 7 10 14 17
21	1̄·5543	5563	5583	5602	5621	5641	5660	5679	5698	5717	3 6 10 13 16
22	1̄·5736	5754	5773	5792	5810	5828	5847	5865	5883	5901	3 6 9 12 15
23	1̄·5919	5937	5954	5972	5990	6007	6024	6042	6059	6076	3 6 9 12 15
24	1̄·6093	6110	6127	6144	6161	6177	6194	6210	6227	6243	3 6 8 11 14
25	1̄·6259	6276	6292	6308	6324	6340	6356	6371	6387	6403	3 5 8 11 13
26	1̄·6418	6434	6449	6465	6480	6495	6510	6526	6541	6556	3 5 8 10 13
27	1̄·6570	6585	6600	6615	6629	6644	6659	6673	6687	6702	2 5 7 10 12
28	1̄·6716	6730	6744	6759	6773	6787	6801	6814	6828	6842	2 5 7 9 12
29	1̄·6856	6869	6883	6896	6910	6923	6937	6950	6963	6977	2 4 7 9 11
30	1̄·6990	7003	7016	7029	7042	7055	7068	7080	7093	7106	2 4 6 9 11
31	1̄·7118	7131	7144	7156	7168	7181	7193	7205	7218	7230	2 4 6 8 10
32	1̄·7242	7254	7266	7278	7290	7302	7314	7326	7338	7349	2 4 6 8 10
33	1̄·7361	7373	7384	7396	7407	7419	7430	7442	7453	7464	2 4 6 8 10
34	1̄·7476	7487	7498	7509	7520	7531	7542	7553	7564	7575	2 4 6 7 9
35	1̄·7586	7597	7607	7618	7629	7640	7650	7661	7671	7682	2 4 5 7 9
36	1̄·7692	7703	7713	7723	7734	7744	7754	7764	7774	7785	2 3 5 7 9
37	1̄·7795	7805	7815	7825	7835	7844	7854	7864	7874	7884	2 3 5 7 8
38	1̄·7893	7903	7913	7922	7932	7941	7951	7960	7970	7979	2 3 5 6 8
39	1̄·7989	7998	8007	8017	8026	8035	8044	8053	8063	8072	2 3 5 6 8
40	1̄·8081	8090	8099	8108	8117	8125	8134	8143	8152	8161	1 3 4 6 7
41	1̄·8169	8178	8187	8195	8204	8213	8221	8230	8238	8247	1 3 4 6 7
42	1̄·8255	8264	8272	8280	8289	8297	8305	8313	8322	8330	1 3 4 5 7
43	1̄·8338	8346	8354	8362	8370	8378	8386	8394	8402	8410	1 3 4 5 7
44	1̄·8418	8426	8433	8441	8449	8457	8464	8472	8480	8487	1 3 4 5 6

Figures in **bold type** show changes of integer

LOGARITHMS OF SINES

°	0′	6′	12′	18′	24′	30′	36′	42′	48′	54′	1′	2′	3′	4′	5′
45	$\bar{1}$·8495	8502	8510	8517	8525	8532	8540	8547	8555	8562	1	2	4	5	6
46	$\bar{1}$·8569	8577	8584	8591	8598	8606	8613	8620	8627	8634	1	2	4	5	6
47	$\bar{1}$·8641	8648	8655	8662	8669	8676	8683	8690	8697	8704	1	2	3	5	6
48	$\bar{1}$·8711	8718	8724	8731	8738	8745	8751	8758	8765	8771	1	2	3	4	6
49	$\bar{1}$·8778	8784	8791	8797	8804	8810	8817	8823	8830	8836	1	2	3	4	5
50	$\bar{1}$·8843	8849	8855	8862	8868	8874	8880	8887	8893	8899	1	2	3	4	5
51	$\bar{1}$·8905	8911	8917	8923	8929	8935	8941	8947	8953	8959	1	2	3	4	5
52	$\bar{1}$·8965	8971	8977	8983	8989	8995	9000	9006	9012	9018	1	2	3	4	5
53	$\bar{1}$·9023	9029	9035	9041	9046	9052	9057	9063	9069	9074	1	2	3	4	5
54	$\bar{1}$·9080	9085	9091	9096	9101	9107	9112	9118	9123	9128	1	2	3	4	5
55	$\bar{1}$·9134	9139	9144	9149	9155	9160	9165	9170	9175	9181	1	2	3	3	4
56	$\bar{1}$·9186	9191	9196	9201	9206	9211	9216	9221	9226	9231	1	2	3	3	4
57	$\bar{1}$·9236	9241	9246	9251	9255	9260	9265	9270	9275	9279	1	2	2	3	4
58	$\bar{1}$·9284	9289	9294	9298	9303	9308	9312	9317	9322	9326	1	2	2	3	4
59	$\bar{1}$·9331	9335	9340	9344	9349	9353	9358	9362	9367	9371	1	1	2	3	4
60	$\bar{1}$·9375	9380	9384	9388	9393	9397	9401	9406	9410	9414	1	1	2	3	4
61	$\bar{1}$·9418	9422	9427	9431	9435	9439	9443	9447	9451	9455	1	1	2	3	3
62	$\bar{1}$·9459	9463	9467	9471	9475	9479	9483	9487	9491	9495	1	1	2	3	3
63	$\bar{1}$·9499	9503	9506	9510	9514	9518	9522	9525	9529	9533	1	1	2	3	3
64	$\bar{1}$·9537	9540	9544	9548	9551	9555	9558	9562	9566	9569	1	1	2	2	3
65	$\bar{1}$·9573	9576	9580	9583	9587	9590	9594	9597	9601	9604	1	1	2	2	3
66	$\bar{1}$·9607	9611	9614	9617	9621	9624	9627	9631	9634	9637	1	1	2	2	3
67	$\bar{1}$·9640	9643	9647	9650	9653	9656	9659	9662	9666	9669	1	1	2	2	3
68	$\bar{1}$·9672	9675	9678	9681	9684	9687	9690	9693	9696	9699	0	1	1	2	2
69	$\bar{1}$·9702	9704	9707	9710	9713	9716	9719	9722	9724	9727	0	1	1	2	2
70	$\bar{1}$·9730	9733	9735	9738	9741	9743	9746	9749	9751	9754	0	1	1	2	2
71	$\bar{1}$·9757	9759	9762	9764	9767	9770	9772	9775	9777	9780	0	1	1	2	2
72	$\bar{1}$·9782	9785	9787	9789	9792	9794	9797	9799	9801	9804	0	1	1	2	2
73	$\bar{1}$·9806	9808	9811	9813	9815	9817	9820	9822	9824	9826	0	1	1	2	2
74	$\bar{1}$·9828	9831	9833	9835	9837	9839	9841	9843	9845	9847	0	1	1	1	2
75	$\bar{1}$·9849	9851	9853	9855	9857	9859	9861	9863	9865	9867	0	1	1	1	2
76	$\bar{1}$·9869	9871	9873	9875	9876	9878	9880	9882	9884	9885	0	1	1	1	2
77	$\bar{1}$·9887	9889	9891	9892	9894	9896	9897	9899	9901	9902	0	1	1	1	1
78	$\bar{1}$·9904	9906	9907	9909	9910	9912	9913	9915	9916	9918	0	1	1	1	1
79	$\bar{1}$·9919	9921	9922	9924	9925	9927	9928	9929	9931	9932	0	0	1	1	1
80	$\bar{1}$·9934	9935	9936	9937	9939	9940	9941	9943	9944	9945	0	0	1	1	1
81	$\bar{1}$·9946	9947	9949	9950	9951	9952	9953	9954	9955	9956	0	0	1	1	1
82	$\bar{1}$·9958	9959	9960	9961	9962	9963	9964	9965	9966	9967	0	0	1	1	1
83	$\bar{1}$·9968	9968	9969	9970	9971	9972	9973	9974	9975	9975	0	0	0	1	1
84	$\bar{1}$·9976	9977	9978	9978	9979	9980	9981	9981	9982	9983	0	0	0	0	1
85	$\bar{1}$·9983	9984	9985	9985	9986	9987	9987	9988	9988	9989	0	0	0	0	0
86	$\bar{1}$·9989	9990	9990	9991	9991	9992	9992	9993	9993	9994	0	0	0	0	0
87	$\bar{1}$·9994	9994	9995	9995	9996	9996	9996	9996	9997	9997	0	0	0	0	0
88	$\bar{1}$·9997	9998	9998	9998	9998	9999	9999	9999	9999	9999	0	0	0	0	0
89	$\bar{1}$·9999	9999	**0000**	**0000**	**0000**	**0000**	**0000**	**0000**	**0000**	**0000**	0	0	0	0	0

Figures in **bold type** show changes of integer

LOGARITHMS OF COSINES

SUBTRACT

°	0′	6′	12′	18′	24′	30′	36′	42′	48′	54′	1′	2′	3′	4′	5′
0	0·0000	0000	0000	0000	0000	0000	0000	0000	0000	$\bar{1}$·9999	0	0	0	0	0
1	$\bar{1}$·9999	9999	9999	9999	9999	9999	9998	9998	9998	9998	0	0	0	0	0
2	$\bar{1}$·9997	9997	9997	9997	9996	9996	9996	9995	9995	9994	0	0	0	0	0
3	$\bar{1}$·9994	9994	9993	9993	9992	9992	9991	9991	9990	9990	0	0	0	0	0
4	$\bar{1}$·9989	9989	9988	9988	9987	9987	9986	9985	9985	9984	0	0	0	0	0
5	$\bar{1}$·9983	9983	9982	9981	9981	9980	9979	9978	9978	9977	0	0	0	0	1
6	$\bar{1}$·9976	9975	9975	9974	9973	9972	9971	9970	9969	9968	0	0	0	1	1
7	$\bar{1}$·9968	9967	9966	9965	9964	9963	9962	9961	9960	9959	0	0	1	1	1
8	$\bar{1}$·9958	9956	9955	9954	9953	9952	9951	9950	9949	9947	0	0	1	1	1
9	$\bar{1}$·9946	9945	9944	9943	9941	9940	9939	9937	9936	9935	0	0	1	1	1
10	$\bar{1}$·9934	9932	9931	9929	9928	9927	9925	9924	9922	9921	0	0	1	1	1
11	$\bar{1}$·9919	9918	9916	9915	9913	9912	9910	9909	9907	9906	0	1	1	1	1
12	$\bar{1}$·9904	9902	9901	9899	9897	9896	9894	9892	9891	9889	0	1	1	1	1
13	$\bar{1}$·9887	9885	9884	9882	9880	9878	9876	9875	9873	9871	0	1	1	1	2
14	$\bar{1}$·9869	9867	9865	9863	9861	9859	9857	9855	9853	9851	0	1	1	1	2
15	$\bar{1}$·9849	9847	9845	9843	9841	9839	9837	9835	9833	9831	0	1	1	1	2
16	$\bar{1}$·9828	9826	9824	9822	9820	9817	9815	9813	9811	9808	0	1	1	1	2
17	$\bar{1}$·9806	9804	9801	9799	9797	9794	9792	9789	9787	9785	0	1	1	2	2
18	$\bar{1}$·9782	9780	9777	9775	9772	9770	9767	9764	9762	9759	0	1	1	2	2
19	$\bar{1}$·9757	9754	9751	9749	9746	9743	9741	9738	9735	9733	0	1	1	2	2
20	$\bar{1}$·9730	9727	9724	9722	9719	9716	9713	9710	9707	9704	0	1	1	2	2
21	$\bar{1}$·9702	9699	9696	9693	9690	9687	9684	9681	9678	9675	0	1	2	2	2
22	$\bar{1}$·9672	9669	9666	9662	9659	9656	9653	9650	9647	9643	1	1	2	2	2
23	$\bar{1}$·9640	9637	9634	9631	9627	9624	9621	9617	9614	9611	1	1	2	2	3
24	$\bar{1}$·9607	9604	9601	9597	9594	9590	9587	9583	9580	9576	1	1	2	2	3
25	$\bar{1}$·9573	9569	9566	9562	9558	9555	9551	9548	9544	9540	1	1	2	2	3
26	$\bar{1}$·9537	9533	9529	9525	9522	9518	9514	9510	9506	9503	1	1	2	3	3
27	$\bar{1}$·9499	9495	9491	9487	9483	9479	9475	9471	9467	9463	1	1	2	3	3
28	$\bar{1}$·9459	9455	9451	9447	9443	9439	9435	9431	9427	9422	1	1	2	3	3
29	$\bar{1}$·9418	9414	9410	9406	9401	9397	9393	9388	9384	9380	1	1	2	3	4
30	$\bar{1}$·9375	9371	9367	9362	9358	9353	9349	9344	9340	9335	1	1	2	3	4
31	$\bar{1}$·9331	9326	9322	9317	9312	9308	9303	9298	9294	9289	1	2	2	3	4
32	$\bar{1}$·9284	9279	9275	9270	9265	9260	9255	9251	9246	9241	1	2	2	3	4
33	$\bar{1}$·9236	9231	9226	9221	9216	9211	9206	9201	9196	9191	1	2	3	3	4
34	$\bar{1}$·9186	9181	9175	9170	9165	9160	9155	9149	9144	9139	1	2	3	3	4
35	$\bar{1}$·9134	9128	9123	9118	9112	9107	9101	9096	9091	9085	1	2	3	4	5
36	$\bar{1}$·9080	9074	9069	9063	9057	9052	9046	9041	9035	9029	1	2	3	4	5
37	$\bar{1}$·9023	9018	9012	9006	9000	8995	8989	8983	8977	8971	1	2	3	4	5
38	$\bar{1}$·8965	8959	8953	8947	8941	8935	8929	8923	8917	8911	1	2	3	4	5
39	$\bar{1}$·8905	8899	8893	8887	8880	8874	8868	8862	8855	8849	1	2	3	4	5
40	$\bar{1}$·8843	8836	8830	8823	8817	8810	8804	8797	8791	8784	1	2	3	4	5
41	$\bar{1}$·8778	8771	8765	8758	8751	8745	8738	8731	8724	8718	1	2	3	4	6
42	$\bar{1}$·8711	8704	8697	8690	8683	8676	8669	8662	8655	8648	1	2	3	5	6
43	$\bar{1}$·8641	8634	8627	8620	8613	8606	8598	8591	8584	8577	1	2	4	5	6
44	$\bar{1}$·8569	8562	8555	8547	8540	8532	8525	8517	8510	8502	1	2	4	5	6

SUBTRACT

LOGARITHMS OF COSINES

°	0'	6'	12'	18'	24'	30'	36'	42'	48'	54'	1' 2' 3' 4' 5'
45	$\bar{1}$·8495	8487	8480	8472	8464	8457	8449	8441	8433	8426	1 3 4 5 6
46	$\bar{1}$·8418	8410	8402	8394	8386	8378	8370	8362	8354	8346	1 3 4 5 7
47	$\bar{1}$·8338	8330	8322	8313	8305	8297	8289	8280	8272	8264	1 3 4 6 7
48	$\bar{1}$·8255	8247	8238	8230	8221	8213	8204	8195	8187	8178	1 3 4 6 7
49	$\bar{1}$·8169	8161	8152	8143	8134	8125	8117	8108	8099	8090	1 3 4 6 7
50	$\bar{1}$·8081	8072	8063	8053	8044	8035	8026	8017	8007	7998	2 3 5 6 8
51	$\bar{1}$·7989	7979	7970	7960	7951	7941	7932	7922	7913	7903	2 3 5 6 8
52	$\bar{1}$·7893	7884	7874	7864	7854	7844	7835	7825	7815	7805	2 3 5 7 8
53	$\bar{1}$·7795	7785	7774	7764	7754	7744	7734	7723	7713	7703	2 3 5 7 9
54	$\bar{1}$·7692	7682	7671	7661	7650	7640	7629	7618	7607	7597	2 4 5 7 9
55	$\bar{1}$·7586	7575	7564	7553	7542	7531	7520	7509	7498	7487	2 4 6 7 9
56	$\bar{1}$·7476	7464	7453	7442	7430	7419	7407	7396	7384	7373	2 4 6 8 10
57	$\bar{1}$·7361	7349	7338	7326	7314	7302	7290	7278	7266	7254	2 4 6 8 10
58	$\bar{1}$·7242	7230	7218	7205	7193	7181	7168	7156	7144	7131	2 4 6 8 10
59	$\bar{1}$·7118	7106	7093	7080	7068	7055	7042	7029	7016	7003	2 4 6 9 11
60	$\bar{1}$·6990	6977	6963	6950	6937	6923	6910	6896	6883	6869	2 4 7 9 11
61	$\bar{1}$·6856	6842	6828	6814	6801	6787	6773	6759	6744	6730	2 5 7 9 12
62	$\bar{1}$·6716	6702	6687	6673	6659	6644	6629	6615	6600	6585	2 5 7 10 12
63	$\bar{1}$·6570	6556	6541	6526	6510	6495	6480	6465	6449	6434	3 5 8 10 13
64	$\bar{1}$·6418	6403	6387	6371	6356	6340	6324	6308	6292	6276	3 5 8 11 13
65	$\bar{1}$·6259	6243	6227	6210	6194	6177	6161	6144	6127	6110	3 6 8 11 14
66	$\bar{1}$·6093	6076	6059	6042	6024	6007	5990	5972	5954	5937	3 6 9 12 15
67	$\bar{1}$·5919	5901	5883	5865	5847	5828	5810	5792	5773	5754	3 6 9 12 15
68	$\bar{1}$·5736	5717	5698	5679	5660	5641	5621	5602	5583	5563	3 6 10 13 16
69	$\bar{1}$·5543	5523	5504	5484	5463	5443	5423	5402	5382	5361	3 7 10 14 17
70	$\bar{1}$·5341	5320	5299	5278	5256	5235	5213	5192	5170	5148	4 7 11 14 18
71	$\bar{1}$·5126	5104	5082	5060	5037	5015	4992	4969	4946	4923	4 8 11 15 19
72	$\bar{1}$·4900	4876	4853	4829	4805	4781	4757	4733	4709	4684	4 8 12 16 20
73	$\bar{1}$·4659	4634	4609	4584	4559	4533	4508	4482	4456	4430	4 9 13 17 21
74	$\bar{1}$·4403	4377	4350	4323	4296	4269	4242	4214	4186	4158	5 9 14 18 23
75	$\bar{1}$·4130	4102	4073	4044	4015	3986	3957	3927	3897	3867	5 10 15 20 24
76	$\bar{1}$·3837	3806	3775	3745	3713	3682	3650	3618	3586	3554	5 11 16 21 26
77	$\bar{1}$·3521	3488	3455	3421	3387	3353	3319	3284	3250	3214	6 11 17 23 28
78	$\bar{1}$·3179	3143	3107	3070	3034	2997	2959	2921	2883	2845	6 12 19 25 31
79	$\bar{1}$·2806	2767	2727	2687	2647	2606	2565	2524	2482	2439	7 14 20 27 34
80	$\bar{1}$·2397	2353	2310	2266	2221	2176	2131	2085	2038	1991	8 15 23 30 38
81	$\bar{1}$·1943	1895	1847	1797	1747	1697	1646	1594	1542	1489	8 17 25 34 42
82	$\bar{1}$·1436	1381	1326	1271	1214	1157	1099	1040	0981	0920	10 19 29 38 48
83	$\bar{1}$·0859	0797	0734	0670	0605	0539	0472	0403	0334	0264	11 22 33 44 55
84	$\bar{1}$·0192	0120	0046	**9970**	**9894**	**9816**	**9736**	**9655**	**9573**	**9489**	13 26 39 52 65
85	$\bar{2}$·9403	9315	9226	9135	9042	8946	8849	8749	8647	8543	16 32 48 64 80
86	$\bar{2}$·8436	8326	8213	8098	7979	7857	7731	7602	7468	7330	
87	$\bar{2}$·7188	7041	6889	6731	6567	6397	6220	6035	5842	5640	Use
88	$\bar{2}$·5428	5206	4971	4723	4459	4179	3880	3558	3210	2832	Interpolation
89	$\bar{2}$·242	$\bar{2}$·196	$\bar{2}$·145	$\bar{2}$·087	$\bar{2}$·020	$\bar{3}$·941	$\bar{3}$·844	$\bar{3}$·719	$\bar{3}$·543	$\bar{3}$·242	

Figures in **bold type** show changes of integer

LOGARITHMS OF TANGENTS

°	0′	6′	12′	18′	24′	30′	36′	42′	48′	54′	1′ 2′ 3′ 4′ 5′
0	Inf. Neg.	$\overline{3}$·242	$\overline{3}$·543	$\overline{3}$·719	$\overline{3}$·844	$\overline{3}$·941	$\overline{2}$·020	$\overline{2}$·087	$\overline{2}$·145	$\overline{2}$·196	Use
1	$\overline{2}$·2419	2833	3211	3559	3881	4181	4461	4725	4973	5208	interpolation
2	$\overline{2}$·5431	5643	5845	6038	6223	6401	6571	6736	6894	7046	
3	$\overline{2}$·7194	7337	7475	7609	7739	7865	7988	8107	8223	8336	
4	$\overline{2}$·8446	8554	8659	8762	8862	8960	9056	9150	9241	9331	16 32 48 64 81
5	$\overline{2}$·9420	9506	9591	9674	9756	9836	9915	9992	**0068**	**0143**	13 26 40 53 66
6	$\overline{1}$·0216	0289	0360	0430	0499	0567	0633	0699	0764	0828	11 22 34 45 56
7	$\overline{1}$·0891	0954	1015	1076	1135	1194	1252	1310	1367	1423	10 20 29 39 49
8	$\overline{1}$·1478	1533	1587	1640	1693	1745	1797	1848	1898	1948	9 17 26 35 43
9	$\overline{1}$·1997	2046	2094	2142	2189	2236	2282	2328	2374	2419	8 16 23 31 39
10	$\overline{1}$·2463	2507	2551	2594	2637	2680	2722	2764	2805	2846	7 14 21 28 35
11	$\overline{1}$·2887	2927	2967	3006	3046	3085	3123	3162	3200	3237	6 13 19 26 32
12	$\overline{1}$·3275	3312	3349	3385	3422	3458	3493	3529	3564	3599	6 12 18 24 30
13	$\overline{1}$·3634	3668	3702	3736	3770	3804	3837	3870	3903	3935	6 11 17 22 28
14	$\overline{1}$·3968	4000	4032	4064	4095	4127	4158	4189	4220	4250	5 10 16 21 26
15	$\overline{1}$·4281	4311	4341	4371	4400	4430	4459	4488	4517	4546	5 10 15 20 25
16	$\overline{1}$·4575	4603	4632	4660	4688	4716	4744	4771	4799	4826	5 9 14 19 23
17	$\overline{1}$·4853	4880	4907	4934	4961	4987	5014	5040	5066	5092	4 9 13 18 22
18	$\overline{1}$·5118	5143	5169	5195	5220	5245	5270	5295	5320	5345	4 8 13 17 21
19	$\overline{1}$·5370	5394	5419	5443	5467	5491	5516	5539	5563	5587	4 8 12 16 20
20	$\overline{1}$·5611	5634	5658	5681	5704	5727	5750	5773	5796	5819	4 8 12 15 19
21	$\overline{1}$·5842	5864	5887	5909	5932	5954	5976	5998	6020	6042	4 7 11 15 19
22	$\overline{1}$·6064	6086	6108	6129	6151	6172	6194	6215	6236	6257	4 7 11 14 18
23	$\overline{1}$·6279	6300	6321	6341	6362	6383	6404	6424	6445	6465	3 7 10 14 17
24	$\overline{1}$·6486	6506	6527	6547	6567	6587	6607	6627	6647	6667	3 7 10 13 17
25	$\overline{1}$·6687	6706	6726	6746	6765	6785	6804	6824	6843	6863	3 7 10 13 16
26	$\overline{1}$·6882	6901	6920	6939	6958	6977	6996	7015	7034	7053	3 6 9 13 16
27	$\overline{1}$·7072	7090	7109	7128	7146	7165	7183	7202	7220	7238	3 6 9 12 15
28	$\overline{1}$·7257	7275	7293	7311	7330	7348	7366	7384	7402	7420	3 6 9 12 15
29	$\overline{1}$·7438	7455	7473	7491	7509	7526	7544	7562	7579	7597	3 6 9 12 15
30	$\overline{1}$·7614	7632	7649	7667	7684	7701	7719	7736	7753	7771	3 6 9 12 14
31	$\overline{1}$·7788	7805	7822	7839	7856	7873	7890	7907	7924	7941	3 6 9 11 14
32	$\overline{1}$·7958	7975	7992	8008	8025	8042	8059	8075	8092	8109	3 6 8 11 14
33	$\overline{1}$·8125	8142	8158	8175	8191	8208	8224	8241	8257	8274	3 5 8 11 14
34	$\overline{1}$·8290	8306	8323	8339	8355	8371	8388	8404	8420	8436	3 5 8 11 13
35	$\overline{1}$·8452	8468	8484	8501	8517	8533	8549	8565	8581	8597	3 5 8 11 13
36	$\overline{1}$·8613	8629	8644	8660	8676	8692	8708	8724	8740	8755	3 5 8 11 13
37	$\overline{1}$·8771	8787	8803	8818	8834	8850	8865	8881	8897	8912	3 5 8 10 13
38	$\overline{1}$·8928	8944	8959	8975	8990	9006	9022	9037	9053	9068	3 5 8 10 13
39	$\overline{1}$·9084	9099	9115	9130	9146	9161	9176	9192	9207	9223	3 5 8 10 13
40	$\overline{1}$·9238	9254	9269	9284	9300	9315	9330	9346	9361	9376	3 5 8 10 13
41	$\overline{1}$·9392	9407	9422	9438	9453	9468	9483	9499	9514	9529	3 5 8 10 13
42	$\overline{1}$·9544	9560	9575	9590	9605	9621	9636	9651	9666	9681	3 5 8 10 13
43	$\overline{1}$·9697	9712	9727	9742	9757	9772	9788	9803	9818	9833	3 5 8 10 13
44	$\overline{1}$·9848	9864	9879	9894	9909	9924	9939	9955	9970	9985	3 5 8 10 13

Figures in **bold type** show changes of integer

LOGARITHMS OF TANGENTS

°	0′	6′	12′	18′	24′	30′	36′	42′	48′	54′	1′ 2′ 3′ 4′ 5′
45	0·0000	0015	0030	0045	0061	0076	0091	0106	0121	0136	3 5 8 10 13
46	0·0152	0167	0182	0197	0212	0228	0243	0258	0273	0288	3 5 8 10 13
47	0·0303	0319	0334	0349	0364	0379	0395	0410	0425	0440	3 5 8 10 13
48	0·0456	0471	0486	0501	0517	0532	0547	0562	0578	0593	3 5 8 10 13
49	0·0608	0624	0639	0654	0670	0685	0700	0716	0731	0746	3 5 8 10 13
50	0·0762	0777	0793	0808	0824	0839	0854	0870	0885	0901	3 5 8 10 13
51	0·0916	0932	0947	0963	0978	0994	1010	1025	1041	1056	3 5 8 10 13
52	0·1072	1088	1103	1119	1135	1150	1166	1182	1197	1213	3 5 8 10 13
53	0·1229	1245	1260	1276	1292	1308	1324	1340	1356	371	3 5 8 11 13
54	0·1387	1403	1419	1435	1451	1467	1483	1499	1516	1532	3 5 8 11 13
55	0·1548	1564	1580	1596	1612	1629	1645	1661	1677	1694	3 5 8 11 14
56	0·1710	1726	1743	1759	1776	1792	1809	1825	1842	1858	3 5 8 11 14
57	0·1875	1891	1908	1925	1941	1958	1975	1992	2008	2025	3 6 8 11 14
58	0·2042	2059	2076	2093	2110	2127	2144	2161	2178	2195	3 6 9 11 14
59	0·2212	2229	2247	2264	2281	2299	2316	2333	2351	2368	3 6 9 12 14
60	0·2386	2403	2421	2438	2456	2474	2491	2509	2527	2545	3 6 9 12 15
61	0·2562	2580	2598	2616	2634	2652	2670	2689	2707	2725	3 6 9 12 15
62	0·2743	2762	2780	2798	2817	2835	2854	2872	2891	2910	3 6 9 12 15
63	0·2928	2947	2966	2985	3004	3023	3042	3061	3080	3099	3 6 9 13 16
64	0·3118	3137	3157	3176	3196	3215	3235	3254	3274	3294	3 6 10 13 16
65	0·3313	3333	3353	3373	3393	3413	3433	3453	3473	3494	3 7 10 13 17
66	0·3514	3535	3555	3576	3596	3617	3638	3659	3679	3700	3 7 10 14 17
67	0·3721	3743	3764	3785	3806	3828	3849	3871	3892	3914	4 7 11 14 18
68	0·3936	3958	3980	4002	4024	4046	4068	4091	4113	4136	4 7 11 15 19
69	0·4158	4181	4204	4227	4250	4273	4296	4319	4342	4366	4 8 12 15 19
70	0·4389	4413	4437	4461	4484	4509	4533	4557	4581	4606	4 8 12 16 20
71	0·4630	4655	4680	4705	4730	4755	4780	4805	4831	4857	4 8 13 17 21
72	0·4882	4908	4934	4960	4986	5013	5039	5066	5093	5120	4 9 13 18 22
73	0·5147	5174	5201	5229	5256	5284	5312	5340	5368	5397	5 9 14 19 23
74	0·5425	5454	5483	5512	5541	5570	5600	5629	5659	5689	5 10 15 20 25
75	0·5719	5750	5780	5811	5842	5873	5905	5936	5968	6000	5 10 16 21 26
76	0·6032	6065	6097	6130	6163	6196	6230	6264	6298	6332	6 11 17 22 28
77	0·6366	6401	6436	6471	6507	6542	6578	6615	6651	6688	6 12 18 24 30
78	0·6725	6763	6800	6838	6877	6915	6954	6994	7033	7073	6 13 19 26 32
79	0·7113	7154	7195	7236	7278	7320	7363	7406	7449	7493	7 14 21 28 35
80	0·7537	7581	7626	7672	7718	7764	7811	7858	7906	7954	8 16 23 31 39
81	0·8003	8052	8102	8152	8203	8255	8307	8360	8413	8467	9 17 26 35 43
82	0·8522	8577	8633	8690	8748	8806	8865	8924	8985	9046	10 20 29 39 49
83	0·9109	9172	9236	9301	9367	9433	9501	9570	9640	9711	11 22 34 45 56
84	0·9784	9857	9932	**0008**	**0085**	**0164**	**0244**	**0326**	**0409**	**0494**	13 26 40 53 66
85	1·0580	0669	0759	0850	0944	1040	1138	1238	1341	1446	16 32 48 64 81
86	1·1554	1664	1777	1893	2012	2135	2261	2391	2525	2663	
87	1·2806	2954	3106	3264	3429	3599	3777	3962	4155	4357	Use
88	1·4569	4792	5027	5275	5539	5819	6119	6441	6789	7167	interpolation
89	1·758	1·804	1·855	1·913	1·980	2·059	2·156	2·281	2·457	2·758	

Figures in **bold type** show changes of integer

LOGARITHMS OF COSECANTS

SUBTRACT

°	0'	6'	12'	18'	24'	30'	36'	42'	48'	54'	1'	2'	3'	4'	5'
0	Inf. Pos.	2·758	2·457	2·281	2·156	2·059	1·980	1·913	1·855	1·804					
1	1·7581	7168	6790	6442	6120	5821	5541	5277	5029	4794		Use			
2	1·4572	4360	4158	3965	3780	3603	3433	3269	3111	2959		interpolation			
3	1·2812	2670	2532	2398	2269	2143	2021	1902	1787	1674					
4	1·1564	1457	1353	1251	1151	1054	0958	0865	0774	0685	16	32	48	64	80
5	1·0597	0511	0427	0345	0264	0184	0106	0030	**9954**	**9880**	13	26	39	52	65
6	0·9808	9736	9666	9597	9528	9461	9395	9330	9266	9203	11	22	33	44	55
7	0·9141	9080	9019	8960	8901	8843	8786	8729	8674	8619	10	19	29	38	48
8	0·8564	8511	8458	8406	8354	8303	8253	8203	8153	8105	8	17	25	34	42
9	0·8057	8009	7962	7915	7869	7824	7779	7734	7690	7647	8	15	23	30	38
10	0·7603	7561	7518	7476	7435	7394	7353	7313	7273	7233	7	14	20	27	34
11	0·7194	7155	7117	7079	7041	7003	6966	6930	6893	6857	6	12	19	25	31
12	0·6821	6786	6750	6716	6681	6647	6613	6579	6545	6512	6	11	17	23	28
13	0·6479	6446	6414	6382	6350	6318	6287	6255	6225	6194	5	11	16	21	26
14	0·6163	6133	6103	6073	6043	6014	5985	5956	5927	5898	5	10	15	20	24
15	0·5870	5842	5814	5786	5758	5731	5704	5677	5650	5623	5	9	14	18	23
16	0·5597	5570	5544	5518	5492	5467	5441	5416	5391	5366	4	9	13	17	21
17	0·5341	5316	5291	5267	5243	5219	5195	5171	5147	5124	4	8	12	16	20
18	0·5100	5077	5054	5031	5008	4985	4963	4940	4918	4896	4	8	11	15	19
19	0·4874	4852	4830	4808	4787	4765	4744	4722	4701	4680	4	7	11	14	18
20	0·4659	4639	4618	4598	4577	4557	4537	4516	4496	4477	3	7	10	14	17
21	0·4457	4437	4417	4398	4379	4359	4340	4321	4302	4283	3	6	10	13	16
22	0·4264	4246	4227	4208	4190	4172	4153	4135	4117	4099	3	6	9	12	15
23	0·4081	4063	4046	4028	4010	3993	3976	3958	3941	3924	3	6	9	12	15
24	0·3907	3890	3873	3856	3839	3823	3806	3790	3773	3757	3	6	8	11	14
25	0·3741	3724	3708	3692	3676	3660	3644	3629	3613	3597	3	5	8	11	13
26	0·3582	3566	3551	3535	3520	3505	3490	3474	3459	3444	3	5	8	10	13
27	0·3430	3415	3400	3385	3371	3356	3341	3327	3313	3298	2	5	7	10	12
28	0·3284	3270	3256	3241	3227	3213	3199	3186	3172	3158	2	5	7	9	12
29	0·3144	3131	3117	3104	3090	3077	3063	3050	3037	3023	2	4	7	9	11
30	0·3010	2997	2984	2971	2958	2945	2932	2920	2907	2894	2	4	6	9	11
31	0·2882	2869	2856	2844	2832	2819	2807	2795	2782	2770	2	4	6	8	10
32	0·2758	2746	2734	2722	2710	2698	2686	2674	2662	2651	2	4	6	8	10
33	0·2639	2627	2616	2604	2593	2581	2570	2558	2547	2536	2	4	6	8	10
34	0·2524	2513	2502	2491	2480	2469	2458	2447	2436	2425	2	4	6	7	9
35	0·2414	2403	2393	2382	2371	2360	2350	2339	2329	2318	2	4	5	7	9
36	0·2308	2297	2287	2277	2266	2256	2246	2236	2226	2215	2	3	5	7	9
37	0·2205	2195	2185	2175	2165	2156	2146	2136	2126	2116	2	3	5	7	8
38	0·2107	2097	2087	2078	2068	2059	2049	2040	2030	2021	2	3	5	6	8
39	0·2011	2002	1993	1983	1974	1965	1956	1947	1937	1928	2	3	5	6	8
40	0·1919	1910	1901	1892	1883	1875	1866	1857	1848	1839	1	3	4	6	7
41	0·1831	1822	1813	1805	1796	1787	1779	1770	1762	1753	1	3	4	6	7
42	0·1745	1736	1728	1720	1711	1703	1695	1687	1678	1670	1	3	4	5	7
43	0·1662	1654	1646	1638	1630	1622	1614	1606	1598	1590	1	3	4	5	7
44	0·1582	1574	1567	1559	1551	1543	1536	1528	1520	1513	1	3	4	5	6

SUBTRACT

Figures in **bold type** show changes of integer.

LOGARITHMS OF COSECANTS

°	0′	6′	12′	18′	24′	30′	36′	42′	48′	54′	1′	2′	3′	4′	5′
45	0·1505	1498	1490	1483	1475	1468	1460	1453	1445	1438	1	2	4	5	6
46	0·1431	1423	1416	1409	1402	1394	1387	1380	1373	1366	1	2	4	5	6
47	0·1359	1352	1345	1338	1331	1324	1317	1310	1303	1296	1	2	3	5	6
48	0·1289	1282	1276	1269	1262	1255	1249	1242	1235	1229	1	2	3	4	6
49	0·1222	1216	1209	1203	1196	1190	1183	1177	1170	1164	1	2	3	4	5
50	0·1157	1151	1145	1138	1132	1126	1120	1113	1107	1101	1	2	3	4	5
51	0·1095	1089	1083	1077	1071	1065	1059	1053	1047	1041	1	2	3	4	5
52	0·1035	1029	1023	1017	1011	1005	1000	0994	0988	0982	1	2	3	4	5
53	0·0977	0971	0965	0959	0954	0948	0943	0937	0931	0926	1	2	3	4	5
54	0·0920	0915	0909	0904	0899	0893	0888	0882	0877	0872	1	2	3	4	5
55	0·0866	0861	0856	0851	0845	0840	0835	0830	0825	0819	1	2	3	3	4
56	0·0814	0809	0804	0799	0794	0789	0784	0779	0774	0769	1	2	3	3	4
57	0·0764	0759	0754	0749	0745	0740	0735	0730	0725	0721	1	2	2	3	4
58	0·0716	0711	0706	0702	0697	0692	0688	0683	0678	0674	1	2	2	3	4
59	0·0669	0665	0660	0656	0651	0647	0642	0638	0633	0629	1	1	2	3	4
60	0·0625	0620	0616	0612	0607	0603	0599	0594	0590	0586	1	1	2	3	4
61	0·0582	0578	0573	0569	0565	0561	0557	0553	0549	0545	1	1	2	3	3
62	0·0541	0537	0533	0529	0525	0521	0517	0513	0509	0505	1	1	2	3	3
63	0·0501	0497	0494	0490	0486	0482	0478	0475	0471	0467	1	1	2	3	3
64	0·0463	0460	0456	0452	0449	0445	0442	0438	0434	0431	1	1	2	2	3
65	0·0427	0424	0420	0417	0413	0410	0406	0403	0399	0396	1	1	2	2	3
66	0·0393	0389	0386	0383	0379	0376	0373	0369	0366	0363	1	1	2	2	3
67	0·0360	0357	0353	0350	0347	0344	0341	0338	0334	0331	1	1	2	2	3
68	0·0328	0325	0322	0319	0316	0313	0310	0307	0304	0301	0	1	1	2	2
69	0·0298	0296	0293	0290	0287	0284	0281	0278	0276	0273	0	1	1	2	2
70	0·0270	0267	0265	0262	0259	0257	0254	0251	0249	0246	0	1	1	2	2
71	0·0243	0241	0238	0236	0233	0230	0228	0225	0223	0220	0	1	1	2	2
72	0·0218	0215	0213	0211	0208	0206	0203	0201	0199	0196	0	1	1	2	2
73	0·0194	0192	0189	0187	0185	0183	0180	0178	0176	0174	0	1	1	2	2
74	0·0172	0169	0167	0165	0163	0161	0159	0157	0155	0153	0	1	1	1	2
75	0·0151	0149	0147	0145	0143	0141	0139	0137	0135	0133	0	1	1	1	2
76	0·0131	0129	0127	0125	0124	0122	0120	0118	0116	0115	0	1	1	1	2
77	0·0113	0111	0109	0108	0106	0104	0103	0101	0099	0098	0	1	1	1	1
78	0·0096	0094	0093	0091	0090	0088	0087	0085	0084	0082	0	1	1	1	1
79	0·0081	0079	0078	0076	0075	0073	0072	0071	0069	0068	0	0	1	1	1
80	0·0066	0065	0064	0063	0061	0060	0059	0057	0056	0055	0	0	1	1	1
81	0·0054	0053	0051	0050	0049	0048	0047	0046	0045	0044	0	0	1	1	1
82	0·0042	0041	0040	0039	0038	0037	0036	0035	0034	0033	0	0	1	1	1
83	0·0032	0032	0031	0030	0029	0028	0027	0026	0025	0025	0	0	0	1	1
84	0·0024	0023	0022	0022	0021	0020	0019	0019	0018	0017	0	0	0	0	1
85	0·0017	0016	0015	0015	0014	0013	0013	0012	0012	0011	0	0	0	0	0
86	0·0011	0010	0010	0009	0009	0008	0008	0007	0007	0006	0	0	0	0	0
87	0·0006	0006	0005	0005	0004	0004	0004	0004	0003	0003	0	0	0	0	0
88	0·0003	0002	0002	0002	0002	0001	0001	0001	0001	0001	0	0	0	0	0
89	0·0001	0001	0000	0000	0000	0000	0000	0000	0000	0000	0	0	0	0	0

LOGARITHMS OF SECANTS

°	0'	6'	12'	18'	24'	30'	36'	42'	48'	54'	1'	2'	3'	4'	5'
0	0·0000	0000	0000	0000	0000	0000	0000	0000	0000	0001	0	0	0	0	0
1	0·0001	0001	0001	0001	0001	0001	0002	0002	0002	0002	0	0	0	0	0
2	0·0003	0003	0003	0003	0004	0004	0004	0005	0005	0006	0	0	0	0	0
3	0·0006	0006	0007	0007	0008	0008	0009	0009	0010	0010	0	0	0	0	0
4	0·0011	0011	0012	0012	0013	0013	0014	0015	0015	0016	0	0	0	0	0
5	0·0017	0017	0018	0019	0019	0020	0021	0022	0022	0023	0	0	0	0	1
6	0·0024	0025	0025	0026	0027	0028	0029	0030	0031	0032	0	0	0	1	1
7	0·0032	0033	0034	0035	0036	0037	0038	0039	0040	0041	0	0	1	1	1
8	0·0042	0044	0045	0046	0047	0048	0049	0050	0051	0053	0	0	1	1	1
9	0·0054	0055	0056	0057	0059	0060	0061	0063	0064	0065	0	0	1	1	1
10	0·0066	0068	0069	0071	0072	0073	0075	0076	0078	0079	0	0	1	1	1
11	0·0081	0082	0084	0085	0087	0088	0090	0091	0093	0094	0	1	1	1	1
12	0·0096	0098	0099	0101	0103	0104	0106	0108	0109	0111	0	1	1	1	1
13	0·0113	0115	0116	0118	0120	0122	0124	0125	0127	0129	0	1	1	1	2
14	0·0131	0133	0135	0137	0139	0141	0143	0145	0147	0149	0	1	1	1	2
15	0·0151	0153	0155	0157	0159	0161	0163	0165	0167	0169	0	1	1	1	2
16	0·0172	0174	0176	0178	0180	0183	0185	0187	0189	0192	0	1	1	1	2
17	0·0194	0196	0199	0201	0203	0206	0208	0211	0213	0215	0	1	1	2	2
18	0·0218	0220	0223	0225	0228	0230	0233	0236	0238	0241	0	1	1	2	2
19	0·0243	0246	0249	0251	0254	0257	0259	0262	0265	0267	0	1	1	2	2
20	0·0270	0273	0276	0278	0281	0284	0287	0290	0293	0296	0	1	1	2	2
21	0·0298	0301	0304	0307	0310	0313	0316	0319	0322	0325	0	1	1	2	2
22	0·0328	0331	0334	0338	0341	0344	0347	0350	0353	0357	1	1	2	2	3
23	0·0360	0363	0366	0369	0373	0376	0379	0383	0386	0389	1	1	2	2	3
24	0·0393	0396	0399	0403	0406	0410	0413	0417	0420	0424	1	1	2	2	3
25	0·0427	0431	0434	0438	0442	0445	0449	0452	0456	0460	1	1	2	2	3
26	0·0463	0467	0471	0475	0478	0482	0486	0490	0494	0497	1	1	2	3	3
27	0·0501	0505	0509	0513	0517	0521	0525	0529	0533	0537	1	1	2	3	3
28	0·0541	0545	0549	0553	0557	0561	0565	0569	0573	0578	1	1	2	3	3
29	0·0582	0586	0590	0594	0599	0603	0607	0612	0616	0620	1	1	2	3	4
30	0·0625	0629	0633	0638	0642	0647	0651	0656	0660	0665	1	1	2	3	4
31	0·0669	0674	0678	0683	0688	0692	0697	0702	0706	0711	1	2	2	3	4
32	0·0716	0721	0725	0730	0735	0740	0745	0749	0754	0759	1	2	2	3	4
33	0·0764	0769	0774	0779	0784	0789	0794	0799	0804	0809	1	2	3	3	4
34	0·0814	0819	0825	0830	0835	0840	0845	0851	0856	0861	1	2	3	3	4
35	0·0866	0872	0877	0882	0888	0893	0899	0904	0909	0915	1	2	3	4	5
36	0·0920	0926	0931	0937	0943	0948	0954	0959	0965	0971	1	2	3	4	5
37	0·0977	0982	0988	0994	1000	1005	1011	1017	1023	1029	1	2	3	4	5
38	0·1035	1041	1047	1053	1059	1065	1071	1077	1083	1089	1	2	3	4	5
39	0·1095	1101	1107	1113	1120	1126	1132	1138	1145	1151	1	2	3	4	5
40	0·1157	1164	1170	1177	1183	1190	1196	1203	1209	1216	1	2	3	4	5
41	0·1222	1229	1235	1242	1249	1255	1262	1269	1276	1282	1	2	3	4	6
42	0·1289	1296	1303	1310	1317	1324	1331	1338	1345	1352	1	2	3	5	6
43	0·1359	1366	1373	1380	1387	1394	1402	1409	1416	1423	1	2	4	5	6
44	0·1431	1438	1445	1453	1460	1468	1475	1483	1490	1498	1	2	4	5	6

LOGARITHMS OF SECANTS

°	0'	6'	12'	18'	24'	30'	36'	42'	48'	54'	1'	2'	3'	4'	5'
45	0·1505	1513	1520	1528	1536	1543	1551	1559	1567	1574	1	3	4	5	6
46	0·1582	1590	1598	1606	1614	1622	1630	1638	1646	1654	1	3	4	5	7
47	0·1662	1670	1678	1687	1695	1703	1711	1720	1728	1736	1	3	4	6	7
48	0·1745	1753	1762	1770	1779	1787	1796	1805	1813	1822	1	3	4	6	7
49	0·1831	1839	1848	1857	1866	1875	1883	1892	1901	1910	1	3	4	6	7
50	0·1919	1928	1937	1947	1956	1965	1974	1983	1993	2002	2	3	5	6	8
51	0·2011	2021	2030	2040	2049	2059	2068	2078	2087	2097	2	3	5	6	8
52	0·2107	2116	2126	2136	2146	2156	2165	2175	2185	2195	2	3	5	7	8
53	0·2205	2215	2226	2236	2246	2256	2266	2277	2287	2297	2	3	5	7	9
54	0·2308	2318	2329	2339	2350	2360	2371	2382	2393	2403	2	4	5	7	9
55	0·2414	2425	2436	2447	2458	2469	2480	2491	2502	2513	2	4	6	7	9
56	0·2524	2536	2547	2558	2570	2581	2593	2604	2616	2627	2	4	6	8	10
57	0·2639	2651	2662	2674	2686	2698	2710	2722	2734	2746	2	4	6	8	10
58	0·2758	2770	2782	2795	2807	2819	2832	2844	2856	2869	2	4	6	8	10
59	0·2882	2894	2907	2920	2932	2945	2958	2971	2984	2997	2	4	6	9	11
60	0·3010	3023	3037	3050	3063	3077	3090	3104	3117	3131	2	4	7	9	11
61	0·3144	3158	3172	3186	3199	3213	3227	3241	3256	3270	2	5	7	9	12
62	0·3284	3298	3313	3327	3341	3356	3371	3385	3400	3415	2	5	7	10	12
63	0·3430	3444	3459	3474	3490	3505	3520	3535	3551	3566	3	5	8	10	13
64	0·3582	3597	3613	3629	3644	3660	3676	3692	3708	3724	3	5	8	11	13
65	0·3741	3757	3773	3790	3806	3823	3839	3856	3873	3890	3	6	8	11	14
66	0·3907	3924	3941	3958	3976	3993	4010	4028	4046	4063	3	6	9	12	15
67	0·4081	4099	4117	4135	4153	4172	4190	4208	4227	4246	3	6	9	12	15
68	0·4264	4283	4302	4321	4340	4359	4379	4398	4417	4437	3	6	10	13	16
69	0·4457	4477	4496	4516	4537	4557	4577	4598	4618	4639	3	7	10	14	17
70	0·4659	4680	4701	4722	4744	4765	4787	4808	4830	4852	4	7	11	14	18
71	0·4874	4896	4918	4940	4963	4985	5008	5031	5054	5077	4	8	11	15	19
72	0·5100	5124	5147	5171	5195	5219	5243	5267	5291	5316	4	8	12	16	20
73	0·5341	5366	5391	5416	5441	5467	5492	5518	5544	5570	4	9	13	17	21
74	0·5597	5623	5650	5677	5704	5731	5758	5786	5814	5842	5	9	14	18	23
75	0·5870	5898	5927	5956	5985	6014	6043	6073	6103	6133	5	10	15	20	24
76	0·6163	6194	6225	6255	6287	6318	6350	6382	6414	6446	5	11	16	21	26
77	0·6479	6512	6545	6579	6613	6647	6681	6716	6750	6786	6	11	17	23	28
78	0·6821	6857	6893	6930	6966	7003	7041	7079	7117	7155	6	12	19	25	31
79	0·7194	7233	7273	7313	7353	7394	7435	7476	7518	7561	7	14	20	27	34
80	0·7603	7647	7690	7734	7779	7824	7869	7915	7962	8009	8	15	23	30	38
81	0·8057	8105	8153	8203	8253	8303	8354	8406	8458	8511	8	17	25	34	42
82	0·8504	8619	8674	8729	8786	8843	8901	8960	9019	9080	10	19	29	38	48
83	0·9141	9203	9266	9330	9395	9461	9528	9597	9666	9736	11	22	33	44	55
84	0·9808	9880	9954	**0030**	**0106**	**0184**	**0264**	**0345**	**0427**	**0511**	13	26	39	52	65
85	1·0597	0685	0774	0865	0958	1054	1151	1251	1353	1457	16	32	48	64	80
86	1·1564	1674	1787	1902	2021	2143	2269	2398	2532	2670					
87	1·2812	2959	3111	3269	3433	3603	3780	3965	4158	4360			Use		
88	1·4572	4794	5029	5277	5541	5821	6120	6442	6790	7168		interpolation			
89	1·7581	1·804	1·855	1·913	1·980	2·059	2·156	2·281	2·457	2·758					

Figures in **bold type** show changes of integer.

LOGARITHMS OF COTANGENTS

SUBTRACT

°	0'	6'	12'	18'	24'	30'	36'	42'	48'	54'	1'	2'	3'	4'	5'
0	Inf. Pos.	2·758	2·457	2·281	2·156	2·059	1·980	1·913	1·855	1·804					
1	1·7581	7167	6789	6441	6119	5819	5539	5275	5027	4792		Use			
2	1·4569	4357	4155	3962	3777	3599	3429	3264	3106	2954		interpolation			
3	1·2806	2663	2525	2391	2261	2135	2012	1893	1777	1664					
4	1·1554	1446	1341	1238	1138	1040	0944	0850	0759	0669	16	32	48	64	81
5	1·0580	0494	0409	0326	0244	0164	0085	0008	**9932**	**9857**	13	26	40	53	66
6	0·9784	9711	9640	9570	9501	9433	9367	9301	9236	9172	11	22	34	45	56
7	0·9109	9046	8985	8924	8865	8806	8748	8690	8633	8577	10	20	29	39	49
8	0·8522	8467	8413	8360	8307	8255	8203	8152	8102	8052	9	17	26	35	43
9	0·8003	7954	7906	7858	7811	7764	7718	7672	7626	7581	8	16	23	31	39
10	0·7537	7493	7449	7406	7363	7320	7278	7236	7195	7154	7	14	21	28	35
11	0·7113	7073	7033	6994	6954	6915	6877	6838	6800	6763	6	13	19	26	32
12	0·6725	6688	6651	6615	6578	6542	6507	6471	6436	6401	6	12	18	24	30
13	0·6366	6332	6298	6264	6230	6196	6163	6130	6097	6065	6	11	17	22	28
14	0·6032	6000	5968	5936	5905	5873	5842	5811	5780	5750	5	10	16	21	26
15	0·5719	5689	5659	5629	5600	5570	5541	5512	5483	5454	5	10	15	20	25
16	0·5425	5397	5368	5340	5312	5284	5256	5229	5201	5174	5	9	14	19	23
17	0·5147	5120	5093	5066	5039	5013	4986	4960	4934	4908	4	9	13	18	22
18	0·4882	4857	4831	4805	4780	4755	4730	4705	4680	4655	4	8	13	17	21
19	0·4630	4606	4581	4557	4533	4509	4484	4461	4437	4413	4	8	12	16	20
20	0·4389	4366	4342	4319	4296	4273	4250	4227	4204	4181	4	8	12	15	19
21	0·4158	4136	4113	4091	4068	4046	4024	4002	3980	3958	4	7	11	15	19
22	0·3936	3914	3892	3871	3849	3828	3806	3785	3764	3743	4	7	11	14	18
23	0·3721	3700	3679	3659	3638	3617	3596	3576	3555	3535	3	7	10	14	17
24	0·3514	3494	3473	3453	3433	3413	3393	3373	3353	3333	3	7	10	13	17
25	0·3313	3294	3274	3254	3235	3215	3196	3176	3157	3137	3	7	10	13	16
26	0·3118	3099	3080	3061	3042	3023	3004	2985	2966	2947	3	6	9	13	16
27	0·2928	2910	2891	2872	2854	2835	2817	2798	2780	2762	3	6	9	12	15
28	0·2743	2725	2707	2689	2670	2652	2634	2616	2598	2580	3	6	9	12	15
29	0·2562	2545	2527	2509	2491	2474	2456	2438	2421	2403	3	6	9	12	15
30	0·2386	2368	2351	2333	2316	2299	2281	2264	2247	2229	3	6	9	12	14
31	0·2212	2195	2178	2161	2144	2127	2110	2093	2076	2059	3	6	9	11	14
32	0·2042	2025	2008	1992	1975	1958	1941	1925	1908	1891	3	6	8	11	14
33	0·1875	1858	1842	1825	1809	1792	1776	1759	1743	1726	3	5	8	11	14
34	0·1710	1694	1677	1661	1645	1629	1612	1596	1580	1564	3	5	8	11	14
35	0·1548	1532	1516	1499	1483	1467	1451	1435	1419	1403	3	5	8	11	13
36	0·1387	1371	1356	1340	1324	1308	1292	1276	1260	1245	3	5	8	11	13
37	0·1229	1213	1197	1182	1166	1150	1135	1119	1103	1088	3	5	8	10	13
38	0·1072	1056	1041	1025	1010	0994	0978	0963	0947	0932	3	5	8	10	13
39	0·0916	0901	0885	0870	0854	0839	0824	0808	0793	0777	3	5	8	10	13
40	0·0762	0746	0731	0716	0700	0685	0670	0654	0639	0624	3	5	8	10	13
41	0·0608	0593	0578	0562	0547	0532	0517	0501	0486	0471	3	5	8	10	13
42	0·0456	0440	0425	0410	0395	0379	0364	0349	0334	0319	3	5	8	10	13
43	0·0303	0288	0273	0258	0243	0228	0212	0197	0182	0167	3	5	8	10	13
44	0·0152	0136	0121	0106	0091	0076	0061	0045	0030	0015	3	5	8	10	13

SUBTRACT

Figures in **bold type** show changes of integer.

SUBTRACT

°	0′	6′	12′	18′	24′	30′	36′	42′	48′	54′	1′	2′	3′	4′	5′
45	0·0000	**9985**	**9970**	**9955**	**9939**	**9924**	**9909**	**9894**	**9879**	**9864**	3	5	8	10	13
46	$\bar{1}$·9848	9833	9818	9803	9788	9772	9757	9742	9727	9712	3	5	8	10	13
47	$\bar{1}$·9697	9681	9666	9651	9636	9621	9605	9590	9575	9560	3	5	8	10	13
48	$\bar{1}$·9544	9529	9514	9499	9483	9468	9453	9438	9422	9407	3	5	8	10	13
49	$\bar{1}$·9392	9376	9361	9346	9330	9315	9300	9284	9269	9254	3	5	8	10	13
50	$\bar{1}$·9238	9223	9207	9192	9176	9161	9146	9130	9115	9099	3	5	8	10	13
51	$\bar{1}$·9084	9068	9053	9037	9022	9006	8990	8975	8959	8944	3	5	8	10	13
52	$\bar{1}$·8928	8912	8897	8881	8865	8850	8834	8818	8803	8787	3	5	8	10	13
53	$\bar{1}$·8771	8755	8740	8724	8708	8692	8676	8660	8644	8629	3	5	8	11	13
54	$\bar{1}$·8613	8597	8581	8565	8549	8533	8517	8501	8484	8468	3	5	8	11	13
55	$\bar{1}$·8452	8436	8420	8404	8388	8371	8355	8339	8323	8306	3	5	8	11	14
56	$\bar{1}$·8290	8274	8257	8241	8224	8208	8191	8175	8158	8142	3	5	8	11	14
57	$\bar{1}$·8125	8109	8092	8075	8059	8042	8025	8008	7992	7975	3	6	8	11	14
58	$\bar{1}$·7958	7941	7924	7907	7890	7873	7856	7839	7822	7805	3	6	9	11	14
59	$\bar{1}$·7788	7771	7753	7736	7719	7701	7684	7667	7649	7632	3	6	9	12	14
60	$\bar{1}$·7614	7597	7579	7562	7544	7526	7509	7491	7473	7455	3	6	9	12	15
61	$\bar{1}$·7438	7420	7402	7384	7366	7348	7330	7311	7293	7275	3	6	9	12	15
62	$\bar{1}$·7257	7238	7220	7202	7183	7165	7146	7128	7109	7090	3	6	9	12	15
63	$\bar{1}$·7072	7053	7034	7015	6996	6977	6958	6939	6920	6901	3	6	9	13	16
64	$\bar{1}$·6882	6863	6843	6824	6804	6785	6765	6746	6726	6706	3	6	10	13	16
65	$\bar{1}$·6687	6667	6647	6627	6607	6587	6567	6547	6527	6506	3	7	10	13	17
66	$\bar{1}$·6486	6465	6445	6424	6404	6383	6362	6341	6321	6300	3	7	10	14	17
67	$\bar{1}$·6279	6257	6236	6215	6194	6172	6151	6129	6108	6086	4	7	11	14	18
68	$\bar{1}$·6064	6042	6020	5998	5976	5954	5932	5909	5887	5864	4	7	11	15	19
69	$\bar{1}$·5842	5819	5796	5773	5750	5727	5704	5681	5658	5634	4	8	12	15	19
70	$\bar{1}$·5611	5587	5563	5539	5516	5491	5467	5443	5419	5394	4	8	12	16	20
71	$\bar{1}$·5370	5345	5320	5295	5270	5245	5220	5195	5169	5143	4	8	13	17	21
72	$\bar{1}$·5118	5092	5066	5040	5014	4987	4961	4934	4907	4880	4	9	13	18	22
73	$\bar{1}$·4853	4826	4799	4771	4744	4716	4688	4660	4632	4603	5	9	14	19	23
74	$\bar{1}$·4575	4546	4517	4488	4459	4430	4400	4371	4341	4311	5	10	15	20	25
75	$\bar{1}$·4281	4250	4220	4189	4158	4127	4095	4064	4032	4000	5	10	16	21	26
76	$\bar{1}$·3968	3935	3903	3870	3837	3804	3770	3736	3702	3668	6	11	17	22	28
77	$\bar{1}$·3634	3599	3564	3529	3493	3458	3422	3385	3349	3312	6	12	18	24	30
78	$\bar{1}$·3275	3237	3200	3162	3123	3085	3046	3006	2967	2927	6	13	19	26	32
79	$\bar{1}$·2887	2846	2805	2764	2722	2680	2637	2594	2551	2507	7	14	21	28	35
80	$\bar{1}$·2463	2419	2374	2328	2282	2236	2189	2142	2094	2046	8	16	23	31	39
81	$\bar{1}$·1997	1948	1898	1848	1797	1745	1693	1640	1587	1533	9	17	26	35	43
82	$\bar{1}$·1478	1423	1367	1310	1252	1194	1135	1076	1015	0954	10	20	29	39	49
83	$\bar{1}$·0891	0828	0764	0699	0633	0567	0499	0430	0360	0289	11	22	34	45	56
84	$\bar{1}$·0216	0143	0068	**9992**	**9915**	**9836**	**9756**	**9674**	**9591**	**9506**	13	26	40	53	66
85	$\bar{2}$·9420	9331	9241	9150	9056	8960	8862	8762	8659	8554	16	32	48	64	81
86	$\bar{2}$·8446	8336	8223	8107	7988	7865	7739	7609	7475	7337					
87	$\bar{2}$·7194	7046	6894	6736	6571	6401	6223	6038	5845	5643			Use		
88	$\bar{2}$·5431	5208	4973	4725	4461	4181	3881	3559	3211	2833		interpolation			
89	$\bar{2}$·242	$\bar{2}$·196	$\bar{2}$·145	$\bar{2}$·087	$\bar{2}$·020	$\bar{3}$·941	$\bar{3}$·844	$\bar{3}$·719	$\bar{3}$·543	$\bar{3}$·242					

SUBTRACT

Figures in **bold type** show changes of integer.

NATURAL SINES

°	0′	6′	12′	18′	24′	30′	36′	42′	48′	54′	1′	2′	3′	4′	5′
0	·0000	0017	0035	0052	0070	0087	0105	0122	0140	0157	3	6	9	12	15
1	·0175	0192	0209	0227	0244	0262	0279	0297	0314	0332	3	6	9	12	15
2	·0349	0366	0384	0401	0419	0436	0454	0471	0488	0506	3	6	9	12	15
3	·0523	0541	0558	0576	0593	0610	0628	0645	0663	0680	3	6	9	12	15
4	·0698	0715	0732	0750	0767	0785	0802	0819	0837	0854	3	6	9	12	14
5	·0872	0889	0906	0924	0941	0958	0976	0993	1011	1028	3	6	9	12	14
6	·1045	1063	1080	1097	1115	1132	1149	1167	1184	1201	3	6	9	12	14
7	·1219	1236	1253	1271	1288	1305	1323	1340	1357	1374	3	6	9	12	14
8	·1392	1409	1426	1444	1461	1478	1495	1513	1530	1547	3	6	9	12	14
9	·1564	1582	1599	1616	1633	1650	1668	1685	1702	1719	3	6	9	11	14
10	·1736	1754	1771	1788	1805	1822	1840	1857	1874	1891	3	6	9	11	14
11	·1908	1925	1942	1959	1977	1994	2011	2028	2045	2062	3	6	9	11	14
12	·2079	2096	2113	2130	2147	2164	2181	2198	2215	2233	3	6	9	11	14
13	·2250	2267	2284	2300	2317	2334	2351	2368	2385	2402	3	6	8	11	14
14	·2419	2436	2453	2470	2487	2504	2521	2538	2554	2571	3	6	8	11	14
15	·2588	2605	2622	2639	2656	2672	2689	2706	2723	2740	3	6	8	11	14
16	·2756	2773	2790	2807	2823	2840	2857	2874	2890	2907	3	6	8	11	14
17	·2924	2940	2957	2974	2990	3007	3024	3040	3057	3074	3	6	8	11	14
18	·3090	3107	3123	3140	3156	3173	3190	3206	3223	3239	3	6	8	11	14
19	·3256	3272	3289	3305	3322	3338	3355	3371	3387	3404	3	5	8	11	14
20	·3420	3437	3453	3469	3486	3502	3518	3535	3551	3567	3	5	8	11	14
21	·3584	3600	3616	3633	3649	3665	3681	3697	3714	3730	3	5	8	11	14
22	·3746	3762	3778	3795	3811	3827	3843	3859	3875	3891	3	5	8	11	13
23	·3907	3923	3939	3955	3971	3987	4003	4019	4035	4051	3	5	8	11	13
24	·4067	4083	4099	4115	4131	4147	4163	4179	4195	4210	3	5	8	11	13
25	·4226	4242	4258	4274	4289	4305	4321	4337	4352	4368	3	5	8	11	13
26	·4384	4399	4415	4431	4446	4462	4478	4493	4509	4524	3	5	8	10	13
27	·4540	4555	4571	4586	4602	4617	4633	4648	4664	4679	3	5	8	10	13
28	·4695	4710	4726	4741	4756	4772	4787	4802	4818	4833	3	5	8	10	13
29	·4848	4863	4879	4894	4909	4924	4939	4955	4970	4985	3	5	8	10	13
30	·5000	5015	5030	5045	5060	5075	5090	5105	5120	5135	3	5	8	10	13
31	·5150	5165	5180	5195	5210	5225	5240	5255	5270	5284	2	5	7	10	12
32	·5299	5314	5329	5344	5358	5373	5388	5402	5417	5432	2	5	7	10	12
33	·5446	5461	5476	5490	5505	5519	5534	5548	5563	5577	2	5	7	10	12
34	·5592	5606	5621	5635	5650	5664	5678	5693	5707	5721	2	5	7	10	12
35	·5736	5750	5764	5779	5793	5807	5821	5835	5850	5864	2	5	7	10	12
36	·5878	5892	5906	5920	5934	5948	5962	5976	5990	6004	2	5	7	9	12
37	·6018	6032	6046	6060	6074	6088	6101	6115	6129	6143	2	5	7	9	12
38	·6157	6170	6184	6198	6211	6225	6239	6252	6266	6280	2	5	7	9	11
39	·6293	6307	6320	6334	6347	6361	6374	6388	6401	6414	2	4	7	9	11
40	·6428	6441	6455	6468	6481	6494	6508	6521	6534	6547	2	4	7	9	11
41	·6561	6574	6587	6600	6613	6626	6639	6652	6665	6678	2	4	7	9	11
42	·6691	6704	6717	6730	6743	6756	6769	6782	6794	6807	2	4	6	9	11
43	·6820	6833	6845	6858	6871	6884	6896	6909	6921	6934	2	4	6	8	11
44	·6947	6959	6972	6984	6997	7009	7022	7034	7046	7059	2	4	6	8	10

NATURAL SINES

°	0′	6′	12′	18′	24′	30′	36′	42′	48′	54′	1′	2′	3′	4′	5′
45	·7071	7083	7096	7108	7120	7133	7145	7157	7169	7181	2	4	6	8	10
46	·7193	7206	7218	7230	7242	7254	7266	7278	7290	7302	2	4	6	8	10
47	·7314	7325	7337	7349	7361	7373	7385	7396	7408	7420	2	4	6	8	10
48	·7431	7443	7455	7466	7478	7490	7501	7513	7524	7536	2	4	6	8	10
49	·7547	7559	7570	7581	7593	7604	7615	7627	7638	7649	2	4	6	8	9
50	·7660	7672	7683	7694	7705	7716	7727	7738	7749	7760	2	4	6	7	9
51	·7771	7782	7793	7804	7815	7826	7837	7848	7859	7869	2	4	5	7	9
52	·7880	7891	7902	7912	7923	7934	7944	7955	7965	7976	2	4	5	7	9
53	·7986	7997	8007	8018	8028	8039	8049	8059	8070	8080	2	3	5	7	9
54	·8090	8100	8111	8121	8131	8141	8151	8161	8171	8181	2	3	5	7	8
55	·8192	8202	8211	8221	8231	8241	8251	8261	8271	8281	2	3	5	7	8
56	·8290	8300	8310	8320	8329	8339	8348	8358	8368	8377	2	3	5	6	8
57	·8387	8396	8406	8415	8425	8434	8443	8453	8462	8471	2	3	5	6	8
58	·8480	8490	8499	8508	8517	8526	8536	8545	8554	8563	2	3	5	6	8
59	·8572	8581	8590	8599	8607	8616	8625	8634	8643	8652	1	3	4	6	7
60	·8660	8669	8678	8686	8695	8704	8712	8721	8729	8738	1	3	4	6	7
61	·8746	8755	8763	8771	8780	8788	8796	8805	8813	8821	1	3	4	6	7
62	·8829	8838	8846	8854	8862	8870	8878	8886	8894	8902	1	3	4	5	7
63	·8910	8918	8926	8934	8942	8949	8957	8965	8973	8980	1	3	4	5	6
64	·8988	8996	9003	9011	9018	9026	9033	9041	9048	9056	1	3	4	5	6
65	·9063	9070	9078	9085	9092	9100	9107	9114	9121	9128	1	2	4	5	6
66	·9135	9143	9150	9157	9164	9171	9178	9184	9191	9198	1	2	3	5	6
67	·9205	9212	9219	9225	9232	9239	9245	9252	9259	9265	1	2	3	4	6
68	·9272	9278	9285	9291	9298	9304	9311	9317	9323	9330	1	2	3	4	5
69	·9336	9342	9348	9354	9361	9367	9373	9379	9385	9391	1	2	3	4	5
70	·9397	9403	9409	9415	9421	9426	9432	9438	9444	9449	1	2	3	4	5
71	·9455	9461	9466	9472	9478	9483	9489	9494	9500	9505	1	2	3	4	5
72	·9511	9516	9521	9527	9532	9537	9542	9548	9553	9558	1	2	3	4	4
73	·9563	9568	9573	9578	9583	9588	9593	9598	9603	9608	1	2	2	3	4
74	·9613	9617	9622	9627	9632	9636	9641	9646	9650	9655	1	2	2	3	4
75	·9659	9664	9668	9673	9677	9681	9686	9690	9694	9699	1	1	2	3	4
76	·9703	9707	9711	9715	9720	9724	9728	9732	9736	9740	1	1	2	3	3
77	·9744	9748	9751	9755	9759	9763	9767	9770	9774	9778	1	1	2	3	3
78	·9781	9785	9789	9792	9796	9799	9803	9806	9810	9813	1	1	2	2	3
79	·9816	9820	9823	9826	9829	9833	9836	9839	9842	9845	1	1	2	2	3
80	·9848	9851	9854	9857	9860	9863	9866	9869	9871	9874	0	1	1	2	2
81	·9877	9880	9882	9885	9888	9890	9893	9895	9898	9900	0	1	1	2	2
82	·9903	9905	9907	9910	9912	9914	9917	9919	9921	9923	0	1	1	2	2
83	·9925	9928	9930	9932	9934	9936	9938	9940	9942	9943	0	1	1	1	2
84	·9945	9947	9949	9951	9952	9954	9956	9957	9959	9960	0	1	1	1	1
85	·9962	9963	9965	9966	9968	9969	9971	9972	9973	9974	0	0	1	1	1
86	·9976	9977	9978	9979	9980	9981	9982	9983	9984	9985	0	0	1	1	1
87	·9986	9987	9988	9989	9990	9990	9991	9992	9993	9993	0	0	0	0	0
88	·9994	9995	9995	9996	9996	9997	9997	9997	9998	9998	0	0	0	0	0
89	·9998	9999	9999	9999	9999	1·000	1·000	1·000	1·000	1·000	0	0	0	0	0

NATURAL COSINES

°	0′	6′	12′	18′	24′	30′	36′	42′	48′	54′	1′	2′	3′	4′	5′
0	1·000	1·000	1·000	1·000	1·000	1·000	**9999**	**9999**	**9999**	**9999**	0	0	0	0	0
1	·9998	9998	9998	9997	9997	9997	9996	9996	9995	9995	0	0	0	0	0
2	·9994	9993	9993	9992	9991	9990	9990	9989	9988	9987	0	0	0	0	0
3	·9986	9985	9984	9983	9982	9981	9980	9979	9978	9977	0	0	1	1	1
4	·9976	9974	9973	9972	9971	9969	9968	9966	9965	9963	0	0	1	1	1
5	·9962	9960	9959	9957	9956	9954	9952	9951	9949	9947	0	1	1	1	1
6	·9945	9943	9942	9940	9938	9936	9934	9932	9930	9928	0	1	1	1	2
7	·9925	9923	9921	9919	9917	9914	9912	9910	9907	9905	0	1	1	2	2
8	·9903	9900	9898	9895	9893	9890	9888	9885	9882	9880	0	1	1	2	2
9	·9877	9874	9871	9869	9866	9863	9860	9857	9854	9851	0	1	1	2	2
10	·9848	9845	9842	9839	9836	9833	9829	9826	9823	9820	1	1	2	2	3
11	·9816	9813	9810	9806	9803	9799	9796	9792	9789	9785	1	1	2	2	3
12	·9781	9778	9774	9770	9767	9763	9759	9755	9751	9748	1	1	2	3	3
13	·9744	9740	9736	9732	9728	9724	9720	9715	9711	9707	1	1	2	3	3
14	·9703	9699	9694	9690	9686	9681	9677	9673	9668	9664	1	1	2	3	4
15	·9659	9655	9650	9646	9641	9636	9632	9627	9622	9617	1	2	2	3	4
16	·9613	9608	9603	9598	9593	9588	9583	9578	9573	9568	1	2	2	3	4
17	·9563	9558	9553	9548	9542	9537	9532	9527	9521	9516	1	2	3	4	4
18	·9511	9505	9500	9494	9489	9483	9478	9472	9466	9461	1	2	3	4	5
19	·9455	9449	9444	9438	9432	9426	9421	9415	9409	9403	1	2	3	4	5
20	·9397	9391	9385	9379	9373	9367	9361	9354	9348	9342	1	2	3	4	5
21	·9336	9330	9323	9317	9311	9304	9298	9291	9285	9278	1	2	3	4	5
22	·9272	9265	9259	9252	9245	9239	9232	9225	9219	9212	1	2	3	4	6
23	·9205	9198	9191	9184	9178	9171	9164	9157	9150	9143	1	2	3	5	6
24	·9135	9128	9121	9114	9107	9100	9092	9085	9078	9070	1	2	4	5	6
25	·9063	9056	9048	9041	9033	9026	9018	9011	9003	8996	1	3	4	5	6
26	·8988	8980	8973	8965	8957	8949	8942	8934	8926	8918	1	3	4	5	6
27	·8910	8902	8894	8886	8878	8870	8862	8854	8846	8838	1	3	4	5	7
28	·8829	8821	8813	8805	8796	8788	8780	8771	8763	8755	1	3	4	6	7
29	·8746	8738	8729	8721	8712	8704	8695	8686	8678	8669	1	3	4	6	7
30	·8660	8652	8643	8634	8625	8616	8607	8599	8590	8581	1	3	4	6	7
31	·8572	8563	8554	8545	8536	8526	8517	8508	8499	8490	2	3	5	6	8
32	·8480	8471	8462	8453	8443	8434	8425	8415	8406	8396	2	3	5	6	8
33	·8387	8377	8368	8358	8348	8339	8329	8320	8310	8300	2	3	5	6	8
34	·8290	8281	8271	8261	8251	8241	8231	8221	8211	8202	2	3	5	7	8
35	·8192	8181	8171	8161	8151	8141	8131	8121	8111	8100	2	3	5	7	8
36	·8090	8080	8070	8059	8049	8039	8028	8018	8007	7997	2	3	5	7	9
37	·7986	7976	7965	7955	7944	7934	7923	7912	7902	7891	2	4	5	7	9
38	·7880	7869	7859	7848	7837	7826	7815	7804	7793	7782	2	4	5	7	9
39	·7771	7760	7749	7738	7727	7716	7705	7694	7683	7672	2	4	6	7	9
40	·7660	7649	7638	7627	7615	7604	7593	7581	7570	7559	2	4	6	8	9
41	·7547	7536	7524	7513	7501	7490	7478	7466	7455	7443	2	4	6	8	10
42	·7431	7420	7408	7396	7385	7373	7361	7349	7337	7325	2	4	6	8	10
43	·7314	7302	7290	7278	7266	7254	7242	7230	7218	7206	2	4	6	8	10
44	·7193	7181	7169	7157	7145	7133	7120	7108	7096	7083	2	4	6	8	10

Figures in **bold** type show change of integer

NATURAL COSINES

°	0'	6'	12'	18'	24'	30'	36'	42'	48'	54'	1'	2'	3'	4'	5'
45	·7071	7059	7046	7034	7022	7009	6997	6984	6972	6959	2	4	6	8	10
46	·6947	6934	6921	6909	6896	6884	6871	6858	6845	6833	2	4	6	8	11
47	·6820	6807	6794	6782	6769	6756	6743	6730	6717	6704	2	4	6	9	11
48	·6691	6678	6665	6652	6639	6626	6613	6600	6587	6574	2	4	7	9	11
49	·6561	6547	6534	6521	6508	6494	6481	6468	6455	6441	2	4	7	9	11
50	·6428	6414	6401	6388	6374	6361	6347	6334	6320	6307	2	4	7	9	11
51	·6293	6280	6266	6252	6239	6225	6211	6198	6184	6170	2	5	7	9	11
52	·6157	6143	6129	6115	6101	6088	6074	6060	6046	6032	2	5	7	9	12
53	·6018	6004	5990	5976	5962	5948	5934	5920	5906	5892	2	5	7	9	12
54	·5878	5864	5850	5835	5821	5807	5793	5779	5764	5750	2	5	7	9	12
55	·5736	5721	5707	5693	5678	5664	5650	5635	5621	5606	2	5	7	10	12
56	·5592	5577	5563	5548	5534	5519	5505	5490	5476	5461	2	5	7	10	12
57	·5446	5432	5417	5402	5388	5373	5358	5344	5329	5314	2	5	7	10	12
58	·5299	5284	5270	5255	5240	5225	5210	5195	5180	5165	2	5	7	10	12
59	·5150	5135	5120	5105	5090	5075	5060	5045	5030	5015	3	5	8	10	13
60	·5000	4985	4970	4955	4939	4924	4909	4894	4879	4863	3	5	8	10	13
61	·4848	4833	4818	4802	4787	4772	4756	4741	4726	4710	3	5	8	10	13
62	·4695	4679	4664	4648	4633	4617	4602	4586	4571	4555	3	5	8	10	13
63	·4540	4524	4509	4493	4478	4462	4446	4431	4415	4399	3	5	8	10	13
64	·4384	4368	4352	4337	4321	4305	4289	4274	4258	4242	3	5	8	11	13
65	·4226	4210	4195	4179	4163	4147	4131	4115	4099	4083	3	5	8	11	13
66	·4067	4051	4035	4019	4003	3987	3971	3955	3939	3923	3	5	8	11	13
67	·3907	3891	3875	3859	3843	3827	3811	3795	3778	3762	3	5	8	11	13
68	·3746	3730	3714	3697	3681	3665	3649	3633	3616	3600	3	5	8	11	14
69	·3584	3567	3551	3535	3518	3502	3486	3469	3453	3437	3	5	8	11	14
70	·3420	3404	3387	3371	3355	3338	3322	3305	3289	3272	3	5	8	11	14
71	·3256	3239	3223	3206	3190	3173	3156	3140	3123	3107	3	6	8	11	14
72	·3090	3074	3057	3040	3024	3007	2990	2974	2957	2940	3	6	8	11	14
73	·2924	2907	2890	2874	2857	2840	2823	2807	2790	2773	3	6	8	11	14
74	·2756	2740	2723	2706	2689	2672	2656	2639	2622	2605	3	6	8	11	14
75	·2588	2571	2554	2538	2521	2504	2487	2470	2453	2436	3	6	8	11	14
76	·2419	2402	2385	2368	2351	2334	2317	2300	2284	2267	3	6	8	11	14
77	·2250	2233	2215	2198	2181	2164	2147	2130	2113	2096	3	6	9	11	14
78	·2079	2062	2045	2028	2011	1994	1977	1959	1942	1925	3	6	9	11	14
79	·1908	1891	1874	1857	1840	1822	1805	1788	1771	1754	3	6	9	11	14
80	·1736	1719	1702	1685	1668	1650	1633	1616	1599	1582	3	6	9	11	14
81	·1564	1547	1530	1513	1495	1478	1461	1444	1426	1409	3	6	9	12	14
82	·1392	1374	1357	1340	1323	1305	1288	1271	1253	1236	3	6	9	12	14
83	·1219	1201	1184	1167	1149	1132	1115	1097	1080	1063	3	6	9	12	14
84	·1045	1028	1011	0993	0976	0958	0941	0924	0906	0889	3	6	9	12	14
85	·0872	0854	0837	0819	0802	0785	0767	0750	0732	0715	3	6	9	12	14
86	·0698	0680	0663	0645	0628	0610	0593	0576	0558	0541	3	6	9	12	15
87	·0523	0506	0488	0471	0454	0436	0419	0401	0384	0366	3	6	9	12	15
88	·0349	0332	0314	0297	0279	0262	0244	0227	0209	0192	3	6	9	12	15
89	·0175	0157	0140	0122	0105	0087	0070	0052	0035	0017	3	6	9	12	15

NATURAL TANGENTS

°	0′	6′	12′	18′	24′	30′	36′	42′	48′	54′	1′	2′	3′	4′	5′
0	·0000	0017	0035	0052	0070	0087	0105	0122	0140	0157	3	6	9	12	15
1	·0175	0192	0209	0227	0244	0262	0279	0297	0314	0332	3	6	9	12	15
2	·0349	0367	0384	0402	0419	0437	0454	0472	0489	0507	3	6	9	12	15
3	·0524	0542	0559	0577	0594	0612	0629	0647	0664	0682	3	6	9	12	15
4	·0699	0717	0734	0752	0769	0787	0805	0822	0840	0857	3	6	9	12	15
5	·0875	0892	0910	0928	0945	0963	0981	0998	1016	1033	3	6	9	12	15
6	·1051	1069	1086	1104	1122	1139	1157	1175	1192	1210	3	6	9	12	15
7	·1228	1246	1263	1281	1299	1317	1334	1352	1370	1388	3	6	9	12	15
8	·1405	1423	1441	1459	1477	1495	1512	1530	1548	1566	3	6	9	12	15
9	·1584	1602	1620	1638	1655	1673	1691	1709	1727	1745	3	6	9	12	15
10	·1763	1781	1799	1817	1835	1853	1871	1890	1908	1926	3	6	9	12	15
11	·1944	1962	1980	1998	2016	2035	2053	2071	2089	2107	3	6	9	12	15
12	·2126	2144	2162	2180	2199	2217	2235	2254	2272	2290	3	6	9	12	15
13	·2309	2327	2345	2364	2382	2401	2419	2438	2456	2475	3	6	9	12	15
14	·2493	2512	2530	2549	2568	2586	2605	2623	2642	2661	3	6	9	12	16
15	·2679	2698	2717	2736	2754	2773	2792	2811	2830	2849	3	6	9	13	16
16	·2867	2886	2905	2924	2943	2962	2981	3000	3019	3038	3	6	9	13	16
17	·3057	3076	3096	3115	3134	3153	3172	3191	3211	3230	3	6	10	13	16
18	·3249	3269	3288	3307	3327	3346	3365	3385	3404	3424	3	6	10	13	16
19	·3443	3463	3482	3502	3522	3541	3561	3581	3600	3620	3	7	10	13	16
20	·3640	3659	3679	3699	3719	3739	3759	3779	3799	3819	3	7	10	13	17
21	·3839	3859	3879	3899	3919	3939	3959	3979	4000	4020	3	7	10	13	17
22	·4040	4061	4081	4101	4122	4142	4163	4183	4204	4224	3	7	10	14	17
23	·4245	4265	4286	4307	4327	4348	4369	4390	4411	4431	3	7	10	14	17
24	·4452	4473	4494	4515	4536	4557	4578	4599	4621	4642	4	7	11	14	18
25	·4663	4684	4706	4727	4748	4770	4791	4813	4834	4856	4	7	11	14	18
26	·4877	4899	4921	4942	4964	4986	5008	5029	5051	5073	4	7	11	15	18
27	·5095	5117	5139	5161	5184	5206	5228	5250	5272	5295	4	7	11	15	18
28	·5317	5340	5362	5384	5407	5430	5452	5475	5498	5520	4	8	11	15	19
29	·5543	5566	5589	5612	5635	5658	5681	5704	5727	5750	4	8	12	15	19
30	·5774	5797	5820	5844	5867	5890	5914	5938	5961	5985	4	8	12	16	20
31	·6009	6032	6056	6080	6104	6128	6152	6176	6200	6224	4	8	12	16	20
32	·6249	6273	6297	6322	6346	6371	6395	6420	6445	6469	4	8	12	16	20
33	·6494	6519	6544	6569	6594	6619	6644	6669	6694	6720	4	8	13	17	21
34	·6745	6771	6796	6822	6847	6873	6899	6924	6950	6976	4	9	13	17	21
35	·7002	7028	7054	7080	7107	7133	7159	7186	7212	7239	4	9	13	18	22
36	·7265	7292	7319	7346	7373	7400	7427	7454	7481	7508	5	9	14	18	23
37	·7536	7563	7590	7618	7646	7673	7701	7729	7757	7785	5	9	14	18	23
38	·7813	7841	7869	7898	7926	7954	7983	8012	8040	8069	5	9	14	19	24
39	·8098	8127	8156	8185	8214	8243	8273	8302	8332	8361	5	10	15	20	24
40	·8391	8421	8451	8481	8511	8541	8571	8601	8632	8662	5	10	15	20	25
41	·8693	8724	8754	8785	8816	8847	8878	8910	8941	8972	5	10	16	21	26
42	·9004	9036	9067	9099	9131	9163	9195	9228	9260	9293	5	11	16	21	27
43	·9325	9358	9391	9424	9457	9490	9523	9556	9590	9623	6	11	17	22	28
44	·9657	9691	9725	9759	9793	9827	9861	9896	9930	9965	6	11	17	23	29

NATURAL TANGENTS

°	0′	6′	12′	18′	24′	30′	36′	42′	48′	54′	1′	2′	3′	4′	5′
45	1·0000	0035	0070	0105	0141	0176	0212	0247	0283	0319	6	12	18	24	30
46	1·0355	0392	0428	0464	0501	0538	0575	0612	0649	0686	6	12	18	25	31
47	1·0724	0761	0799	0837	0875	0913	0951	0990	1028	1067	6	13	19	25	32
48	1·1106	1145	1184	1224	1263	1303	1343	1383	1423	1463	7	13	20	26	33
49	1·1504	1544	1585	1626	1667	1708	1750	1792	1833	1875	7	14	21	28	34
50	1·1918	1960	2002	2045	2088	2131	2174	2218	2261	2305	7	14	22	29	36
51	1·2349	2393	2437	2482	2527	2572	2617	2662	2708	2753	8	15	23	30	38
52	1·2799	2846	2892	2938	2985	3032	3079	3127	3175	3222	8	16	24	31	39
53	1·3270	3319	3367	3416	3465	3514	3564	3613	3663	3713	8	16	25	33	41
54	1·3764	3814	3865	3916	3968	4019	4071	4124	4176	4229	9	17	26	34	43
55	1·4281	4335	4388	4442	4496	4550	4605	4659	4715	4770	9	18	27	36	45
56	1·4826	4882	4938	4994	5051	5108	5166	5224	5282	5340	10	19	29	38	48
57	1·5399	5458	5517	5577	5637	5697	5757	5818	5880	5941	10	20	30	40	50
58	1·6003	6066	6128	6191	6255	6319	6383	6447	6512	6577	11	21	32	43	53
59	1·6643	6709	6775	6842	6909	6977	7045	7113	7182	7251	11	23	34	45	56
60	1·7321	7391	7461	7532	7603	7675	7747	7820	7893	7966	12	24	36	48	60
61	1·8040	8115	8190	8265	8341	8418	8495	8572	8650	8728	13	26	38	51	64
62	1·8807	8887	8967	9047	9128	9210	9292	9375	9458	9542	14	27	41	55	68
63	1·9626	9711	9797	9883	9970	**0057**	**0145**	**0233**	**0323**	**0413**	15	29	44	58	73
64	2·0503	0594	0686	0778	0872	0965	1060	1155	1251	1348	16	31	47	63	78
65	2·1445	1543	1642	1742	1842	1943	2045	2148	2251	2355	17	34	51	68	85
66	2·2460	2566	2673	2781	2889	2998	3109	3220	3332	3445	18	37	55	73	92
67	2·3559	3673	3789	3906	4023	4142	4262	4383	4504	4627	20	40	60	79	99
68	2·4751	4876	5002	5129	5257	5386	5517	5649	5782	5916	22	43	65	87	108
69	2·6051	6187	6325	6464	6605	6746	6889	7034	7179	7326	24	47	71	95	118
70	2·7475	7625	7776	7929	8083	8239	8397	8556	8716	8878	26	52	78	104	130
71	2·9042	9208	9375	9544	9714	9887	**0061**	**0237**	**0415**	**0595**	29	58	87	116	145
72	3·0777	0961	1146	1334	1524	1716	1910	2106	2305	2506	32	64	96	129	161
73	3·2709	2914	3122	3332	3544	3759	3977	4197	4420	4646	36	72	108	144	180
74	3·4874	5105	5339	5576	5816	6059	6305	6554	6806	7062	41	81	122	163	203
75	3·7321	7583	7848	8118	8391	8667	8947	9232	9520	9812	Use interpolation				
76	4·0108	0408	0713	1022	1335	1653	1976	2303	2635	2972					
77	4·3315	3662	4015	4373	4737	5107	5483	5864	6252	6646					
78	4·7046	7453	7867	8288	8716	9152	9594	**0045**	**0504**	**0970**					
79	5·1446	1929	2422	2924	3435	3955	4486	5026	5578	6140					
80	5·6713	7297	7894	8502	9124	9758	**0405**	**1066**	**1742**	**2432**					
81	6·3138	3859	4596	5350	6122	6912	7720	8548	9395	**0264**					
82	7·1154	2066	3002	3962	4947	5958	6996	8062	9158	**0285**					
83	8·1443	2636	3863	5126	6427	7769	9152	**0579**	**2052**	**3572**					
84	9·5144	6768	8448	10·02	10·20	10·39	10·58	10·78	10·99	11·20					
85	11·43	11·66	11·91	12·16	12·43	12·71	13·00	13·30	13·62	13·95					
86	14·30	14·67	15·06	15·46	15·89	16·35	16·83	17·34	17·89	18·46					
87	19·08	19·74	20·45	21·20	22·02	22·90	23·86	24·90	26·03	27·27					
88	28·64	30·14	31·82	33·69	35·80	38·19	40·92	44·07	47·74	52·08					
89	57·29	63·66	71·62	81·85	95·49	114·6	143·2	191·0	286·5	573·0					

Figures in **bold type** show changes of integer

NATURAL COSECANTS

SUBTRACT

°	0′	6′	12′	18′	24′	30′	36′	42′	48′	54′	1′	2′	3′	4′	5′
0	Inf. Pos.	573·0	286·5	191·0	143·2	114·6	95·49	81·85	71·62	63·66					
1	57·30	52·09	47·75	44·08	40·93	38·20	35·81	33·71	31·84	30·16					
2	28·65	27·29	26·05	24·92	23·88	22·93	22·04	21·23	20·47	19·77					
3	19·11	18·49	17·91	17·37	16·86	16·38	15·93	15·50	15·09	14·70					
4	14·34	13·99	13·65	13·34	13·03	12·75	12·47	12·20	11·95	11·71			Use		
5	11·47	11·25	11·03	10·83	10·63	10·43	10·25	10·07	9·895	9·728		interpolation			
6	9·567	9·411	9·259	9·113	8·971	8·834	3·700	8·571	8·446	8·324					
7	8·206	8·091	7·979	7·870	7·764	7·661	7·561	7·463	7·368	7·276					
8	7·185	7·097	7·011	6·927	6·845	6·765	6·687	6·611	6·537	6·464					
9	6·392	6·323	6·255	6·188	6·123	6·059	5·996	5·935	5·875	5·816					
10	5·7588	7023	6470	5928	5396	4874	4362	3860	3367	2883	86	173	259	345	432
11	5·2408	1942	1484	1034	0593	0159	9732	9313	8901	8496	72	144	216	287	359
12	4·8097	7706	7321	6942	6569	6202	5841	5486	5137	4793	61	121	182	243	304
13	4·4454	4121	3792	3469	3150	2837	2527	2223	1923	1627	52	104	156	208	260
14	4·1336	1048	0765	0486	0211	9939	9672	9408	9147	8890	45	90	135	180	225
15	3·8637	8387	8140	7897	7657	7420	7186	6955	6727	6502	39	79	118	157	196
16	3·6280	6060	5843	5629	5418	5209	5003	4799	4598	4399	35	69	104	138	173
17	3·4203	4009	3817	3628	3440	3255	3072	2891	2712	2535	31	61	92	123	153
18	3·2361	2188	2017	1848	1681	1515	1352	1190	1030	0872	27	55	82	110	137
19	3·0716	0561	0407	0256	0106	9957	9811	9665	9521	9379	25	49	74	99	123
20	2·9238	9099	8960	8824	8688	8555	8422	8291	8161	8032	22	44	67	89	111
21	2·7904	7778	7653	7529	7407	7285	7165	7046	6927	6811	20	40	60	81	101
22	2·6695	6580	6466	6354	6242	6131	6022	5913	5805	5699	18	37	55	73	92
23	2·5593	5488	5384	5282	5180	5078	4978	4879	4780	4683	17	34	50	67	84
24	2·4586	4490	4395	4300	4207	4114	4022	3931	3841	3751	15	31	46	62	77
25	2·3662	3574	3486	3400	3314	3228	3144	3060	2976	2894	14	28	42	57	71
26	2·2812	2730	2650	2570	2490	2412	2333	2256	2179	2103	13	26	39	52	65
27	2·2027	1952	1877	1803	1730	1657	1584	1513	1441	1371	12	24	36	48	60
28	2·1301	1231	1162	1093	1025	0957	0890	0824	0757	0692	11	22	34	45	56
29	2·0627	0562	0498	0434	0371	0308	0245	0183	0122	0061	10	21	31	42	52
30	2·0000	9940	9880	9821	9762	9703	9645	9587	9530	9473	10	19	29	39	49
31	1·9416	9360	9304	9249	9194	9139	9084	9031	8977	8924	9	18	27	36	45
32	1·8871	8818	8766	8714	8663	8612	8561	8510	8460	8410	8	17	25	34	42
33	1·8361	8312	8263	8214	8166	8118	8070	8023	7976	7929	8	16	24	32	40
34	1·7883	7837	7791	7745	7700	7655	7610	7566	7522	7478	7	15	22	30	37
35	1·7434	7391	7348	7305	7263	7221	7179	7137	7095	7054	7	14	21	28	35
36	1·7013	6972	6932	6892	6852	6812	6772	6733	6694	6655	7	13	20	26	33
37	1·6616	6578	6540	6502	6464	6427	6390	6353	6316	6279	6	12	19	25	31
38	1·6243	6207	6171	6135	6099	6064	6029	5994	5959	5925	6	12	18	24	29
39	1·5890	5856	5822	5788	5755	5721	5688	5655	5622	5590	6	11	17	22	28
40	1·5557	5525	5493	5461	5429	5398	5366	5335	5304	5273	5	10	16	21	26
41	1·5243	5212	5182	5151	5121	5092	5062	5032	5003	4974	5	10	15	20	25
42	1·4945	4916	4887	4859	4830	4802	4774	4746	4718	4690	5	9	14	19	23
43	1·4663	4635	4608	4581	4554	4527	4501	4474	4448	4422	4	9	13	18	22
44	1·4396	4370	4344	4318	4293	4267	4242	4217	4192	4167	4	8	13	17	21

SUBTRACT

Figures in bold type show changes of integer.

NATURAL COSECANTS

SUBTRACT

°	0′	6′	12′	18′	24′	30′	36′	42′	48′	54′	1′	2′	3′	4′	5′
45	1·4142	4118	4093	4069	4044	4020	3996	3972	3949	3925	4	8	12	16	20
46	1·3902	3878	3855	3832	3809	3786	3763	3741	3718	3696	4	8	11	15	19
47	1·3673	3651	3629	3607	3585	3563	3542	3520	3499	3478	4	7	11	14	18
48	1·3456	3435	3414	3393	3373	3352	3331	3311	3291	3270	3	7	10	14	17
49	1·3250	3230	3210	3190	3171	3151	3131	3112	3093	3073	3	7	10	13	16
50	1·3054	3035	3016	2997	2978	2960	2941	2923	2904	2886	3	6	9	12	15
51	1·2868	2849	2831	2813	2796	2778	2760	2742	2725	2708	3	6	9	12	15
52	1·2690	2673	2656	2639	2622	2605	2588	2571	2554	2538	3	6	8	11	14
53	1·2521	2505	2489	2472	2456	2440	2424	2408	2392	2376	3	5	8	11	13
54	1·2361	2345	2329	2314	2299	2283	2268	2253	2238	2223	3	5	8	10	13
55	1·2208	2193	2178	2163	2149	2134	2120	2105	2091	2076	2	5	7	10	12
56	1·2062	2048	2034	2020	2006	1992	1978	1964	1951	1937	2	5	7	9	11
57	1·1924	1910	1897	1883	1870	1857	1844	1831	1818	1805	2	4	7	9	11
58	1·1792	1779	1766	1753	1741	1728	1716	1703	1691	1679	2	4	6	8	10
59	1·1666	1654	1642	1630	1618	1606	1594	1582	1570	1559	2	4	6	8	10
60	1·1547	1535	1524	1512	1501	1490	1478	1467	1456	1445	2	4	6	8	9
61	1·1434	1423	1412	1401	1390	1379	1368	1357	1347	1336	2	4	5	7	9
62	1·1326	1315	1305	1294	1284	1274	1264	1253	1243	1233	2	3	5	7	9
63	1·1223	1213	1203	1194	1184	1174	1164	1155	1145	1136	2	3	5	6	8
64	1·1126	1117	1107	1098	1089	1079	1070	1061	1052	1043	2	3	5	6	8
65	1·1034	1025	1016	1007	0998	0989	0981	0972	0963	0955	1	3	4	6	7
66	1·0946	0938	0929	0921	0913	0904	0896	0888	0880	0872	1	3	4	5	7
67	1·0864	0856	0848	0840	0832	0824	0816	0808	0801	0793	1	3	4	5	7
68	1·0785	0778	0770	0763	0755	0748	0740	0733	0726	0719	1	2	4	5	6
69	1·0711	0704	0697	0690	0683	0676	0669	0662	0655	0649	1	2	3	5	6
70	1·0642	0635	0628	0622	0615	0608	0602	0595	0589	0583	1	2	3	4	5
71	1·0576	0570	0564	0557	0551	0545	0539	0533	0527	0521	1	2	3	4	5
72	1·0515	0509	0503	0497	0491	0485	0480	0474	0468	0463	1	2	3	4	5
73	1·0457	0451	0446	0440	0435	0429	0424	0419	0413	0408	1	2	3	4	4
74	1·0403	0398	0393	0388	0382	0377	0372	0367	0363	0358	1	2	2	3	4
75	1·0353	0348	0343	0338	0334	0329	0324	0320	0315	0311	1	2	2	3	4
76	1·0306	0302	0297	0293	0288	0284	0280	0276	0271	0267	1	1	2	3	4
77	1·0263	0259	0255	0251	0247	0243	0239	0235	0231	0227	1	1	2	3	3
78	1·0223	0220	0216	0212	0209	0205	0201	0198	0194	0191	1	1	2	2	3
79	1·0187	0184	0180	0177	0174	0170	0167	0164	0161	0157	1	1	2	2	3
80	1·0154	0151	0148	0145	0142	0139	0136	0133	0130	0127	0	1	1	2	2
81	1·0125	0122	0119	0116	0114	0111	0108	0106	0103	0101	0	1	1	2	2
82	1·0098	0096	0093	0091	0089	0086	0084	0082	0079	0077	0	1	1	2	2
83	1·0075	0073	0071	0069	0067	0065	0063	0061	0059	0057	0	1	1	1	2
84	1·0055	0053	0051	0050	0048	0046	0045	0043	0041	0040	0	1	1	1	1
85	1·0038	0037	0035	0034	0032	0031	0030	0028	0027	0026	0	0	1	1	1
86	1·0024	0023	0022	0021	0020	0019	0018	0017	0016	0015	0	0	0	1	1
87	1·0014	0013	0012	0011	0010	0010	0009	0008	0007	0007	0	0	0	1	1
88	1·0006	0006	0005	0004	0004	0003	0003	0003	0002	0002	0	0	0	0	0
89	1·0002	0001	0001	0001	0001	0000	0000	0000	0000	0000	0	0	0	0	0

SUBTRACT

NATURAL SECANTS

°	0′	6′	12′	18′	24′	30′	36′	42′	48′	54′	1′	2′	3′	4′	5′
0	1·0000	0000	0000	0000	0000	0000	0001	0001	0001	0001	0	0	0	0	0
1	1·0002	0002	0002	0003	0003	0003	0004	0004	0005	0006	0	0	0	0	0
2	1·0006	0007	0007	0008	0009	0010	0010	0011	0012	0013	0	0	0	1	1
3	1·0014	0015	0016	0017	0018	0019	0020	0021	0022	0023	0	0	0	1	1
4	1·0024	0026	0027	0028	0030	0031	0032	0034	0035	0037	0	0	1	1	1
5	1·0038	0040	0041	0043	0045	0046	0048	0050	0051	0053	0	1	1	1	1
6	1·0055	0057	0059	0061	0063	0065	0067	0069	0071	0073	0	1	1	1	2
7	1·0075	0077	0079	0082	0084	0086	0089	0091	0093	0096	0	1	1	2	2
8	1·0098	0101	0103	0106	0108	0111	0114	0116	0119	0122	0	1	1	2	2
9	1·0125	0127	0130	0133	0136	0139	0142	0145	0148	0151	0	1	1	2	2
10	1·0154	0157	0161	0164	0167	0170	0174	0177	0180	0184	1	1	2	2	3
11	1·0187	0191	0194	0198	0201	0205	0209	0212	0216	0220	1	1	2	2	3
12	1·0223	0227	0231	0235	0239	0243	0247	0251	0255	0259	1	1	2	3	3
13	1·0263	0267	0271	0276	0280	0284	0288	0293	0297	0302	1	1	2	3	4
14	1·0306	0311	0315	0320	0324	0329	0334	0338	0343	0348	1	2	2	3	4
15	1·0353	0358	0363	0367	0372	0377	0382	0388	0393	0398	1	2	2	3	4
16	1·0403	0408	0413	0419	0424	0429	0435	0440	0446	0451	1	2	3	4	4
17	1·0457	0463	0468	0474	0480	0485	0491	0497	0503	0509	1	2	3	4	5
18	1·0515	0521	0527	0533	0539	0545	0551	0557	0564	0570	1	2	3	4	5
19	1·0576	0583	0589	0595	0602	0608	0615	0622	0628	0635	1	2	3	4	5
20	1·0642	0649	0655	0662	0669	0676	0683	0690	0697	0704	1	2	3	5	6
21	1·0711	0719	0726	0733	0740	0748	0755	0763	0770	0778	1	2	4	5	6
22	1·0785	0793	0801	0808	0816	0824	0832	0840	0848	0856	1	3	4	5	7
23	1·0864	0872	0880	0888	0896	0904	0913	0921	0929	0938	1	3	4	5	7
24	1·0946	0955	0963	0972	0981	0989	0998	1007	1016	1025	1	3	4	6	7
25	1·1034	1043	1052	1061	1070	1079	1089	1098	1107	1117	2	3	5	6	8
26	1·1126	1136	1145	1155	1164	1174	1184	1194	1203	1213	2	3	5	6	8
27	1·1223	1233	1243	1253	1264	1274	1284	1294	1305	1315	2	3	5	7	9
28	1·1326	1336	1347	1357	1368	1379	1390	1401	1412	1423	2	4	5	7	9
29	1·1434	1445	1456	1467	1478	1490	1501	1512	1524	1535	2	4	6	8	9
30	1·1547	1559	1570	1582	1594	1606	1618	1630	1642	1654	2	4	6	8	10
31	1·1666	1679	1691	1703	1716	1728	1741	1753	1766	1779	2	4	6	8	10
32	1·1792	1805	1818	1831	1844	1857	1870	1883	1897	1910	2	4	7	9	11
33	1·1924	1937	1951	1964	1978	1992	2006	2020	2034	2048	2	5	7	9	11
34	1·2062	2076	2091	2105	2120	2134	2149	2163	2178	2193	2	5	7	10	12
35	1·2208	2223	2238	2253	2268	2283	2299	2314	2329	2345	3	5	8	10	13
36	1·2361	2376	2392	2408	2424	2440	2456	2472	2489	2505	3	5	8	11	13
37	1·2521	2538	2554	2571	2588	2605	2622	2639	2656	2673	3	6	8	11	14
38	1·2690	2708	2725	2742	2760	2778	2796	2813	2831	2849	3	6	9	12	15
39	1·2868	2886	2904	2923	2941	2960	2978	2997	3016	3035	3	6	9	12	15
40	1·3054	3073	3093	3112	3131	3151	3171	3190	3210	3230	3	7	10	13	16
41	1·3250	3270	3291	3311	3331	3352	3373	3393	3414	3435	3	7	10	14	17
42	1·3456	3478	3499	3520	3542	3563	3585	3607	3629	3651	4	7	11	14	18
43	1·3673	3696	3718	3741	3763	3786	3809	3832	3855	3878	4	8	11	15	19
44	1·3902	3925	3949	3972	3996	4020	4044	4069	4093	4118	4	8	12	16	20

NATURAL SECANTS

°	0'	6'	12'	18'	24'	30'	36'	42'	48'	54'	1'	2'	3'	4'	5'
45	1·4142	4167	4192	4217	4242	4267	4293	4318	4344	4370	4	8	13	17	21
46	1·4396	4422	4448	4474	4501	4527	4554	4581	4608	4635	4	9	13	18	22
47	1·4663	4690	4718	4746	4774	4802	4830	4859	4887	4916	5	9	14	19	23
48	1·4945	4974	5003	5032	5062	5092	5121	5151	5182	5212	5	10	15	20	25
49	1·5243	5273	5304	5335	5366	5398	5429	5461	5493	5525	5	10	16	21	26
50	1·5557	5590	5622	5655	5688	5721	5755	5788	5822	5856	6	11	17	22	28
51	1·5890	5925	5959	5994	6029	6064	6099	6135	6171	6207	6	12	18	24	29
52	1·6243	6279	6316	6353	6390	6427	6464	6502	6540	6578	6	12	19	25	31
53	1·6616	6655	6694	6733	6772	6812	6852	6892	6932	6972	7	13	20	26	33
54	1·7013	7054	7095	7137	7179	7221	7263	7305	7348	7391	7	14	21	28	35
55	1·7434	7478	7522	7566	7610	7655	7700	7745	7791	7837	7	15	22	30	37
56	1·7883	7929	7976	8023	8070	8118	8166	8214	8263	8312	8	16	24	32	40
57	1·8361	8410	8460	8510	8561	8612	8663	8714	8766	8818	8	17	25	34	42
58	1·8871	8924	8977	9031	9084	9139	9194	9249	9304	9360	9	18	27	36	45
59	1·9416	9473	9530	9587	9645	9703	9762	9821	9880	9940	10	19	29	39	49
60	2·0000	0061	0122	0183	0245	0308	0371	0434	0498	0562	10	21	31	42	52
61	2·0627	0692	0757	0824	0890	0957	1025	1093	1162	1231	11	22	34	45	56
62	2·1301	1371	1441	1513	1584	1657	1730	1803	1877	1952	12	24	36	48	60
63	2·2027	2103	2179	2256	2333	2412	2490	2570	2650	2730	13	26	39	52	65
64	2·2812	2894	2976	3060	3144	3228	3314	3400	3486	3574	14	28	42	57	71
65	2·3662	3751	3841	3931	4022	4114	4207	4300	4395	4490	15	31	46	62	77
66	2·4586	4683	4780	4879	4978	5078	5180	5282	5384	5488	17	34	50	67	84
67	2·5593	5699	5805	5913	6022	6131	6242	6354	6466	6580	18	37	55	73	92
68	2·6695	6811	6927	7046	7165	7285	7407	7529	7653	7778	20	40	60	81	101
69	2·7904	8032	8161	8291	8422	8555	8688	8824	8960	9099	22	44	67	89	111
70	2·9238	9379	9521	9665	9811	9957	**0106**	**0256**	**0407**	**0561**	25	49	74	99	123
71	3·0716	0872	1030	1190	1352	1515	1681	1848	2017	2188	27	55	82	110	137
72	3·2361	2535	2712	2891	3072	3255	3440	3628	3817	4009	31	61	92	123	153
73	3·4203	4399	4598	4799	5003	5209	5418	5629	5843	6060	35	69	104	138	173
74	3·6280	6502	6727	6955	7186	7420	7657	7897	8140	8387	39	79	118	157	196
75	3·8637	8890	9147	9408	9672	9939	**0211**	**0486**	**0765**	**1048**	45	90	135	180	225
76	4·1336	1627	1923	2223	2527	2837	3150	3469	3792	4121	52	104	156	208	260
77	4·4454	4793	5137	5486	5841	6202	6569	6942	7321	7706	61	121	182	243	304
78	4·8097	8496	8901	9313	9732	**0159**	**0593**	**1034**	**1484**	**1942**	72	144	216	287	359
79	5·2408	2883	3367	3860	4362	4874	5396	5928	6470	7023	86	173	259	345	432
80	5·759	5·816	5·875	5·935	5·996	6·059	6·123	6·188	6·255	6·323					
81	6·392	6·464	6·537	6·611	6·687	6·765	6·845	6·927	7·011	7·097					
82	7·185	7·276	7·368	7·463	7·561	7·661	7·764	7·870	7·979	8·091					
83	8·206	8·324	8·446	8·571	8·700	8·834	8·971	9·113	9·259	9·411					
84	9·567	9·728	9·895	10·07	10·25	10·43	10·63	10·83	11·03	11·25		Use			
85	11·47	11·71	11·95	12·20	12·47	12·75	13·03	13·34	13·65	13·99		interpolation			
86	14·34	14·70	15·09	15·50	15·93	16·38	16·86	17·37	17·91	18·49					
87	19·11	19·77	20·47	21·23	22·04	22·93	23·88	24·92	26·05	27·29					
88	28·65	30·16	31·84	33·71	35·81	38·20	40·93	44·08	47·75	52·09					
89	57·30	63·66	71·62	81·85	95·49	114·6	143·2	191·0	286·5	573·0					

Figures in **bold type** show changes of integer.

NATURAL COTANGENTS

SUBTRACT

°	0'	6'	12'	18'	24'	30'	36'	42'	48'	54'	1'	2'	3'	4'	5'
0	Inf. Pos.	573·0	286·5	191·0	143·2	114·6	95·49	81·85	71·62	63·66					
1	57·29	52·08	47·74	44·07	40·92	38·19	35·80	33·69	31·82	30·14					
2	28·64	27·27	26·03	24·90	23·86	22·90	22·02	21·20	20·45	19·74		Use			
3	19·08	18·46	17·89	17·34	16·83	16·35	15·89	15·46	15·06	14·67					
4	14·30	13·95	13·62	13·30	13·00	12·71	12·43	12·16	11·91	11·66		interpolation			
5	11·43	11·20	10·99	10·78	10·58	10·39	10·20	10·02	9·845	9·677					
6	9·514	9·357	9·205	9·058	8·915	8·777	8·643	8·513	8·386	8·264					
7	8·144	8·028	7·916	7·806	7·700	7·596	7·495	7·396	7·300	7·207					
8	7·115	7·026	6·940	6·855	6·772	6·691	6·612	6·535	6·460	6·386					
9	6·314	6·243	6·174	6·107	6·041	5·976	5·912	5·850	5·789	5·730					
10	5·6713	6140	5578	5026	4486	3955	3435	2924	2422	1929	88	176	263	351	439
11	5·1446	0970	0504	0045	9594	9152	8716	8288	7867	7453	73	147	220	293	367
12	4·7046	6646	6252	5864	5483	5107	4737	4373	4015	3662	62	124	187	249	311
13	4·3315	2972	2635	2303	1976	1653	1335	1022	0713	0408	53	107	160	214	267
14	4·0108	9812	9520	9232	8947	8667	8391	8118	7848	7583	46	93	139	186	232
15	3·7321	7062	6806	6554	6305	6059	5816	5576	5339	5105	41	81	122	163	203
16	3·4874	4646	4420	4197	3977	3759	3544	3332	3122	2914	36	72	108	144	180
17	3·2709	2506	2305	2106	1910	1716	1524	1334	1146	0961	32	64	96	129	161
18	3·0777	0595	0415	0237	0061	9887	9714	9544	9375	9208	29	58	87	116	145
19	2·9042	8878	8716	8556	8397	8239	8083	7929	7776	7625	26	52	78	104	130
20	2·7475	7326	7179	7034	6889	6746	6605	6464	6325	6187	24	47	71	95	118
21	2·6051	5916	5782	5649	5517	5386	5257	5129	5002	4876	22	43	65	87	108
22	2·4751	4627	4504	4383	4262	4142	4023	3906	3789	3673	20	40	60	79	99
23	2·3559	3445	3332	3220	3109	2998	2889	2781	2673	2566	18	37	55	73	92
24	2·2460	2355	2251	2148	2045	1943	1842	1742	1642	1543	17	34	51	68	85
25	2·1445	1348	1251	1155	1060	0965	0872	0778	0686	0594	16	31	47	63	78
26	2·0503	0413	0323	0233	0145	0057	9970	9883	9797	9711	15	29	44	58	73
27	1·9626	9542	9458	9375	9292	9210	9128	9047	8967	8887	14	27	41	55	68
28	1·8807	8728	8650	8572	8495	8418	8341	8265	8190	8115	13	26	38	51	64
29	1·8040	7966	7893	7820	7747	7675	7603	7532	7461	7391	12	24	36	48	60
30	1·7321	7251	7182	7113	7045	6977	6909	6842	6775	6709	11	23	34	45	56
31	1·6643	6577	6512	6447	6383	6319	6255	6191	6128	6066	11	21	32	43	53
32	1·6003	5941	5880	5818	5757	5697	5637	5577	5517	5458	10	20	30	40	50
33	1·5399	5340	5282	5224	5166	5108	5051	4994	4938	4882	10	19	29	38	48
34	1·4826	4770	4715	4659	4605	4550	4496	4442	4388	4335	9	18	27	36	45
35	1·4281	4229	4176	4124	4071	4019	3968	3916	3865	3814	9	17	26	34	43
36	1·3764	3713	3663	3613	3564	3514	3465	3416	3367	3319	8	16	25	33	41
37	1·3270	3222	3175	3127	3079	3032	2985	2938	2892	2846	8	16	24	31	39
38	1·2799	2753	2708	2662	2617	2572	2527	2482	2437	2393	8	15	23	30	38
39	1·2349	2305	2261	2218	2174	2131	2088	2045	2002	1960	7	14	22	29	36
40	1·1918	1875	1833	1792	1750	1708	1667	1626	1585	1544	7	14	21	28	34
41	1·1504	1463	1423	1383	1343	1303	1263	1224	1184	1145	7	13	20	26	33
42	1·1106	1067	1028	0990	0951	0913	0875	0837	0799	0761	6	13	19	25	32
43	1·0724	0686	0649	0612	0575	0538	0501	0464	0428	0392	6	12	18	25	31
44	1·0355	0319	0283	0247	0212	0176	0141	0105	0070	0035	6	12	18	24	30

SUBTRACT

Figures in **bold type** show changes of integer.

NATURAL COTANGENTS

°	0′	6′	12′	18′	24′	30′	36′	42′	48′	54′	1′	2′	3′	4′	5′
45	1·0000	**9965**	**9930**	**9896**	**9861**	**9827**	**9793**	**9759**	**9725**	**9691**	6	11	17	23	29
46	0·9657	9623	9590	9556	9523	9490	9457	9424	9391	9358	6	11	17	22	28
47	0·9325	9293	9260	9228	9195	9163	9131	9099	9067	9036	5	11	16	21	27
48	0·9004	8972	8941	8910	8878	8847	8816	8785	8754	8724	5	10	16	21	26
49	0·8693	8662	8632	8601	8571	8541	8511	8481	8451	8421	5	10	15	20	25
50	0·8391	8361	8332	8302	8273	8243	8214	8185	8156	8127	5	10	15	20	24
51	0·8098	8069	8040	8012	7983	7954	7926	7898	7869	7841	5	9	14	19	24
52	0·7813	7785	7757	7729	7701	7673	7646	7618	7590	7563	5	9	14	18	23
53	0·7536	7508	7481	7454	7427	7400	7373	7346	7319	7292	5	9	14	18	23
54	0·7265	7239	7212	7186	7159	7133	7107	7080	7054	7028	4	9	13	18	22
55	0·7002	6976	6950	6924	6899	6873	6847	6822	6796	6771	4	9	13	17	21
56	0·6745	6720	6694	6669	6644	6619	6594	6569	6544	6519	4	8	13	17	21
57	0·6494	6469	6445	6420	6395	6371	6346	6322	6297	6273	4	8	12	16	20
58	0·6249	6224	6200	6176	6152	6128	6104	6080	6056	6032	4	8	12	16	20
59	0·6009	5985	5961	5938	5914	5890	5867	5844	5820	5797	4	8	12	16	20
60	0·5774	5750	5727	5704	5681	5658	5635	5612	5589	5566	4	8	12	15	19
61	0·5543	5520	5498	5475	5452	5430	5407	5384	5362	5340	4	8	11	15	19
62	0·5317	5295	5272	5250	5228	5206	5184	5161	5139	5117	4	7	11	15	18
63	0·5095	5073	5051	5029	5008	4986	4964	4942	4921	4899	4	7	11	15	18
64	0·4877	4856	4834	4813	4791	4770	4748	4727	4706	4684	4	7	11	14	18
65	0·4663	4642	4621	4599	4578	4557	4536	4515	4494	4473	4	7	11	14	18
66	0·4452	4431	4411	4390	4369	4348	4327	4307	4286	4265	3	7	10	14	17
67	0·4245	4224	4204	4183	4163	4142	4122	4101	4081	4061	3	7	10	14	17
68	0·4040	4020	4000	3979	3959	3939	3919	3899	3879	3859	3	7	10	13	17
69	0·3839	3819	3799	3779	3759	3739	3719	3699	3679	3659	3	7	10	13	17
70	0·3640	3620	3600	3581	3561	3541	3522	3502	3482	3463	3	7	10	13	16
71	0·3443	3424	3404	3385	3365	3346	3327	3307	3288	3269	3	6	10	13	16
72	0·3249	3230	3211	3191	3172	3153	3134	3115	3096	3076	3	6	10	13	16
73	0·3057	3038	3019	3000	2981	2962	2943	2924	2905	2886	3	6	9	13	16
74	0·2867	2849	2830	2811	2792	2773	2754	2736	2717	2698	3	6	9	13	16
75	0·2679	2661	2642	2623	2605	2586	2568	2549	2530	2512	3	6	9	12	16
76	0·2493	2475	2456	2438	2419	2401	2382	2364	2345	2327	3	6	9	12	15
77	0·2309	2290	2272	2254	2235	2217	2199	2180	2162	2144	3	6	9	12	15
78	0·2126	2107	2089	2071	2053	2035	2016	1998	1980	1962	3	6	9	12	15
79	0·1944	1926	1908	1890	1871	1853	1835	1817	1799	1781	3	6	9	12	15
80	0·1763	1745	1727	1709	1691	1673	1655	1638	1620	1602	3	6	9	12	15
81	0·1584	1566	1548	1530	1512	1495	1477	1459	1441	1423	3	6	9	12	15
82	0·1405	1388	1370	1352	1334	1317	1299	1281	1263	1246	3	6	9	12	15
83	0·1228	1210	1192	1175	1157	1139	1122	1104	1086	1069	3	6	9	12	15
84	0·1051	1033	1016	0998	0981	0963	0945	0928	0910	0892	3	6	9	12	15
85	0·0875	0857	0840	0822	0805	0787	0769	0752	0734	0717	3	6	9	12	15
86	0·0699	0682	0664	0647	0629	0612	0594	0577	0559	0542	3	6	9	12	15
87	0·0524	0507	0489	0472	0454	0437	0419	0402	0384	0367	3	6	9	12	15
88	0·0349	0332	0314	0297	0279	0262	0244	0227	0209	0192	3	6	9	12	15
89	0·0175	0157	0140	0122	0105	0087	0070	0052	0035	0017	3	6	9	12	15

Figures in **bold type** show changes of integer.

PART A: ALGEBRA

EXERCISE 1

If $a = 2$, $b = 1$, $c = 3$, $x = 0$, $y = 5$, $z = 10$, find the value of:

1. $2a$ **2.** $6c$ **3.** $3xy$ **4.** $2cz$

5. ab **6.** cy **7.** $2axy$ **8.** $4abc$

9. $\dfrac{a}{2}$ **10.** $\dfrac{b}{4}$ **11.** $\dfrac{y}{z}$ **12.** $\dfrac{x}{b}$

If $p = 3$, $q = 1$, $r = 2$, $s = 4$, $t = 0$, find the value of:

13. $\dfrac{qr}{2} + \dfrac{rs}{4}$ **14.** $\dfrac{2p + 2s}{3q + 2r}$ **15.** $\dfrac{pq}{r + s}$

16. $\dfrac{p + q + 8}{s}$ **17.** $\dfrac{p}{q} + \dfrac{r}{q} + \dfrac{s}{r}$ **18.** $\dfrac{2s}{4r} - \dfrac{3q}{p} + \dfrac{2t}{1}$

19. $\dfrac{rs}{pq} \times 1\frac{1}{2}$ **20.** $\dfrac{q + s}{3} \times \dfrac{1}{p + r}$ **21.** $\dfrac{p}{s} \div \dfrac{q + r}{s - t}$

EXERCISE 2

Write out the following using symbols; some are simple terms, others are expressions or equations.

1. Add 3 to x.   **2.** Subtract 6 from p.   **3.** Multiply b by 4.

4. Divide m by 2.   **5.** Six times S.   **6.** Add a to twice b.

7. The difference between a and b is equal to nine.

8. The product of x and y is added to the product of b and c.

9. Take x from 6, the result is 2.

10. x is equal to twice y.

11. Two is divided into p. The quotient is equal to six.

12. Three minus a is equal to a minus five.

EXERCISE 3

Write down the shorter forms for the following expressions. If no shorter form is possible, write "NO SHORTER FORM".

1. $a + a + a$ **2.** $x + x + x + x$ **3.** $p + p + p + p + p$

4. $5a + 2a - 7a$ **5.** $4c + 2c + c$ **6.** $f + f - f$

7. $2x + x + x - 2x - x + 3x$ **8.** $5y + y + 2y - 4y - 3y$

9. $t - 4t - 2t + 3t + 4t$ **10.** $6b - b - 2b - 3b + 2b$

11. $3b - 2d - d - 3d + 4b$ **12.** $3n - 2p - 2n + 4p$

13. $7r - 3s - 4s - 2r - 3r$ **14.** $4 - 3k - 4l$

15. $2f + 3g - 4h + 1$ **16.** $2r + r - s + 3r - 2s + 3s$

17. $s + t - s + t - s + 2s$ **18.** $4k + 2 - l + m - 4$

19. $3p + 4q - 5r$ **20.** $2p + 3q - p - 2q - p - q$

1

MAKING EXPRESSIONS: PROBLEMS

EXERCISE 4

1. In a class there are 16 girls and 20 boys. How many pupils are there altogether?

2. In a class there are G girls and B boys. How many pupils are there altogether?

3. In a morning I travelled x miles, and in the afternoon, y miles. How far did I travel altogether?

4. I made a journey of 38 miles, travelling t miles by train. I completed the journey by bus. How far did I travel by bus?

5. A boy is 14 years old now; how old was he x years ago?

6. A girl will be z years old in three years time. How old is she now?

7. How many pence are there: (i) in one pound; (ii) in six pounds; (iii) in S pounds; (iv) in $6S$ pounds; (v) in $\dfrac{S}{2}$ pounds; (vi) in $2\frac{1}{2}S$ pounds?

8. How many halfpennies are there (i) in 1p; (ii) in 3p; (iii) in xp; (iv) in $3x$p?

9. A boy has x marbles; his friend has half this number. How many has the second boy? How many have they altogether?

10. Six cakes are to be shared equally among four girls. How many will each receive? How many will each receive: (i) when x cakes are shared equally among y girls; (ii) when $3x$ cakes are shared equally among $2y$ girls; (iii) when $6x$ cakes are shared equally among $4y$ girls?

ADDITION OF COMPOUND EXPRESSIONS

$2x$ is called a **simple** expression because it contains only one term.

$\left.\begin{array}{l} 2x + 3y \\ 3a + 2b - 3c \end{array}\right\}$ These are called **compound** expressions because each contains *more than one* term.

$\left.\begin{array}{l} 2x + 3y \\ 3a - 5c \end{array}\right\}$ These are called **binomial** expressions or binomials because each contains *two* terms.

$\left.\begin{array}{l} 3a + 2b - 3c \\ x - 3y + 2z \end{array}\right\}$ These are called **trinomial** expressions or trinomials because each contains *three* terms.

Add:

$$
\begin{array}{r}
a + 2b + c \\
a + 3b + 2c \\
2a + 2b + 3c \\
\hline
4a + 7b + 6c
\end{array}
$$

Add:

$$
\begin{array}{r}
2x + y + 3z \\
x + 3y + 2z \\
3x + 2y + z \\
\hline
6x + 6y + 6z
\end{array}
$$

NOTE. Compound expressions are usually written in alphabetical order, for convenience. It is sometimes necessary to rearrange the terms to make the alphabetical order possible. Numerals are usually written last.

EXERCISE 5

Find the sum of the following expressions:

1. $2a + 4b + c$; $3a + 2b + 4c$; $a + 3b + 2c$

2. $2x + 3z + 4y$; $2y + 2x + 2z$; $y + z + x$

3. $m + n - p$; $2m - 2n + 3p$; $2m + 3n + 4p$

4. $2r - 3s + 4t$; $3r + 2s - t$; $-2r + s - 2t$

5. $q - 2p - 3r$; $4p + 2r - 2q$; $r + p + q$

6. $x + 2y + 4$; $3x - 3y - 2$; $4y - 2 + x$

7. $2u - w + v$; $2w - u + v$; $-2v - u - w$

8. $3ab + 2cz + 4lm$; $2ab$; $4cz + 2lm$; $3lm$

9. $4xy - 2yz + 3$; $3xy + 3yz - 2$; $xy + 4yz + 1$

10. $2by - 3cz + ax$; $2ax + 4cz - 3by$; $cz - ax + by$

POWERS AND MULTIPLICATION

When a term is multiplied by itself a number of times the **product** is called a **power** of the original term. It is expressed by writing the number of **factors** at the top right of the term concerned.

Thus: $\qquad 3 \times 3 = 3^2 \qquad$ This is called the second power of 3.

$\qquad 5 \times 5 \times 5 = 5^3 \qquad$ This is called the third power of 5.

$a \times a \times a \times a = a^4 \qquad$ This is called the fourth power of a.

The small numbers written at the top right of a term are called **index** numbers.

NOTE

1. x^2 may be read as, "x to the power 2". It is usually called "x **squared**".

2. x^3 may be read as, "x to the power 3". What do you think it is usually called?

3. x^4 may be read as, "x to the power 4" or simply as, "x to the fourth". There is no short form as in x^2 and x^3—can you say why?

4. When a term is expressed in the *first* power, the index figure 1 is omitted, thus: $\qquad a^1$ is written simply as a

You will remember that the coefficient 1 is also omitted, thus:

$\qquad 1a^1$ is written simply as a

5. REMEMBER $\qquad 3x$ means $3 \times x$

$\qquad\qquad\qquad\quad x^3$ means $x \times x \times x$

EXERCISE 6

Simplify the following:

1. $2a \times 2b \times 0$ $\qquad$ **2.** $xy \times xy \times 1$ $\qquad$ **3.** $x \times xy \times y$

4. $x \times xy \times 2y$ $\qquad$ **5.** $2x \times x \times y^2$ $\qquad$ **6.** $2 \times x^2 \times y^2$

7. $a \times 2b \times 3c$ $\qquad$ **8.** $x^2 \times y^3 \times z^4$ $\qquad$ **9.** $2x \times 3y^2 \times 4z^3$

10. $2abc \times 3a$ $\qquad$ **11.** $3xyz \times 2xy$ $\qquad$ **12.** $3xyz \times 3xyz$

ROOTS

Finding the roots of terms or expressions is the reverse of finding powers. That is, a power of a term is obtained when the term is multiplied by itself a number of times, but a root is that factor of a term which has been multiplied by itself to provide the term given.

Thus: a to the power $2 = a^2$

The square root of $a^2 = a$

Another symbol is employed to denote a root, it is $\sqrt{}$. This symbol is called the **radical sign.**

$\sqrt{}$ means the second root, usually called the **square root**
$\sqrt[3]{}$ means the third root, usually called the **cube root**
$\sqrt[4]{}$ means the fourth root

Thus: $\sqrt{16} = 4$ because $4 \times 4 = 4^2 = 16$
           $\sqrt{a^2} = a$ because $a \times a = a^2$
           $\sqrt[3]{8} = 2$ because $2 \times 2 \times 2 = 2^3 = 8$
           $\sqrt[3]{x^3} = x$ because $x \times x \times x = x^3$
           $\sqrt[4]{81} = 3$ because $3 \times 3 \times 3 \times 3 = 3^4 = 81$
           $\sqrt[4]{y^4} = y$ because $y \times y \times y \times y = y^4$

EXERCISE 7

Simplify the following:

1. $\sqrt{9}$	**2.** $\sqrt{81}$	**3.** $\sqrt{64}$	**4.** $\sqrt{4}$
5. $\sqrt{144}$	**6.** $\sqrt{x^2}$	**7.** $\sqrt[3]{y^3}$	**8.** $\sqrt[3]{64}$
9. $\sqrt{144b^2}$	**10.** $\sqrt[4]{a^4}$	**11.** $\sqrt{16a^2}$	**12.** $\sqrt{4x^2}$
13. $\sqrt[3]{8a^3}$	**14.** $\sqrt{1}$	**15.** $\sqrt[4]{16x^4}$	**16.** $\sqrt[5]{0}$

DIVISION

In algebra simple cases of division may be treated in the same way as similar examples in arithmetic; that is they can be expressed as fractions and the numerator and denominator may be divided by a common factor (the process of cancelling).

Thus: $8a \div 2 = \dfrac{\overset{4}{8a}}{\underset{1}{2}} = \dfrac{4 \times a}{1} = 4a$

          $4ab \div 2b = \dfrac{\overset{2}{4}a\overset{1}{b}}{\underset{1}{2}\underset{1}{b}} = \dfrac{2 \times a}{1 \times 1} = 2a$

          $6x^2 \div 2x = \dfrac{6x^2}{2x} = \dfrac{\overset{3}{6} \times \overset{1}{x} \times x}{\underset{1}{2} \times \underset{1}{x}} = \dfrac{3 \times 1 \times x}{1 \times 1} = 3x$

EXERCISE 8

Simplify the following:

1. $3y \div y$
2. $ay \div y$
3. $8l \div 4$
4. $8n \div 4n$
5. $12a \div 6a$
6. $9y \div 3y$
7. $8ab \div 2ab$
8. $9xy \div 3xy$
9. $4cd \div 2cd$
10. $4ay \div a$
11. $4ay \div 4a$
12. $4ay \div 4y$
13. $6x^2 \div 3$
14. $6x^2 \div x^2$
15. $6x^2 \div 3x^2$
16. $4ax^2 \div 2ax^2$
17. $16a^2x^2 \div 4ax$
18. $8a^2x^2 \div 2a^2x$

ORDER OF TERMS

It is usual to arrange terms as follows:

1. In alphabetical order when possible.

2. Terms using the same letter symbol are usually placed in *descending order of powers*, numerical terms last.

Thus: $$2a^4 + 3a^3 - 5a^2 + a - 4$$

3. Sometimes the terms may be placed in *ascending order*, such an arrangement may avoid starting an expression with the minus sign. The numerical term must then be placed first.

Thus: $$6 + a + 3a^2 + 4a^3 - 2a^4$$

NOTE. In neither arrangement do the coefficients of the terms affect the order in which the terms are placed.

EXERCISE 9

Simplify the following expressions and arrange them in *descending* order of powers:

1. $a - 2a^2 - 3a + 4a^2$
2. $x - 2x + 3x^2 + 4x$
3. $x^2 + 2 - 2x + 3$
4. $3 - 2b^2 + 4b + 6b^2$
5. $y + 1 - 2y + 3 + y^2 + 3y$
6. $l^2 - 2l + l^3 + 3l - 4l^2$
7. $c^3 - c - 2c^3 + c^2 + 2c + 4c^3$
8. $a^3b + 2a^2b^2 + 2ab^3 + a^3b$
9. $3z - z + 2 - z + 3 + z^2$
10. $p^2 + 4 - p - 3 + p^2 + 1$
11. $2t + t^3 - t^2 + 2t - t^2$
12. $6s^3 - 2s^2 + s + s^3 + 2s$

Simplify the following expressions and arrange them in *ascending* order of powers:

13. $a^2 - 3a^3 + 2a^2 - a + 4$
14. $y - 2y + 1 - y^2 - 2 + 3$
15. $2b - 3b^3 + b + b^3 - b^2$
16. $x^3 + 2x - x + 3x^2 - x^3 + 1$
17. $e^3 + 3 - 2e^2 + e + 2e^2$
18. $2 - g^4 + 2g - g^3 + 2g^2 - 2$
19. $a^2b + a^3b + ab + ab$
20. $a^3b + a^2b^2 + ab^3 + 3$
21. $a^2b - 2 + ab^2 - 2a^2b + 6$
22. $3a^3 - ab - 4a^2 + 3ab$
23. $l^2m^2 + 2l^3 - m + 2l^2m^2 - m$
24. $l^3 + m - 2m^2 - 3l^2 + 2m$

MAKING EXPRESSIONS

Consider this example:

Add 3 to 2

We **may** write *either* 2 + 3 *or* 3 + 2.
Both forms give the result 5.
But according to the instruction, 2 must already be present for 3
to be added to it, therefore 2 + 3 is the *correct answer*.

EXERCISE 10

Make expressions from the following instructions:

1. The sum of p and q is divided by their difference.
2. The square root of m is subtracted from the cube of a.
3. The product of x, y and z is divided by their sum.
4. Multiply a by 2 and add b to the result.
5. x is divided by y, y is divided by z, the results are added.
6. The square of t is multiplied by 5.
7. The cube root of a is added to the square of b.
8. The difference between p and q is divided by their sum.
9. From the cube of x take the square of x, add x to the result.
10. The square root of c is subtracted from the sum of the cube root
   of b and the fourth root of a.

MULTIPLICATION AND DIVISION: COEFFICIENTS

In both multiplication and division of terms the numerical coefficients
are treated as they would be in arithmetic.

REMEMBER

1. *In multiplication* the numerical coefficients are multiplied together.
Letter terms which are the same have their indices added.

Thus: $2ax^2 \times 3a^2x^2 = 6a^3x^4$

2. *In division* the numerical coefficient of the dividend is divided by
the numerical coefficient of the divisor. Where letter terms are the
same the index number of the divisor is subtracted from the index
number of the dividend.

Thus: $6b^2y^3 \div 2by^2 = 3by$

The following exercise contains a few unusual examples of this type:
$a^{\frac{1}{2}} \times a^{1\frac{1}{2}}$; we must still add the index numbers and thus obtain the
result a^2. We shall be able to give *meaning* to $a^{\frac{1}{2}}$ or $a^{1\frac{1}{2}}$ later on, in the
meantime try to develop skill in *using* the index numbers.

EXERCISE 11

1. $4a \times 2a^4$
2. $3e^2 \times 2d$
3. $4de^2 \div 6de$
4. $8d^2e^2 \div 4de^2$
5. $2a^2y \times 3ay^2$
6. $6p^3q^2 \div 2pq$
7. $9yz^4 \div 9yz^3$
8. $12x^3y^3 \div 4y^2$
9. $e^2f \times 3fg^2$
10. $pq \times 2pq^2$
11. $3a^4 \times 3a^4$
12. $2cd \times cd^3$
13. $k^{10} \div k^5$
14. $a^{\frac{1}{2}} \times a^{\frac{1}{2}}$
15. $b^{\frac{1}{4}} \times b^{\frac{3}{4}}$
16. $x^{2\frac{1}{2}} \times x^{1\frac{1}{2}}$
17. $h^5 \times h^2$
18. $3kl^2 \times mn^2$
19. $3f^2 \div 2f$
20. $x^{2\frac{1}{2}} \div x^{\frac{1}{2}}$
21. $x^{1\frac{1}{2}} \div x^{\frac{1}{2}}$
22. $2a^{2\frac{1}{2}} \div a^{\frac{1}{2}}$
23. $2a \times 2a \times 2a$
24. $n^3 \div n^3$
25. $a^4c^3z^5 \div a^2c^2z^2$
26. $4x^{2\frac{1}{2}} \div 2x^{\frac{1}{4}}$
27. $3b^{\frac{1}{2}} \times 2b^{\frac{1}{2}} \div 6b$
28. $a^{\frac{1}{2}} \times a^{\frac{1}{2}} \div 2a$
29. $2x^5y^5 \div 3x^2y^2$
30. $a^2 \times a^4 \times a^6$
31. $2xy^2 \times 3x^2y \times xy$
32. $\dfrac{x}{y} \times \dfrac{y}{z} \times \dfrac{z}{a}$
33. $\dfrac{x^2}{a^2} \div \dfrac{x^2}{a^2}$
34. $\dfrac{4a}{2} \div \dfrac{2a}{4}$
35. $x \times 2x^2 \times 3x^3$
36. $\dfrac{a^2 \times a^3 \times a^4}{a^7}$

PROBLEMS

EXERCISE 12

1. I have two boxes, each contains p oranges. If I sell q oranges, how many shall I have left?

2. y girls are to be divided into two teams to play rounders. How many girls in each team? What kind of number must y be, for the teams to be fairly matched?

3. A train journey costs p pence. What is the fare for three people? How much for one person for four journeys? How much for x persons for t journeys?

4. A tank holds x gal of water. How many times can a 2-gal bucket be filled from the tank?

5. A boy has m marbles; one friend has twice this number; a second friend has only half the number. How many marbles have the three boys altogether?

6. How many inches: (i) in 1 ft; (ii) in F ft?

7. How many feet: (i) in 1 yd; (ii) in Y yd?

8. How many inches in Y yd?

9. How many yards in 1 chain? How many chains in Y yd?

10. How many yards in 1 mile? How many miles in Y yd?

11. In fig. A: 1 call the area of the rectangle A. What does A equal?

fig. A: 1

12. In fig. A: 1, what is the value of A; (i) when $l = 18$, $b = 12$; (ii) when $l = 3\frac{1}{2}$, $b = 2\frac{1}{2}$?

13. In fig. A: 1 if the perimeter is P, what is P equal to?

14. What is the value of P, when $l = 8\frac{1}{2}$, $b = 4\frac{1}{2}$?

DIRECTED NUMBERS: ADDITION AND SUBTRACTION

31° C means 31° above 0° C.

−35° C means 35° below 0° C.

You will observe that it is necessary to draw *particular* attention to those readings *less* than zero, and we do so by using the minus sign. The sign is not used, however, in the sense of "subtracting" or "taking away", instead it indicates the **direction** of a reading in relation to 0° C.

−35° C would be called a **negative** quantity.

Those readings *above* zero are regarded as *normal* and no special attention is given to the sign, but in fact 92° C could be written as +92° C. It can be called a **positive** quantity, and the quantities used in everyday life are of this type. E.g.,

A second-hand car is advertised at £500
(NOT +£500)

QUESTIONS

1. What would be your reaction to an advertisement in which the price was given as −£500?

2. Is zero (0) a positive or negative quantity?

Perhaps we could summarize these results as follows:

+ means ⟶ the NORMAL procedure.

− means ⟵ the REVERSE (or opposite) to normal procedure.

With these statements in mind, let us give meaning to the following examples.

(*a*) **4 + (−2)**

(i) Having no sign attached to it, the number 4 is a normal positive value.

(ii) Next we have to add (−2), but the sign + can be read as "do the normal thing to that which follows"; thus "do the normal thing" with (−2) is to "subtract 2" from 4.

∴ 4 + (−2) means 4 − 2 = 2

(*b*) **6 − (+3)**

(i) The number 6 is a normal positive value.

(ii) −(+3) means "reverse the normal procedure" with (+3) giving us −3.

∴ 6 − (+3) means 6 − 3 = 3

Similarly: $+(+2)$ means "do the normal thing" to $(+2)$

$$\therefore \ +(+2) = +2$$

and $\qquad -(-3)$ means "do the opposite thing" to (-3)

$$\therefore \ -(-3) = +3$$

The complete set of sign combinations and their values is as follows:

$+(+)$ **means** $+$ ⎫ **Like signs give**
$-(-)$ **means** $+$ ⎭ **"plus"**

$+(-)$ **means** $-$ ⎫ **Unlike signs give**
$-(+)$ **means** $-$ ⎭ **"minus"**

Deal with each pair of signs using the appropriate phrase "do the normal thing" or "do the opposite thing" and you will better understand the results, which are best learnt from the statements at the right. (**Known as "The Rule of Signs".**)

EXERCISE 13

Write down the values of the following:

1. $+(+2)$	**2.** $+(-2)$	**3.** $-(+2)$
4. $+(-3)$	**5.** $-(+3)$	**6.** $-(-4)$
7. $+(+1)$	**8.** $-(+0)$	**9.** $+(-5)$
10. $-(+7)$	**11.** $+(+9)$	**12.** $+(-0)$
13. $8 + (-3)$	**14.** $3-(-8)$	**15.** $8 - (-3)$
16. $7 - (+4)$	**17.** $4 - (-7)$	**18.** $4 + (-7)$
19. $9 + (-5)$	**20.** $9 - (+5)$	**21.** $5 - (-9)$
22. $5 + (-9)$	**23.** $2 - (+0)$	**24.** $0 - (+2)$
25. $2 - (+2)$	**26.** $(-2) + (+2)$	**27.** $(-3) - (-3)$
28. $(+4) - (+4)$	**29.** $(-6) + (-2)$	**30.** $(-2) + (-6)$
31. $(-1) - (-0)$	**32.** $(-0) + (-1)$	**33.** $(-1) - (-1)$
34. $(-7) + (+7)$	**35.** $(-2) + (+4)$	**36.** $(+5) - (+3)$
37. $(+x) + (+x)$	**38.** $(+x) - (+x)$	**39.** $(-x) + (-x)$
40. $(-2x) - (-3x)$	**41.** $(+3x) + (-2x)$	**42.** $(+2x) - (+3x)$
43. $(-6y) + (+8y)$	**44.** $(+4a) + (-4a)$	**45.** $(0) - (-3b)$
46. $(+3t) + (-0)$	**47.** $9 + (-3x)$	**48.** $(3x) - (+9)$

NOTE. From the previous exercise and reading matter you will see that the use of brackets enables us to express certain symbols more clearly. By employing such devices it is possible to avoid ambiguous situations in mathematics.

MULTIPLICATION OF DIRECTED NUMBERS

Consider the temperature of a beaker of water being heated steadily so that the temperature changes are equal, in equal periods of time. We will assume that in 1 minute the temperature rises $2°$ C and continues to do so every minute.

Then in 10 minutes time (that is $+10$) with the temperature rising $2°$ C each minute (that is $+2$) the change in temperature is given by:

RESULT (i) $$(+10) \times (+2) = (+20)$$

That is $20°$ C **above** the present temperature.

But 6 minutes ago (that is -6) the change in temperature is given by:

RESULT (ii) $$(-6) \times (+2) = (-12)$$

That is $12°$ C **below** the present temperature.

Assume that the water is now allowed to cool and the temperature falls by $2°$ C (that is -2) every minute.

Then in 10 minutes time (that is $+10$) the change in temperature is given by:

RESULT (iii) $$(+10) \times (-2) = (-20)$$

That is $20°$ C **below** the present temperature.

But 6 minutes ago (that is -6) the change in temperature is given by:

RESULT (iv) $$(-6) \times (-2) = (+12)$$

That is $12°$ C **above** the present temperature.

You may need to read these statements several times in order to satisfy yourself that the results are what *you* would expect when carrying out a similar experiment.

We can summarize the results of these explanations as follows:

$$(+) \times (+) = +$$
$$(-) \times (-) = +$$
$$(+) \times (-) = -$$
$$(-) \times (+) = -$$

and again we see that the **Rule of Signs** is applicable.

Like signs give "plus".
Unlike signs give "minus".

EXERCISE 14

Write down the values of the following:

1. $(-2) \times (+2)$	**2.** $(-0) \times (-1)$	**3.** $(+4) \times (+4)$
4. $(-6) \times (-2)$	**5.** $(-2) \times (+4)$	**6.** $(-1) \times (-0)$
7. $4 \times (-7)$	**8.** $4 \times (+7)$	**9.** $9 \times (-5)$
10. $9 \times (+5)$	**11.** $5 \times (-9)$	**12.** $0 \times (+5)$
13. $(+x) \times (+x)$	**14.** $(+x) \times (-x)$	**15.** $(-x) \times (-x)$
16. $(-2x) \times (-3x)$	**17.** $(+3x) \times (-2x)$	**18.** $(+2x) \times (+3x)$
19. $(-6y) \times (+8y)$	**20.** $(+4a) \times (-4a)$	**21.** $(0) \times (-3b)$
22. $(3t) \times (-0)$	**23.** $9 \times (-3x)$	**24.** $3x \times (+9)$

NOTE. Example 24 in the previous exercise should suggest to you a good reason for avoiding the use of the multiplication sign ($\times$) in algebra. In its place we sometimes use a full stop, thus: $2x . 3x$, but a better device is the use of brackets thus: $(2x)(3x)$. Once more we see how the use of brackets enables us to express things clearly.

DIVISION OF DIRECTED NUMBERS

If the temperature of a quantity of heated water is to be $20°$ C higher than at present (that is $+20$) in 10 minutes time (that is $+10$), then the temperature change per minute is given by:

RESULT (i) $(+20) \div (+10) = (+2)$ A **rise** of $2°$ C per minute.

But if the temperature was $12°$ C lower than at present (that is -12) at a time 6 minutes ago (that is -6), then the temperature change per minute is given by:

RESULT (ii) $(-12) \div (-6) = (+2)$ A **rise** of $2°$ C per minute.

If the temperature in 10 minutes time (that is $+10$) is to be $20°$ C lower than at present (that is -20), then the temperature change per minute is given by:

RESULT (iii) $(-20) \div (+10) = (-2)$ A **fall** of $2°$ C per minute

If the temperature 6 minutes ago (that is -6) was $12°$ C higher than at present (that is $+12$), then the temperature change per minute is given by:

RESULT (iv) $(+12) \div (-6) = (-2)$ A **fall** of $2°$ C per minute.

When you are satisfied that these statements are correct you will **see** that the results can be summarized as follows:

$$(+) \div (+) = +$$
$$(-) \div (-) = +$$
$$(+) \div (-) = -$$
$$(-) \div (+) = -$$

and once more the **Rule of Signs** is applicable. In particular, note that $(+2) \div (-4)$ will simplify to $-\frac{1}{2}$.

EXERCISE 15

Write down the values of the following:

1. $(+2) \div (+1)$ **2.** $(+1) \div (+2)$ **3.** $(+4) \div (-2)$

4. $(-2) \div (+4)$ **5.** $(-4) \div (-1)$ **6.** $(-1) \div (-4)$

7. $(0) \div (-1)$ **8.** $(-1) \div (+1)$ **9.** $(+3) \div (-2)$

10. $(ax) \div (-x)$ **11.** $(-ax) \div (a)$ **12.** $(-a) \div (-a)$

13. $(-xy) \div (xy)$ **14.** $(xy) \div (-y)$ **15.** $(-x) \div (-2)$

16. $\dfrac{4ab}{2b}$ **17.** $\dfrac{2ab}{4ab}$ **18.** $\dfrac{9a^2b}{3a}$ **19.** $\dfrac{-4a^2b}{2b}$

20. $\dfrac{2b^2}{-4b}$ **21.** $\dfrac{-9a^2b}{-3a}$ **22.** $\dfrac{6a^2b^2c}{-2ab^2}$ **23.** $\dfrac{-x^3y^2z}{x^2y}$

NOTE. Do not forget that directed numbers are sometimes described as positive $(+)$ or negative $(-)$ numbers.

MISCELLANEOUS EXAMPLES

EXERCISE 16

Simplify the following:

1. $6 + (-2)$ **2.** $(6)(-2)$ **3.** $6 \div (-2)$

4. $2 - (-6)$ **5.** $(-2)(-6)$ **6.** $(-6) \div (-3)$

7. $a^2 \div a$ **8.** $(-2a)^2$ **9.** $a^2 \div (-a)$

10. $4 - 2(-1)$ **11.** $0 - (-3)$ **12.** $(-a)(2a)$

13. $x^2 + (-x)$ **14.** $(-a)^3$ **15.** $(-3)^4$

16. $(4x^3) \div (2x^2)$ **17.** $(-4x^3) \div (-2x^2)$ **18.** $4x^3 - (-2x^2)$

19. $(-2a)(-3b)$ **20.** $(-0) + (-1)$ **21.** $(-a) - (+0)$

22. $(-x)^2 - (-x^2)$ **23.** $(-2x) \div (-4x)$ **24.** $(-4x)(-2x)$

25. $\dfrac{2x(-x)^2}{(-x)(-x)(-x)}$ **26.** $\dfrac{(-2a) + (-4a)}{6(-a)}$ **27.** $\dfrac{(-a^2)(-a^3)}{2a(a^4)}$

28. $\left(\dfrac{-a}{b}\right) \times \left(\dfrac{-b^2}{a^2}\right)$ **29.** $\left(\dfrac{-a}{b}\right)\left(\dfrac{a}{b}\right)\left(\dfrac{a}{-b}\right)$ **30.** $\left(\dfrac{-2}{4}\right) \div \left(\dfrac{16}{-4}\right)$

MORE ABOUT COEFFICIENTS

IN earlier work on coefficients it was sufficient to identify the numeral or letter preceding a term, but now we must take account of the sign attached to the term.

EXERCISE 17 (*Oral*)

Read off the coefficients of the following, state whether they are numerical or literal:

1. $2a$	**2.** $-3b$	**3.** $+6a$	**4.** $-4c$	**5.** $4x$
6. $4c$	**7.** $-9y$	**8.** $-5y$	**9.** $+3y$	**10.** $3x$
11. $-16c$	**12.** $-x$	**13.** ax	**14.** $-bx$	**15.** b
16. bc	**17.** $-c$	**18.** c	**19.** $-xy$	**20.** xyz
21. xy	**22.** $-abc$	**23.** $-bc$	**24.** $-b$	**25.** $2ab$
26. $-2b$	**27.** $-2xy$	**28.** $16xy$	**29.** $-16x$	**30.** $-16xy$

USE OF BRACKETS

EXAMPLE 1. *The sum of x and y is added to the sum of 2x and 2y.*

This is expressed thus: $(2x + 2y) + (x + y)$

EXAMPLE 2. *The sum of x and y is subtracted from the sum of 2x and 2y.*

This is expressed thus: $(2x + 2y) - (x + y)$

EXAMPLE 3. *The difference between x and y is added to the difference between 2x and 2y.*

This is expressed thus: $(2x - 2y) + (x - y)$

EXAMPLE 4. *The difference between x and y is subtracted from the difference between 2x and 2y.*

This is expressed thus: $(2x - 2y) - (x - y)$

In none of these cases can we first simplify the contents of the brackets. We must, therefore, remove the brackets, paying careful attention to the signs.

EXAMPLE 1. $(2x + 2y) + (x + y) = 2x + 2y + x + y$
$$= 3x + 3y \; Ans.$$

EXAMPLE 2. $(2x + 2y) - (x + y) = 2x + 2y - x - y$
$$= x + y \; Ans.$$

EXAMPLE 3. $(2x - 2y) + (x - y) = 2x - 2y + x - y$
$$= 3x - 3y \; Ans.$$

EXAMPLE 4. $(2x - 2y) - (x - y) = 2x - 2y - x + y$
$$= x - y \; Ans.$$

From these results we can conclude:

(i) If a + sign precedes a bracket the signs of the terms within the bracket remain unchanged when the bracket is removed.

(ii) If a — sign precedes a bracket the sign of each term within the bracket is changed when the bracket is removed.

EXERCISE 18

Write expressions for the following, simplifying the expressions by removing brackets and collecting like terms where possible.

1. The sum of a and b is added to $2a$.

2. The difference between a and b is added to $2b$.

3. The sum of $2a$ and $2b$ is subtracted from a.

4. The difference between $2a$ and $2b$ is added to $2a$.

5. The difference between $2a$ and $2b$ is subtracted from $2a$.

6. To the sum of x and y is added the sum of $3x$ and $4y$.

7. From the sum of $3x$ and $4y$ is subtracted the difference between x and y.

8. To the difference between x and y is added the sum of $2x$ and $3y$.

9. From the difference between $3x$ and $2y$ is subtracted the sum of $2x$ and $3y$.

10. To the sum of $3a$ and b is added the difference between a and $3b$.

11. $2y$ is added to $2x$, y is subtracted from $3x$. The results are added and $3y$ is subtracted from the expression.

12. $3a$ and $5b$ are added together; from the result is subtracted the sum of $2a$, $4b$ and $3c$.

13. The difference between a and $3b$ is added to the sum of a and $3b$; the result is divided by 2.

14. The sum of 3 and 2 is added to the difference between 3 and 2.

ADDITION AND SUBTRACTION OF COMPOUND EXPRESSIONS

The use of brackets makes it possible to carry out addition or subtraction of expressions set out **horizontally** instead of **vertically**, as we usually do when adding or subtracting sums of money.

EXAMPLE 1. $(2a + 3b + c) + (a + 2b - 3c) + (3a - 4b + 2c)$
$$= 2a + 3b + c + a + 2b - 3c + 3a - 4b + 2c$$
$$= 6a + b \; Ans.$$

EXAMPLE 2. $(3x - 4y + 2z) - (x - 4y + z)$
$$= 3x - 4y + 2z - x + 4y - z$$
$$= 2x + z \; Ans.$$

EXAMPLE 3. $(2a^2 + 2ab - 2b^2) + (a^2 - 3ab + b^2) - (a^2 - ab + b^2)$
$$= 2a^2 + 2ab - 2b^2 + a^2 - 3ab + b^2 - a^2 + ab - b^2$$
$$= 2a^2 - 2b^2 \; Ans.$$

EXERCISE 19

Add the following expressions:

1. $2m - 3n + 3p$; $3m - 2n - 3p$; $-4m + 3n - 4p$
2. $r + 2s - 4t$; $3r + 2s - t$; $-2r - 3s + 2t$
3. $q - 2p + 3r$; $-4p + 2r + 2q$; $r + 2p - q$
4. $2x + 3y - 4$; $-3y + 3x + 2$; $4y - 2 - x$
5. $3u - w + 2v$; $2w - u + v$; $-2v - u - w$
6. $pq + qr$; $pq + rs$; $qr - rs$; $2pq - 2qr$
7. $ab + cz$; $2lm - pq$; $2cz + pq$; $-2ab - 2lm$
8. $-2a + b$; $-2b + c$; $-2c + d$; $-2d + e$
9. Subtract $a^2 - ab + 2b$ from $2a^2 - 2b$
10. Subtract $a^2 - ab + 2$ from $a^2 - ab - 2$
11. Subtract $-a^3 + a^2 + 4$ from a
12. Subtract a from $a^2 - a + 4$
13. From $x^2 - 2y + 3$ subtract $x^2 - x - 3$
14. From $-a + ab - y$ subtract $a + ab + y$
15. From $a^2 - 2ab + b^2$ subtract $a^2 - 2ab + b^2$
16. From $a^3 - 2a^2$ subtract $2a^2 - 3a + 4$
17. What must be added to $x^2 - 2xy$ to make $-3x^2 + 2xy + 2y^2$?
18. What must be added to $-3x^2 + 2xy + 2y^2$ to make $x^2 - 2xy$?
19. What must be subtracted from $3x^2$ to make $x^2 - 2xy$?
20. What must be subtracted from $x^2 - 2xy$ to make $3x^2$?

MULTIPLICATION OF COMPOUND EXPRESSIONS

A further use for brackets is to assist the process of multiplication.

EXAMPLE 1.

$$4(a - b)$$
$$= 4a - 4b \ Ans.$$

EXAMPLE 2.

$$a(a^2 - b)$$
$$= a^3 - ab \ Ans.$$

EXAMPLE 3.

$$-6(2x - 3y)$$
$$= -12x + 18y$$
or $18y - 12x \ Ans.$ (Why?)

EXAMPLE 4.

$$-3a(2a - 4)$$
$$= -6a^2 + 12a$$
or $12a - 6a^2 \ Ans.$

The following is a good routine to adopt during the manipulation of algebraic expressions: (i) **signs**; (ii) **numerals**; (iii) **letters**; (iv) **powers**.

When we have to multiply expressions each containing two or more terms we can employ a similar method to that already shown.

NOTE. Pairs of brackets without a sign separating them represent multiplication, thus: (   )(   )

EXAMPLE 5.

$(a + 2)(a + 3)$
$= a(a + 3) + 2(a + 3)$
$= a^2 + 3a + 2a + 6$
$= a^2 + 5a + 6$ *Ans.*

EXAMPLE 6.

$(x - 4)(x^3 - 2x^2 + 1)$
$= x(x^3 - 2x^2 + 1) - 4(x^3 - 2x^2 + 1)$
$= x^4 - 2x^3 + x - 4x^3 + 8x^2 - 4$
$= x^4 - 6x^3 + 8x^2 + x - 4$ *Ans.*

STAGES IN CALCULATION

(i) Break the first bracket into component terms, each with its appropriate sign.

(ii) Multiply the second bracket by each component from (i); care with signs.

(iii) Collect like terms.

NOTE. It is advisable to place the *shorter* expression first, since by using this as the multiplier we shall carry out the calculation more conveniently.

EXERCISE 20 (1–12 *Oral*)

Simplify the following:

1. $2(a + b)$
2. $4(x + 2)$
3. $-2(x + y)$
4. $-4(a - b)$
5. $a(3 + b)$
6. $-a(3 - b)$
7. $4(a + 2b + 3c)$
8. $-3(2x - 3y - 4z)$
9. $2a(a^3 - a^2 + a)$
10. $-3x(2x^3 - 4x^2 - x)$
11. $2a^2(3a + ab + 1)$
12. $-3x^2(x^2 + 2xy + y^2)$
13. $(a + b)(a + b)$
14. $(a + b)(a - b)$
15. $(a - b)(a - b)$
16. $(a + 2)(a + 3)$
17. $(a - 2)(a + 3)$
18. $(a + 2b)(a - 2b)$
19. $(2a - b)(2a - b)$
20. $(a - 2b)(2a - b)$
21. $(a + b)(x + y)$
22. $(2a + 1)(a - 2)$
23. $(a - b)(x - y)$
24. $(a^2 + a)(a^2 + a)$
25. $(2a^2 + 3a)(3a^2 - 2a)$
26. $(a^2c + b)(c^2 - ac^2)$
27. $(x - y)^2$
28. $(2a + 1)^2$
29. $(a + 1)(a^2 + a + 1)$
30. $(a - 1)(a^2 + a + 1)$
31. $(a + 1)(a^2 - a - 1)$
32. $(a - 1)(a^2 - a - 1)$
33. $(a + 2)(a^2 + 4a - 6)$
34. $(a - 3)(a^2 - 6a + 5)$
35. $(2a + 3)(2a^2 - a + 4)$
36. $(3a - 4)(3a^2 + 4a - 6)$
37. $(x + y)(x^2 + x + 1)$
38. $(2x - 2y)(2x^2 + 2y^2 - 2)$
39. $(x^2 + x)(x^2 + x + 3)$
40. $(x^2 - x)(y^2 - 2y + 3)$
41. $(x + 1)(x^3 - x^2 + x - 4)$
42. $(2x - 2)(3x^3 + 2x^2 - x + 1)$
43. $(2x - y + 2)(2 + y - 2)$
44. $(x + 2y - 1)(x - 2y + 1)$
45. $(x + 2y + 3)(2x - 3y - 4)$
46. $(a + b + c)(a - b - c)$

DIVISION OF COMPOUND EXPRESSIONS

When dividing a *compound* expression by a *simple* expression we can set the work down as a fraction, but we must remember that **each** term of the dividend must be divided by the divisor.

Thus: $\dfrac{6x^2 + 4x^2y - 2xy^2}{2x} = \underline{\underline{3x + 2xy - y^2}}$ *Ans.*

EXERCISE 21

Simplify the following:

1. $\dfrac{a + ab + ac}{a}$ 2. $\dfrac{abc - abd - abe}{ab}$ 3. $\dfrac{2x - 3xy + 4xz}{x}$

4. $\dfrac{4x + 6xy - 8xz}{-2x}$ 5. $\dfrac{2ax - 4a^2x^2 + 6a^3x^3}{2ax}$ 6. $\dfrac{a^2x - ax^2 - 2ax^3}{-ax}$

7. $\dfrac{8x^3y^2 - 12x^2y^3 + 16x^2y^2}{4x^2y^2}$ 8. $\dfrac{12a^3b^2c + 18a^2b^3c^2 - 36a^2b^2}{6a^2b^2}$

9. $\dfrac{a^2 + b^2 + c^2}{d^2}$ 10. $\dfrac{10pqr + 20p^2q^2r^2 - 30p^3q^3r}{10pqr}$

When dividing one compound expression by another certain precautions are necessary:

1. **Both** the dividend and the divisor must be arranged with the terms in the same order—usually in descending powers of some common term.

2. The **first term only** of the dividend is divided by the **first term** of the divisor. The result gives the first term of the quotient.

3. The **whole** divisor is now multiplied by the quotient. This product is set down below the dividend and subtracted. Great care is needed at this stage due to the use of directed numbers.

4. Terms are now brought down from the dividend and the process described above is repeated, as in arithmetic, until all the terms from the dividend have been used.

EXAMPLE 1. *Divide* $6x^3 - 17x^2 + 15x - 4$ *by* $3x - 4$

$$\begin{array}{r} 2x^2 - 3x + 1 \\ 3x - 4\overline{)6x^3 - 17x^2 + 15x - 4} \end{array}$$

* **Care** $\quad \dfrac{6x^3 - 8x^2}{}$

$\qquad\qquad -9x^2 + 15x$

Care $\quad \dfrac{-9x^2 + 12x}{}$

$\qquad\qquad\qquad 3x - 4$

Ans. $= \underline{\underline{2x^2 - 3x + 1}} \qquad \dfrac{3x - 4}{}$

* This operation requires care—we can regard the subtraction as follows:

$(6x^3 - 17x^2) - (6x^3 - 8x^2)$
$= 6x^3 - 17x^2 - 6x^3 + 8x^2$
$= \underline{-9x^2}$

The following rule may help you during the subtraction steps:

Change the sign of the lower line (mentally) and add.

In the actual division steps remember: (i) **signs**; (ii) **numerals**; (iii) **letters**; (iv) **powers**.

EXAMPLE 2. *Divide $2a^5 - 2a$ by $a + a^2$*

First arrange the terms in the same order in both dividend and divisor, allowing spaces for missing terms.

$$
\begin{array}{r}
2a^3 - 2a^2 + 2a - 2 \\
a^2 + a)\overline{2a^5 \qquad\qquad\qquad\qquad -2a} \\
\underline{2a^5 + 2a^4} \\
-2a^4 \\
\underline{-2a^4 - 2a^3} \\
2a^3 \\
\underline{2a^3 + 2a^2} \\
-2a^2 - 2a \\
-2a^2 - 2a
\end{array}
$$

Ans. $= \underline{\underline{2a^3 - 2a^2 + 2a - 2}}$

By spreading out the terms in the dividend in this way we can bring down any term as the need arises.

EXERCISE 22

Divide:

1. $x^2 + 6x + 9$ by $x + 3$
2. $x^2 + 7x + 12$ by $x + 3$
3. $x^2 + 8x + 15$ by $x + 3$
4. $x^2 + 9x + 20$ by $x + 4$
5. $6x^2 + 7x + 2$ by $2x + 1$
6. $10x^2 + 23x + 12$ by $2x + 3$
7. $35x^2 - x - 70$ by $5x + 7$
8. $49a^2 - 16$ by $7a - 4$
9. $12l^2 + 6l - 36$ by $3l + 6$
10. $30p^2 - 61p + 30$ by $6p - 5$
11. $16r^2 - 1$ by $4r + 1$
12. $36y^2 - 138y + 120$ by $9y - 12$
13. $6x^3 + 3x^2 - 6x - 3$ by $3x + 3$
14. $a^3 + 2a^2b + 2ab^2 + b^3$ by $a + b$
15. $2x^4 + 12x^3 + 14x^2 - 8x$ by $x + 4$
16. $6x^4 - 10x^3 + 6x^2 - 4x + 2$ by $3x^3 - 2x^2 + x - 1$
17. $x^4 + x^3 - 2x^2 + 3x - 1$ by $x^2 + 2x - 1$
18. $6x^4 + x^3 + 14x^2 - 11x + 2$ by $2x^3 + x^2 + 5x - 2$
19. $8x^4 - 18x^3 + 13x^2 - 8x + 3$ by $4x^3 - 3x^2 + 2x - 1$
20. $a^6 + a^5 - a^4 + a^3 - 2a^2$ by $a^2 + 2a$
21. $a^3 + b^3$ by $a + b$
22. $x^2 - y^2$ by $x - y$
23. $a^3 - b^3$ by $a - b$
24. $a^3 + b^3 - 2a^2b$ by $a - b$
25. $4a^5 + 18a^3 - 486a$ by $3a + a^2$
26. $x^4 - y^4$ by $x + y$
27. $a^4 - b^4$ by $a - b$
28. $6a^5b - 96ab^5$ by $2a + 4b$
29. $a^4b - ab^4$ by $a - b$
30. $x^6 - y^6$ by $x + y$

REMAINDERS

EXAMPLE. *Divide $2a^3 + 7a^2 + 8a + 6$ by $2a + 3$. Give quotient and remainder. What must be added to the dividend to avoid a remainder?*

$$
\begin{array}{r}
a^2 + 2a + 1 \\
2a + 3 \overline{)2a^3 + 7a^2 + 8a + 6} \\
\underline{2a^3 + 3a^2} \\
4a^2 + 8a \\
\underline{4a^2 + 6a} \\
2a + 6 \\
\underline{2a + 3} \\
\text{Rem.} = +3
\end{array}
$$

Ans. $= a^2 + 2a + 1$ rem. 3. -3 must be added.

EXERCISE 23

In each of the following give the quotient and remainder and state what must be **added** to the dividend to avoid a remainder.

Divide:

1. $x^3 + 3x^2 + 3x - 1$ by $x + 1$
2. $x^3 - 4x^2 + 5x$ by $x - 3$
3. $4x^3 + 4x^2 - x + 1$ by $2x + 1$
4. $2x^3 + 6x^2 - 26x + 6$ by $2x - 4$
5. $3x^4 + 12x^3 + 9x^2 - 3x - 3$ by $3x + 6$
6. $2a^3 - 3a^2b + 4ab - 2ab^2$ by $a^2 - 2ab + 2b$
7. $a^2 + 2ab + b^2$ by $a + b + c$
8. $2x^3 - 2x^2 - 2x + 3x^2y - 2xy + 2y$ by $2x + 3y$
9. a^4 by $a + 1$
10. $x^5 - 1$ by $x^4 - x^3 + x^2 - x + 1$

EXERCISE 24: PROBLEMS

1. I have a marbles in one pocket and b marbles in another.

  (i) How many marbles have I altogether?
 (ii) John has twice as many, how many has he?
(iii) How many have we altogether?

2. A box weighs 2 lb when empty. What is the total weight if it contains x lb of apples? What is the total weight of 6 boxes of apples?

3. A barrel will hold G gal, I pour in g gal; how many gallons are required to fill it?

4. I have n planks, each x ft long. I cut 2 ft from each plank, what is the total length of timber left?

5. I have *b* boxes, each contains *p* oranges. I remove *q* oranges from each box, how many oranges are left: (i) in each box; (ii) altogether?

6. I have a number of coins, *x* of them are 50p coins and 6 are 10p coins.

  (i) How many coins have I altogether?

 (ii) How much money have I in pence?

(iii) How much money have I in pounds?

7. A fence is supported by posts *f* ft apart.

  (i) How long is a fence containing 60 posts?

 (ii) How many posts are required in a fence 60 ft long?

8. A man works *H* hours per day for 5 days and 4 hours on Saturday. He is paid *r* pence per hour. What is his weekly wage: (i) in pence; (ii) in pounds?

9. What is the area of the rectangle in fig. A:2? Dimensions are in inches.

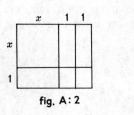

fig. A:2

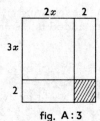

fig. A:3

10. (i) What is the area of the complete rectangle in fig. A: 3?

 (ii) What is the area of the shaded portion?

(iii) What is the area of the remainder?

FURTHER WORK WITH POWERS AND ROOTS

$(a^2bc^3)^3$ may be written as $a^2bc^3 \cdot a^2bc^3 \cdot a^2bc^3$

$$\therefore (a^2bc^3)^3 = \underline{a^6b^3c^9} \ Ans.$$

Since $(\frac{1}{2})^2$ clearly gives us $\frac{1}{4}$, it should be easy for us to understand that $(\frac{1}{2})^2$ may be written as $\frac{1^2}{2^2}$, which gives us the same answer: $\frac{1}{4}$.

Thus: $\qquad \left(\dfrac{x^2}{y^3}\right)^2 = \dfrac{(x^2)^2}{(y^3)^2} = \underline{\underline{\dfrac{x^4}{y^6}}} \ Ans.$

Similarly: $\qquad \sqrt{\dfrac{a^2}{b^4}} = \dfrac{\sqrt{a^2}}{\sqrt{b^4}} = \underline{\underline{\dfrac{a}{b^2}}} \ Ans.$

EXERCISE 25

Simplify the following:

1. $(-xy)^4$ 2. $(2ab)^3$ 3. $(-2a)^4$
4. $(3a^2)^2$ 5. $(-x^2)^2$ 6. $(-3x)^3$
7. $(3ab)^2$ 8. $(2a^2b)^2$ 9. $(a^2b)^3$
10. $(2ab^3)^2$ 11. $(2a^2b^3)^3$ 12. $(3a^3b^2)^2$
13. $(a^3b^3)^3$ 14. $(-3a^2b^2)^4$ 15. $(4a^2x^3y^4)^3$
16. $\sqrt{a^2b^2}$ 17. $\sqrt[3]{x^3y^3}$ 18. $\sqrt[4]{a^4y^4}$
19. $\sqrt{4a^2}$ 20. $\sqrt[3]{27x^3}$ 21. $\sqrt[4]{16x^4}$
22. $\sqrt{9a^4}$ 23. $\sqrt[3]{8x^6}$ 24. $\sqrt[4]{81a^8}$
25. $\sqrt{16a^4x^2}$ 26. $\sqrt[3]{64a^3b^6}$ 27. $\sqrt{81a^8}$
28. $\sqrt[4]{16a^8b^4c^{16}}$ 29. $\sqrt[3]{125a^6x^9}$ 30. $\sqrt[4]{a^4b^8c^{12}}$
31. $\sqrt{(a^2b^2c^2)^3}$ 32. $\sqrt[3]{(a^3b^3c^3)^2}$ 33. $(\sqrt{a^4})^3$
34. $(\frac{1}{4})^3$ 35. $\left(\frac{4a^2}{a}\right)^2$ 36. $\left(\frac{a^2b}{b^2c}\right)^3$
37. $\left(\frac{ax^2y^3}{a^2x^4y^2}\right)^3$ 38. $(\frac{3}{7})^2$ 39. $\left(\frac{4k^2lm^3}{8kl^2m^2}\right)^2$
40. $(\frac{3}{6})^4$ 41. $\left(\frac{3xyz^3}{3x^2yz^2}\right)^3$ 42. $(8\frac{1}{27})^3$
43. $\sqrt[3]{\frac{a^3b^3c^6}{a^3b^6c^3}}$ 44. $\sqrt{\frac{9a^2b^2c^4}{16c^4x^2y^2}}$ 45. $\sqrt[4]{\frac{a^2x^5}{a^2x^5}}$
46. $\sqrt{\frac{a^3b^5c^7}{ab^3c}}$ 47. $\sqrt[3]{\frac{x^6y^9}{a^9b^6}}$ 48. $\sqrt{\frac{18a^4}{2a^2}}$
49. $\sqrt{\frac{4a^2b^2}{9x^2y^4}}$ 50. $\sqrt[4]{\frac{81a^4y^{16}}{16a^8y^{12}}}$ 51. $\sqrt[3]{\frac{a^4b^6}{ay^3}}$

MORE ABOUT BRACKETS

The brackets used are these: ( ), { }, [ ].

Another device having the same function as brackets is a line called a **vinculum**, this is placed above the symbols which are to be associated as though in brackets.

REMEMBER

1. The contents of a pair of brackets can be thought of as a single term.

2. Each separate term within a pair of brackets must be operated on in the same way. Thus:

(*a*) If a coefficient is placed before a bracket it indicates that each term within the bracket must be multiplied by that coefficient.

(*b*) Signs before brackets can be regarded as coefficients.

3. A useful procedure is: "signs, numbers, letters and powers".

EXAMPLE 1. *Simplify the following:*

$$22a + 3[2a - 4\{a - 2(a - 3a) + a\} - a]$$
$$= 22a + 3[2a - 4\{a - 2a + 6a + a\} - a]$$
$$= 22a + 3[2a - 4\{6a\} - a]$$
$$= 22a + 3[2a - 24a - a]$$
$$= 22a + 3[-23a]$$
$$= 22a - 69a$$
$$= -47a \text{ Ans.}$$

EXAMPLE 2. *Simplify the following:*

$$2[x + 3\{2 + 6(2 - \overline{x + 2})\}]$$
$$= 2[x + 3\{2 + 6(2 - x - 2)\}]$$
$$= 2[x + 3\{2 + 6(-x)\}]$$
$$= 2[x + 3\{2 - 6x\}]$$
$$= 2[x + 6 - 18x]$$
$$= 2[6 - 17x]$$
$$= 12 - 34x \text{ Ans.}$$

EXERCISE 26

Simplify the following:

1. $a + [-a - \{a + 2a(-1)\}]$
2. $2x - 2(x + y) + 2[x + 2\{x + y\}]$
3. $2(a + b) - 3\{2a - b\} + 4[a - a + b]$
4. $\{a - (a + b)\} - (-b)$
5. $3p - [p + \{p - (p - \overline{p + 1})\}]$
6. $2[x + y - (2 - y)] - [2x - 3\{x - (x - y)\}]$
7. $x - \overline{x - y + z} + 2\{x - (y - z)\}$
8. $[(c - d) - (c - d) + d] - [-e + f - (e + f)]$
9. $2k - \{-l - (k + l)\} - [-k - \{-(k + l) + l\}]$
10. $p + q(-1) - 1\{p - (p + q) + (2p - q)\}$
11. $a[a - \{-a(a + b) - a^2\} - ab]$
12. $2a - a(a + 1) - 2\{a - 3(a + 1)\}$
13. $x - [-x - \{x - (-\overline{x - 1}) - x\} + x]$
14. $x^2[2 + x\{x - 2(x^2 - x)\}]$
15. $2ax\{a - (ax - x + 1)\} - x(a + 1)$
16. $\{a^2 - a(a - \overline{a - y})\} - \{a(a - \overline{a^2 - y})\}$
17. $-d - [d - \{e - d - (d - e) - d\} - e]$
18. $2[4 - 2\{6 - 3(3 + 2)\} + 5(-4)]$
19. $m - n(m + n) - [m - n^2 + m\{m - n\} - 2n]$
20. $5p - 5[p - \{q - (q + r)\}]$
21. $3x[2 + 4\{x^2 - 2x + 1\} - 3x(x - 2)]$
22. $[2a + 3\{b + (a - c)\}] - [2\{a + (b - c)\}]$
23. $x - 3\{x - 2(y - z)\} - \{-x - 3(x + y - z)\}$
24. $-\overline{a + b - c} - \overline{b - c} + 2(a + b - c)$
25. $ax[a - x(a + x) - a\{a - a(a + x)\}]$

SIMPLE EQUATIONS

A simple equation is one which, in its simplest form, contains no power of the unknown higher than the first (i.e., only a or x NOT a^2 or x^3, etc.). For this reason they are sometimes known as **equations of the first degree.**

EXAMPLE 1. *Solve the equation* $7x - 3x + 4x = 9 + 11 + 4$
Collecting like terms: $8x = 24$
Dividing each side by 8: $x = 3$ *Ans.*

EXAMPLE 2. *Solve the equation* $13x - 9x + 2x = 19 - 21$
Collecting like terms: $6x = -2$
Dividing both sides by 6: $x = -\dfrac{2^1}{6_3}$
$\therefore \; x = -\frac{1}{3}$ *Ans.*

EXERCISE 27 (1–20 *Oral*)

Solve the following equations:

1. $3x = 6$ **2.** $4x = 16$ **3.** $2x = 10$
4. $2x = 8$ **5.** $5x = 15$ **6.** $3x = 18$
7. $4x = 20$ **8.** $6x = 6$ **9.** $3x = 0$
10. $14x = 70$ **11.** $x + 2x = 15$ **12.** $3x = 9 + 3$
13. $2x - x = 3$ **14.** $4x = 11 - 3$
15. $2x + 3x + x = 18 + 6$ **16.** $x + 5x = 5 + 7$
17. $x + 5x + 7x = 20 + 8 + 11$ **18.** $3x + 9x = 18 + 18$
19. $3x + 4x - 5x = 9 - 4 - 3$ **20.** $7x - 9x + 15x = 17 + 9$
21. $8x - 5x + 2x + x = 3 - 9 + 5 + 7$
22. $x + 2x - 3x + 4x = 11 - 9 + 2$
23. $5x - 9x + 11x = 7 - 10 - 4$
24. $5x + 6x - 10x = 25 + 17 - 45$
25. $3x - 5x + 6x = 15 + 9 - 6$
26. $x + 7x - 5x = 17 - 9 + 7$
27. $4x - 6x + 7x = 21 - 19 + 14$
28. $-6x + 11x + 2x = 13 - 21 - 6$
29. $3x - 13x + 20x = -15 - 7 - 3$
30. $-5x - 2x + 9x = 16 + 7 - 24$
31. $17x - 12x + 11x = 15 + 1 - 4$
32. $23x - 5x + 6x - 12x = 3 + 9 - 26$
33. $-16x + 9x + 11x = 2 + 3 - 15$
34. $7x + 2x + 9x = -37 + 9 + 11 - 1$
35. $2x - 21x + 33x + x = 43 - 21 - 2$
36. $3x - 4x + 6x + 11x = 12 + 16 + 10$
37. $5x + 9x - 3x + 2x = 47 - 15 - 6$
38. $7x - 13x + 17x = 200 - 79$
39. $13x + 14x - 2x = 72 - 17 - 10$
40. $14x - 17x + 19x = 21 + 19 - 43$

TRANSPOSING TERMS

Moving terms from one side of an equation to the other is called **transposing** terms, and it is common practice to arrange the unknowns on the left of the equation. We shall refer to the "left-hand side" of the equation as L.H.S. and to the "right-hand side" as R.H.S.

EXAMPLE 1. $3x + 2 = 17$

Subtract 2 from both sides: $3x = 17 - 2$
$$\therefore\ 3x = 15$$
Divide both sides by 3: $\underline{\underline{x = 5}}$ *Ans.*

CHECK. When $x = 5$

$$\begin{aligned} \text{L.H.S.} &= 3x + 2 \qquad & \text{R.H.S.} &= 17 \\ &= 15 + 2 \\ &= 17 \end{aligned}$$

$$\therefore\ \underline{\underline{\text{L.H.S.} = \text{R.H.S.}}}$$

EXAMPLE 2. $9x = 3x + 18$

Subtract $3x$ from both sides: $9x - 3x = 18$
$$\therefore\ 6x = 18$$
Divide both sides by 6: $\underline{\underline{x = 3}}$ *Ans.*

CHECK. When $x = 3$

$$\begin{aligned} \text{L.H.S.} &= 9x \qquad & \text{R.H.S.} &= 3x + 18 \\ &= 27 & &= 9 + 18 \\ & & &= 27 \end{aligned}$$

$$\therefore\ \underline{\underline{\text{L.H.S.} = \text{R.H.S.}}}$$

EXAMPLE 3. $2x + 3 - 5 = 7 - 4x + x$

Collect like terms: $2x - 2 = 7 - 3x$
Add 2 to both sides: $2x = 7 - 3x + 2$
Collect like terms: $2x = 9 - 3x$
Add $3x$ to both sides: $5x = 9$
Divide both sides by 5: $\underline{\underline{x = 1\tfrac{4}{5}}}$ *Ans.*

CHECK. When $x = 1\tfrac{4}{5}$

$$\begin{aligned} \text{L.H.S.} &= 2x - 2 \qquad & \text{R.H.S.} &= 7 - 3x \\ &= 3\tfrac{3}{5} - 2 & &= 7 - 5\tfrac{2}{5} \\ &= 1\tfrac{3}{5} & &= 1\tfrac{3}{5} \end{aligned}$$

$$\therefore\ \underline{\underline{\text{L.H.S.} = \text{R.H.S.}}}$$

NOTE. Our solutions must always be given in the form: $x =$, NOT $3x =$, NOR $-x =$.

If the latter solution appears, the procedure is as follows: $-x = 3$. Multiply both sides by -1: $\therefore x = -3$ *Ans.*

"Unwanted" terms do not actually disappear from an equation. In fact, they reappear on the other side but with their signs changed. Since this happens in *every* case, we can establish a rule:

"Any term may be transposed from one side of an equation to the other providing its sign is changed."

Use the rule with care and understanding!

EXERCISE 28

Solve the following equations (*check your answers*):

1. $3x + 2 = 14$ 2. $5x + 6 = 16$
3. $2x + 3 = 6$ 4. $4x + 4 = 6$
5. $7x - 2 = 5$ 6. $3x - 5 = 10$
7. $5x - 6 = 4$ 8. $9x - 1 = 8$
9. $4x - 4 = 4$ 10. $11x - 18 = 4$
11. $2x = 5 - 3x$ 12. $3x = 8 - x$
13. $3x = 5 - 2x$ 14. $9x = 20 - x$
15. $8x = 27 - x$ 16. $13x = 22 - 9x$
17. $7x = 27 + 4x$ 18. $6x = 15 + 3x$
19. $4x + 3 = 3x + 1$ 20. $3x - 6 = 2x - 7$
21. $5x - 9 = 4x - 11$ 22. $6x + 7 = 4x + 6$
23. $2x - 3 + 4 - 5x + 7 = 8 - 4x + 2 - 3x$
24. $5x + 6 - 7x + 2 - 3x = 7x + 16 - 11x - 14$
25. $9 - 4x + 3x - 7 - x = 13 - 9x + 4 + 2x$
26. $14x - 16 + 31 - 16x = 27x - 14 - 16x + 21 - x$
27. $13 - 9x - 7x - 21 = 17 + 31x + 14 - 38x$
28. $21x + 17 - 13x - 29 = 23 - 9x + 16x - 35$

SIMPLE EQUATIONS: BRACKETS

NOTE. Each side of the equation must be simplified by removing brackets before transposing terms.

EXAMPLE. *Solve the equation:* $2\{x - 2(3 - x) - (x - 1)\} = 2$

Remove inner brackets: $2\{x - 6 + 2x - x + 1\} = 2$
Collect like terms: $2\{2x - 5\} = 2$
Remove brackets: $4x - 10 = 2$
Add 10 to each side: $4x = 12$
Divide both sides by 4: $\therefore x = 3$ *Ans.*

Check the solution by substituting for x in first line.

Exercise 29

Solve the following equations (*check your answers*):

1. $2(x + 1) = x + 8$ 2. $3x = 2(5 - x)$
3. $2 = 6(1 - 2x)$ 4. $3(2x - 5) = 9$
5. $2x + 3 = 4(6 - x)$ 6. $x - 5 = 2(x - 3)$
7. $2(2x - 1) = 3(x + 2)$ 8. $5(1 + x) = 3(2x + 3)$
9. $4(3 - 2x) = 3(6 - 4x)$ 10. $7(x - 3) = 3(7 + x)$
11. $6(2x + 7) = 7(x + 6)$ 12. $2(6x + 7) = 6(7 - 6x)$
13. $2(2x + 1) + 3(x - 2) + 4(6 - x) = 3(16 - 2x) - 1$
14. $2\{2 + 2(x - 2) - 3(3 - x) + 1\} = 0$
15. $4(2 - x) - 7(x - 3) - 4 = x + 1$
16. $6 - \{3x - (4x + 3) + 4\} = 10$
17. $37 - 4(5x - 4 + 2x) = -3$
18. $3(x + 4) - 4(3 - x) = 2(x + 2) - 8(7 + x)$
19. $5x + 2\{3 - (3 + x) + 6x\} = 2(6 + x) + 1$
20. $20(4x + 3) - 8(3x + 4) - 16 = 15(x + 3) - 3x$

EXAMPLE. *Solve the equation:* $(x + 2)(x - 3) = (x + 2)(x + 4)$

Form the products:

$$x(x - 3) + 2(x - 3) = x(x + 4) + 2(x + 4)$$
$$x^2 - 3x + 2x - 6 = x^2 + 4x + 2x + 8$$

Collect like terms: $x^2 - x - 6 = x^2 + 6x + 8$

Subtract x^2 from both sides: $-x - 6 = 6x + 8$

Transpose terms: $-7x = 14$

Divide both sides by -7:* $x = -2$ *Ans.*

CHECK. Substitute for x in first line—care needed!

NOTE. * **Note this method—it avoids the need to multiply both sides by -1.**

Exercise 30

Solve the following equations (*check your answers*):

1. $(x + 1)(x + 2) = (x - 3)(x - 4)$
2. $(x + 1)(x + 3) = (x - 3)(x - 5)$
3. $(x - 2)(x - 5) = (x + 4)(x + 3)$
4. $(x - 3)(x + 7) = (x + 1)(x - 9)$
5. $(2x - 1)(2x + 3) = (4x + 4)(x - 2)$
6. $(2x + 5)(x - 4) = (x + 6)(2x - 7)$
7. $(x - 1)^2 = (x + 3)(x - 1)$
8. $(x + 3)^2 = (x - 5)^2$
9. $(2x + 1)^2 - 1 = 4(x^2 - x + 3)$
10. $(2x - 3)^2 = (x - 3)(4x - 2)$

11. $(3x - 3)(2x + 4) = (2x - 1)^2 + 2(x^2 + 2x + 3)$

12. $(x + 1)(x - 2) - 3(x^2 - 2x + 4) = (2x - 4)(x - 4) - 4(x^2 - 1)$

FURTHER WORK WITH BRACKETS

Consider this expression: $4(2x + 3)(3x - 2)$

Clearly the product of the two binomials is to be multiplied by 4 and **the brackets must not be removed** until this product has been obtained. The setting down is made easier if we introduce additional brackets thus:

$$4\{(2x + 3)(3x - 2)\}$$
$$= 4\{2x(3x - 2) + 3(3x - 2)\}$$
$$= 4\{6x^2 - 4x + 9x - 6\}$$
$$= 4\{6x^2 + 5x - 6\}$$
$$= 24x^2 + 20x - 24 \text{ } Ans.$$

EXERCISE 31

Simplify the following:

1. $2(x + 1)(x + 2)$

2. $4(x - 2)(x + 3)$

3. $6(x + 2)(x - 2)$

4. $3(x - 1)(x + 1)$

5. $5(x - 3)(x + 4)$

6. $8(x + 2)(x + 3)$

7. $3(2x + 1)(x - 2)$

8. $5(3x - 2)(x - 1)$

9. $4(2x - 2)(3x + 1)$

10. $2(4x + 1)(4x - 1)$

11. $x(x + 1)(x - 1)$

12. $x(2x + 1)(x + 1)$

13. $x(2x + 3)(3x + 2)$

14. $x(4x - 1)(x + 4)$

15. $2x(x + 2)(x - 2)$

16. $2x(2x + 1)(2x - 1)$

17. $3x(3x + 2)(x + 1)$

18. $3ax(a + x)(a - x)$

19. $2a^2x(a + 1)(x - 1)$

20. $a^2x^2(a + x)^2$

EXERCISE 32

Solve the following equations (*check your answers*):

1. $2(x + 1)(x + 2) = 4(x - 2)(x + 3) - 2x^2$

2. $6(x + 2)(x - 2) - 3(x^2 - x) = 3(x - 1)(x + 1)$

3. $5(x - 3)(x + 4) + 3(x^2 + 1) = 8(x + 2)(x + 3)$

4. $5(3x - 2)(x - 1) = 3(2x + 1)(x - 2) + (3x + 2)(3x - 2)$

5. $4(2x - 2)(3x + 1) = 2(4x + 1)(4x - 1) - 8(x^2 - 1)$

6. $x(x + 1)(x - 1) + 4 = x(2x + 1)(x + 1) - x(x^2 + 3x)$

7. $x(2x + 3)(3x + 2) - 2x^2(x - 1) = x(4x - 1)(x + 4) + 30$

8. $2x(x + 2)(x - 2) + 12 = 2x(2x + 1)(2x - 1) - 6(x^3 - 1)$

9. $3x(3x + 2)(x + 1) + 3x(7x - 1) = 9(x^2 - 1)(x + 4)$

10. $3ax(a + x)(a - x) = 3a(a^2 - x^3 + 2a^2x)$

MAKING EXPRESSIONS
EXERCISE 33

1. One number is x, another is y. What is: (i) their sum; (ii) their difference; (iii) their product?

2. The sum of two numbers is 16 and one of them is x; what is the other?

3. The difference between two numbers is 12 and the larger is y; what is the other?

4. The sum of two numbers is $3a$ and one of them is $2b$; what is the other?

5. The difference between two numbers is $4x$ and the smaller is $3y$; what is the other?

6. x and y are the two factors of a number; what is the number?

7. A number x has a factor y; what is the other factor?

8. There are three numbers each equal to x: (i) What is their sum? (ii) What is their product?

9. There are three consecutive numbers of which 6 is the smallest; what are the others?

10. There are four consecutive numbers of which 6 is the largest; what are the others?

11. There are three consecutive numbers of which a is the smallest; what are the others?

12. There are four consecutive numbers of which x is the largest; what are the others?

13. If x is the middle of three consecutive numbers what are the other two?

14. n is the smallest of three consecutive even numbers; what are the others?

15. b is the largest of three consecutive odd numbers; what are the others?

16. x is a number, when it is halved the result is 6; what is the number?

17. x is a number, when it is halved the result is 18; what is the number?

18. x is a number, when it is doubled the result is 16; what is the number?

19. x is a number, thrice it equals 27; what is the number?

20. x is a number, when 2 is added the result is 7; what is the number?

21. x is a number, when 5 is subtracted the result is 4; what is the number?

22. x is a number, it is doubled and 3 is added, the result is 11; what is the number?

SIMPLE EQUATIONS: PROBLEMS

In problems we must use letters to represent our unknown numbers, and it is most important to **begin by stating precisely what a letter stands for.**

EXAMPLE. *The sum of two numbers is* 12, *their difference is* 6. *Find the numbers.*
Let x be the smaller number.
Then $(12 - x)$ is the larger number.
Their difference will be $(12 - x) - x$.

$$\therefore (12 - x) - x = 6$$

Remove brackets:	$12 - x - x = 6$
Collect like terms:	$12 - 2x = 6$
Add $2x$ to both sides:	$12 = 2x + 6$ *
Subtract 6 from both sides:	$6 = 2x$
Divide both sides by 2:	$x = 3$

The numbers are 3 and 9.

NOTE. * Whilst it is necessary to finish our solution with the letter symbol on the L.H.S., this device avoids the introduction of a minus sign attached to the unknown.

EXERCISE 34

1. A number is multiplied by 3 and 5 is subtracted, the result is 13; find the number.

2. A number is multiplied by 7 and 15 is subtracted, the result is −1; find the number.

3. The sum of two consecutive numbers is 27; find the numbers.

4. The sum of two consecutive numbers is 43; find the numbers.

5. The difference between two consecutive numbers is equal to half the value of the smaller; find the larger.

6. The sum of two numbers is 15, their difference is 7; find the numbers.

7. The sum of two numbers is 28, their difference is 4; find the numbers.

8. Divide 36 into two numbers whose difference is 12.

9. Divide 47 into two numbers whose difference is 23.

10. The sum of three consecutive numbers is 36; find the numbers.

11. The sum of three consecutive numbers is 57; find the numbers.

12. The sum of two consecutive even numbers is 54; find the numbers.

13. The sum of two consecutive odd numbers is 36; find the numbers.

14. 23 is added to a number, this sum is multiplied by 4, the result is 104; find the number.

Exercise 35

1. Share 36p between A, B and C, so that A may have 6p more than B, who has 3p more than C.

2. Share £25 between A, B and C, so that A's share is twice B's, whose share is three times C's.

3. Share 70p between A, B and C, so that A's share is half B's, whose share is four times C's.

4. A, who has twice as much money as B, gives half his money to C, who has 50p already; if they have £2 altogether, what did each have originally?

5. A, who has four times as much money as B, gives all his money to B, who now has £2·50. What did each have originally?

6. A board is twice as long as it is wide, the perimeter is 60 in. Find the area in in².

7. A man is twice the age of his son, the sum of their ages is 75 years. How old is each?

8. In 10 years a father will be twice the age of his son, the sum of their ages will be 60 years. How old is the son now?

9. A is 10 years older than B, in 6 years A will be twice as old as B. Find their present ages.

10. A is 12 years older than B, who is 6 years older than C, their combined age is 60 years; find their separate ages.

11. A is 12 years older than B and 6 years older than C, their combined age is 60 years; find their separate ages.

12. A bag contains 50 coins consisting of 2p and 10p denominations, they amount to £2·60. Find how many of each denomination.

13. A packet contains 60 bank-notes consisting of £5 and £1 notes, they amount to £100. Find how many of each denomination.

14. A fertilizer consists of two chemicals A and B. Chemical A costs 10p a pound, chemical B costs 25p a pound. If 112 lb of fertilizer costs £24, find the quantities of A and B in pounds.

15. A bag contains coins to the value of £1. The coins consist of 5p, 2p and 1p coins. There are three times as many 5p as there are 1p, and the 2p are five less in number than the 5p. Find how many coins there are of each kind.

16. A man purchases a quantity of cloth at 50p a yard. He keeps 25 yd and sells the rest at 70p a yard, making a profit of 50p. How much cloth did he purchase?

fig. A : 4

17. The diagram (fig. A: 4) shows a lawn, dimensions in yards. The length of the lawn is increased by 3 yd and the width by 4 yd, the area is thus increased by 57 yd². Find the original dimensions.

FRACTIONS: EXPRESSIONS

EXAMPLE 1. *Simplify* $\dfrac{a+b}{3} - \dfrac{2a-2b}{4} + 3\frac{1}{2}$

(i) Change $3\frac{1}{2}$ to improper fraction: $\dfrac{a+b}{3} - \dfrac{2a-2b}{4} + \dfrac{7}{2}$

(ii) Find L.C.M. of denominators (12) and proceed as in arithmetic:

$$= \frac{4(a+b) - 3(2a-2b) + 42}{12}$$

(iii) Clear brackets: $= \dfrac{4a + 4b - 6a + 6b + 42}{12}$

(iv) Collect like terms: $= \dfrac{-2a + 10b + 42}{12}$

(v) Cancel to lowest terms: $= \dfrac{-a + 5b + 21}{6}$ *Ans.*

NOTE

Stage (ii) Regard the numerators as single terms and *introduce brackets* to show the necessary multiplication. **Do not work out brackets until next stage.**

Stage (v) In order to cancel, the number in mind must be **a factor of every term in the denominator and numerator.**

EXERCISE 36

Simplify the following:

1. $\dfrac{2a}{3} + \dfrac{3a}{2} + a$ 2. $\dfrac{3a}{4} + \dfrac{5a}{3} - 2a$ 3. $\dfrac{3a}{5} + \dfrac{5a}{2} - 4a$

4. $2\frac{1}{2}a + 3\frac{1}{4}a - 1\frac{2}{3}a$ 5. $\frac{1}{4}a + \frac{2}{3}a - \frac{1}{2}a$ 6. $\frac{3}{4}a - \frac{5}{6}a + \frac{2}{3}a$

7. $\dfrac{4}{a} - \dfrac{2}{a} + \dfrac{3}{a}$ 8. $\dfrac{2}{3a} + \dfrac{3}{2a} - \dfrac{1}{a}$ 9. $\dfrac{3}{4a} + \dfrac{5}{2a} - \dfrac{2}{3a}$

10. $\dfrac{a}{b} + \dfrac{b}{a} - 1$ 11. $\dfrac{2a}{b} + \dfrac{4b}{a} - \frac{2}{3}$ 12. $\dfrac{a}{b} + \dfrac{b}{c} - \dfrac{c}{a}$

13. $\dfrac{a}{2b} - \dfrac{3b}{2a} + \frac{1}{2}$ 14. $\dfrac{5a}{2b} + \dfrac{5a}{4b} + \dfrac{7a}{8b}$ 15. $\dfrac{2}{ab} + \dfrac{3}{bc} - \dfrac{4}{ac}$

16. $\dfrac{a+b}{2} + \dfrac{b+c}{3} + \dfrac{a+c}{1}$ 17. $\dfrac{a-b}{3} + \dfrac{a+2b}{4} - \dfrac{2a-b}{2}$

18. $\dfrac{3a+b}{4} + \dfrac{a-3b}{3} - \dfrac{4a-4b}{6}$ 19. $\dfrac{2a-5b}{4} - \dfrac{3a+2b}{2} + \dfrac{2b}{8}$

20. $\dfrac{2a+2b}{a^2} + \dfrac{3a-4b}{b^2} + 2$ 21. $\dfrac{2a^3 + 3b^3}{a^3} - \dfrac{3a^3 + 2b^3}{b^3}$

22. $\dfrac{2(a+2b)}{3b} - \dfrac{3(a-b)}{2a} + 2$ 23. $\dfrac{5(a+2b)}{3} - \dfrac{4(a-b)}{6} - 2$

FRACTIONS: EQUATIONS

EXAMPLE 1. *Solve the equation* $\dfrac{2a}{3} + \dfrac{3a}{2} - \dfrac{a}{1} = \dfrac{5a}{4} - \dfrac{5a}{6} + \dfrac{2}{1}$

(i) Find the L.C.M. of **all** the denominators:

$$\frac{8a + 18a - 12a}{12} = \frac{15a - 10a + 24}{12}$$

(ii) Multiply both sides of the equation by 12:

$$\frac{8a + 18a - \cancel{12a}}{\cancel{12}} \times \frac{\overset{1}{\cancel{12}}}{1} = \frac{15a - 10a + 24}{\cancel{12}} \times \frac{\overset{1}{\cancel{12}}}{1}$$

(iii) The purpose of stage (ii) was to remove the denominators from **both sides**, with experience we can go straight to:

$$8a + 18a - 12a = 15a - 10a + 24$$

(iv) Collect terms: $\qquad 14a = 5a + 24$

(v) Subtract $5a$ from both sides:

$$9a = 24$$

(vi) Divide both sides by 9: $\quad \underline{\underline{a = 2\tfrac{2}{3} \; Ans.}}$

EXERCISE 37

Solve the following equations:

1. $\dfrac{a}{2} - \dfrac{a}{3} = \dfrac{1}{5}$

2. $\dfrac{2b}{3} + \dfrac{3b}{2} = 2\tfrac{1}{6}$

3. $3x + 2\cdot57 = 6\cdot32$

4. $0\cdot25x - 0\cdot75 = 0$

5. $\tfrac{3}{4}a + 1\tfrac{2}{3}a - 2 = \tfrac{1}{2}a - \tfrac{3}{4}a + 6$

6. $\tfrac{3}{4}x - 2\tfrac{1}{3} = \tfrac{2}{3}x - \tfrac{1}{2}x$

7. $\dfrac{3}{a} + \dfrac{3}{2a} - 3 = \dfrac{4}{3a} - \dfrac{1}{a} + 2$

8. $\dfrac{1}{4a} + \dfrac{5}{2a} + \dfrac{2}{3} = \dfrac{5}{6} + \dfrac{2}{3a} - \dfrac{4}{9a}$

9. $\dfrac{x+2}{3} - \dfrac{3x+2}{6} = \dfrac{2x+3}{9} + \dfrac{x-3}{2}$

10. $\dfrac{a+2}{a} + \dfrac{2a-5}{3a} = \dfrac{3a-5}{6a} + \dfrac{2a-3}{a}$

The experience gained in the previous exercise should enable us to omit one of the early stages in setting down, namely stage (i) as shown in the worked example above. The necessary calculation being carried out mentally.

You must understand why such an omission is permitted:

We select an L.C.M. of the denominators such that it is common to both sides of the equation. Subsequently, we multiply both sides by this L.C.M. If this is to be the ultimate purpose of the L.C.M. let us carry out the multiplication immediately the calculation begins.

EXAMPLE 2. *Solve the equation:* $\frac{2x}{5} + 1\frac{1}{3}x - 3\frac{4}{5} = \frac{1}{3}x + \frac{2}{5}$

Rewrite as follows: $\quad \frac{2x}{5} + \frac{4x}{3} - \frac{19}{5} = \frac{x}{3} + \frac{2}{5}$

Multiply both sides by L.C.M. (15):

$$6x + 20x - 57 = 5x + 6$$

Collect like terms: $\qquad 26x - 57 = 5x + 6$

Transpose terms: $\qquad\qquad 21x = 63$

Divide both sides by 21: $\qquad \underline{\underline{x = 3}}$ *Ans.*

EXAMPLE 3. *Solve the equation:*

$$\frac{a + 2}{6a} - \frac{3(2a - 1)}{8a} = \frac{4(3a - 2)}{12a} - \frac{3a + 2}{24a}$$

(i) Multiply both sides by L.C.M. (24*a*):

$$4(a + 2) - 9(2a - 1) = 8(3a - 2) - 1(3a + 2)$$

(ii) Remove brackets: $4a + 8 - 18a + 9 = 24a - 16 - 3a - 2$

(iii) Collect like terms: $\qquad\quad 17 - 14a = 21a - 18$

(iv) Transpose terms: $\qquad\qquad\quad -35a = -35$

(v) Divide both sides by -35: $\qquad \underline{\underline{a = 1}}$ *Ans.*

NOTE. The brackets MUST be retained in stage (i); note the use of 1 in the last expression. This is not essential, but the practice can help to avoid mistakes.

EXERCISE 38

Solve the following equations:

1. $\frac{2x}{3} + \frac{3x}{4} = 4\frac{1}{4}$ **2.** $\frac{3x}{5} - \frac{x}{3} = 1\frac{1}{15}$ **3.** $\frac{1}{2}x + \frac{3}{8}x = 1\frac{1}{4}$

4. $\frac{5}{6}x - \frac{3}{4} = \frac{1}{3}x$ **5.** $\frac{4x}{7} - \frac{3}{4} - \frac{5x}{14} = 0$ **6.** $\frac{1}{2}x - \frac{8}{9} - \frac{2}{3}x = 0$

7. $\frac{7}{8} - \frac{3x}{4} = \frac{x}{2} - \frac{5}{16}$ **8.** $\frac{5}{9x} + \frac{14}{27x} = \frac{5}{6x} + \frac{1}{2}$ **9.** $\frac{4}{5} - \frac{2}{3}x = \frac{4}{15}x - \frac{2}{3}$

10. $\frac{1}{2} - \frac{x}{6} = \frac{4}{9x} - \frac{x}{6}$ **11.** $\frac{3x - 5}{8} + 3 = \frac{5x - 3}{5}$

12. $\frac{x + 1}{2} = \frac{3x + 4}{7} - 1$ **13.** $\frac{2x + 3}{2x} + 2 = \frac{3x + 2}{3x} + 3$

14. $\frac{2x^2 + x}{6x} = \frac{x - 1}{3} - \frac{2}{x}$

15. $\frac{x + 1}{2} + \frac{x + 2}{3} - \frac{3x - 2}{4} = \frac{4x + 3}{6} + \frac{x - 4}{12}$

16. $\frac{3x - 2}{2x} - \frac{x + 3}{x} + \frac{4x + 1}{3x} = \frac{x - 2}{6x} + \frac{6}{x} - \frac{2}{3}$

CROSS-MULTIPLICATION

Some economy of thought is possible in equations, **but only in those** which contain single fractions on each side.

EXAMPLE 1. *Solve the equation:* $\dfrac{x+2}{3} = \dfrac{3x+5}{2}$

We could proceed to multiply both sides by L.C.M. (6) giving us:

$$2(x+2) = 3(3x+5)$$

The need for a common denominator is no longer necessary in such cases, instead we can **"cross-multiply"** thus:

$$\frac{x+2}{3} \times \frac{3x+5}{2}$$

Giving:
$$2(x+2) = 3(3x+5)$$
$$2x+4 = 9x+15, \text{ etc.}$$

EXAMPLE 2. *Solve the equation:* $\dfrac{x+2}{2x-3} = \dfrac{x-4}{2x+3}$

Cross-multiply: $(2x+3)(x+2) = (2x-3)(x-4)$
Form products: $2x(x+2)+3(x+2) = 2x(x-4)-3(x-4)$
Remove brackets: $2x^2+4x+3x+6 = 2x^2-8x-3x+12$
Collect terms: $2x^2+7x+6 = 2x^2-11x+12$
Transpose, etc. $18x = 6$
Divide both sides by 18: $x = \tfrac{1}{3}$ *Ans.*

EXERCISE 39 (1–12 *Oral*)

Solve the following equations:

1. $\dfrac{x}{3} = \dfrac{1}{2}$ **2.** $\dfrac{3x}{2} = 4$ **3.** $\dfrac{x}{2} = 1$ **4.** $\dfrac{2x}{5} = 2$

5. $4x = \tfrac{3}{2}$ **6.** $\dfrac{1}{2x} = \dfrac{4}{3}$ **7.** $\dfrac{x}{4} = \dfrac{2}{3}$ **8.** $\dfrac{3}{4x} = 2$

9. $\dfrac{2}{5x} = \dfrac{2}{5}$ **10.** $\dfrac{2}{5x} = \dfrac{3}{5}$ **11.** $2\tfrac{1}{2} = \dfrac{x}{2}$ **12.** $\dfrac{2x}{3} = 3\tfrac{1}{3}$

13. $\dfrac{a+1}{2} = \dfrac{2+a}{3}$ **14.** $\dfrac{a-3}{3} = \dfrac{4-2a}{4}$ **15.** $\dfrac{3a-5}{2} = \dfrac{5a+3}{4}$

16. $\dfrac{2-4a}{3} = \dfrac{2a-4}{5}$ **17.** $\dfrac{a+3}{5} = \dfrac{4-2a}{3}$ **18.** $\dfrac{6a-4}{3} = \dfrac{9a-7}{4}$

19. $\dfrac{x+1}{x-2} = \dfrac{x-4}{x+3}$ **20.** $\dfrac{x-5}{x+2} = \dfrac{x+4}{x-3}$ **21.** $\dfrac{4x-3}{2-2x} = \dfrac{6x+2}{3-3x}$

22. $\dfrac{4x-5}{2x+7} = \dfrac{8x-3}{4x+4}$ **23.** $\dfrac{8x+4}{4x-5} = \dfrac{6x-7}{3x-5}$ **24.** $\dfrac{a+1}{a-1} = \dfrac{a-1}{a+1}$

EXERCISE 40

1. Find a number such that by adding 1 and dividing the sum by 8 the result is the same as subtracting 1 and dividing by 4.

2. Find a number such that by subtracting 1 and dividing by 5 the result is the same as adding 1 and dividing by 6.

3. Find a number such that adding 2 and dividing by 4 gives the same result as dividing the number by 3.

4. The numerator of a fraction is 3 less than the denominator, if 1 is subtracted from both numerator and denominator the result is equal to $\frac{2}{3}$. Find the fraction.

5. A boy cycles a certain distance from X to Y at 10 mile/h, he returns at 12 mile/h, the total time taken is 1 hr 50 min. Find the distance XY.

6. A boy walks to school at 4 mile/h and arrives at the proper time, if he had walked at 3 mile/h he would have been 10 min late. Find the distance he walks.

7. A man cycles to work at 15 mile/h and arrives 1 min early, at a speed of 14 mile/h he would have been 1 min late. Find the distance he cycles.

8. Postcards were sold: "Twopence coloured, penny plain." If 24 cards were purchased for 36p, how many were coloured?

9. Plain buttons are $2\frac{1}{2}$p each, fancy ones are 6p each; if I pay $66\frac{1}{2}$p for 14 buttons, how many of each do I purchase?

10. In a guessing game 2 points are awarded for a correct answer and 1 point is deducted for a wrong answer. After 20 guesses a girl has a total of 10 points, how many correct guesses did she make?

11. An arithmetic exercise contains 10 sums; 3 marks are awarded for a completely correct sum and 1 mark for a partly correct sum. For a score of 18 marks, how many sums were completely correct?

12. A wind is blowing at 50 mile/h, an aircraft flying *into* the wind makes half the speed it obtains when the wind is *following*. Find the aircraft's speed in still air.

13. Find two consecutive odd numbers whose squares differ by 16.

14. Chemical A is 36p per lb cheaper than chemical B. $1\frac{1}{2}$ lb of A is mixed with $4\frac{1}{2}$ lb of B to give a mixture costing 41p per lb. Find the cost per lb of A and B.

15. Two trains are approaching each other from towns 140 miles apart, they pass each other after two hours travelling. One train is travelling 10 mile/h faster than the other, find their speeds.

16. Fig. A: 5 represents a rectangle, dimensions in yards. Find: (i) the area in yd²; (ii) the perimeter in ft.

fig. A: 5

TRANSFORMATION OF FORMULAE

A formula is a convenient way of expressing certain mathematical facts about different quantities; it shows the relationship which exists between these quantities. Most formulae are remembered in a particular form, but it may become necessary to express a formula differently, i.e. to "change the subject".

To change the subject of a formula we employ the same techniques as in equations.

HINTS

(i) Remove fractions by multiplying both sides by a suitable L.C.M.

(ii) Collect together, **on one side only**, those terms containing the new subject.

(iii) Carry out necessary operations to leave the new subject only on one side; all other terms on the opposite side.

EXAMPLE 1. *Make K the subject of the formula* $M = \dfrac{5K}{8}$

$$M = \frac{5K}{8}$$

Multiply both sides by 8: $8M = 5K$

Divide both sides by 5: $K = \dfrac{8M}{5}$ *Ans.*

EXAMPLE 2. *Make l the subject of the formula* $A = 6l^2$

$$A = 6l^2$$

Divide both sides by 6: $l^2 = \dfrac{A}{6}$

Square root both sides: $l = \sqrt{\dfrac{A}{6}}$ *Ans.*

CHECK. We can verify the correctness of our new formula by substituting simple values:

From the first formula $A = 6l^2$. If $l = 2$,

   then $A = 24$

From the second formula $l = \sqrt{\dfrac{A}{6}}$. If $A = 24$,

   then $l = \sqrt{4}$

   $\therefore l = 2$

In this way we have proved that the relationship which existed between the components in the first formula remains unchanged in the new arrangement.

Exercise 41

Rearrange the following formulae for the new subject:

1. $A = lb.$ $b = ?$ **2.** $V = lbh.$ $h = ?$

3. $A = \pi r^2.$ $\pi = ?$ **4.** $C = 2\pi r.$ $r = ?$

5. $V = Ah.$ $A = ?$ **6.** $V = \frac{1}{3}Ah.$ $A = ?$

7. $W = VI.$ $V = ?$ **8.** $V = RC.$ $R = ?$

9. $W = 5h - 180.$ $h = ?$ **10.** $S = 2n - 4.$ $n = ?$

11. $P = 2l + 2b.$ $b = ?$ **12.** $S = \frac{1}{2}(a + b + c).$ $a = ?$

13. $R = \frac{1}{2}(24 - T).$ $T = ?$ **14.** $A = I + P.$ $I = ?$

15. $C = \frac{5}{9}(F - 32).$ $F = ?$ **16.** $A = \frac{1}{2}(B \times h).$ $B = ?$

17. $A = h(2l + 2b).$ $l = ?$ **18.** $A = \frac{1}{2}h(a + b).$ $h = ?$

19. $I = \dfrac{PRT}{100}.$ $R = ?$ **20.** $N = \dfrac{20P}{S}.$ $P = ?$

21. $V = \frac{4}{3}\pi r^3.$ $r = ?$ **22.** $A = 4\pi r^2.$ $r = ?$

VARIABLES

Our brief study of a number of simple formulae should enable us to understand that the value of one quantity may often depend upon the value of some other quantity.

EXAMPLE. *The cost of petrol (C) for a car depends upon the quantity Q which we purchased. C might be "cost in pence" whilst Q might be "quantity in gallons".*

There is a direct relationship between these two quantities and we can say that C is proportional to Q. Thus:

$$C \propto Q$$

But if we are told that petrol costs 50 pence per gallon then:

$$C = 50Q$$

Note the introduction of the sign for proportion ($\propto$) and observe the difference between the two statements above. Both C and Q can change their values readily but 50 remains fixed (whilst the price of petrol remains stable).

NOTE

   (i) C and Q are **variables**, 50 is a **constant**.

   (ii) C clearly depends upon Q and is called a **dependent variable**.

   (iii) Q may have any value we choose and is called an **independent variable**.

FUNCTIONS

When the **value of an expression** depends on the variable contained in the expression, it is called **a function of the variable.** That is, the dependent variable is a function of the independent variable.

From the example already given: C is a function of Q.

The most common form of associating two variables is to use the letters x and y, where x is **the independent variable** and y is **the dependent variable.**

In the form $y = 3x + 2$ the value of y will clearly depend on the changing value of x. Therefore y is **a function** of x. This fact is sometimes denoted in symbol form by $y = f(x)$, and whilst this is obviously an equation and may often be treated as such, the important thing to remember is that the one quantity or variable is a function of the other.

For general purposes y is usually a function of x, and the relationship between the two variables is often more conveniently demonstrated by diagrams—namely graphs—which can show the values of the variables at a glance. We usually select particular values for x, and the values of y are calculated according to the nature of the function; the results are tabulated.

EXAMPLE. *If $y = x - 2$, evaluate $f(x)$ for $x = -5$ to $+5$*

Value of x	-5	-4	-3	-2	-1	0	1	2	3	4	5
Value of $f(x)$	-7	-6	-5	-4	-3	-2	-1	0	1	2	3

NOTE. The values of x are given **left to right most negative** to **most positive.**

EXERCISE 42

Calculate the value of $f(x)$ from the following equations, where the value of x is -5 to $+5$. *Tabulate your results:*

1. $y = x$　　　　　　**2.** $y = 2x$　　　　　　**3.** $y = 4x$

4. $y = x^2$　　　　　　**5.** $y = x^3$　　　　　　**6.** $y = 3x^2$

7. $y = 2x^3$　　　　　**8.** $y = \dfrac{x}{2}$　　　　　**9.** $y = \frac{1}{4}x$

10. $y = x + 1$　　　　**11.** $y = x + 3$　　　　**12.** $y = x + 10$

13. $y = 2x + 2$　　　**14.** $y = 3x + 4$　　　**15.** $y = 5x + 1$

16. $y = 4x - 1$　　　**17.** $y = 3x - 4$　　　**18.** $y = 5x - 5$

19. $y = \dfrac{1}{x}$　　　　　**20.** $y = \dfrac{1}{2x}$　　　　**21.** $y = \dfrac{2}{x}$

22. $y = \dfrac{3}{2x}$　　　　**23.** $y = -\dfrac{1}{3x}$　　　**24.** $y = -\dfrac{1}{x^2}$

PLOTTING POINTS

To illustrate the relationship between two variables we can draw graphs, preferably on squared paper, plotting the values of the variables with reference to two lines known as **the axes.** The **vertical axis represents the dependent variable (usually *y*)** and **the horizontal**

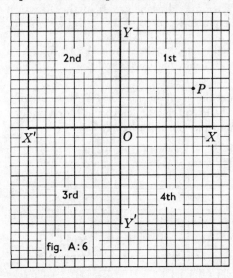

fig. A:6

axis represents the independent variable (usually *x*). The point of interaction of the axes is called the **origin.** (See fig. A: 6.) The axes divide the paper into four parts called the 1st, 2nd, 3rd and 4th **quadrants**; note that they are numbered **anti-clockwise.**

The point *P* in fig. A: 6 is plotted by counting 8 divisions along the *x*-axis (from the origin) and 4 divisions up the *y*-axis (from the origin). The two distances 8 and 4 are known as the **coordinates of *P***; the distance along the *x*-axis is called the **abscissa of *P*,** and the distance along the *y*-axis is called the **ordinate of *P*.**

This information may be given in the form *P*(8, 4), telling us that *P* is plotted with abscissa 8 and ordinate 4. *The abscissa is always given first.*

EXERCISE 43 (*Axes each* 4 *in long*)

1. Plot the points $A(5, 2)$; $B(5, 10)$; $C(5, 17)$; $D(5, -4)$; $E(5, -8)$; $F(5, -15)$; join the points.

2. Plot the points $M(5, 5)$; $N(13, 5)$; $P(20, 5)$; $Q(-3, 5)$; $R(-9, 5)$; $S(-17, 5)$; join the points.

3. Take new axes 4 in long and plot the points $K(20, 0)$; $L(0\ 15)$; $M(-20, 0)$; $N(0, -15)$. Construct the figure *KLMN*; what can you say about it?

4. Take new axes 4 in long and plot the points $R(0, 20)$; $S(-15, -10)$; $T(15, -10)$. Construct the figure *RST*; what can you say about it?

GRAPHS OF FUNCTIONS

Having plotted a number of points against two axes, we shall usually find that the points can be joined by a single line, straight or curved, which we call **a graph**. The graph must be a graph of *something* that is either of a function or of an equation; thus in the equation $y = 3x + 2$, since $y = f(x)$ we may draw **a graph of the function $3x + 2$** or **a graph of the equation $y = 3x + 2$.**

NOTE. The axes do not need to employ the same scale (see fig. A: 7), but try to use all the space available and devise scales which are easy to measure.

Let us proceed to construct the graph of $y = 3x + 2$ for values of x from -3 to $+3$. First we shall require a table of values for x and y. The graph can now be drawn as shown in fig. A: 7.

Table of values when $y = 3x + 2$ ($x = -3$ to $+3$):

x	-3	-2	-1	0	1	2	3
$3x$	-9	-6	-3	0	3	6	9
$+2$	$+2$	$+2$	$+2$	$+2$	$+2$	$+2$	$+2$
y	-7	-4	-1	$+2$	$+5$	$+8$	$+11$

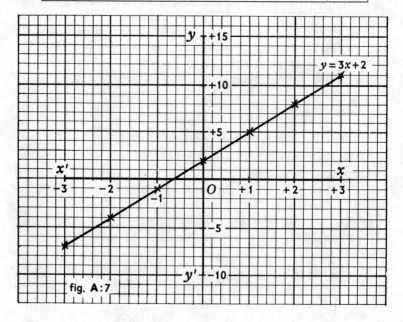

fig. A:7

Since the graph of the function $3x + 2$ is a straight line, we say that $3x + 2$ is a **linear function of x** and the equation $y = 3x + 2$ is **a linear equation.**

The following exercise employs most of the equations from Exercise 42; much time in calculation and tabulation will be saved by using the results of that exercise.

EXERCISE 44

Plot the graphs of the following equations:

1. $y = x$	**2.** $y = 2x$	**3.** $y = 4x$
4. $y = \dfrac{x}{2}$	**5.** $y = \frac{1}{4}x$	**6.** $y = x + 1$
7. $y = x + 3$	**8.** $y = x + 10$	**9.** $y = 2x + 2$
10. $y = 3x + 4$	**11.** $y = 5x + 1$	**12.** $y = 4x - 1$
13. $y = 3x - 4$	**14.** $y = 5x - 5$	**15.** $y = 4$

NOTE. Every simple equation can be reduced to the form $y = mx$ or $y = mx + c$, where m and c are constants. Such equations represent straight lines and can be described as **linear equations.**

INTERPOLATION

From fig. A: 7 find the values of y [that is $f(x)$] when $x = -2\frac{1}{3}$, $x = +2\frac{1}{3}$. What values must x have if $y = 6\frac{1}{2}$, $y = -3$?

The process of inserting intermediate values of a variable is called **interpolation.** Check the results by interpolation with the results obtained by substituting the given values in the equation, and hence solving the equation.

EXERCISE 45

On the same axes, using suitable values for x, plot the following graphs:

1. $2y = 4x - 6$. From your graph find: (i) y when $x = -\frac{1}{2}$; (ii) x when $y = 0$.
2. $y = 2x$. From your graph find: (i) y when $x = 0$; (ii) x when $y = -2\frac{1}{2}$.
3. $3y = 6x + 9$. From your graph find: (i) y when $x = 0$; (ii) x when $y = 1\frac{1}{2}$.

QUESTIONS

   (i) What feature have the three graphs in common?
   (ii) What angle does each graph make with the x-axis? (Use a protractor.)
   (iii) Where does each graph cut the y-axis?

INTERCEPT AND GRADIENT

The point at which a graph line will cut the y-axis is determined by the value of the constant term c in the equation $y = mx + c$. Verify this from the graphs drawn in Exercise 45, thus: $y = 2x - 3$; $y = 2x(+0)$; $y = 2x + 3$.

This is not surprising when you remember that the y-axis is the line for which $x = 0$, and if you substitute for $x = 0$ in the three equations given you will see that $y = c$. Thus at a glance we can obtain **the intercept** of the graph on the y-axis. The intercept on the x-axis is less obvious, but can be obtained by substituting for $y = 0$.

It should now be understood that equations of the form $y = mx$ or $y = mx + c$ will produce straight line-graphs. To produce the straight line we require only **two points** through which to plot the graph, though by selecting three we can ensure that mistakes have been avoided. Convenient points are best obtained by substituting suitable values in the equation to give **integral** values (whole numbers) for x and y.

EXERCISE 46 (Oral)

Where will the following graphs cut the y-axis?

1. $y = x$	**2.** $y = x + 1$	**3.** $y = x - 1$
4. $y = x - 2$	**5.** $y = 2x$	**6.** $y = 2x - 1$
7. $y = 2x + 1$	**8.** $y = 2x - 2$	**9.** $y = 3x$
10. $y = 3x + 1$	**11.** $y = 3x - 1$	**12.** $y = 3x + 2$
13. $y = 3x - 3$	**14.** $y = 3x + 3$	**15.** $y = x + 3$
16. $y = 2x - 3$	**17.** $2y = x$	**18.** $3y = 4x + 6$
19. $y = x - \frac{1}{2}$	**20.** $2y = 3x + 3$	**21.** $2y = x + 1$
22. $2y = 4x - 4$	**23.** $3y = x$	**24.** $3y = 2x + 3$
25. $3y = 3$	**26.** $2y = x - \frac{1}{2}$	**27.** $4y = 8$
28. $4y = 3x + 10$		

From your construction of the graphs:
$$y = 2x - 3; \quad y = 2x; \quad y = 2x + 3$$
you will have observed that they are parallel, they are at the same angle to the x-axis. You verified this by using a protractor, but in fact the slope or *gradient* of a linear graph is measured by *the tangent ratio* of the angle formed with the x-axis. **To obtain this ratio we must employ the scales used on the two axes.**

Fig. A: 8 shows the graph of $y = 2x$. Since it is *a linear graph*, $\tan \theta$ will have the same value at any point along the graph line. This enables us to select points which give convenient values for the tangent ratio. Thus:

$$\tan \theta_1 = \tfrac{8}{4} \quad \therefore \tan \theta = 2$$
$$\tan \theta_2 = \tfrac{4}{2} \quad \therefore \tan \theta = 2$$

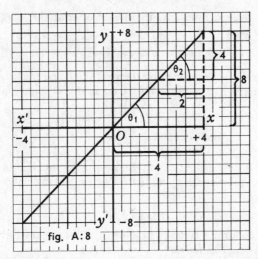

fig. A:8

QUESTION. Is there any connection between tan θ and the coefficient of x in the equation $y = 2x$?

Fig. A:9 shows the graph of $y = 2x + 3$. This is of the form $y = mx + c$, where $c = 3$ gives the intercept on the y-axis.

$$\tan \theta_1 = \tfrac{4}{2} \qquad \therefore \tan \theta = 2$$
$$\tan \theta_2 = \frac{7}{3 \cdot 5} \qquad \therefore \tan \theta = 2$$

Note that tan θ_2 is less convenient than tan θ_1 which is obtained by using *integral values*.

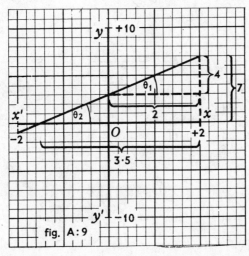

fig. A:9

From the two examples given it should be clear that on a graph of $y = mx$ or $y = mx + c$

$$m = \tan \theta$$

where $\angle \theta$ is the angle formed by the graph line and the axis $X'OX$.

On the graph of $y = mx + c$ let us consider a point $P(x, y)$; then

$$m = \tan \theta$$

and $$\tan \theta = \frac{y - c}{x}$$

Verify this from figs. A: 8 and A: 9 using a number of different points and employing the appropriate abscissa and ordinate.

EXERCISE 47

From the following sketches produce equations of the form $y = mx$ or $y = mx + c$. *The diagrams are not to scale.*

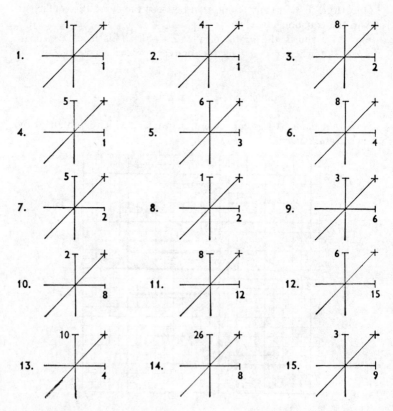

16.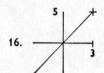

17.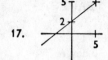

18.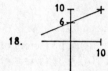

19.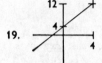

20.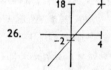

21.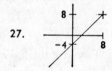

22.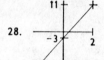

23.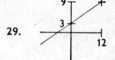

24.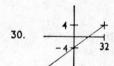

25.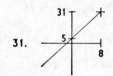

26.

27.

28.

29.

30.

31.

32.

EXERCISE 48

Make *sketches* of the following functions:

1. $y = 3x$

2. $y = 3x + 4$

3. $y = 3x - 2$

4. $y = \dfrac{x}{3}$

5. $2y = 5x$

6. $3y = 4x$

7. $\dfrac{y}{2} = x$

8. $\dfrac{y}{3} = x + 1$

9. $\dfrac{y}{4} = x - 2$

10. $y = 2\tfrac{1}{2}x + 2$

11. $3y = 4x + 6$

12. $\tfrac{2}{3}y = x - 2$

FACTORS

EXAMPLES

(i) The factors of $2a$ are 2 and a $\qquad 2a = 2 \cdot a$

(ii) The factors of ab are a and b. $\qquad ab = a \cdot b$

(iii) The factors of x^3 are x and x and x. Thus: $x^3 = x \cdot x \cdot x$

EXERCISE 49 (*Oral*)

Express the following as factors in their simplest terms:

1. $3x$ **2.** $5a$ **3.** xy **4.** $2xy$

5. axy **6.** b^2 **7.** $3b^2$ **8.** $2ab^2$

9. a^2b^2 **10.** $5a^2b^2$ **11.** $7ax^2y^3$ **12.** $a^2x^2y^2$

EXAMPLES

(i) $\qquad 2a + ab = a(2 + b)$

(ii) $\qquad a^2 + ab = a(a + b)$

(iii) $\qquad a^2b - ab^2 = ab(a - b)$

(iv) $\qquad 2a^3b^2 - a^2b = a^2b(2ab - 1)$

Is the next example correct? If not, give the correct answer.

(v) $\quad l^2m^3n^4 - lm^2n^5 \overset{?}{=} lm^2n^4(lmn - n^2)$

EXERCISE 50

Factorize the following, *where possible*:

1. $a + 2ab$ **2.** $2 + 4a$ **3.** $4a + 4b$

4. $a + b$ **5.** $ac - ab$ **6.** $ax + ay$

7. $2ac - 2xy$ **8.** $3acx - 6cxy$ **9.** $a^2 + b^2$

10. $2a^2 + 2b^2$ **11.** $a^2x + b^2x$ **12.** $2x^2y - 2xy^2$

13. $p^3 - p^4$ **14.** $3mn^2 + 15m^2n$ **15.** $3m^2n^2 - 15mn$

16. $3m^2 + 15n^2$ **17.** $3m^2 - 5n^2$ **18.** $3mn^2 + 15m^2n^3$

19. $2p + 4q - 6r$ **20.** $pq + pr + ps$

21. $2pq - 4pr + 6ps$ **22.** $2pq - 4qr - 6rs$

23. $4x - 8x^2 + 10x^3$ **24.** $4ax^2 + 8a^2x - 16ax$

REVISION: MULTIPLICATION

EXERCISE 51

Multiply the following:

1. $a(a - b)$ **2.** $ax(x + 1)$ **3.** $x^2y^2(a + b)$

4. $(a + b)(x - y)$ **5.** $(a - x)(b + y)$ **6.** $(a + y)(b + x)$

7. $(a - y)(b - x)$ **8.** $(a + b)(2p + q)$ **9.** $(a - b)(q - 2p)$

10. $(2a - b)(l + m)$ **11.** $(2a + 2b)(3l - 3m)$ **12.** $(x + y)(p^2 + q^2)$

13. $(x^2 - y^2)(a - b)$ **14.** $(ax + y)(b + c)$ **15.** $(a - xy)(b - c)$

16. $(ax + by)(a - b)$ **17.** $(ax - by)(x + y)$

18. $(ax + bx)(ay - by)$ **19.** $(a + 3)(a + 2)$

20. $(x + y)(x - y)$ **21.** $(c + 1)(c - 1)$

22. $(a + b)(a + b)$ **23.** $(a - b)(a - b)$

COMPOUND FACTORS

EXAMPLES (i) $\qquad ax + bx = x(a + b)$

(ii) $\quad a(x + y) + b(x + y) = (x + y)(a + b)$

NOTE. x is a simple factor, $(a + b)$ and $(x + y)$ are compound factors. Extract the common factor first.

EXERCISE 52

Factorize the following, *where possible*:

1. $x(a + b) + y(a + b)$
2. $x(a - b) - y(a - b)$
3. $x(a + b) - y(a + b)$
4. $x(a + b) + y(a - b)$
5. $a(x + y) + b(x - y)$
6. $2(p + q) - a(p + q)$
7. $b(l - m) + 3(l - m)$
8. $ab(c + d) - xy(c + d)$
9. $a^2(r^2 - s) + b^2(r^2 - s)$
10. $2a(a + b) - b^2(a - b)$
11. $3(y - z) - (y - z)t$
12. $(x^2 + y^2)a + (x^2 + y^2)b$
13. $(l + m^2)x + y(l + m^2)$
14. $(l^2 + m)c - d(l + m^2)$
15. $ab(c + d) - ab$
16. $a^2 + a^2(c - d)$
17. $x^2 + y^2(a - b)$
18. $a^2(p - q) + (p - q)$
19. $z^2(a + 2b) - (a - 2b)$
20. $b^3(a^2 - b) + b^3(a + b^2)$

GROUPING TERMS

EXAMPLE. *Factorize* $2a^3 + a^2b^2 - 4ab - 2b^3$

$$2a^3 + a^2b^2 - 4ab - 2b^3 = (2a^3 + a^2b^2) - (4ab + 2b^3) \text{ *}$$
$$= a^2(2a + b^2) - 2b(2a + b^2)$$
$$Ans. = \underline{\underline{(2a + b^2)(a^2 - 2b)}}$$

* NOTE. The effect of introducing brackets must be given careful thought, particularly in regard to the signs. Thus:

$$-4ab - 2b^3 = -(4ab + 2b^3)$$

EXERCISE 53

Factorize the following expressions, rearranging the terms if necessary:

1. $ac + bc + ad + bd$
2. $cx + cy + dx + dy$
3. $2x + 2y + cx + cy$
4. $2x + ax + 2y + ay$
5. $a^2 + ax + 2a + 2x$
6. $x^2 + 6x + ax + 6a$
7. $2a + 9ab + 3b + 6a^2$
8. $4ax + 6bx + 6ay + 9by$
9. $ax - by - ay + bx$
10. $ay - bx + ax - by$
11. $bc + a^2 - ac - ab$
12. $bx - ax + by - ay$
13. $ax - ay + x - y$
14 $ax - b + bx - a$
15. $xy - x - y^2 + y$
16. $ax + y^2 - ay - xy$
17. $a^3 + x^3 + a^2x + ax^2$
18. $y^3 + x^3 + xy + x^2y^2$
19. $x^4 - xy^3 - x^3y + y^4$
20. $a^4b - 1 - a^4 + b$
21. $a^2 + ab + ac + ax + bx + cx$
22. $a^2 + ab - ac + a + b - c$

THE PRODUCT OF TWO BINOMIAL EXPRESSIONS

EXAMPLE
$$(a + 2)(a + 3)$$
$$= a(a + 3) + 2(a + 3)$$
$$= a^2 + 3a + 2a + 6$$
$$Ans. = a^2 + 5a + 6$$

This operation may be carried out mentally as follows:

Stage 1. $(a + 2)(a + 3)$ Multiply $+a$ by $+a$, giving $+a^2$

Stage 2. $(a + 2)(a + 3)$ Multiply $+a$ by $+2$, and $+a$ by $+3$, giving $+2a$ and $+3a$, making $+5a$

Stage 3. $(a + 2)(a + 3)$ Multiply $+2$ by $+3$, giving $+6$

The "partial products" can be written down as they are obtained and thus provide the complete product: $a^2 + 5a + 6$

The process may be summarized thus:

1. The product of the two "lefts".
2. The product of the two "inners" plus the product of the two "outers".
3. The product of the two "rights".

NOTE. Particular care with the signs is necessary when combining the "two inners" with the "two outers". (Stage 2.)

EXERCISE 54

Give the products of the following *by inspection*:

1. $(a + 1)(a + 2)$	**2.** $(x + 1)(x + 2)$	**3.** $(x + 2)(x + 3)$
4. $(a + 1)(a + 3)$	**5.** $(x + 3)(x + 1)$	**6.** $(x + 2)(x + 2)$
7. $(a + 1)(a + 4)$	**8.** $(x + 2)(x + 4)$	**9.** $(x + 1)(x + 4)$
10. $(x + 4)(x + 3)$	**11.** $(a + 5)(a + 1)$	**12.** $(a + 2)(a + 4)$
13. $(a + 2)(a + 5)$	**14.** $(5 + x)(3 + x)$	**15.** $(4 + a)(3 + a)$
16. $(a + 1)(a - 2)$	**17.** $(a - 1)(a + 2)$	**18.** $(x - 2)(x + 3)$
19. $(x - 3)(x - 2)$	**20.** $(a - 1)(a - 3)$	**21.** $(a + 3)(a - 1)$
22. $(x - 1)(x - 2)$	**23.** $(x + 1)(x - 4)$	**24.** $(x - 2)(x + 4)$
25. $(x - 1)(x - 4)$	**26.** $(x + 3)(x - 4)$	**27.** $(a - 5)(a + 1)$
28. $(a - 2)(a - 4)$	**29.** $(a + 2)(a - 5)$	**30.** $(5 + x)(3 - x)$
31. $(4 - a)(3 - a)$	**32.** $(a + 1)(a - 1)$	**33.** $(2a + 3)(3a + 2)$
34. $(2a + 2)(2a - 2)$	**35.** $(a^2 - 3)(2a^2 - 1)$	**36.** $(2a + 3b)(3a - 2b)$
37. $(2a + 1)(2a + 2)$	**38.** $(2a - 1)(2a - 1)$	**39.** $(2a + 1)(2a - 1)$
40. $(a + b)(a + b)$	**41.** $(a - b)(a - b)$	**42.** $(a + b)(a - b)$
43. $(2a + 3)(a + 4)$	**44.** $(x - 4)(3x - 2)$	**45.** $(2x + 2)(3x - 4)$
46. $(3x + 2)(2x - 3)$	**47.** $(3x + 3)(3x - 3)$	**48.** $(2x + y)(x + 2y)$

49. $(2x + y)(2x - y)$ **50.** $(2 + 2a)(3 + 3a)$ **51.** $(3 - 2a)(2 - 3a)$
52. $(3 + 2a)(3 - 2a)$ **53.** $(2 + 3a)(3 - 2a)$ **54.** $(3 + 2a)(2a - 3)$
55. $(x - 2y)(3y + x)$ **56.** $(a + b)(x + y)$ **57.** $(a^2 + 1)(a^2 - 1)$
58. $(1 - x^3)(1 + x^3)$ **59.** $(2a^2 + bx^2)(3a^2 - 2bx^2)$
60. $(a + b^2c^3)(a^2 - b^3c^4)$ **61.** $(a + 1)^2$
62. $(2a + 3b)(4a - 5b)$ **63.** $(2x - 3)^2$
64. $(4x + 5y)(4x - 5y)$ **65.** $(a + b)^2$
66. $(3y - 4z)(4y - 3z)$ **67.** $(2x - y)^2$
68. $(3a + 5b)(3a - 5b)$ **69.** $(3a + 3b)^2$
70. $(5a^2 + b)(a + 5b)$ **71.** $(a^2 - 1)^2$ **72.** $(3x^2 - 2y^2)^2$

GEOMETRICAL ILLUSTRATION OF PRODUCTS

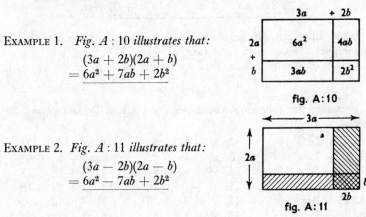

EXAMPLE 1.   *Fig. A* : 10 *illustrates that:*
$$(3a + 2b)(2a + b)$$
$$= 6a^2 + 7ab + 2b^2$$

fig. A:10

EXAMPLE 2. *Fig. A* : 11 *illustrates that:*
$$(3a - 2b)(2a - b)$$
$$= 6a^2 - 7ab + 2b^2$$

fig. A:11

QUESTIONS (Fig. A : 11)

(i) How many times have we subtracted the area of the small rectangle measuring $2b$ by b?

(ii) Why is the third term of the product $+2b^2$?

EXERCISE 55

Draw figures to illustrate the following and give the products:

1. $a(a + 1)$	**2.** $a(a + b)$	**3.** $a(a - 1)$
4. $a(a - b)$	**5.** $2a(a + 1)$	**6.** $2a(a + b)$
7. $2a(a - 1)$	**8.** $2a(a - b)$	**9.** $2a(a + b + c)$
10. $(a + b)^2$	**11.** $(2a + b)(a + 2b)$	**12.** $(a + b)(2a + 2b)$
13. $(2x + y)(x + 3y)$	**14.** $(2x + 2y)^2$	**15.** $(a - 1)^2$
16. $(a - b)^2$	**17.** $(2x - 3y)^2$	**18.** $(2a + b)(2a - b)$

FACTORIZING TRINOMIAL EXPRESSIONS

We have seen that the product of two binomial expressions frequently simplifies to three terms, that is a **trinomial expression.** By regarding

factorizing as the reverse process of multiplication we can devise a
method for factorizing trinomials.

EXAMPLE. *Factorize* $2x^2 + x - 3$

Stage 1.   (   )(   )

Stages 2 and 3.   (i) $(x + \quad)(2x - \quad)$
                or   (ii) $(x - \quad)(2x + \quad)$

Stage 4.   (i) $(x + 3)(2x - 1)$ $+5x$ is incorrect
                         (ii) $(x + 1)(2x - 3)$ $-x$ is incorrect
                         (iii) $(x - 3)(2x + 1)$ $-5x$ is incorrect
                         (iv) $(x - 1)(2x + 3)$ $+x$ is correct

$$\therefore \ 2x^2 + x - 3 = (x - 1)(2x + 3) \ \textit{Ans.}$$

NOTE. In practice we can avoid the incorrect solutions by considering
the required factors with great care.

Check, by multiplication, that the factors give the correct product.

EXERCISE 56

Factorize the following where possible. *Check by multiplying the
factors.*

1. $x^2 + 2x + 1$	**2.** $x^2 + 4x + 4$	**3.** $x^2 + 3x + 2$
4. $x^2 + 5x + 6$	**5.** $x^2 + 4x + 3$	**6.** $x^2 + 5x + 4$
7. $x^2 + 7x + 12$	**8.** $x^2 + 6x + 8$	**9.** $x^2 + 8x + 15$
10. $x^2 + 7x + 10$	**11.** $x^2 + 6x + 5$	**12.** $x^2 + 8x + 16$
13. $x^2 + 10x + 25$	**14.** $x^2 + 9x + 20$	**15.** $x^2 + 10x + 16$
16. $x^2 - 2x + 1$	**17.** $x^2 - 4x + 4$	**18.** $x^2 - 3x + 2$
19. $x^2 - 4x + 3$	**20.** $x^2 - 5x + 6$	**21.** $x^2 - 6x + 9$
22. $x^2 - 5x + 4$	**23.** $x^2 - 6x + 8$	**24.** $x^2 - 7x + 12$
25. $x^2 - 8x + 16$	**26.** $x^2 - 10x + 25$	**27.** $x^2 - 12x + 36$
28. $x^2 - 7x + 6$	**29.** $x^2 - 7x + 10$	**30.** $x^2 - 11x + 10$
31. $x^2 + 2x - 3$	**32.** $x^2 + 3x - 4$	**33.** $x^2 + 4x - 5$
34. $2x^2 - 5x - 3$	**35.** $2x^2 + 7x + 3$	**36.** $2x^2 + x - 3$
37. $2x^2 - x - 3$	**38.** $2x^2 - x + 1$	**39.** $2x^2 - x - 1$
40. $2x^2 + x - 1$	**41.** $2x^2 - 3x + 1$	**42.** $2x^2 + 3x + 1$
43. $2x^2 + 7x - 6$	**44.** $2x^2 + 7x + 6$	**45.** $2x^2 + 8x + 6$
46. $2x^2 + x - 6$	**47.** $2x^2 - x - 6$	**48.** $2x^2 - 4x - 6$
49. $2x^2 + 4x + 5$	**50.** $2x^2 + 4x - 6$	**51.** $3x^2 + 7x + 2$
52. $3x^2 + 5x + 2$	**53.** $3x^2 + 5x - 2$	**54.** $3x^2 + 2x + 5$
55. $3x^2 - x - 2$	**56.** $3x^2 + x - 2$	**57.** $3x^2 - 5x + 2$
58. $3x^2 - 5x - 2$	**59.** $3x^2 - 7x + 2$	**60.** $3x^2 + 11x - 6$
61. $3x^2 + 11x + 6$	**62.** $3x^2 + 9x + 6$	**63.** $3x^2 - 11x + 6$
64. $3x^2 - 9x + 6$	**65.** $3x^2 - 7x - 6$	**66.** $3x^2 + 7x - 6$
67. $3x^2 - 3x + 8$	**68.** $x^2 + 5x - 6$	**69.** $3x^2 - 9x - 6$

FURTHER POINTS ON FACTORIZING

EXAMPLE 1. *Factorize* $3ax + 3ay + 3bx + 3by$

$$3ax + 3ay + 3bx + 3by = 3(ax + ay + bx + by)$$
$$= 3[(ax + ay) + (bx + by)]$$
$$= 3[a(x + y) + b(x + y)]$$
$$Ans. = \underline{\underline{3(x + y)(a + b)}}$$

EXAMPLE 2. *Factorize* $24 - 4x - 4x^2$

$$24 - 4x - 4x^2 = 4(6 - x - x^2)$$
$$= 4[(2 - x)(3 + x)]$$
$$Ans. = \underline{\underline{4(2 - x)(3 + x)}}$$

EXERCISE 57

Factorize where possible:

1. $3ax - 3by + 3bx - 3ay$
2. $24a^2 + 20a - 24$
3. $4b^2 - 6ab + 2a^2$
4. $3a^2 - a + 5$
5. $ax + by + cx + dy$
6. $16a^2 + 8ab^2 + b^4$
7. $16a^2 + 40ab + 16b^2$
8. $3a^2 + bc + c^2$
9. $5x^2 + 25x - 10$
10. $9ax + 4by - 6ay - 6bx$
11. $72 - 6b^2 - 6b^4$
12. $4a^2c - 8a^2c^2 - 4ac^2$
13. $12a^2 + 25ab + 12b^2$
14. $9a^2 - 4ab + 4b^2$
15. $27a^2 - 27ab - 12b^2$
16. $6x^3y - 6x^2y^2 - 36xy^3$
17. $ax + x + ay + y$
18. $3a + 2b + 5x$
19. $4a^2 - 35a - 9$
20. $ab^2 + bx - by^2 + bz$
21. $6x^2 + 17xy + 12y^2$
22. $6x^2 + 73xy + 12y^2$
23. $6x^2 + 54xy + 12y^2$
24. $a^3 - a^2 - 6a$
25. $a^2 + ab - a - b$
26. $4a^2 + 3a + b^2$
27. $12 - 3x - x^2$
28. $6 + 3x - 2x^2$
29. $12 - 2x - 4x^2$
30. $6ax + 6by - 4bx - 9ay$

SOME IMPORTANT PRODUCTS AND FACTORS

$$\text{(i) } (a + b)^2 = (a + b)(a + b) = a^2 + 2ab + b^2$$
$$\text{(ii) } (a - b)^2 = (a - b)(a - b) = a^2 - 2ab + b^2$$
$$\text{(iii) } \qquad\quad (a + b)(a - b) = a^2 - b^2$$

(i) The square of the sum of two terms is equal to **'the square of the first plus twice the product plus the square of the second'**.

(ii) The square of the difference of two terms is equal to **'the square of the first minus twice the product plus the square of the second'**.

(iii) The product of the sum and difference of two terms is equal to 'the square of the first minus the square of the second' or **'the difference of the two squares'**.

EXAMPLES   (i) $4x^2 + 12xy + 9y^2 = (2x + 3y)(2x + 3y)$
$$= (2x + 3y)^2$$

         (ii) $9y^2 - 30y + 25 = (3y - 5)(3y - 5)$
$$= (3y - 5)^2$$

         (iii) $4x^2y^2 - 9 = (2xy + 3)(2xy - 3)$

NOTE. It is clearly an advantage to recognize the third example as **'the difference of two squares'.** Hence the factors will be 'the square root of the first **plus** the square root of the second' multiplied by 'the square root of the first **minus** the square root of the second'.

EXERCISE 58

Give the products of the following:

1. $(x + y)(x + y)$ **2.** $(x - y)(x - y)$ **3.** $(x + y)(x - y)$
4. $(a + 1)^2$ **5.** $(a - 1)^2$ **6.** $(a + 1)(a - 1)$
7. $(a + 2)^2$ **8.** $(x + 2)(x - 2)$ **9.** $(x + 2)^2$
10. $(2x + 1)^2$ **11.** $(2x + 1)(2x - 1)$ **12.** $(2x - 1)^2$
13. $(2a + 2)(2a - 2)$ **14.** $(2x - y)^2$ **15.** $(x + 2y)^2$
16. $(x + 2y)(x - 2y)$ **17.** $(x + 3)^2$ **18.** $(3 - x)^2$
19. $(3x - 2y)^2$ **20.** $(3 + x)(3 - x)$ **21.** $(1 + x^3)(1 - x^3)$

Factorize the following:

22. $x^2 - y^2$ **23.** $a^2 + 4a + 4$ **24.** $x^2 - 6x + 9$
25. $4x^2 - y^2$ **26.** $9x^2 - 25$ **27.** $a^2 + 4ab + 4b^2$
28. $9a^2 - 6ab + b^2$ **29.** $9a^2 - b^2$ **30.** $16 - a^6$
31. $16 - 8y^2 + y^4$ **32.** $y^2 - 8y + 16$ **33.** $a^2 - 100$
34. $a^2x^2 + 2ax + 1$ **35.** $9y^2 + 24yz + 16z^2$ **36.** $1 - t^4$
37. $4 - t^2$ **38.** $4 - 4t + t^2$ **39.** $l^4 + 2pl^2 + p^2$
40. $p^2 - 2lp + l^2$ **41.** $4s^2 - 12st + 9t^2$ **42.** $9r^2 - 4s^2$

Write down the products of the following:

43. $(2a + 2b)^2$ **44.** $(2a + 3b)(2a - 3b)$ **45.** $(3x^2 - 1)^2$
46. $(a^4 + 1)(a^4 - 1)$ **47.** $(6 + ax)^2$ **48.** $(1 + ax)(1 - ax)$
49. $(5 - a^2)^2$ **50.** $(5 - 3a)^2$ **51.** $(ax + 10)(ax - 10)$
52. $(5x + 4y)^2$ **53.** $(1 + y^4)(1 - y^4)$ **54.** $(ax + bx)^2$
55. $(a^2 + x^3)^2$ **56.** $(4 - 3t)^2$ **57.** $(a^3 - b^2)^2$
58. $(a^3 + b^2)(a^3 - b^2)$ **59.** $(10t + s^2)(10t - s^2)$ **60.** $(a^4 + 6a^2)^2$

Factorize the following, *where possible*:

61. $36t^2 - 9$ **62.** $x^4 + 14x + 49$ **63.** $a^6 - 6a^3 + 9$
64. $16 - 8t^2 + t^4$ **65.** $14a^2 - y^2$ **66.** $121x^2 - 9y^2$
67. $a^2t^4 + 2at^2 + 1$ **68.** $a^5t^2 - 1$ **69.** $9 - 12a + 4a^2$
70. $c^4 - 2c^2d^4 + d^8$ **71.** $d^8 - c^6 + 4$ **72.** $a^2 + 1$
73. $81 - 144a + 64a^2$ **74.** $144x^4 - 100y^2$ **75.** $9a^8 + 12a^6 + 4a^4$

FURTHER WORK WITH THE DIFFERENCE OF TWO SQUARES

EXAMPLE. *Evaluate* $(21)^2 - (19)^2$

$$(21)^2 - (19)^2 = (21 + 19)(21 - 19)$$
$$= 40 \times 2$$
$$Ans. = \underline{\underline{80}}$$

EXERCISE 59

Evaluate the following:

1. $(31)^2 - (29)^2$ **2.** $(41)^2 - (39)^2$ **3.** $(55)^2 - (45)^2$
4. $(105)^2 - (95)^2$ **5.** $(51)^2 - (49)^2$ **6.** $(61)^2 - (59)^2$
7. $(52)^2 - (48)^2$ **8.** $(71)^2 - (69)^2$ **9.** $(51)^2 - (29)^2$
10. $(61)^2 - (39)^2$ **11.** $(89)^2 - (11)^2$ **12.** $(101)^2 - 1$
13. $(164)^2 - (36)^2$ **14.** $(6 \cdot 6)^2 - (3 \cdot 4)^2$ **15.** $(8 \cdot 9)^2 - (1 \cdot 1)^2$

HARDER FACTORS

EXAMPLE 1.

$$(a + b)^2 - c^2 = [(a + b) + c][(a + b) - c]$$
$$= \underline{(a + b + c)(a + b - c)}$$

EXAMPLE 2.

$$a^2 - (a - b)^2 = [a + (a - b)][a - (a - b)]$$
$$= (a + a - b)(a - a + b)$$
$$= \underline{(2a - b)b \text{ or } b(2a - b)}$$

The process of factorizing is completed only when the result is in the form of a multiplication; the product of the factors should be equivalent to the value of the original expression.

EXERCISE 60

Factorize the following, *where possible*:

1. $(a - b)^2 - c^2$ **2.** $(x + y)^2 - y^2$ **3.** $a^2 + (a - b)^2$
4. $p^4 - 1$ **5.** $27 - 12a^2$ **6.** $a^2 + (b + c)^2$
7. $(a^2 + b^2) + (a + b)^2$ **8.** $(a - b)^2 - (b + c)^2$
9. $(a + b)^2 + (a + c)^2$ **10.** $16x + (4 - x)^2$
11. $(a - b)^2 - a^2$ **12.** $81 + 9x^2$
13. $(2x + y)^2 - (x + 2y)^2$ **14.** $16 + a^4$
15. $4(a + b)^2 - a^2$ **16.** $(a + b)^2 - b^2$
17. $\frac{1}{16} - y^4$ **18.** $8x^2 - 18y^2$
19. $(a + x)^2 + (a + x)^2$ **20.** $16 - (4 + x)^2$
21. $x^4y^8 - z^{12}$ **22.** $7 - 40x^2$
23. $81x^2 - 9$ **24.** $1 - x^4$
25. $9 - (a + b)^2$ **26.** $(x + y)^4 + z^4$

MISCELLANEOUS EXAMPLES

EXERCISE 61

Factorize the following, *where possible*:

1. $36a^2xy^2 + 9ax^2y$
2. $48a^2 + 96ab + 48b^2$
3. $a^4b^3c^2 - a^2b$
4. $14p^2 + 12pq + 7q^2$
5. $2a(a + b) + 3b(a + b)$
6. $112a^4 - 7$
7. $x^3 - x(y - 3)^2$
8. $4(a + b)^2 - 9(a - b)^2$
9. $15x^2 - 60$
10. $2x(x + y) - 2y(x + y)$
11. $8x^2 - 9y^2$
12. $25a^2b^2 + 20ab + 4$
13. $2ax + by - ay - 2bx$
14. $(4 + x)^2 - (4 + y)^2$
15. $36a^3x^2 - 9ay^2$
16. $10a^2 - 15a - 10$
17. $a^2x - b^2x$
18. $16x^2 + 25y^2$
19. $21 - x - 2x^2$
20. $2x + ay + ax + 2y$
21. $4 + (a - b)^2$
22. $40x^2 - 25y^2$
23. $\frac{9}{16} - x^2$
24. $9a^2 - 12ab + 4b^2$

HIGHEST COMMON FACTOR

The **highest common factor** (H.C.F.) of a given number of expressions is the largest factor which will divide exactly into each of the given expressions.

EXAMPLE. The H.C.F. of $2x^2y^3z^4$, $4x^3y^2z$, $6x^2y^2$ is $\underline{2x^2y^2}$

To obtain the H.C.F. of two or more compound expressions we must first factorize each of the expressions as follows:

EXAMPLE. *Find the H.C.F. of* $x^2 + 2xy + y^2$, $x^2 - y^2$

$$x^2 + 2xy + y^2 = (x + y)(x + y)$$
$$x^2 - y^2 = (x + y)(x - y)$$
$$\therefore \underline{\underline{\text{H.C.F.} = x + y}}$$

EXERCISE 62

Find the H.C.F. of the following:

1. ax^2, ay^2
2. a^2x, x^2y
3. ax^2, a^2x
4. ax^2y, a^2y
5. a^2xy^2, ax^2y
6. a^2x^2y, a^2xy^2
7. $2ax^2$, $3a^2x$
8. $3x^2y^2$, $4x^2y^2$
9. $3ay$, $5bx$
10. $3ab$, $6bz$
11. $4a^2bc$, $6ab^2c$
12. $3xy^2z^2$, $9y^3z$
13. $12a^5x^4$, $16a^4x^4$, $8a^4x^5$
14. $30b^5c^4$, $15b^3c^4$, $45b^3c^3$
15. $14p^4qr$, $7p^3r$, $2q^4s$
16. $28qr^2$, $14r^2s$, $42r^3t$
17. $10l^2m^4$, $15l^3m^5$, $30l^4m^4$
18. $30a^5x^4z^3$, $15a^4x^3z^5$, $45a^4x^3z^2$
19. $3a - 2b$, $2a + 3b$
20. $x^2 + xy$, $y^2 + xy$
21. $2ax^2 + 2x^2$, $2ax^2 - 2x^2$
22. $6a^2x + 9a^2y$, $8a^2x + 12a^2y$
23. $6a^2x + 9a^2y$, $8b^2x - 12b^2y$
24. $4ax^3y^2 + 4x^3y^2$, $4ax^2y^3 + 4x^2y^3$

25. $x^2 - xy, x^2 - y^2$ **26.** $a^2 + ab, a^2 + 2ab + b^2$

27. $a + ax, x^2 + 2x + 1$ **28.** $4a - 4x, 2a^2 - 2x^2$

29. $x^4 + x^3y, x^4 - x^2y^2$ **30.** $a^2 + 2ab + b^2, a^2 - 2ab + b^2$

31. $a^3x + a^2x + x, a^3y + a^2y + y$ **32.** $x^4 - 1, 2x^3 - 2x$

LOWEST COMMON MULTIPLE

The **lowest common multiple** (L.C.M.) of a given number of expressions is the smallest expression which is exactly divisible by each of the given expressions.

EXAMPLE. The L.C.M. of a^2x, $4ax^3$, $12a^3y$ is $\underline{12a^3x^3y}$

To obtain the L.C.M. of two or more compound expressions we must first factorize each of the expressions as follows:

EXAMPLE. *Find the L.C.M. of* $x^2 - 1$, $x^2 + 2x + 1$
$$x^2 - 1 = (x + 1)(x - 1)$$
$$x^2 + 2x + 1 = (x + 1)(x + 1)$$
$$\therefore \quad \underline{\text{L.C.M.} = (x + 1)(x + 1)(x - 1)}$$

NOTE. The L.C.M. must contain **all** the factors of **each** expression.

EXERCISE 63

Find the L.C.M. of the following:

1. ab^2, a^2b, ab **2.** ab, bc, ac **3.** $2a, 3a, 4a$

4. $2a, 3b, 4c$ **5.** $2a^2, 3bc, ac^2$ **6.** $2a^3, ab^3, 3a^2b^2$

7. $a^2b, ab^2, 3c^2$ **8.** $2x^2, ax^3, bx$ **9.** ab^3, a^2b^2, a^3b

10. $a^4xy^3, a^2x^3y^2$ **11.** $3x^2y^3, 4y^2z^3$ **12.** $4a^3z, az^3, 5y^2$

13. $3y, 4z, 5x$ **14.** $3y^2, 4y^3, yz^2$ **15.** $2a^2bc, 9b^3c$

16. a^2bc, ab^3c, b^2c^2 **17.** $3x^2, 5a^2x, 9a^3$ **18.** $2a^2b^2, 2$

19. p^2q^2, r^2s^2, t^2 **20.** $5l, 4m, 3n$ **21.** l^2m^2, m^3n^3, n^4p^5

22. $2x, 3x, 4x^3$ **23.** ax^2y, bxy^2, cxy^3 **24.** n^4, p^4, n^5p^5

25. $a^2 + ab, 3a^2 + 3ab$ **26.** $2a - 2b, 5a - 5b$

27. $a^3 + a^2b, a^4 + a^3b$ **28.** $2x^3 - 2x^2y, 3x^2 - 3xy$

29. $a^2x + abx, aby + b^2y$ **30.** $2a + 2b, 2a - 2b$

31. $a^2 + ab, ab - b^2$ **32.** $2a + b, a + 2b$

33. $ax + 2ay, 2bx + by$ **34.** $a^3x + a^2x, a^2x^2 + a^2x$

35. $ax^3 + ax^2, a^2x^2 - a^2x$ **36.** $a^2 + a^3, x^2 + x^3$

37. $a^2 - 1, a^2 - 2a + 1$ **38.** $x^2 + 3x + 2, x^2 + 2x + 1$

39. $2x - x^2, 4 - 4x + x^2$ **40.** $12a^2 + 12a + 3, 20a^2 + 20a + 5$

41. $a^2 + 5a + 6, a^2 + 6a + 8$ **42.** $2x^2 + x - 6, 2x^2 - 7x + 6$

43. $2x^2 - x - 1, 2x^2 + x - 1$ **44.** $2x^2 - x - 3, 2x^2 - 5x + 3$

45. $ax + x - a - 1, ax + x + a + 1$

46. $6 - 2x - 3a + ax, 9 - 3x - 3a + ax$

47. $3a^4 + 9a^3 + 6a^2, 2a^5 + 2a^4 - 4a^3$

48. $6ax^3 - 3ax^4 - 3x^4 + 6x^3, 10ax^2 - 5ax^3 + 5x^3 - 10x^2$

FRACTIONS

EXAMPLE
$$\frac{\overset{2}{\cancel{4}}\overset{1}{\cancel{x^2}}\overset{y}{\cancel{y^3}}}{\underset{3}{\cancel{6}}\underset{x}{\cancel{x^3}}\underset{1}{\cancel{y^2}}} = \frac{2\,.\,1\,.\,y}{3\,.\,x\,.\,1} = \frac{2y}{3x}$$

EXERCISE 64

Simplify the following:

1. $\dfrac{2a^2x^3}{3x^3y^2}$　　2. $\dfrac{5a^2b^2}{6a^2b^2}$　　3. $\dfrac{4x^2yz^3}{16xy^2z^2}$　　4. $\dfrac{3ab}{4xy}$

5. $\dfrac{12p^3q^2}{18q^2r^3}$　　6. $\dfrac{l^2m^5n^3}{l^4m^2n^5}$　　7. $\dfrac{r^5s^2t^3}{r^3s^2t^2}$　　8. $\dfrac{14a^5x^2z^4}{42a^3x^3z^4}$

9. $\dfrac{36k^4l^2m^3}{27k^3l^3m^4}$　　10. $\dfrac{a^5b^3cd^2}{ab^5c^3d}$　　11. $\dfrac{p^3q^3r}{p^4q^2r}$　　12. $\dfrac{a^2b^3x^4}{a^2c^4z^3}$

HARDER FRACTIONS

EXAMPLE　　*Simplify* $\dfrac{2a^2 + 7a + 3}{a^2 + 7a + 12} \div \dfrac{6a^2 + 7a + 2}{6a^2 + 28a + 16}$

$$\text{EXPRESSION} = \frac{\overset{1}{\cancel{(2a+1)}}\overset{1}{\cancel{(a+3)}}}{\cancel{(a+3)}(a+4)} \times \frac{2(3a^2 + 14a + 8)}{\underset{1}{\cancel{(2a+1)}}(3a+2)}$$

$$= \frac{2\overset{1}{\cancel{(a+4)}}\overset{1}{\cancel{(3a+2)}}}{\underset{1}{\cancel{(a+4)}}\underset{1}{\cancel{(3a+2)}}} \quad Ans. = \underline{\underline{2}}$$

NOTE. Factorize all numerators and denominators then cancel.

EXERCISE 65

Simplify the following, *where possible*:

1. $\dfrac{2a + 2b}{3a + 3b}$　　2. $\dfrac{ax + ay}{2ax + 2ay}$　　3. $\dfrac{ax + bx}{ax - bx}$

4. $\dfrac{ax^2 - bx^2}{a^2 + bx^2}$　　5. $\dfrac{3x^3 + 3x}{4x^2y + 4y}$　　6. $\dfrac{a^3b^2 - a^2b^2}{a^2b^2 - a^3b^2}$

7. $\dfrac{ab + ay}{ab + by}$　　8. $\dfrac{a + b}{a^2 - b^2}$　　9. $\dfrac{x - 1}{x^2 - 1}$

10. $\dfrac{a^3 + a^2b}{a^2 - b^2} \times \dfrac{a - b}{a}$　　11. $\dfrac{a^2 + 2ab + b^2}{3a^2 - 3b^2} \times \dfrac{a - b}{a + b}$

12. $\dfrac{2x^4y^2z^3}{6a^5b^7c^3} \div \dfrac{4x^5yz}{9a^7b^4c}$　　13. $\dfrac{3a^2x^2y + 3ax^2y}{2axy + 2ax} \times \dfrac{4y + 4}{9axy + 9xy}$

14. $\dfrac{5a^4 + 5a^3b}{4a^2x - 4a^2y} \div \dfrac{15ax^2 + 15bx^2}{12x^2 - 12xy}$　　15. $\dfrac{(x^2 - 2x + 1) - (x^2 - 1)}{(x - 1)^2 - (x^2 - 6x + 5)}$

16. $\dfrac{p^4 - 1}{4p} \times \dfrac{16p^2}{4p^3 + 4p}$ **17.** $\dfrac{2ax^2 - 2ax}{ax - 1 + x - a} \div \dfrac{3ax - 3x}{6a^2 - 6}$

18. $\dfrac{2ax + 2ay}{6x^3 + 12x^2} \times \dfrac{3x^3 - 3x^2y}{2ax + 6a} \times \dfrac{x^2 + 5x + 6}{x^2 - y^2}$

19. $\dfrac{ax^3 + x^3}{3a^2 - 3a} \times \dfrac{3ax^2 - 3ax}{a^2 + 2a + 1} \times \dfrac{3a^2 - 3}{x^5 - x^4}$

20. $\dfrac{a^2 + 3a + 2}{a - 1 - ax + x} \times \dfrac{x^2 - 1}{a^2 - a - 2} \times \dfrac{a^2 - 3a + 2}{ax + 2x + a + 2}$

ADDITION AND SUBTRACTION OF FRACTIONS

EXAMPLE. *Express as a single fraction*

$$\frac{a}{a + b} + \frac{b}{a - b} + \frac{2ab}{a^2 - b^2}$$

$$\text{EXPRESSION} = \frac{a(a - b) + b(a + b) + 2ab}{(a + b)(a - b)}$$

$$= \frac{a^2 - ab + ab + b^2 + 2ab}{(a + b)(a - b)} = \frac{a^2 + 2ab + b^2}{(a + b)(a - b)}$$

$$= \frac{\overset{1}{\cancel{(a + b)}}(a + b)}{\underset{1}{\cancel{(a + b)}}(a - b)} = \underline{\underline{\frac{a + b}{a - b}}} \; \textit{Ans.}$$

EXERCISE 66

Simplify the following, *leaving the numerator and denominator in factor form*:

1. $\dfrac{2}{a - 1} + \dfrac{1}{a + 2}$ **2.** $\dfrac{a}{b - a} - \dfrac{b}{a - b}$

3. $\dfrac{a + b}{a - b} - 1$ **4.** $\dfrac{ab}{a - b} - b$

5. $\dfrac{x + 1}{x} + \dfrac{3x - 2}{2x} - \dfrac{2}{3}$ **6.** $\dfrac{a + 2}{6} - \dfrac{5a}{12} + \dfrac{a - 3}{3}$

7. $\dfrac{1}{a + 2} + \dfrac{2}{a - 1} - \dfrac{3}{a}$ **8.** $\dfrac{2}{x^4 - 1} + \dfrac{1}{1 - x^2}$

9. $\dfrac{2a + 4b}{a - b} - \dfrac{2a^2 + 10ab}{a^2 - b^2}$ **10.** $\dfrac{a + 2}{4a - 8} - \dfrac{a - 2}{4a + 8} - \dfrac{2a}{a^2 - 4}$

11. $\dfrac{2}{x + 1} - \dfrac{3}{x + 2} + \dfrac{1}{x + 3}$ **12.** $\dfrac{1}{x + y} - \dfrac{1}{x - y} - \dfrac{2x}{y^2 - x^2}$

13. $\dfrac{x - 2}{x^2 - 5x + 4} - \dfrac{x - 3}{x^2 - 3x + 2} + \dfrac{x}{x^2 - 6x + 8}$

14. $\dfrac{a + b}{(a - 1)(b - 1)} + \dfrac{a + 1}{(a - b)(1 - b)} + \dfrac{b + 1}{(b - a)(1 - a)}$

15. $\dfrac{x}{2x + y} + \dfrac{y}{x + 2y} + \dfrac{x^2 + xy + y^2}{2x^2 + 5xy + 2y^2}$

16. $\dfrac{a^3}{a^2 + ab} - \dfrac{b^3}{a^2 - ab} + \dfrac{a^3b + ab^3}{a^3 - ab^2}$

17. $\dfrac{2x^2}{9x^2 - 20x + 4} - \dfrac{5x}{9x^2 + 16x - 4} - \dfrac{2}{x^2 - 4}$

18. $\dfrac{2a^4b - a^2}{a^2b} - \dfrac{2ab^2 + b^2}{ab^2} + \dfrac{2a + 2b}{2ab}$

HARDER EQUATIONS

EXAMPLE 1. *Solve* $\qquad \dfrac{3x + 4}{5} = \dfrac{2x - 5}{3}$

(i) Multiply throughout by L.C.M. 15: $\quad 3(3x + 4) = 5(2x - 5)$

(ii) Clear brackets, etc.: $\qquad\qquad\qquad 9x + 12 = 10x - 25$

$$37 = x$$
$$x = 37 \ Ans.$$

NOTE. This is an example in which **"cross-multiplying"** may be used.

You are reminded that this method applies only when each side of the equation contains a single fraction.

EXAMPLE 2. *Solve* $\qquad \dfrac{x + 4}{2x - 3} + \dfrac{x - 3}{x + 1} = \dfrac{(3x + 4)(x - 3)}{2x^2 - x - 3}$

(i) Factorize denominator on R.H.S.:

$$\frac{x + 4}{2x - 3} + \frac{x - 3}{x + 1} = \frac{(3x + 4)(x - 3)}{(2x - 3)(x + 1)}$$

(ii) Multiply throughout by L.C.M. $(2x - 3)(x + 1)$:

$$(x + 1)(x + 4) + (2x - 3)(x - 3) = (3x + 4)(x - 3)$$
$$(x^2 + 5x + 4) + (2x^2 - 9x + 9) = (3x^2 - 5x - 12)$$

(iii) Clear brackets:

$$x^2 + 5x + 4 + 2x^2 - 9x + 9 = 3x^2 - 5x - 12$$
$$3x^2 - 4x + 13 = 3x^2 - 5x - 12$$
$$x = -25 \ Ans.$$

NOTE

(i) When simplifying **expressions** containing fractions we retain a common denominator.

(ii) When solving **equations** containing fractions we eliminate the denominators by multiplying throughout by the L.C.M. of the denominators.

Solve the following equations:

1. $\dfrac{3x-2}{6} + \dfrac{4}{x+1} = \dfrac{2}{3x+3} + \dfrac{x+3}{2}$

2. $\dfrac{3}{x+1} = \dfrac{5}{x+2}$

3. $\frac{4}{5}(5x+2) - \frac{2}{3}(2x+3) = \frac{7}{10}(6x-5)$

4. $\dfrac{x+2}{x-1} = \dfrac{x+3}{x-4}$

5. $\dfrac{2x^2 - 5x - 3}{2x^2 + 9x + 4} = \dfrac{2x^2 + 7x - 4}{2x^2 - 7x + 3}$

6. $\dfrac{4}{x} + \dfrac{2}{x^2} = \dfrac{5}{x} + \dfrac{1}{x^2}$

7. $\dfrac{1}{x+1} + \dfrac{1}{x+2} = \dfrac{2}{x+3}$

8. $\dfrac{2x+1}{2x+3} + \dfrac{2x-1}{x+1} = 3$

9. $\dfrac{3x(x+1)}{(2x-1)(x-2)} = 1\frac{1}{2}$

10. $\dfrac{2x-1}{x} = 2 \cdot 5 - \dfrac{0 \cdot 5x}{x+1}$

11. $\dfrac{x^2 + 5x + 6}{x^2 + 3x + 2} = \dfrac{x+1}{x+3}$

12. $\dfrac{3}{x-3} = \dfrac{5}{x-2} - \dfrac{2}{x}$

13. $\dfrac{x+1}{x+2} + \dfrac{x+1}{x-2} = \dfrac{5x^2+1}{2x^2-8} - \dfrac{1}{2}$

14. $\dfrac{6x+5}{3x-2} = \dfrac{3x+1}{x+4} - 1$

15. $\dfrac{2}{x^2+3x+2} - \dfrac{3}{x^2+4x+3} = \dfrac{1}{x^2+5x+6}$

16. $\dfrac{3x+2}{3} - \dfrac{2(x+3)}{x+2} = \dfrac{x-1}{3} + \dfrac{2x^2}{3x+6}$

17. $\dfrac{(x+2)(x-1)}{2} - \dfrac{(x+1)(x-2)}{3} = \dfrac{x^2}{6}$

18. $\dfrac{x+3}{x^2+2x-8} + \dfrac{x+2}{x^2+x-12} = \dfrac{2(x-3)}{x^2-5x+6}$

FACTORS AND DIVISION

Given that $(2x - 1)$ is a factor of $(6x^3 - 7x^2 + 1)$ the second factor may be found by the process of division. The quotient should now be examined for further factors if the original expression is to be given completely in factor form.

Given that the first expression is a factor of the second, factorize the latter completely:

1. $3x + 1$; $6x^3 - 7x^2 + 1$
2. $x - 1$; $6x^3 - 7x^2 + 1$
3. $x + 1$; $x^3 + x^2 - x - 1$
4. $x + 1$; $x^3 - x^2 - x + 1$
5. $x - 1$; $x^3 - 3x^2 + 3x - 1$
6. $x + 1$; $x^3 + 3x^2 + 3x + 1$
7. $x + 1$; $x^3 + 6x^2 + 11x + 6$
8. $x - 1$; $x^3 - 9x^2 + 23x - 15$
9. $2x + 1$; $8x^3 + 4x^2 - 2x - 1$
10. $x^4 - 16$; $x^8 - 32x^4 + 256$
11. $a + b$; $a^3 + b^3$
12. $a - b$; $a^3 - b^3$

THE SUM AND DIFFERENCE OF TWO CUBES

$$a^3 + b^3 = (a + b)(a^2 - ab + b^2)$$
$$a^3 - b^3 = (a - b)(a^2 + ab + b^2)$$

EXAMPLE. *Factorize* $64x^3 - 8y^3$

Since $a^3 - b^3 = (a - b)(a^2 + ab + b^2)$
Then $64x^3 - 8y^3 = 8(8x^3 - y^3)$
       $Ans. = \underline{8(2x - y)(4x^2 + 2xy + y^2)}$

EXERCISE 69

Factorize the following, *where possible*:

1. $a^3 + b^6$ 2. $x^6 - y^3$ 3. $a^3 + a^2$
4. $4x^3 + 32y^3$ 5. $64 + z^3$ 6. $27p^3 - 8q^3$
7. $8b^3 - 1$ 8. $1 + 1000a^3$ 9. $a^2 - b^2$
10. $a^2 + b^2$ 11. $125x^3 + 216y^3$ 12. $1 - x^3y^6$
13. $(a + 1)^3 + (a - 1)^3$ 14. $(a + b)^3 - (a - b)^3$

EXPANSIONS AND COEFFICIENTS

The product of two expressions may sometimes be referred to as an **expansion** of the two expressions.

EXAMPLE. *Write down the coefficient of x^2 in the expansion of*
$$(2x - 3)(3x^2 - 4x + 2)$$

The term containing x^2 will be given by the product of $+2x$ and $-4x$ and also -3 and $+3x^2$, i.e., $-8x^2 - 9x^2$ giving $-17x^2$.

$$\therefore \underline{\text{Coefficient of } x^2 = -17}$$

EXERCISE 70

Write down the expansions of the following:

1. $(x + 1)(x^2 + x + 1)$ 2. $(x - 1)(x^2 - x - 1)$
3. $(1 + x)(x + 1 - x^2)$ 4. $(x - 1)(1 + x^2 + x)$
5. $(2x + 1)(x^2 + 3 - x)$ 6. $(x - 3)(3x - 4 + x^2)$
7. $(2x - 4)(3x^2 - 4x + 2)$ 8. $(x + 4)(2x^2 - 4 - x)$

Write down the coefficient of x^2 in the following expansions:

9. $(2x - 1)(x^2 + 2x - 1)$ 10. $(1 - 2x)(4 + x - 2x^2)$
11. $(2x + 4)(3x^2 - 3 - 3x)$ 12. $(3x - 2)(3 - x^2 - 2x)$
13. $(x - 5)(5x^2 - 3x + 1)$ 14. $(3 - 2x)(3x - x^2 + 1)$

Write down the coefficient of x in the following expansions:

15. $(x - 1)(3x^2 - 4x + 1)$ 16. $(4 - 3x)(1 + x^2 - x)$
17. $(2x - 3)(3x + x^2 - 4)$ 18. $(4x - 3)(x^2 - x - 1)$
19. $(x + 4)(4 - x + x^2)$ 20. $(x + y)(x^2 + y^2 + 1)$

GRAPHICAL SOLUTION OF SIMULTANEOUS EQUATIONS

$$2x + 3y = 18 \quad\text{————————}\quad ①$$
$$x + 2y = 11 \quad\text{————————}\quad ②$$

We can construct a table of values for each of the equations, this will be made easier if we **express y as a function of x,** i.e., $y = f(x)$, where **x is the independent variable** and **y is the dependent variable.**

From ① $\quad y = \dfrac{18 - 2x}{3}$ From ② $\quad y = \dfrac{11 - x}{2}$

$2x + 3y = 18$

x	1	2	3	4	5	6	7	etc.
y	$5\frac{1}{3}$	$4\frac{2}{3}$	4	$3\frac{1}{3}$	$2\frac{2}{3}$	2	$1\frac{1}{3}$	etc.

$x + 2y = 11$

x	1	2	3	4	5	6	7	etc.
y	5	$4\frac{1}{2}$	4	$3\frac{1}{2}$	3	$2\frac{1}{2}$	2	etc.

We now plot a graph for each equation from the tables of values (see fig. A : 12). (Only 3 points required, see p. 42.)

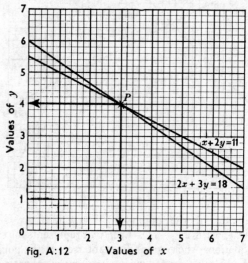

fig. A:12 Values of x

The two graph lines intersect at point P whose coordinates are (3, 4). You should remember that the x-coordinate (**abscissa**) is given first and the y-coordinate (**ordinate**) is given second.

The values $x = 3$, $y = 4$ provide the common solution to the pair of simultaneous equations $2x + 3y = 18$ and $x + 2y = 11$.

NOTE. Check the solution by substituting in the equations.

When two or more equations are true for the same values of the unknowns they are called **Simultaneous Equations**.

EXERCISE 71

Solve the following simultaneous equations graphically. *Check your solutions.*

1. $x + 3y = 5$ **2.** $2x - y = 0$ **3.** $3x + 2y = 12$
   $3x + y = 7$ $2x + y = 4$ $x + y = 5$
4. $8x - y = 4$ **5.** $x + 2y = 9$ **6.** $3y - x = 8$
   $y - 4x = 0$ $x - 2y = 1$ $y - 3x = 0$

The solutions of the examples in the previous exercise were all positive integral values. Such will not always be the case, consider fig. A : 13.

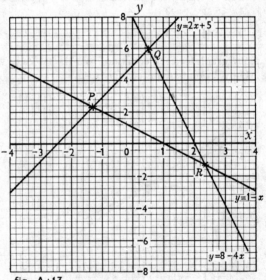

fig. A : 13

The coordinates of the points of intersection are as follows: $P(-1\frac{1}{3}, 2\frac{1}{3})$; $Q(\frac{1}{3}, 6)$; $R(2\frac{1}{3}, -1\frac{1}{3})$. Each pair represents the solution for that pair of equations whose graph lines intersect at the points named.

From fig. A : 13 solve the following simultaneous equations:

(i) $y - 2x = 5$ (ii) $y - 5 = 2x$ (iii) $y + 4x = 8$
    $x + y = 1$ $4x + y = 8$ $y + x = 1$

EXERCISE 72

Taking values of x from -4 to $+4$ draw graphs to solve the following. *Use a new pair of axes for each graph.*

1. $x + y = 0$
$2x - y = 6$

2. $x = y - 4$
$2y = 7 + x$

3. $x - y = 1$
$3x - 2y = 0$

4. $x + 4y = 7$
$4y - x = 1$

5. $2x - y = 2$
$4x + y = 13$

6. $2x + y = 2$
$4x + y = 7$

7. $y - x = 6\frac{1}{2}$
$5x + 2y = -1$

8. $2y - x = 1$
$2y + x = 8$

9. $y - 2x = 7$
$4x + y = -\frac{1}{2}$

SOLUTION OF SIMULTANEOUS EQUATIONS BY CALCULATION

FIRST METHOD—EQUATING EXPRESSIONS FOR ONE OF THE UNKNOWNS

EXAMPLE. *Solve the equations:*

$$y - 2x = 8\frac{1}{2} \quad\text{————}\quad ①$$
$$y = 2\frac{1}{2} + \frac{1}{2}x \quad\text{——}\quad ②$$

From ①
$$y = 8\frac{1}{2} + 2x \quad\text{———}\quad ③$$

Equating ② and ③:
$$2\frac{1}{2} + \frac{1}{2}x = 8\frac{1}{2} + 2x$$

$$\therefore \frac{5}{2} + \frac{x}{2} = \frac{17}{2} + \frac{2x}{1}$$

Multiply throughout by L.C.M. 2:
$$5 + x = 17 + 4x$$
$$-12 = 3x$$
$$x = -4$$

Substitute for x in ①:
$$y + 8 = 8\frac{1}{2}$$
$$y = \frac{1}{2}$$

Ans. $\underline{\underline{x = -4, \ y = \frac{1}{2}}}$

CHECK in ②:
L.H.S. $= \frac{1}{2}$ R.H.S. $= 2\frac{1}{2} - 2 = \frac{1}{2}$
$$\therefore \text{ L.H.S.} = \text{R.H.S.}$$

EXERCISE 73

Solve the following equations. *Check your answers.*

1. $2x - y = 3$
$3x - 2y = 4$

2. $2x - y = 4$
$3x - 2y = 5$

3. $3x - y = 0$
$y - x = 2$

4. $y - x = 2$
$y + x = 0$

5. $9 - 2y = -13x$
$13 + 5x = -4y$

6. $y - x = 3$
$3 - y = x$

7. $3x - y = 0$
$2y - 4x = 1$

8. $x - 2y = 6$
$x + 2y = 4$

9. $7 + x = y$
$2y + x = 5$

10. $y + 2x = 1$
$3y - 4x = 28$

11. $y - 4x = 6$
$3y + 16x = 11$

12. $x + 4 = y$
$3x - 16y = 1$

13. $x - 2y = 5$
$8y + 6x = 15$

14. $5x + 7y = 35$
$7x - 5y = 49$

15. $4x - 9y = 9$
$5y + 2x = 33$

16. $4x - 5y = 65$
$5x + 4y = 30$

17. $5x + 7y = 2$
$7x + 10y = 4$

18. $5x - 5y = 1$
$2x - 3y = 0$

SECOND METHOD—SUBSTITUTING AN EXPRESSION FOR ONE OF THE UNKNOWNS

EXAMPLE. *Solve the equations:*

$$x + 8 = 3y \quad\text{———}\quad ①$$
$$3x + 2y = 9 \quad\text{———}\quad ②$$

From ① $x = 3y - 8$
Substitute for x in ②: $3(3y - 8) + 2y = 9$
$9y - 24 + 2y = 9$
$11y = 33$
$y = 3$
Substitute for y in ②: $3x + 6 = 9$
$3x = 3$
$x = 1$

Ans. $x = 1, y = 3$

CHECK in ①:
L.H.S. $= 1 + 8 = 9$ R.H.S. $= 9$
$\therefore$ L.H.S. $=$ R.H.S.

EXERCISE 74

Solve the following equations. *Check your answers.*

1. $3x - 4y = 2$
$x + 3y = 5$

2. $3y - 5x = 4$
$5x - y = 2$

3. $x - y = 1$
$2x + 3y = 12$

4. $y - 2x = 1$
$3y + x = 17$

5. $3x - 4y = 8$
$x + 5y = 9$

6. $y - 3x = 1$
$x + 9y = 37$

7. $6x - y = 10$
$6y - x = 10$

8. $4x - y = 5$
$5y - 4x = 7$

9. $3x - 10y = 5$
$2x + 3y = 13$

10. $5x + 8 = 4y$
$5y - 4x = 19$

11. $7x + 6y = 17$
$8y - 6x = 38$

12. $7y - 9x = 13$
$8x + 14 = 5y$

13. $20y + 2x = 2$
$2y - 10x = 41$

14. $4x - 5y = 35$
$4y + 18x = 25$

15. $8y - 10x = 2$
$10y - 8x = 16$

16. $8y + 4x = 4$
$2y - 16x = 35$

17. $9x - 9y = 16$
$5x - 3y = 10$

18. $6y - 7x = 5$
$6x - 7y = \frac{41}{42}$

19. $\frac{x}{2} - \frac{y}{4} = 2$
$\frac{2x}{5} + \frac{y}{6} = 6$

20. $\frac{x}{4} - \frac{y}{3} = 0$
$\frac{x}{2} - \frac{y}{2} = 1$

21. $x + \frac{3y}{2} = 4$
$y + \frac{2x}{5} = 4$

THIRD METHOD—ADDITION/SUBTRACTION METHOD

EXAMPLE. *Solve the equations:*

$$20x + 21y = 24 \quad \text{———————} \quad ①$$
$$25x + 28y = 37 \quad \text{———————} \quad ②$$

Multiply ① by 4 $80x + 84y = 96 \quad \text{———————} \quad ③$
and ② by 3: $75x + 84y = 111 \quad \text{———————} \quad ④$
Subtract ④ from ③: $5x = -15$
                $x = -3$

* Multiply ① by 5 $100x + 105y = 120 \quad \text{———————} \quad ⑤$
and ② by 4: $100x + 112y = 148 \quad \text{———————} \quad ⑥$
Subtract ⑤ from ⑥: $7y = 28$
                $y = 4$

Ans. $\underline{\underline{x = -3,\ y = 4}}$

CHECK in ①: L.H.S. $= -60 + 84 = 24 = $ R.H.S.
CHECK in ②: L.H.S. $= -75 + 112 = 37 = $ R.H.S.

NOTE

   * The second unknown may be found by elimination instead of the usual substitution. This method is often more convenient, particularly if the value of the first unknown is found to be an awkward fraction.

EXERCISE 75

Solve the following equations. *Check your answers.*

1. $x + 3y = 7$ 2. $2x + 3y = 13$ 3. $x + 5y = 8$
    $2x + 5y = 12$ $3x + y = 9$ $2x - 5y = 1$

4. $3x - 2y = 2$ 5. $x + 2y = 1$ 6. $2x - 3y = 12$
    $x + 3y = 8$ $2y - x = 3$ $3x - 2y = 13$

7. $3x + 2y = 11$ 8. $5y - 4x = 1$ 9. $7x - 3y = 14$
    $2x - 6y = 0$ $3x - 10y = 18$ $5x + 4y = 53$

10. $8y - 5x = 20$ 11. $3x + 22y = 55$ 12. $6x + 5y = 46$
    $8x - 5y = 46$ $4x + 33y = 77$ $3x + 10y = 38$

13. $12x + 9y = 6$ 14. $35y - 36x = 30$ 15. $20x - 27y = 14$
    $8x - 6y = 20$ $42x - 30y = 30$ $28x + 9y = 4$

16. $5x + \frac{1}{2}y = 12$ 17. $2\frac{1}{2}x - \frac{5}{8}y = 25$ 18. $3 \cdot 2x - 0 \cdot 7y = 0 \cdot 47$
    $2\frac{1}{2}x - y = 1$ $1\frac{1}{3}y - \frac{3}{4}x = 20$ $2 \cdot 4x + 0 \cdot 35y = 0 \cdot 265$

19. $\dfrac{3x - 2y}{3} = \dfrac{1}{3}$ 20. $\dfrac{x + y}{2} + \dfrac{x - y}{3} = 2\frac{1}{6}$

    $3y - 2x = 2x + 1$ $\dfrac{x + y}{3} - \dfrac{x - y}{2} = 2\frac{1}{6}$

SIMULTANEOUS EQUATIONS WITH THREE UNKNOWNS

EXAMPLE. *Solve the equations:*

$$2x + 3y + 2z = 16 \quad\text{———————} \quad ①$$
$$3x - 2y - 3z = 4 \quad\text{———————} \quad ②$$
$$4x - 3y - 3z = 3 \quad\text{———————} \quad ③$$

Add ① and ③: $6x - z = 19 \quad\text{———————} \quad ④$

Multiply ① by 2 $4x + 6y + 4z = 32 \quad\text{———————} \quad ⑤$
and ② by 3: $9x - 6y - 9z = 12 \quad\text{———————} \quad ⑥$

Add ⑤ and ⑥: $13x - 5z = 44 \quad\text{———————} \quad ⑦$

NOTE. We have eliminated y from ④ and ⑦, now proceed with these equations as previously.

EXERCISE 76

Solve the following equations. *Check your answers.*

1. $x + y + z = 6$
$x - y + z = 2$
$x + y - z = 0$

2. $x + y + z = 6$
$x - y + z = 2$
$x + y - z = 4$

3. $2x - 3z + y = -7$
$2z + 3y - x = -2$
$3x - y + 2z = 14$

4. $2x + y + z = 8$
$x - 2y + z = 7$
$x + y - 2z = 1$

5. $x + y + z = 5$
$y - x + z = 1$
$x + y - z = 5$

6. $4x - 15y + 6z = 15$
$6x - 20y + 3z = 8$
$8x - 10y + 9z = 4$

7. $z + y = x + \frac{1}{4}$
$x - 2z = 3y$
$5x - 6y = 4z$

8. $2x = z - y$
$3z = 2x - 3y$
$3 + y = z - x$

9. $x + y = z$
$3y - 2z = x + 1$
$9x + z = 3y + 2$

EXERCISE 77 (*Miscellaneous*)

Solve the following equations:

1. $x + 7y = -16$
$x - 8 = y$

2. $54x - 95y = 13$
$81x - 76y = 86$

3. $0 \cdot 4x - 1 \cdot 3y = 0 \cdot 72$
$0 \cdot 4y + 1 \cdot 3x = 0 \cdot 49$

4. $4x = 2(z - y)$
$3y = x + z$
$2z = 4y + 2$

5. $x + y = 3$
$y + z = 1$
$z + x = 12$

6. $\dfrac{2x + 1}{3} - \dfrac{3y - 4}{5} = 1\frac{4}{15}$
$2x + 3y = 2$

7. $\dfrac{x}{4} + \dfrac{y}{3} = 4$
$\dfrac{24}{x} - \dfrac{18}{y} = 0$

8. $8y - 3x = 3 - 5z$
$2y + x = 4 + z$
$6y + 3x + z = 0$

9. $\dfrac{2x + 3y}{4} = \dfrac{2y + x}{3} = \dfrac{y + 4}{3}$

10. $14x + 9y = 6x + y = 15$

11. $x + 2y + z = 3y + z - 4 = 6x - z + 1 = 0$

PROBLEMS LEADING TO SIMULTANEOUS EQUATIONS

NOTE

(i) The solution should begin with a precise definition of the unknowns.

(ii) The data from the question must be set down in the form of accurate algebraic expressions before equations can be constructed.

(iii) Great care must be taken to ensure that all the statements (and hence the equations) employ the same units.

(iv) The solution of a problem is not complete until the answer has been expressed in the correct form.

EXERCISE 78

1. Find two numbers whose sum is 12 and whose difference is 4.

2. Find two numbers whose sum is 56 and whose difference is 16.

3. Find two numbers whose sum is 72 and such that five times the smaller equals the larger.

4. Find two numbers whose difference is equal to two more than the smaller and whose sum is 44.

5. Find two numbers such that the smaller is equal to $\frac{2}{5}$ of the larger and the larger is five less than twice the smaller.

6. Find two numbers such that the smaller is equal to $\frac{2}{3}$ of the larger and $\frac{5}{9}$ of the larger exceeds $\frac{2}{3}$ of the smaller by three.

7. Find a fraction such that if the numerator and denominator are each increased by one the fraction reduces to $\frac{4}{5}$ and if the numerator and denominator are each decreased by one the fraction reduces to $\frac{2}{3}$.

8. Find a fraction such that if the denominator is diminished by two the fraction reduces to unity and if the numerator is increased by seven the fraction reduces to two.

9. If the numerator of a fraction is increased by unity and the denominator decreased by unity the fraction reduces to $\frac{3}{4}$. If the numerator is decreased by unity and the denominator increased by the same amount the fraction reduces to $\frac{2}{5}$. Find the fraction.

10. A fraction is such that if three is subtracted from the numerator it becomes equal to half the denominator and if the denominator is decreased by unity the fraction reduces to unity. Find the fraction.

11. A number of two digits is such that the units digit is one more than the tens digit. If the sum of the digits is added to the number, the digits would be reversed. Find the number.

12. A number of two digits is such that the tens digit is equal to a third of the units digit and such that the sum of the digits plus their difference is $1\frac{1}{2}$ less than half the number. Find the number.

13. Find a number of two digits such that the tens digit is twice the units digit and if the digits are reversed the result is twelve more than three times the sum of the digits.

14. A number consists of three digits whose sum is ten. The units digit is equal to the sum of the other two and the tens digit is two less than the units digit. Find the number.

15. A number consists of three digits such that the tens digit is two more than the units digit and three less than the hundreds digit. If the number is reduced by unity and the result divided by the sum of the original digits the quotient is 57. Find the number.

16. 3 kg of tea and 2 kg of coffee cost £5·10; 2 kg of tea and 3 kg of coffee each of the same brand as before cost £5·40. Find the cost of each per kilogramme.

17. The ratio of two sums of money is 7 : 5. If the larger sum of money is increased by £15 the ratio becomes 2 : 1. Find the sums of money.

18. Four metres of material A and eight metres of material B cost £5·20. Eight metres of material A and four metres of material B cost £4·28. Find the cost of each per metre.

19. A bag contains some 10p and 2p coins. If there were two more 2p coins the bag would contain £5, but if there were half the number of 10p coins the bag would contain only £2·66. Find the number of each.

20. Two boys each had a number of marbles. One said to the other, 'If you gave me half your marbles, I should have three times as many as I have now.' The other boy replied, 'If you gave me two of your marbles, I should have five times as many as you had left.' How many marbles had each?

21. Find the present ages of a father and son such that half the sum of their ages is equal to the difference between their ages now and in fifteen years time the father will be twice as old as his son.

22. Find the perimeter and area of the rectangle in fig. A : 14. The dimensions are in metres.

$4x - y$

$7x + 3y$ $1 - 2y$

$8x + 2y$

fig. A:14

23. At a school concert admission prices were as follows: Children 10p, Adults 40p, Reserved 50p. The proceeds from the sale of children's tickets were half that from the sale of reserved tickets. The number of adults' tickets sold was three times the number of children's tickets. The total proceeds amounted to £120. How many tickets were sold?

24. In a farmyard there are a number of cows and ducks. Between them there are 20 heads and 56 feet. How many cows and ducks are there?

25. A stationer buys black lead pencils at 20p per dozen and coloured pencils at 30p per dozen. He sells three-quarters of his stock of each for a total of £20·25. The rest he sells for £5·25 at 16p per dozen for the black and 28p per dozen for the coloured, making a total profit of £1·50. How many of each type of pencil were purchased?

26. Two towns are joined by road and rail. If an average speed of 30 mile/h is maintained on each route the journey by rail takes forty minutes longer, but if the road speed is reduced by 5 mile/h, the journey by road takes an hour longer than by rail. Find the distances by road and rail.

27. Two cars approach each other from opposite directions along the same route joining two distant towns. The difference in their speeds is 12 mile/h and the faster car travels 40 miles more than the slower. If each travels for the same period of time, how long will they take to meet and what is the distance between the towns if the slower car takes 8 hours to complete the whole journey?

LITERAL SIMPLE EQUATIONS

EXAMPLE. *Solve* $\dfrac{a(b-x)}{b} - ab = a(b-2x)$

Multiply throughout by b: $a(b-x) - ab^2 = ab(b-2x)$

Clear brackets: $ab - ax - ab^2 = ab^2 - 2abx$

Collect x terms on L.H.S.: $2abx - ax = ab^2 + ab^2 - ab$

Factorize L.H.S. to isolate x: $x(2ab - a) = 2ab^2 - ab$

Divide throughout by $(2ab - a)$ and factorize: $x = \dfrac{ab(2b-1)}{a(2b-1)}$

Reduce to lowest terms: $\underline{\underline{x = b \ Ans.}}$

EXERCISE 79

Solve the following equations:

1. $a = b + x$ **2.** $b = x - a$ **3.** $ab - x = 0$

4. $ax + bx = a + b$ **5.** $a = bx$ **6.** $ax - b = c$

7. $(x + a)(x + b) + 1 = (a + 1)(b + 1) + x^2$

8. $\dfrac{a - b}{x} = \dfrac{a - b}{a + b}$ **9.** $\dfrac{a}{x} + \dfrac{a}{b} = \dfrac{a + b}{x}$

10. $a(x - 1) = b(x + 1)$ **11.** $a(b + x) - b(a - x) = a + b$

12. $\dfrac{a - x}{3} + \dfrac{b + x}{2} = \dfrac{x}{4}$ **13.** $\dfrac{a - b}{a - x} = \dfrac{a - c}{a + x} + \dfrac{2a(x - b)}{a^2 - x^2}$

14. $(a + b)(c + x) + (a + c)(b + x) = (b + c)(a + x)$

15. $a(b + c + x) - b(a + c - x) = c(a - b - x) + 1$

16. $(a - x)(b - x) - (a + x)(b + x) = (a - b)^2 - (a + b)^2$

17. $\dfrac{ax - b}{bc} + \dfrac{bx - c}{ac} = \dfrac{a - cx}{ab} = \dfrac{a^2 - ab + b^2 - bc + c^2 - ac}{abc}$

LITERAL SIMULTANEOUS EQUATIONS

EXAMPLE 1. *Solve the equations*

$$x + y = 3a + b \quad\text{——————} \quad ①$$
$$x - y = a - 3b \quad\text{——————} \quad ②$$

Add ① and ②: $\qquad 2x = 4a - 2b$

$$\therefore\ x = 2a - b$$

Subtract ② from ①: $\qquad 2y = 2a + 4b$

$$\therefore\ y = a + 2b$$

Ans. $x = 2a - b,\ y = a + 2b$

EXERCISE 80

Solve the following equations:

1. $x + y = 2a$
   $x - y = 2b$
2. $x + y = 2a + b + c$
   $x - y = b - c$
3. $2x + 3y = 2a + 2b$
   $2x - 3y = 0$
4. $x + y = 4a$
   $x - y = 2b$
5. $a + b = x + y - 2$
   $a - x = b - y$
6. $2x + 3y = a + b + 3$
   $a - b = 2x - 3y + 1$
7. $2b = 3x - 5y$
   $3a = 6x + 5y - b$
8. $10b = 5a + 3x - 4y$
   $4x = 10a + 5b - 3y$
9. $b = \dfrac{3y - 2x - 5c}{a}$
   $c = 5ab - 3x - 2y$
10. $a = \dfrac{x + y - cd - bd}{b + c}$
    $d = \dfrac{x - y - ab + ac}{c - b}$

THE USE OF RECIPROCALS

EXERCISE 81

Solve the following equations:

1. $\dfrac{1}{x} + \dfrac{1}{y} = \dfrac{3}{4}$
   $\dfrac{1}{x} - \dfrac{1}{y} = \dfrac{1}{4}$
2. $\dfrac{2}{x} + \dfrac{1}{y} = \dfrac{2}{3}$
   $\dfrac{1}{y} - \dfrac{2}{x} = 0$
3. $\dfrac{1}{x} + \dfrac{1}{y} = 5$
   $\dfrac{1}{y} - \dfrac{1}{x} = 1$
4. $\dfrac{1}{x} - \dfrac{1}{y} = 3$
   $\dfrac{3}{x} + \dfrac{1}{y} = 1$
5. $\dfrac{3}{x} + \dfrac{2}{y} = 2$
   $\dfrac{3}{y} - \dfrac{2}{x} = \dfrac{5}{6}$
6. $\dfrac{3}{x} - \dfrac{4}{y} = 25$
   $\dfrac{4}{x} + \dfrac{3}{y} = 0$
7. $\dfrac{1}{y} - \dfrac{1}{x} = \dfrac{4}{15}$
   $\dfrac{5}{y} - \dfrac{3}{x} = 4$
8. $\dfrac{3}{2x} + \dfrac{2}{y} = 1\tfrac{4}{5}$
   $\dfrac{2}{x} - \dfrac{3}{2y} = \dfrac{11}{15}$
9. $\dfrac{4}{3x} - \dfrac{3}{4y} = 1\tfrac{14}{15}$
   $\dfrac{3}{4x} - \dfrac{4}{3y} = 1\tfrac{49}{60}$

TRANSFORMATION OF FORMULAE

EXAMPLE. $V = \frac{4}{3}\pi r^3$. *Find r.*

(i) Clear fractions: $\qquad 3V = 4\pi r^3$

(ii) Divide throughout by 4π to leave r terms on one side only:

$$r^3 = \frac{3V}{4\pi}$$

(iii) Cube root both sides: $\quad r = \sqrt[3]{\left(\frac{3V}{4\pi}\right)} \quad Ans.$

EXERCISE 82

Find r in 1–6:

1. $d = 2r$ $\qquad$ 2. $C = 2\pi r$ $\qquad$ 3. $A = \pi r^2$
4. $A = 2\pi rh$ $\qquad$ 5. $V = \pi r^2 h$ $\qquad$ 6. $V = \frac{1}{3}\pi r^2 h$

7. $E = \dfrac{La}{Fb}$, find: (i) L; (ii) b $\qquad$ 8. $T = 2\pi\sqrt{\left(\dfrac{l}{g}\right)}$, find: (i) l: (ii) g

9. $X = \dfrac{4\pi D}{K}$, find: (i) K; (ii) D $\qquad$ 10. $z = \dfrac{ay^2}{by^2 + bx}$, find y

11. $H = \dfrac{4\pi IN}{10l}$, find: (i) l; (ii) I $\qquad$ 12. $\dfrac{1}{a^2} - \dfrac{1}{b^2} = \dfrac{1}{c^2}$, find a

13. $C = \dfrac{AKN}{4000\pi d}$, find: (i) d; (ii) K $\qquad$ 14. $\dfrac{1}{R} = \dfrac{1}{r_1} + \dfrac{1}{r_2} + \dfrac{1}{r_3}$, find r_1

15. $F = \dfrac{Q_1 Q_2}{Kd^2}$, find: (i) Q_1; (ii) d $\qquad$ 16. $f = \dfrac{1}{2\pi\sqrt{(LC)}}$, find L

17. $I = \dfrac{PRT}{100}$, find: (i) P; (ii) T $\qquad$ 18. $P = \dfrac{RE^2}{(R + b)^2}$, find E

19. $S = \frac{1}{2}at^2$, find: (i) a; (ii) t $\qquad$ 20. $s = ut + \frac{1}{2}at^2$, find: (i) u; (ii) a
21. $v = u + at$, find: (i) u; (ii) t
22. $v^2 = u^2 + 2as$, find: (i) a; (ii) u
23. $S = 2\pi r^2 + 2\pi rh$, find h
24. $A = \frac{1}{2}h(a + b)$, find: (i) h; (ii) a
25. $E = \dfrac{L(R - r)}{2R}$, find: (i) L; (ii) r

26. $\dfrac{a}{b} + \dfrac{b}{c} = \dfrac{x}{y}$, find c $\qquad$ 27. $a = K + P\sqrt{\left(\dfrac{P + T}{PT}\right)}$, find T

28. $A = P + I$, where $I = \dfrac{PRT}{100}$, find: (i) R; (ii) P

29. $S = \sqrt{(R - r)^2 + H^2}$, find H

30. $I = \sqrt{a^2 + \dfrac{b^2}{2}}$, find: (i) b; (ii) a

SYMBOLIC EXPRESSION

EXERCISE 83

1. How many metres in: (i) x km; (ii) y dam; (iii) H dm; (iv) F cm?

2. From a length of cloth T m long, x pieces are cut, each of length T cm. Express the length remaining in: (i) cm; (ii) m.

3. An article costs S pence; how many for: (i) £P; (ii) p pence?

4. How long will it take to travel D miles at V mile/h: (i) in hours; (ii) in min?

5. Express V km/h in metres per second.

6. A cyclist travels S miles, his average speed for the first half of the journey is u mile/h and for the second half his speed is 2 mile/h faster. Express in its simplest form the total time taken in hours.

7. An article is sold for S pence at a loss of $l\%$ of the cost price. Express the cost price in £.

8. If S pounds are spent on articles costing P pence each and P pounds on articles costing S pence each, find the average cost per article in pence. Give the expression in its simplest form.

9. A grocer purchased x tonnes of flour at £y a tonne and sold it at 10p per kg. Express his profit in pounds as a fraction in its simplest form.

10. If a man walks x miles at y mile/h and y miles at x mile/h, express the average speed in mile/h in its simplest form.

11. The cost of M kl of milk is s pounds. If the price is increased by $\frac{1}{2}$p per litre, how many kilolitres can now be bought for s pounds?

12. In a group of x boys 3 are each x years of age, half the remainder are each 6 months older and the others are each 3 months younger. Find an expression, in its simplest form, for the average age of the group in years.

13. The average weight of n people is M stone and the average weight of the H heaviest people is S stone. Give an expression, in its simplest form, for the average weight in stones of the remainder.

14. The expenses for running a car are: Licence £x, Insurance £$(x + 8)$, Repairs £$4x$, Petrol £$\dfrac{x}{50}$ per gallon and the car does 30 miles per gallon. Find the average cost per head in £ for a family of four if the car travels 15 000 miles in a year.

15. A cyclist travels uphill at an average speed of $(u - 2)$ mile/h, but on the return journey his average speed is $(u + 2)$ mile/h. Find the average speed in mile/h for the whole journey of S miles.

16. A wholesaler purchases an article catalogued at £x and is allowed a discount of $x\%$ on the catalogue price. He resells the article at a profit of $x\%$ on his purchase price. Find his selling price.

QUADRATIC EQUATIONS

POINT 1.

The square root of a positive quantity may be given either a positive or negative sign.

POINT 2.

If the product of two or more quantities is zero then at least one of those quantities must itself be equal to 0.

EXAMPLE. *Solve* $x(x + 2) - 2(x^2 - x) = 4(x - 3)$

Clear brackets: $x^2 + 2x - 2x^2 + 2x = 4x - 12$

$$x^2 = 12$$
$$x = \sqrt{12}$$
$$\underline{\underline{x = \pm 3 \cdot 464 \quad Ans.}}$$

EXERCISE 84

Solve the following equations:

1. $x^2 = 9$ **2.** $x^2 = 25$ **3.** $x^2 = 64$ **4.** $x^2 = 121$

5. $2x^2 = 72$ **6.** $2x^2 = 98$ **7.** $3x^2 = 300$ **8.** $4x^2 = 576$

9. $x^2 = \frac{4}{9}$ **10.** $x^2 = 6\frac{1}{4}$ **11.** $4x^2 = 9$ **12.** $49x^2 = 36$

13. $x^2 + 2x - 2 = 2(x + 1)$ **14.** $x(2x + 4) = 4(x - 1) + 22$

15. $(x + 1)(x + 2) = 3x(1 - x) + 66$

16. $(2x - 3)(x + 5) = 7(x + 5)$ **17.** $\dfrac{x + 1}{2x - 2} = \dfrac{2x + 2}{x - 1}$

18. $(2x - 5)(x + 3) = (x + 3)(x - 2)$

19. $(3x - 1)(x - 4) = (x - 5)(4x + 9) - 2x$ **20.** $\dfrac{3x - 1}{2x - 5} = \dfrac{x - 6}{2x - 5}$

21. $\dfrac{x + 1}{x + 2} + \dfrac{x + 2}{x - 1} = \dfrac{4x + 11}{x^2 + x - 2}$ **22.** $\dfrac{x + 3}{3} - \dfrac{x - 2}{x} = \dfrac{x^2 - 1}{x}$

23. $\dfrac{2x + 3}{x + 1} - \dfrac{3x - 2}{x - 1} + \dfrac{9}{x^2 - 1} = 0$ **24.** $\dfrac{x + 4}{4} = \dfrac{x + 2}{2x} + \dfrac{1}{2}$

SOLUTION BY FACTORS

EXAMPLE. *Solve* $x^2 - 5x + 6 = 0$

Factorize L.H.S.: $(x - 2)(x - 3) = 0$

Then *either* $x - 2 = 0$ and $x = 2$

or $x - 3 = 0$ and $x = 3$ *

   Ans. $\underline{x = 2 \text{ or } 3}$

CHECK   When $x = 2$; L.H.S. $= 4 - 10 + 6 = 0 =$ R.H.S.

         When $x = 3$; L.H.S. $= 9 - 15 + 6 = 0 =$ R.H.S.

QUESTION. From the above example why must either $x - 2 = 0$ or $x - 3 = 0$?

EXERCISE 85

Solve the following equations. *Check your solution.*

1. $x^2 - 4x + 4 = 0$ **2.** $x^2 - 4x + 3 = 0$
3. $x^2 - 3x + 2 = 0$ **4.** $x^2 + 2x - 3 = 0$
5. $x^2 - 2x - 3 = 0$ **6.** $x^2 + 5x + 6 = 0$
7. $x^2 - x - 6 = 0$ **8.** $x^2 + 6x + 9 = 0$
9. $x^2 - 5x + 4 = 0$ **10.** $x^2 + 4x + 4 = 0$
11. $x^2 - 2x = 0$ **12.** $x^2 - 5x = 0$
13. $x^2 + 3x = 0$ **14.** $x^2 + 7x = 0$
15. $x^3 - x^2 = 0$ **16.** $x^3 + x^2 = 0$
17. $2x^2 - 5x + 2 = 0$ **18.** $3x^2 + 2x - 1 = 0$
19. $6x^2 - 5x + 1 = 0$ **20.** $3x^2 + 5x - 12 = 0$
21. $3x^2 + 4x - 4 = 0$ **22.** $4x^2 - 4x + 1 = 0$
23. $x^2 - 1 = 0$ **24.** $x^2 - 9 = 0$ **25.** $x^2 - 25 = 0$
26. $4x^2 - 16 = 0$ **27.** $9x^2 - 1 = 0$ **28.** $16x^2 - 9 = 0$
29. $2x^2 + x = 15$ **30.** $3x^2 - 10x = 8$ **31.** $x^2 = 2x + 15$
32. $6x^2 = 5 - 7x$ **33.** $6x^2 + 6 = 13x$ **34.** $4x^2 = 14x$

NOTE. If the coefficients of the terms of a quadratic equation contain a common factor the equation should first be divided throughout by that common factor.

EXAMPLE. *Solve* $9x^2 - 15x = 6$
  Express equation thus: $9x^2 - 15x - 6 = 0$
  Divide both sides by 3: $3x^2 - 5x - 2 = 0$
  Factorize: $(3x + 1)(x - 2) = 0$
  Then *either* $3x + 1 = 0$ and $x = -\frac{1}{3}$
       or $x - 2 = 0$ and $x = 2$*
       Ans. $\underline{x = 2 \text{ or } -\frac{1}{3}}$

SHORTER METHOD

With care we may now omit this (*) stage.

(i) The factor $(x - 2)$ gives the root $+2$; note the change of sign.
(ii) The factor $(3x + 1)$ gives the root $-\frac{1}{3}$; note the change of sign and the numerical term 1 is divided by 3, the coefficient of x.

EXERCISE 86

Give the roots of the equations whose factors are as follows:

1. $(x - 1)(x - 2) = 0$ **2.** $(x - 1)(x + 1) = 0$
3. $(x + 1)(x + 1) = 0$ **4.** $(x + 1)(x - 2) = 0$
5. $(x - 1)(x + 2) = 0$ **6.** $(x + 2)(x - 3) = 0$
7. $(x + 3)(x - 2) = 0$ **8.** $(x + 1)(x - 3) = 0$
9. $(x - 1)(x + 3) = 0$ **10.** $(x - 4)(x - 4) = 0$

11. $x(x - 5) = 0$ **12.** $x(x + 2) = 0$

13. $x^2(x - 6) = 0$ **14.** $x^2(x + 7) = 0$

15. $x(2x - 1) = 0$ **16.** $x(3x - 4) = 0$

17. $x(2x - 4) = 0$ **18.** $x(2x + 3) = 0$

19. $(2x - 1)(x - 2) = 0$ **20.** $(2x + 1)(x + 2) = 0$

21. $(2x - 3)(3x - 2) = 0$ **22.** $(2x + 5)(2x + 3) = 0$

23. $(3x - 1)(3x + 5) = 0$ **24.** $(4x - 7)(6x + 5) = 0$

If we know the two roots of a quadratic equation we can, by reversing the procedure, obtain the equation to which the roots belong.

EXAMPLE. *Find the quadratic equation whose roots are 5 and* $-\frac{1}{2}$.

The factors must be $(x - 5)(2x + 1)$

$\therefore$ The equation is $(x - 5)(2x + 1) = 0$

Ans. $2x^2 - 9x - 5 = 0$

EXERCISE 87

Form the equations whose roots are:

1. $1, 2$ **2.** $2, 3$ **3.** $-3, 2$ **4.** $-2, -2$

5. $4, 4$ **6.** $-1, +1$ **7.** $\frac{1}{2}, 1$ **8.** $\frac{1}{4}, 2$

9. $\frac{1}{3}, -1$ **10.** $\frac{1}{2}, \frac{1}{2}$ **11.** $\frac{2}{3}, -\frac{2}{3}$ **12.** $2\frac{3}{4}, -1\frac{1}{3}$

13. $0, 1$ **14.** $0, -2$ **15.** $0, -\frac{1}{4}$ **16.** $0, 0, \frac{1}{2}$

17. $\frac{2}{3}, 1\frac{1}{4}$ **18.** $2\frac{1}{2}, -2\frac{1}{2}$ **19.** $0, 2\frac{1}{4}$ **20.** $1\frac{1}{2}, -1\frac{1}{4}$

MISCELLANEOUS EXAMPLES

EXERCISE 88

Solve the following equations:

1. $ax^2 - 5ax + 6a = 0$ **2.** $16x^2 - 16 = 0$

3. $2(x^2 - 1) + x(x + 3) = 2x(x + 1)$ **4.** $6x^2 + 12x - 18 = 0$

5. $3x(x + 5) - 5(x^2 + 1) = 7x + 1$ **6.** $12x^2 - 7x - 12 = 0$

7. $\dfrac{1}{x + 1} - \dfrac{2}{x + 2} - \dfrac{3}{x - 3} = \dfrac{2}{x^2 + 3x + 2}$ **8.** $8x^2 - 2 = 0$

9. $(2x - 3)^2 - (x + 4)^2 = 4(2 - 5x)$ **10.** $9x^2 - 24x - 9 = 0$

11. $4x(x - 5) - 3(x^2 + 9) = 2x(x - 4)$ **12.** $8x^2 + 20x + 8 = 0$

13. $(3x + 2)^2 - (2x + 3)^2 = 10$ **14.** $4x^2 = 50x - 100$

15. $(x + 1)^2 + (x + 2)^2 = (x + 3)^2$ **16.** $7x^2 + 48x = 7$

17. $(2x + 1)^2 - (x + 2)^2 = (x + 1)(x - 1)$ **18.** $18x^2 = 19x + 12$

19. $(1 + x)(1 - x) + 2x(x - 4) = 2(1 - 4x)$ **20.** $51x^2 = 17x$

21. $(3x + 1)^2 - (3x - 1)^2 = x^2$ **22.** $6x^2 - 3x = 0$

23. $\dfrac{x^2 + 3}{2} + \dfrac{x^2 - 1}{3} = 2x$ **24.** $\dfrac{x - 4}{3x} + \dfrac{2x^2 - 3}{4x} = \frac{3}{4}$

PROBLEMS LEADING TO QUADRATIC EQUATIONS

Exercise 89

Give positive values only.

1. If the square of a number is reduced by seven times the number the result is zero. Find the number.

2. The square of a number is reduced by twice the number and the result is 3. Find the number.

3. A fraction is such that it is equal to twice its square. Find the fraction.

4. The product of two consecutive even numbers is equal to eight times the larger. Find the numbers.

5. The difference between a number and its square is thirty. Find the number.

6. If thirty times a number is increased by 32 the result is equal to twice the square of the number. Find the number.

7. One number is twice the other, and their product is equal to six times the larger. Find the numbers.

8. The sum of the squares of two consecutive odd numbers is 290. Find the numbers.

9. The difference between the squares of two consecutive numbers is 19. Find the numbers.

10. The difference between two numbers is three. If three times the larger is divided by the square of the smaller, the quotient is equal to $3\frac{3}{4}$. Find the numbers.

11. A fraction is such that the reciprocal of its square is equal to $2\frac{1}{4}$. Find the fraction.

12. The sum of two numbers is twelve. If the square of the smaller is subtracted from their product the result is ten. Find the numbers. (Two pairs.)

13. The area of a rectangle is 108 in² and the length is three times the breadth. Find the length and breadth.

14. The denominator of a fraction is three more than the numerator. If each is increased by four, the fraction is increased by $\frac{1}{8}$. Find the original fraction.

15. A motorist travels 160 miles. If he decreases his average speed by 2 mile/h the journey takes 20 min longer. Find his faster average speed.

16. The width of a room is 3 ft less than the length and the area is 180 ft². Find the dimensions of the room.

17. If the price of eggs were to decrease by 6p per dozen two more could be purchased for the same sum of 60 pence. Find the present price of one egg.

18. A cyclist travels 60 miles. If he decreased his speed by 2 mile/h he would take one hour longer. Find the original time taken.

19. *ABCD* and *WXYZ* are squares. $AW = WX = BX$. If the difference in area of the two squares is 12 in², find the length of *WX*. (Fig. A : 15.)

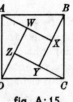

fig. A: 15

20. A merchant buys a length of cloth for £50. He could have purchased 25 yd more of a cheaper cloth at 10p per yard less. How many yards of the better cloth did he buy?

21. A path 4 ft wide is constructed around a square lawn. If the area of the lawn is 356 ft² more than the area of the path, find the length of the side of the lawn.

22. The back wheels of a cart are 4 ft greater in circumference than those at the front, which make eight revolutions more in a distance of 120 ft. Find the circumference of a front and back wheel.

23. A tank is supplied by two pipes, the larger of which can fill the tank in 10 min less than the smaller. Running together, the pipes can fill the tank in 12 min. Find the times taken by each pipe separately to fill the tank.

24. The hypotenuse of a right-angled triangle is 1 in longer than one of the perpendicular sides and 3 in more than twice the other perpendicular side. Find the three sides.

25. A number of people were to share a football pool dividend of £12,800, but four of them had their entries declared void due to late posting. As a result, the rest each received £160 more as their share of the prize. How many actually received a share?

26. A rectangle is such that if it is increased to a square on the longer side the area is increased by 50%. If the rectangle is reduced to a square on the shorter side the area is reduced by 33⅓%. If the difference in area of the two squares is 20 in², find the dimensions of the rectangle.

27. A tank is supplied by two pipes, one of which takes 6 min longer than the other to fill the tank. Find the time taken by each pipe on its own if they fill the bath together in 7 min 12 sec.

28. A stationer buys a number of items for £2·10 and sells each of them at a profit of 6p. After a month he had sold all but four of them and had exactly recovered his outlay. Find the number of items purchased.

29. A man sprints, at an average speed of 100 yd in 10 sec, along two adjacent sides of a rectangular plot of ground in 11½ sec. Maintaining the same average speed, he sprints along a diagonal of the plot in 8½ sec. Find the dimensions of the plot.

30. *ABCDEF* is a regular polygon centre *O*. *AOB* is an equilateral triangle. If the area of the figure is $24\sqrt{3}$ in², find the length of *AB*.

SURDS

NUMBERS which can be expressed exactly by integers (whole numbers) or precise fractions are called **rational numbers.** Numbers which cannot be expressed in this way are called **irrational numbers,** π is an excellent example.

If the root of a number is irrational the root is called a **surd.**

NOTE. In this work $\sqrt{4}$ will refer to the **positive root only.**

EXAMPLE. *Simplify* $\sqrt{3} + \sqrt{12}$

$$\sqrt{3} + \sqrt{12} = \sqrt{3} + \sqrt{4 \times 3} = \sqrt{3} + 2\sqrt{3}$$
$$= \underline{\underline{3\sqrt{3}}} \; Ans.$$

EXERCISE 90

Express the following as surds in their simplest form:

1. $\sqrt{12}$	**2.** $\sqrt{18}$	**3.** $\sqrt{20}$	**4.** $\sqrt{24}$
5. $\sqrt{27}$	**6.** $\sqrt{32}$	**7.** $\sqrt{45}$	**8.** $\sqrt{48}$
9. $\sqrt{50}$	**10.** $\sqrt{54}$	**11.** $\sqrt{2} \times \sqrt{12}$	**12.** $\sqrt{5} \times \sqrt{10}$
13. $\sqrt{9} \div \sqrt{12}$	**14.** $\sqrt{18} \div \sqrt{12}$	**15.** $\sqrt{6} \div \sqrt{18}$	
16. $\sqrt{30} \div \sqrt{54}$	**17.** $\sqrt{3} + \sqrt{12}$	**18.** $\sqrt{12} \div \sqrt{24}$	
19. $\sqrt{5} + \sqrt{20}$	**20.** $\sqrt{12} + \sqrt{12}$	**21.** $\sqrt{2} + \sqrt{8}$	
22. $\sqrt{18} + \sqrt{32}$	**23.** $\sqrt{54} - \sqrt{24}$	**24.** $\sqrt{72} - \sqrt{50}$	
25. $\sqrt{98} - \sqrt{50}$	**26.** $\sqrt{112} + \sqrt{63}$	**27.** $\sqrt{112} - \sqrt{7}$	

RATIONAL DENOMINATORS

EXAMPLE. *Evaluate* $\dfrac{2}{\sqrt{5} + \sqrt{3}}$

Multiply top and bottom by $(\sqrt{5} - \sqrt{3})$:

$$\frac{2(\sqrt{5} - \sqrt{3})}{(\sqrt{5} + \sqrt{3})(\sqrt{5} - \sqrt{3})} = \frac{2(\sqrt{5} - \sqrt{3})}{(\sqrt{5})^2 - (\sqrt{3})^2} = \frac{2(\sqrt{5} - \sqrt{3})}{5 - 3}$$
$$= \frac{2(\sqrt{5} - \sqrt{3})}{2} = \sqrt{5} - \sqrt{3} = 2 \cdot 236 - 1 \cdot 732$$
$$Ans. = \underline{\underline{0 \cdot 504}}$$

EXERCISE 91

Express the following with rational denominators:

1. $\dfrac{1}{\sqrt{8}}$	**2.** $\dfrac{1}{\sqrt{12}}$	**3.** $\dfrac{1}{\sqrt{20}}$	**4.** $\dfrac{3}{\sqrt{27}}$
5. $\dfrac{2}{\sqrt{18}}$	**6.** $\dfrac{6}{\sqrt{20}}$	**7.** $\dfrac{8}{\sqrt{24}}$	**8.** $\dfrac{8}{\sqrt{32}}$
9. $\dfrac{5}{\sqrt{45}}$	**10.** $\dfrac{15}{\sqrt{75}}$	**11.** $\dfrac{1}{\sqrt{2} + 1}$	**12.** $\dfrac{2}{\sqrt{3} + 1}$

13. $\dfrac{1}{2 - \sqrt{3}}$ 14. $\dfrac{1}{\sqrt{5} - 2}$ 15. $\dfrac{10}{3 - \sqrt{4}}$ 16. $\dfrac{1}{\sqrt{3} + \sqrt{2}}$

Simplify:

17. $(\sqrt{3} + \sqrt{2})(\sqrt{3} - \sqrt{2})$ 18. $(\sqrt{8} + \sqrt{5})(\sqrt{8} - \sqrt{5})$
19. $(\sqrt{3} + \sqrt{3})^2$ 20. $(\sqrt{5} + \sqrt{3})^2$ 21. $(\sqrt{6} - \sqrt{5})^2$

Simplify and evaluate:

22. $\dfrac{10}{\sqrt{50}}$ 23. $\dfrac{18}{\sqrt{54}}$ 24. $\dfrac{6}{\sqrt{72}}$ 25. $\dfrac{28}{\sqrt{98}}$

26. $\dfrac{3}{\sqrt{6} - \sqrt{3}}$ 27. $\dfrac{\sqrt{5} - \sqrt{2}}{\sqrt{5} + \sqrt{2}}$ 28. $\dfrac{2\sqrt{3} + 3\sqrt{2}}{2\sqrt{3} + 2\sqrt{2}}$

PERFECT SQUARES

$$(x + y)^2 = x^2 + 2xy + y^2 \qquad\qquad (x - y)^2 = x^2 - 2xy + y^2$$

REMEMBER

The square of the first term: twice the product of the terms: the square of the second term.

NOTE. The 'product of the terms' must take account of the signs possessed by each term.

EXERCISE 92

Give the expansions of the following:

1. $(a + b)^2$ 2. $(a + 2)^2$ 3. $(2 - a)^2$ 4. $(a - x)^2$
5. $(3 + y)^2$ 6. $(y - 4)^2$ 7. $(2x - y)^2$ 8. $(x - 2y)^2$
9. $(2x + 3y)^2$ 10. $(2a + b)^2$ 11. $(3a - 2b)^2$ 12. $(2x - 3y)^2$
13. $(x - 4y)^2$ 14. $(4x - y)^2$ 15. $(4 - y)^2$ 16. $(x + 4)^2$

The trinomials which result from the above expansions are all **'perfect squares'**, since they can be expressed in the form $(a + b)^2$.

If $4a^2 + 20ab + 9b^2$ is a perfect square it must be the square of $(2a + 3b)$, since these are the square roots of the first and last terms.

But 'twice the product of the roots' should give $+12ab$ and **not** $+20ab$. Therefore the expression is not a perfect square.

EXERCISE 93

Which of the following are perfect squares?

1. $a^2 + 4ab + b^2$ 2. $a^2 + 2a + 1$ 3. $1 - 2x + x^2$
4. $x^2 - xy + y^2$ 5. $x^2 + 4x + 2y^2$ 6. $4a^2 - 4a + 1$
7. $x^2 - 6x + 9$ 8. $x^2 + 6x - 9$ 9. $9 - 12y + 4y^2$
10. $y^2 - 4xy - 4x^2$ 11. $4x^2 + 2xy + y^2$
12. $4a^2 + 12ab + 9b^2$ 13. $9x^2 - 12xy + 4y^2$
14. $x^4 + 2x^2 + 1$ 15. $x^2 - 2xy^3 + y^6$
16. $4x^2 + 40x + 25$ 17. $x^4 - 2x^2y + 4y^2$
18. $9a^2 - 6ab - b^2$ 19. $9x^2 + 24xy + 16y^2$

COMPLETING THE SQUARE GEOMETRICALLY

fig. A:16

Fig. A : 16 represents a rectangle of dimensions x by $(x + 8)$, i.e., area $= x^2 + 8x$.

By halving the rectangle (8 by x) and rearranging the rectangles (4 by x) as shown in fig. A : 17, we see that a square 4 by 4 is required to 'complete the square', whose area will then be $(x + 4)^2$. Thus we must add 16.

fig. A:17

Exercise 94

Find by geometrical construction what must be added to the following to complete the square:

1. $x^2 + 6x$	**2.** $x^2 + 10x$	**3.** $x^2 + 16x$	**4.** $4x^2 + 12x$
5. $4x^2 + 20x$	**6.** $9x^2 + 24x$	**7.** $16x^2 + 40x$	**8.** $x^2 - 4x$

THE RULE FOR COMPLETING THE SQUARE

For expressions of the type $x^2 + 8x$ we may complete the square by **'adding the square of half the coefficient of x'.** This may be verified from the diagrams above.

We must also remind ourselves that there are **two square roots, one positive the other negative.**

NOTE. The rule 'add the square of half the coefficient of x' can be employed only when the coefficient of x^2 is unity.

Exercise 95

What must be added to each of the following to make the expression a perfect square? Give the two square roots of each expression.

1. $x^2 + 6x$	**2.** $x^2 - 4x$	**3.** $x^2 - 8x$	**4.** $x^2 + 10x$
5. $x^2 + 16x$	**6.** $x^2 - 12x$	**7.** $x^2 - 6x$	**8.** $x^2 + 4x$
9. $x^2 + 8x$	**10.** $x^2 - 10x$	**11.** $x^2 - 16x$	**12.** $x^2 + 12x$
13. $4x^2 - 20x$	**14.** $16x^2 + 56x$	**15.** $9x^2 - 48x$	**16.** $4x^2 - 12x$
17. $25x^2 + 40x$	**18.** $64x^2 - 16x$	**19.** $a^2 + 2ab$	**20.** $4x^2 - 12xy$
21. $x^2 + 4x + 2$		**22.** $x^2 - 6x + 3$	**23.** $x^2 - 8x + 10$
24. $x^2 + 6x - 3$		**25.** $x^2 - 4x + 8$	**26.** $x^2 + 2x + 1$
27. $x^2 - 2x - 1$		**28.** $x^2 - 2x + 4$	**29.** $x^2 + x$
30. $4x^2 - 4x + 4$		**31.** $x^2 - 5x$	**32.** $9x^2 + 12x + 4$
33. $4x^2 + 12x - 5$		**34.** $x^2 + 1\frac{1}{2}x$	**35.** $16x^2 - 8xy - y^2$
36. $x^2 - x + 1$		**37.** $x^2 - 1\frac{1}{3}x - 1$	**38.** $x^2 + \frac{1}{2}x + \frac{1}{16}$
39. $4x^2 - 1\frac{1}{3}x$		**40.** $9x^2 - 1\frac{1}{2}x + 1$	

SOLUTION OF QUADRATIC EQUATIONS BY COMPLETING THE SQUARE

When using the method of 'completing the square', begin by reducing the coefficient of x^2 to unity by dividing **both sides** of the equation by that coefficient. (*Where necessary.*)

EXAMPLE. *Solve $9x^2 - 24x + 7 = 0$, giving the roots correct to three significant figures.*

Rearrange the equation: $\qquad\qquad 9x^2 - 24x = -7$

* Divide both sides by 9: $\qquad\qquad x^2 - \frac{24}{9}x = -\frac{7}{9}$

To complete the square on L.H.S. add $(-\frac{12}{9})^2$ to both sides:
$$x^2 - \frac{8}{3}x + (\tfrac{4}{3})^2 = (\tfrac{4}{3})^2 - \tfrac{7}{9}$$

Factorize L.H.S.: $\qquad\qquad (x - \tfrac{4}{3})^2 = \tfrac{16}{9} - \tfrac{7}{9}$

Square root both sides: $\qquad\qquad x - 1\tfrac{1}{3} = \pm\sqrt{1}$

Add $1\tfrac{1}{3}$ to both sides: $\therefore\ x = 1\tfrac{1}{3} + 1$ or $1\tfrac{1}{3} - 1$

$\qquad\qquad$ *Ans.* $x = 2\cdot33$ or $0\cdot333$ $\quad$ (*Corr. to 3 sig. fig.*)

NOTE

(i) 'Completing the square' is obviously more involved than 'factorizing', and should be used only when the latter method fails to provide a solution.

(ii) The instruction 'Solve the equation and *give the roots correct to three significant figures*' is a good indication that the quadratic will not factorize.

EXERCISE 96

Solve the following equations by an appropriate method:

1. $x^2 - 4x - 5 = 0$ $\quad$ 2. $x^2 - 2x - 1 = 0$ $\quad$ 3. $x^2 + 2x - 2 = 0$
4. $x^2 + 4x + 4 = 0$ $\quad$ 5. $x^2 + x = 6$ $\quad$ 6. $x^2 + 6x = 1$
7. $x^2 - 6x + 6 = 0$ $\quad$ 8. $x^2 + 4x = 8$ $\quad$ 9. $x^2 + 4x = 4$
10. $x^2 - 4x + 2 = 0$ $\quad$ 11. $4x^2 = 1$ $\quad$ 12. $x^2 - 2x - 1 = 3$
13. $x^2 + 2x = 15$ $\quad$ 14. $2x^2 = x + 21$ $\quad$ 15. $x^2 + 8x = 4$
16. $x^2 + 10x = 25$ $\quad$ 17. $x^2 - 14x\ 1 = 0$ $\quad$ 18. $x^2 - 20x + 95 = 0$

Solve the following equations, giving the roots correct to three significant figures:

19. $4x^2 + 4x - 3 = 0$ $\qquad\qquad$ 20. $4x^2 - 4x - 8 = 0$
21. $4x^2 - 12x + 5 = 0$ $\qquad\qquad$ 22. $9x^2 + 6x - 3 = 0$
23. $4x^2 + 12x + 5 = 0$ $\qquad\qquad$ 24. $x^2 + 8x + 14 = 0$
25. $4x^2 + 16x + 14 = 0$ $\qquad\qquad$ 26. $9x^2 - 12x - 1 = 0$
27. $x^2 - 8x + 10 = 0$ $\qquad\qquad$ 28. $9x^2 - 6x - 9 = 0$
29. $x^2 - 6x - 3 = 0$ $\qquad\qquad$ 30. $x^2 + 14x + 47 = 0$
31. $4x^2 + 4x - 14 = 0$ $\qquad\qquad$ 32. $4x^2 - 4x - 11 = 0$
33. $4x^2 - 12x + 2 = 0$ $\qquad\qquad$ 34. $9x^2 - 12x + 1 = 0$
35. $9x^2 + 6x - 2 = 0$ $\qquad\qquad$ 36. $x^2 + 18x + 70 = 0$

SOLUTION OF QUADRATIC EQUATIONS BY FORMULA

From the numerous examples we have met it should be clear that any quadratic equation can be expressed in the following general form:

$$ax^2 + bx + c = 0$$

where the coefficients a, b and the constant term c represent **any numerical value** (positive or negative).

By the method of 'completing the square' it is possible to solve this literal equation for x in terms of a, b and c.

NOTE. **The formula should be learnt by heart:**

$$x = \frac{-b \pm \sqrt{b^2 - 4ac}}{2a}$$

EXAMPLE 1. *Solve the equation* $36x^2 - 84x + 40 = 0$

Divide both sides by the common factor 4:

$$9x^2 - 21x + 10 = 0$$

$$x = \frac{-b \pm \sqrt{b^2 - 4ac}}{2a} \text{ and } a = 9, b = -21, c = 10$$

$$\therefore x = \frac{21 \pm \sqrt{(-21)^2 - 4(9 \cdot 10)}}{18} = \frac{21 \pm \sqrt{441 - 360}}{18}$$

$$x = \frac{21 \pm \sqrt{81}}{18} = \frac{21 + 9}{18} \text{ or } \frac{21 - 9}{18}$$

$$\therefore x = \frac{30}{18} \text{ or } \frac{12}{18} = \frac{5}{3} \text{ or } \frac{2}{3} \quad \textit{(Check result by factors)}$$

Ans. $\underline{x = 1 \cdot 67 \text{ or } 0 \cdot 667}$ *(Corr. to 3 sig. fig.)*

EXERCISE 97

Solve the following equations, giving the roots correct to two decimal places:

1. $x^2 - 3x - 5 = 0$ **2.** $x^2 - 2x - 1 = 0$
3. $2x^2 - x - 2 = 0$ **4.** $x^2 - 2x - 2 = 0$
5. $x^2 - 3x + 1 = 0$ **6.** $3x^2 + 2x - 2 = 0$
7. $2x^2 + 3x - 1 = 0$ **8.** $x^2 - 4x + 1 = 0$
9. $x^2 + 5x + 3 = 0$ **10.** $x^2 + 4x - 7 = 0$
11. $9x^2 - 15x + 3 = 0$ **12.** $3x^2 - 5x - 6 = 0$
13. $4x^2 + 3x - 2 = 0$ **14.** $4x^2 - 16x + 14 = 0$
15. $5x^2 - x - 1 = 0$ **16.** $5x^2 + 5x - 35 = 0$
17. $8x^2 - 6x - 3 = 0$ **18.** $8x^2 - 20x + 12\frac{1}{2} = 0$
19. $3x^2 - 2\frac{1}{4}x - 1\frac{1}{3} = 0$ **20.** $2\frac{1}{2}x^2 + 6\frac{2}{3}x + 3\frac{1}{2} = 0$

EXAMPLE 2. *Solve the equation* $3x^2 + 4x + 5 = 0$

After substitution: $$x = \frac{-4 \pm \sqrt{16 - 4(+15)}}{6}$$

$$\therefore x = \frac{-4 \pm \sqrt{16 - 60}}{6} = \frac{-4 \pm \sqrt{-44}}{6}$$

Since we cannot obtain the square root of (-44) we cannot (at present) solve the equation.

Ans. The equation has no real roots.

NOTE

(i) If $b^2 - 4ac$ is negative, the quadratic equation has **no roots.**

(ii) If $b^2 - 4ac$ is equal to zero, i.e., $b^2 = 4ac$, the quadratic equation has **two equal roots.**

(iii) **The solution of quadratic equations should first be attempted by factors, if this method fails the use of the formula is probably less tedious than 'completing the square'.**

MISCELLANEOUS QUADRATICS

EXERCISE 98

Solve the following equations, where possible, using the most appropriate method and giving the roots correct to two decimal places where necessary:

1. $x^2 - x = 0$ **2.** $6x^2 - 5x - 6 = 0$

3. $x^2 + 10x + 22 = 0$ **4.** $x^2 - 2x + 5 = 0$

5. $x^3 - x = 0$ **6.** $3x^2 - 5x + 2 = 0$

7. $14x^2 - 51x + 7 = 0$ **8.** $48x^2 + 42x = 45$

9. $9x^2 + 6x = 16$ **10.** $4x^2 - 3x + 5 = 0$

11. $x^2 - 10x + 25 = 25$ **12.** $x^2 - 10x + 25 = 49$

13. $x^2 - 16x + 56 = 0$ **14.** $16x^2 - 8x - 29 = 0$

15. $4x^2 - 7x + 1 = 0$ **16.** $x(x + 1) + 3(x - 2) = 4(2x - 1)$

17. $2(x^2 - 3) = x(x + 1)$ **18.** $x(x^2 - 1) - 3(x^2 - 3x) = x(4x - 2)$

19. $2x(x + 4) = 3(2 - x^2)$ **20.** $\dfrac{x - 3}{x} - \dfrac{2x - 1}{3} = \dfrac{x + 2}{6x}$

21. $\dfrac{x^2 + 4}{3x} + \dfrac{x - 5}{4} = \dfrac{x^2 - 3x + 7}{6x}$ **22.** $\dfrac{x - 1}{x} - \dfrac{5}{x^2} = \dfrac{2x - 3}{3x}$

23. $\dfrac{2x - 4}{x} - \dfrac{4x - 3}{x^2} = \dfrac{5 - 8x}{4x} + \dfrac{3x - 13}{4x^2}$

24. $\dfrac{3x^2}{2} - \dfrac{9x}{4} + \dfrac{5}{6} = 0$

GRAPHICAL SOLUTION OF QUADRATICS

The expression $2x^2 + x - 15$ is a **'quadratic function'** of x, since the value of the whole expression will depend upon the value assigned to x. By employing the notation $y = f(x)$:

$y = 2x^2 + x - 15$

x	-4	-3	-2	-1	0	1	2	3	4	$-\frac{1}{2}$	$+\frac{1}{2}$
$2x^2$	32	18	8	2	0	2	8	18	32	$\frac{1}{2}$	$\frac{1}{2}$
-15	-15	-15	-15	-15	-15	-15	-15	-15	-15		
y	13	0	-9	-14	-15	-12	-5	6	21	*	*

Before attempting to draw the graph line, appropriate values for x should be added to the table where the turning-point occurs.

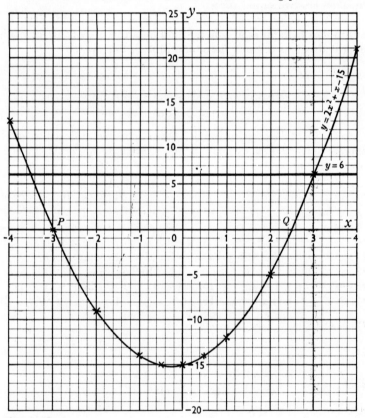

fig. A:18

USING THE GRAPH

To solve the equation $2x^2 + x - 15 = 0$ we require to find those values of x which make the expression equal to zero. Having drawn the graph of $y = 2x^2 + x - 15$, we now have to find the values of x for which $y = 0$.

Referring to the graph we see that $y = 0$ for the **whole length of the x-axis**, therefore the function $2x^2 + x - 15$ will equal 0 where the curve cuts the x-axis. (Points P and Q.)

Read off the values of x at P and Q. These are the roots of the equation $2x^2 + x - 15 = 0$.

This example has been chosen because it will factorize, and thus we can check our graphical solution.

$$(2x \quad)(x \quad) = 0 \quad \therefore \ x = ? \text{ or } ?$$

ORAL DISCUSSION (Answers at back)

(i) Is the graph symmetrical about the y-axis?

(ii) Through what value of x should a line be drawn parallel to the y-axis to provide an 'axis of symmetry'?

(iii) Has y (i.e., the function $2x^2 + x - 15$) a maximum value? If so, what is it?

(iv) Has y [i.e., $f(x)$] a minimum value? If so, what is it?

(v) For what values of x is $f(x)$ positive? (Remember $>$ means greater than, etc.)

(vi) For what values of x is $f(x)$ negative?

(vii) For what values of x is $f(x) = 0$?

(viii) For what values of x is $f(x) = 6$? Check your answer by solving $2x^2 + x - 15 = 6$. (Using factors.)

(ix) For what values of x is $f(x) = -5$? Check your solution.

(x) For what values of x is $f(x) = -9$? Check your solution.

NOTE. The last three questions should suggest a method for solving equations of the type

$$2x^2 + x - 21 = 0, \text{ i.e., } 2x^2 + x - 15 = 6$$

Draw the graph of $y = 6$ (viz., a straight line through $y = 6$ and parallel to x-axis).

Where the two graphs intersect read off values of x.

QUESTIONS

(xi) How could our graph be used to solve:

(a) $2x^2 + x - 10 = 0$ (b) $2x^2 + x - 6 = 0$
(c) $2x^2 + x + 1 = 0$?

(xii) From (c) above, does the graph of $y = -16$ cut the parabola? Can we solve the equation $2x^2 + x + 1 = 0$?

EXERCISE 99

1. Draw the graph of $y = 2x^2 + x - 6$, taking values of x from -4 to $+4$. From your graph answer the following:

(*a*) For what values of x does: (i) $f(x) = 0$; (ii) $f(x) = 4$; (iii) $f(x) = -4$?

(*b*) What are the coordinates of the turning-point?

(*c*) For what values of x is $f(x)$ negative?

(*d*) Use your graph to solve:

    (i) $2x^2 + x - 28 = 0$ (ii) $2x^2 + x = 0$
    (iii) $2x^2 + x + 1 = 0$

2. Draw the graph of $3x^2 + 7x - 10$, taking values of x from -4 to $+2$. From your graph answer the following:

(*a*) For what values of x does: (i) $f(x) = -10$; (ii) $f(x) = 0$; (iii) $f(x) = -14$?

(*b*) Through what value of x should a line be drawn parallel to the y-axis to provide an axis of symmetry?

(*c*) Has $f(x)$: (i) a maximum value; (ii) a minimum value? Give the value where applicable.

(*d*) For what values of x is $f(x)$ positive?

(*e*) Use your graph to solve: (i) $3x^2 + 7x - 15 = 0$; (ii) $3x^2 + 7x - 5 = 0$; (iii) $3x^2 + 7x + 5 = 0$.

3. Draw the graph of $4x^2 - 9x$ taking values of x from -2 to $+4$. From your graph answer the following:

(*a*) Through what value of x should a line be drawn parallel to the y-axis to provide an axis of symmetry?

(*b*) Use your graph to solve the following equations:

    (i) $4x^2 - 9x = 0$ (ii) $4x^2 - 9x = 25$
    (iii) $4x^2 - 9x - 15 = 0$

(*c*) After what value does $f(x)$ cease to have any roots?

4. Draw the graph of $y = 9 + 3x - 2x^2$ taking values of x from $-2\frac{1}{2}$ to $+4$. From your graph answer the following:

(*a*) Has $f(x)$ a maximum or minimum value? What is it? What is the corresponding value of x?

(*b*) For what values of x is $f(x)$: (i) positive; (ii) negative?

(*c*) Use your graph to solve:

    (i) $9 + 3x - 2x^2 = 0$ (ii) $9 + 3x - 2x^2 = -11$
    (iii) $2 + 3x - 2x^2 = 0$ (iv) $2x^2 - 3x - 9 = 0$
    (v) $2x^2 - 3x - 14 = 0$ (vi) $2x^2 - 3x - 7 = 0$

5. Draw the graph of $4 - 9x - 3x^2$, taking values of x from -4 to $+1$ using half values of x. From your graph answer the following:

(a) What is the greatest value of $f(x)$?
(b) Has $f(x)$ a minimum value?
(c) Between what values of x is $f(x)$ positive?
(d) Solve the equations:

    (i) $3x^2 + 9x - 4 = 0$ (ii) $3x^2 + 9x + 6\frac{3}{4} = 0$
    (iii) $x^2 + 3x = 0$ (iv) $6\frac{1}{2} - 9x - 3x^2 = 0$

(e) Form the equation whose roots are -4 and $+1$.

(HINT. Draw the graph of $y = ?$ which gives the roots. $\therefore f(x) = ?$ Simplify the equation.)

6. Draw the graph of $y = x^3 - 7x$ using half values of x from -3 to $+3$. Use 1 in per unit for the horizontal scale and 1 in per two units for the vertical scale. The resulting curve is the graph of a **cubic function**, note the shape. From your graph answer the following:

(a) How many 'turning-points' are there? Give the coordinates of each.

(b) Has $f(x)$ a maximum or minimum value?

(c) For what values of x is $f(x)$ positive?

(d) For what values of x is $f(x)$ negative?

(e) How many roots has the equation $x^3 - 7x = 5$? What are the roots?

(f) Use the graph to solve the following:

    (i) $x^3 - 7x = 0$ (ii) $x^3 - 7x + 5 = 0$
    (iii) $x^3 - 7x = 7\cdot5$ (iv) $x^3 - 7x + 7\frac{1}{2} = 0$

VARIATIONS OF QUADRATIC GRAPHS

A. The graph of $y = 2x^2 + x$

Compare fig. A : 19 with fig. A : 18, and note that in the latter the graph crossed the y-axis at $y = -15$. In fig. A : 19 the intercept on the y-axis is zero. Why?

What is the constant term? To solve $2x^2 + x - 15 = 0$ we rearrange the equation to give $2x^2 + x = 15$, and having drawn the graph of $y = 2x^2 + x$, we must now draw the graph of $y = 15$. What are the roots of:

    (i) $2x^2 + x - 15 = 0$
    (ii) $2x^2 + x = 0$
    (iii) $2x^2 + x = 6$
    (iv) $2x^2 + x + 5 = 0$
    (v) $2x^2 + x + 9 = 0$?

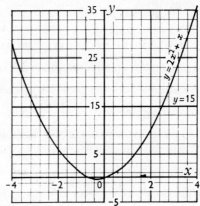

fig. A:19

B. The graph of $y = 2x^2$

The graph will appear as shown in fig. A : 20. Compare with fig. A : 19. The intercept on the y-axis is still zero. Why? Which of the graphs is symmetrical about the y-axis?

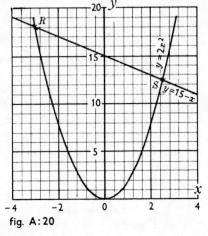
fig. A: 20

In order to solve the equation $2x^2 + x - 15 = 0$ we re-arrange to give $2x^2 = 15 - x$, and having drawn the graph of $y = 2x^2$ we now require the graph of $y = 15 - x$ (see fig. A : 20). Note that $y = 15 - x$ is no longer a line parallel to the x-axis. What is the value of the intercept on the y-axis (of $y = 15 - x$) and why?

The roots of the equation $2x^2 = 15 - x$ (i.e., $2x^2 + x - 15 = 0$) are given by the points of intersection of the two graph lines. What are the values of x at points R and S?

How would you solve the equation:

(i) $2x^2 + x - 3 = 0$ (ii) $2x^2 - x - 15 = 0$
(iii) $2x^2 - x + 15 = 0$?

Draw the graph of $y = 2x^2$ as in fig. A : 20 and use it to solve the equations above.

Has the equation $2x^2 - x + 15 = 0$ any roots?

C. The graph of $y = x^2$

Draw the graph of $y = x^2$ from -4 to $+4$ using half values of x.

How shall we use the graph to solve $2x^2 + x - 15 = 0$? By re-arranging the equation we obtain $2x^2 = 15 - x$ and then $x^2 = \dfrac{15 - x}{2}$. We already have the graph of $y = x^2$ and must therefore proceed to draw the graph of $y = \dfrac{15 - x}{2}$, since this is a **linear** function, only two points are required to plot the graph (three points provide a check). Where the two graphs intersect we can obtain those values of x which satisfy the two functions $y = x^2$ and $y = \dfrac{15 - x}{2}$, and thus satisfy the equation $x^2 = \dfrac{15 - x}{2}$ or $2x^2 + x - 15 = 0$.

What are the roots of the equation $2x^2 + x - 15 = 0$?

Conclusion

We have now examined a variety of methods for solving the equation $2x^2 + x - 15 = 0$.

These methods may be applied to the graphical solution of simultaneous linear/quadratic equations.

EXAMPLE. *Solve the equations* $y = x^2$, $9x - y = 20$

Rearranging the second equation gives $y = 9x - 20$
We draw the graphs of $y = x^2$ and $y = 9x - 20$
Where the graphs intersect we can obtain the values of x and y.

NOTE. Since $y = x^2$ and $y = 9x - 20$
Then $x^2 = 9x - 20$ and $x^2 - 9x + 20 = 0$
Thus the values of x obtained in the simultaneous equations above will give the roots of the quadratic $x^2 - 9x + 20 = 0$
In this case we can check by factors:

$$(x - 5)(x - 4) = 0$$
$$\therefore \ x = 4 \text{ or } 5$$

In the simultaneous equations:

When $x = 4$, $y = 16$; when $x = 5$, $y = 25$

EXERCISE 100

1. Draw the graph of $y = x^2$ with values of x from -7 to $+7$. Use $\frac{1}{2}$ in per unit on the x-axis and $\frac{1}{2}$ in per five units on the y-axis. Use the graph for the following:

(*a*) Find the roots of:

    (i) $x^2 = 15$;   (ii) $x^2 = 23$;   (iii) $x^2 = 34$;   (iv) $x^2 = 45$

(*b*) Solve: (i) $x^2 - 5x = 0$ (ii) $x^2 - 6 \cdot 4x = 0$
                 (iii) $x^2 + 4x = 0$ (iv) $2x^2 + 11x = 0$

(*c*) Solve: (i) $x^2 - x - 6 = 0$ (ii) $x^2 - 3x - 10 = 0$
                 (iii) $x^2 + x - 20 = 0$ (iv) $x^2 + 2x - 25 = 0$

(*d*) What quadratic equations could be solved by superimposing the following graphs on $y = x^2$?

    (i) $y = 3 - 1\frac{1}{2}x$;   (ii) $y = \frac{2}{3} - x$;   (iii) $y = x$

(*e*) Solve graphically the equations referred to in (*d*).

(*f*) Using the graph of $y = x^2$ together with another suitable linear graph solve:

    (i) $2x^2 - 4x - 5 = 0$;   (ii) $\dfrac{x^2 + 3x}{2} = 4\frac{2}{3}$

2. Taking values of x from 0 to 5, solve graphically the simultaneous equations $x^2 = y - 5$, $5x = y - 2$. What quadratic equation is solved by the combined use of these graphs and what are the roots of the quadratic?

3. Using quarter values of x from -1 to $+1\frac{1}{2}$, solve graphically the simultaneous equations $x^2 = \dfrac{2y - x}{2}$, $4y - 5x = 2$. Use these graphs to solve $4x^2 - 3x - 2 = 0$.

4. Draw the graph of $\frac{1}{2}(3 - x)(2 + x)$ for values of x from -2 to $+3$. From your graph:

(a) Solve: (i) $(3 - x)(2 + x) = 4$; (ii) $\frac{1}{4}(3 - x)(2 + x) = \frac{1}{2}$.

(b) For what value of x is $f(x)$ a maximum?

(c) What is the maximum value of $f(x)$?

5. Using half values of x from -2 to $+2$, draw the graphs of $y = \dfrac{6}{x^2 + x + 1}$ and $y = \dfrac{2x}{3}$.

What cubic equation in x can be solved from these two graphs? (Give the equation in its simplest form.) From the graphs find the value of one root of the cubic equation.

6. On the same axes, with values of x from -2 to $+4$, draw the graphs of $y = \dfrac{3x^2}{x + 5}$ and $y = \dfrac{x + 7\frac{1}{2}}{3}$. For what range of values is $\dfrac{3x^2}{x + 5} < 2$? Use your graphs to solve the equation

$$16x^2 - 25x - 75 = 0$$

7. Using half values of x from -2 to $+4$, draw the graphs of $\dfrac{x^3}{x + 3}$ and $\dfrac{4x + 1}{2}$.

(a) For what positive values of x is $\dfrac{4x + 1}{2}$ greater than $\dfrac{x^3}{x + 3}$?

(b) Use your graph to solve the equation $2x^3 = 4x^2 + 13x + 3$.

8. Draw the graph of $y = \dfrac{x^2 - 4}{\bullet x + 4}$ using values of x from $-2\frac{1}{2}$ to $+4$. (a) For what range of values of x is y negative?

(b) Solve: (i) $\dfrac{x^2 - 4}{x + 4} = 1$; (ii) $\dfrac{x^2 - 4}{x + 4} = \dfrac{x}{4}$

9. It is required to make a wooden box of base area 180 in². If the box is x in long, prove that the four sides will require a total length of wood measuring $\dfrac{2x^2 + 360}{x}$ in. (Neglect thickness of wood, etc.)

If the total length of wood is y in, draw a graph showing the relationship between x and y for values of x between 10 and 17. (Begin y-axis at 53 and x-axis at 10.) From your graph find:

(i) The dimensions of the box using a total length of 54 in of wood for the sides.

(ii) (*a*) What is the shortest length of wood from which the sides of the box can be made?

(*b*) What are the dimensions of the box in this case?

10. A lidless box is made from a sheet of metal having squares removed at the four corners and the remaining material bent up to form the four sides and bottom of the box.

If the box (open at the top) is a perfect cube and the sheet of metal from which it is made measures x in by x in, show that:

(i) The volume of the box is $\dfrac{x^3}{27}$ in³.

(ii) The surface area of the metal used in the box is $\dfrac{5x^2}{9}$ in².

Draw the graphs of $y = \dfrac{5x^2}{9}$ and $y = \dfrac{x^3}{27}$ for values of x from 3 to 8, using the same pair of axes for x and y.

Use these graphs to find:

(*a*) The volume and area of metal for a box cut from a sheet:

(i) $4\frac{1}{2}$ in square; and (ii) $7\frac{1}{2}$ in square.

(*b*) The area of metal for a box of volume $6\frac{1}{4}$ in³.

(*c*) The volume of a box with an area of metal 23·5 in².

GENERAL SOLUTION OF SIMULTANEOUS LINEAR/QUADRATIC EQUATIONS

EXAMPLE. *Solve the simultaneous equations:*

$$x + y = 3 \quad\text{———————}\quad ①$$
$$x^2 - 2xy + y^2 = 1 \quad\text{———————}\quad ②$$

From ① $x = 3 - y$, substitute for x in ②:

$$(3 - y)^2 - 2y(3 - y) + y^2 = 1$$

Clear brackets:

$$9 - 6y + y^2 - 6y + 2y^2 + y^2 = 1$$

Simplify: $\qquad\qquad 4y^2 - 12y + 8 = 0$

Divide both sides by 4: $\quad y^2 - 3y + 2 = 0$

Factorize: $\qquad\qquad (y - 2)(y - 1) = 0$

$$\therefore\ y = 1 \text{ or } 2$$

Substitute for each value of y in linear equation: $x = 3 - y$

When	$y = 1$	When	$y = 2$
	$x = 3 - 1$		$x = 3 - 2$
$\therefore$	$x = 2$	$\therefore$	$x = 1$

Ans. $x = 2, y = 1$ or $x = 1, y = 2$

NOTE.

(i) Before obtaining an expression for x or y and then substituting, careful thought should be given to the possible work involved.

(*a*) If the expression for x or y is in fraction form the smaller denominator would usually be preferred.

(*b*) The unknown requiring least substitution would usually be selected for elimination.

Consider the equations carefully before you begin work on them.

(ii) The solution of the quadratic may require the use of the general formula.

EXERCISE 101

Solve the following simultaneous equations:

1. $x = 4y$
   $4xy = 1$

2. $y = 4x$
   $xy = 1$

3. $x + 9y = 0$
   $xy + 1 = 0$

4. $3x + 4y = 0$
   $3xy + 1 = 0$

5. $2y - x = 4$
   $xy = 6$

6. $2x - 3y = 4$
   $xy = 2$

7. $2x - y = 0$
   $x^2 + xy = 3$

8. $x - 8y = 0$
   $x^2 - 2xy = 3$

9. $2x + 3y = 0$
   $x^2 - 3xy = 3$

10. $x + y = 0$
    $2y^2 - 3x = 2$

11. $x = y$
    $x^2 + y^2 = 8$

12. $x = 2y$
    $x^2 + 4y^2 = 2$

13. $2x = y$
    $5x^2 - y^2 = 9$

14. $x = 3y$
    $x^2 + 9y^2 = 2$

15. $x + 3y = 0$
    $x^2 - y^2 = 2$

16. $2x + y = 0$
    $x^2 - y^2 + 3 = 0$

17. $2x + 3y = 1$
    $16x^2 = 9y^2$

18. $3x - y = 0$
    $x^2 + \frac{1}{2} = y^2$

19. $2x - y = 2$
    $4x^2 - 2xy - y^2 = 1$

20. $4x - 3y = 4$
    $4x^2 + xy - 9y^2 = 6$

21. $y - x = 1$
    $x^2 + xy - y^2 = 1$

22. $x - 2y = 2$
    $x^2 - 3xy - 6y^2 = 3$

23. $2x - 3y = 1$
    $4x^2 + xy - 9y^2 = 6$

24. $x - 2y = 4$
    $3x^2 + xy - y^2 = 9$

25. $2x + 3y = 7$
    $x^2 - 3xy + 3y^2 = 1$

26. $x + 2y = 0$
    $x^2 + 2xy + 2y^2 = 3\frac{1}{8}$

27. $2x + 3y = 5$
    $4x^2 + xy - 9y^2 = 6$

28. $3y - 2x = 3$.
    $y(2x - y) = 2$

29. $2(x + y) = 1$
    $2x(x - y) = 3$

30. $x - y = 1$
    $(x + y)^2 = 2$

31. $x + y = 1$
    $(x - 2y)(2x - y) = 20$

32. $2x - y = x(x - y) - y(y - x) = 3$

33. $2x - 3y = 1$
    $\dfrac{2x}{y} - \dfrac{2y}{x} = 3$

34. $\dfrac{x + y}{3} - \dfrac{2x - y}{2} = \dfrac{1}{4}$
    $\dfrac{x + y}{x} - \dfrac{x - y}{y} = \dfrac{1}{2}$

FACTORS AND REMAINDERS

The Factor Theorem states that if an expression $f(x)$ is equal to zero when $x = a$, then $(x - a)$ is a factor of the expression.

DISCUSSION. Is $(x + 3)$ a factor of $2x^3 - 9x^2 + 13x - 12$?

When $x = -3$; Expression $= -54 - 81 - 39 - 12 = -186$

It would seem that $(x + 3)$ is not a factor.

Checking by actual division:

$$
\begin{array}{r}
2x^2 - 15x + 58 \\
x + 3{\overline{\smash{\big)}\,2x^3 - 9x^2 + 13x - 12}} \\
\underline{2x^3 + 6x^2\phantom{{}+ 13x - 12}} \\
-15x^2 + 13x\phantom{{}- 12} \\
\underline{-15x^2 - 45x\phantom{{}- 12}} \\
58x - 12 \\
\underline{58x + 174} \\
-186 \ \text{(R.)}
\end{array}
$$

Here we see that the remainder -186 (R.) agrees with the value of the expression when substituting $x = -3$. The existence of a remainder proves that $(x + 3)$ is not a factor.

CONCLUSION

(i) If $(x - a)$ is a factor of an expression, then the expression is equal to zero when substituting $x = a$.

(ii) If $(x - a)$ is not a factor of an expression, then the remainder is given by substituting $x = a$. **For this reason the theorem is sometimes called the Remainder Theorem.**

EXERCISE 102

1. Which of the following are factors of $x^3 + 2x^2 - 5x - 6$?
(i) $x - 1$; (ii) $x + 1$; (iii) $x - 2$; (iv) $x + 2$; (v) $x - 3$; (vi) $x + 3$

2. Find the remainder (R.) when $x^3 - 6x^2 + 11x - 6$ is divided by: (i) $x - 1$; (ii) $x + 1$; (iii) $x - 2$; (iv) $x + 2$; (v) $x - 3$; (vi) $x + 3$

3. Which of the following are factors of $x^3 - x^2 - 4x + 4$?
(i) $x - 1$; (ii) $x + 1$; (iii) $x - 2$; (iv) $x + 2$; (v) $x - 3$

4. Which of the following are factors of $x^3 + 7x^2 + 15x + 9$?
(i) $x - 1$; (ii) $x + 1$; (iii) $x - 2$; (iv) $x - 3$; (v) $x + 3$

5. Which of the following are factors of $x^3 - 2x^2 - 5x + 6$?
(i) $x - 1$; (ii) $x - 2$; (iii) $x + 2$; (iv) $x - 3$; (v) $x + 3$

6. Find the remainder when $x^3 + 3x^2 - 4$ is divided by: (i) $x + 1$; (ii) $x - 2$; (iii) $x + 2$

7. Find the remainder when $x^3 - 4x^2 - 3x + 18$ is divided by: (i) $x - 2$; (ii) $x + 2$; (iii) $x + 3$

8. Find the remainder when $x^3 + 5x^2 + 3x - 9$ is divided by: (i) $x - 1$; (ii) $x - 2$; (iii) $x + 3$

FURTHER USE OF THE FACTOR THEOREM*

* or Remainder Theorem.

EXAMPLE 1. *Factorize completely* $2x^3 - x^2 - 13x - 6$

   (i) If $x = -1$,   Expression $= -2 - 1 + 13 - 6 = 4$
         ∴ $(x + 1)$ is not a factor.

  (ii) If $x = 1$,   Expression $= 2 - 1 - 13 - 6 = -18$
         ∴ $(x - 1)$ is not a factor.

 (iii) If $x = -2$,   Expression $= -16 - 4 + 26 - 6 = 0$
         ∴ $(x + 2)$ is a factor.

$$\begin{array}{r} 2x^2 - 5x - 3 \\ x + 2 \overline{)2x^3 - x^2 - 13x - 6} \end{array}$$

Another factor may be found by division:

∴ Expression $= (x + 2)(2x^2 - 5x - 3)$

    Ans. $= (x + 2)(2x + 1)(x - 3)$

NOTE

It would have been unnecessary to substitute $x = \pm 4$, $x = \pm 5$, etc. Why?

EXAMPLE 2. *For what values of x does* $2x^3 + 3x^2 - 2x - 3$ *disappear?*
The words 'disappear' or 'vanish' are sometimes used, and they mean that the expression is equal to zero.

Thus we have to solve the equation:

$$2x^3 + 3x^2 - 2x - 3 = 0$$

By trial and division, etc., the factors are found to be as follows:

$$(x - 1)(x + 1)(2x + 3) = 0$$
$$∴ \ x = 1 \text{ or } -1 \text{ or } -1\tfrac{1}{2} \ Ans.$$

EXERCISE 103

Factorize the following completely:

1. $x^3 - x^2 + 2x - 2$ **2.** $x^3 + x^2 - 3x - 3$
3. $x^3 - x^2 - x + 1$ **4.** $x^3 - 3x^2 + 3x - 1$
5. $x^3 + 2x^2 + x$ **6.** $x^3 - 3x + 2$
7. $x^3 + 6x^2 + 12x + 8$ **8.** $x^3 - 6x^2 + 12x - 8$
9. $x^3 - 3x^2 - 16x + 48$ **10.** $x^3 - 12x^2 + 48x - 64$
11. $x^3 - 27$ **12.** $x^4 - 10x^3 + 35x^2 - 50x + 24$

For what values of x do the following expressions vanish?

13. $x^3 + x^2 - x - 1$ **14.** $x^3 - 3x^2 + 3x - 1$
15. $x^3 - 6x^2 + 11x - 6$ **16.** $x^3 + 6x^2 + 11x + 6$
17. $2x^3 - 9x^2 + 13x - 6$ **18.** $4x^3 + 4x^2 - x - 1$
19. $4x^3 - 16x^2 + 19x - 6$ **20.** $9x^3 + 27x^2 + 20x + 4$

EXAMPLE 3. *For what values of a and b is $ax^3 + bx^2 - 2x - 3$ divisible both by $(x - 1)$ and $(x + 1)$? Factorize the expression completely.*

If $(x - 1)$ and $(x + 1)$ are factors, then the expression is equal to zero when substituting for $x = 1$ or $x = -1$.

When $x = 1$,	$a + b - 2 - 3 = 0$	①
When $x = -1$,	$-a + b + 2 - 3 = 0$	②
From ①:	$a + b = 5$	③
From ②:	$b - a = 1$	④
Subtract ④ from ③:	$2a = 4$ $a = 2$	
Substitute for a in ④:	$b = 3$	

$$\text{Ans. } a = 2, b = 3$$

The expression now reads $2x^3 + 3x^2 - 2x - 3$

Since two factors are known: Expression $= (x - 1)(x + 1)(\quad ? \quad)$

Since the product of the first terms must give $2x^3$ and the product of the last terms must give -3: the third factor must be $(2x + 3)$

$$\therefore \text{ Expression} = (x - 1)(x + 1)(2x + 3) \text{ Ans.}$$

NOTE. Alternatively, the third factor may be found by division.

EXERCISE 104

1. For what value of k is $(x + 4)$ a factor of $x^3 + 6x^2 + 9x + k$? Factorize completely.

2. If $(2x + 3)$ is a factor of $2x^3 + bx^2 - 5x + 6$, find b and factorize completely.

3. Find the value of k which makes $3x^3 - x^2 - 3x - k$ divisible by $(3x - 1)$ and factorize completely.

4. Find the value of c which makes $4x^3 - x + cx^2 + 3$ divisible by $(2x - 1)$ and factorize completely.

5. Find the value of a which makes $ax^3 - 11x^2 + 6x - 1$ divisible by $(3x - 1)$. For what values of x does the expression vanish?

6. Find the value of b which makes $24x^3 - bx^2 + 9x + 1$ divisible by $(2x + 1)$. For what values of x does the expression disappear?

7. If $x^3 - bx^2 + cx - 6$ is divisible both by $(x - 1)$ and $(x - 2)$, find the values of b and c and the third factor.

8. If $(2x - 1)$ and $(x + 2)$ are both factors of $4x^3 + 8x^2 + cx + k$, find the values of c and k and the third factor.

9. If $ax^3 + 2x^2 + cx + 6$ is divisible both by $(x + 3)$ and $(x - 2)$, find the values of a and c and the third factor.

10. For what values of b and k are $(x - 1)$ and $(2x + 3)$ both factors of $4x^3 - bx^2 - x - k$? Factorize the expression completely.

11. For what values of a and b are $(2x + 1)$ and $(4x + 1)$ both factors of $ax^3 + bx^2 + 9x + 1$? Factorize the expression completely.

SQUARE ROOT OF A POLYNOMIAL

The method for finding the square root of a polynomial is similar to that used in arithmetic.

EXAMPLE. *Find the square root of* $4x^4 - 12x^3 + 17x^2 - 12x + 4$

$$
\begin{array}{r}
2x^2 - 3x + 2 \\
\hline
)4x^4 - 12x^3 + 17x^2 - 12x + 4 \\
4x^4
\end{array}
$$

$4x^2 - 3x$ $\qquad - 12x^3 + 17x^2$
$\qquad\qquad\qquad - 12x^3 + 9x^2$

$4x^2 - 6x + 2$ $\qquad\qquad 8x^2 - 12x + 4$
$\qquad\qquad\qquad\qquad 8x^2 - 12x + 4$

Ans. $\underline{\underline{2x^2 - 3x + 2}}$

STAGES

1. Square root of 1st term $(4x^4)$ gives 1st term in answer $(2x^2)$.

2. Subtract square of answer $(4x^4)$ from 1st term in expression and bring down next *two* terms, i.e., $(-12x^3 + 17x^2)$.

3. Double answer so far and place at L.H.S. $(4x^2)$.

4. Divide 1st term at L.H.S. into 1st term of new expression $(-12x^3 + 17x^2)$ and place quotient in answer $(-3x)$ and also as 2nd term at L.H.S. $(4x^2 - 3x)$.

5. Multiply expression at L.H.S. $(4x^2 - 3x)$ by 2nd term in answer $(-3x)$ giving $(-12x^3 + 9x^2)$.

6. Subtract and bring down next two terms $(8x^2 - 12x + 4)$.

7. Double answer so far and place at L.H.S. $(4x^2 - 6x)$.

8. Repeat from 4.

EXERCISE 105

Find the square roots of the following:

1. $16x^2 + 40x + 25$　　　　**2.** $64x^2 - 112x + 49$
3. $4x^4 - 4x^2 + 1$　　　　　**4.** $x^6 - 4x^3 + 4$
5. $4x^4 - 12x^2 + 9$　　　　**6.** $9x^4 + 12x^2 + 4$
7. $4x^6 - 4x^3 + 1$　　　　　**8.** $x^4 + 2x^3 + 3x^2 + 2x + 1$
9. $x^4 - 2x^3 + 3x^2 - 2x + 1$　　**10.** $x^4 + 4x^3 + 2x^2 - 4x + 1$
11. $x^4 - 4x^3 + 8x^2 - 8x + 4$　　**12.** $4x^4 + 4x^3 - 7x^2 - 4x + 4$
13. $4x^4 - 12x^3 + 5x^2 + 6x + 1$　**14.** $x^6 - 2x^4 + 2x^3 + x^2 - 2x + 1$
15. $x^6 - 4x^5 + 4x^4 - 2x^3 + 4x^2 + 1$
16. $4x^6 + 4x^5 + 13x^4 + 6x^3 + 9x^2$
17. $16x^4 - 40x^3 + 49x^2 - 30x + 9$
18. $9x^6 - 12x^5 + 4x^4 - 24x^3 + 16x^2 + 16$
19. $4x^4 - 4x^3y - 3x^2y^2 + 2xy^3 + y^4$
20. $9x^4 - 12x^3y + 22x^2y^2 - 12xy^3 + 9y^4$

INDICES

From earlier work on logarithms you should remember the following:

(i) $a^3 \times a^2 = a^5$ (ii) $a^4 \div a^2 = a^2$ (iii) $(a^3)^2 = a^6$

(iv) $\sqrt[3]{a^6} = a^2$ (v) $a^0 = 1$ (vi) $a^{\frac{1}{2}} = \sqrt{a}$

(vii) $a^{\frac{3}{4}} = \sqrt[4]{a^3}$ or $(\sqrt[4]{a})^3$ (viii) $a^{-3} = \dfrac{1}{a^3}$

(ix) $a^{-1\frac{1}{2}} = \dfrac{1}{a^{1\frac{1}{2}}} = \dfrac{1}{a^{\frac{3}{2}}} = \dfrac{1}{\sqrt{a^3}}$ or $\dfrac{1}{(\sqrt{a})^3}$

These results enable us to deal with problems on indices.

EXAMPLE 1. *Express with positive indices:*

(i) x^{-2} (ii) x^{-5} (iii) $x^{-\frac{1}{2}}$ (iv) $x^{-2\frac{1}{2}}$

(i) $x^{-2} = \dfrac{1}{x^2}$ *Ans.* (ii) $x^{-5} = \dfrac{1}{x^5}$ *Ans.* (iii) $x^{-\frac{1}{2}} = \dfrac{1}{x^{\frac{1}{2}}} = \dfrac{1}{\sqrt{x}}$ *Ans.*

(iv) $x^{-2\frac{1}{2}} = \dfrac{1}{x^{2\frac{1}{2}}} = \dfrac{1}{x^{\frac{5}{2}}} = \dfrac{1}{\sqrt{x^5}}$ or $\dfrac{1}{(\sqrt{x})^5}$ *Ans.*

EXAMPLE 2. *Evaluate without using tables:*

(i) $8^{\frac{2}{3}}$ (ii) $16^{\frac{3}{4}}$ (iii) $9^{-\frac{1}{2}}$ (iv) $(2\frac{1}{4})^{-1\frac{1}{2}}$

(i) $8^{\frac{2}{3}} = (\sqrt[3]{8})^2 = 4$ *Ans.* (ii) $16^{\frac{3}{4}} = (\sqrt[4]{16})^3 = 8$ *Ans.*

(iii) $9^{-\frac{1}{2}} = \dfrac{1}{9^{\frac{1}{2}}} = \dfrac{1}{\sqrt{9}} = \dfrac{1}{3}$ *Ans.*

(iv) $(2\frac{1}{4})^{-1\frac{1}{2}} = \dfrac{1}{(2\frac{1}{4})^{\frac{3}{2}}} = \dfrac{1}{(\sqrt{\frac{9}{4}})^3} = \dfrac{1}{(\frac{3}{2})^3} = \dfrac{1}{\frac{27}{8}} = \dfrac{1}{1} \times \dfrac{8}{27} = \dfrac{8}{27}$ *Ans.*

EXERCISE 106

Simplify the following without using tables:

1. $4^{\frac{1}{2}}$	2. $27^{\frac{1}{3}}$	3. $16^{\frac{1}{4}}$	4. $9^{\frac{1}{2}}$
5. $64^{\frac{1}{6}}$	6. 8^0	7. $49^{\frac{1}{2}}$	8. $64^{\frac{2}{3}}$
9. $16^{\frac{3}{4}}$	10. $81^{\frac{3}{4}}$	11. $9^{\frac{3}{2}}$	12. $16^{1\frac{1}{4}}$
13. $4^{2\frac{1}{2}}$	14. $32^{\frac{3}{5}}$	15. $25^{1\frac{1}{2}}$	16. 27^0
17. 3^{-2}	18. 4^{-1}	19. $(\frac{1}{2})^{-3}$	20. 8^{-2}
21. 5^{-2}	22. $(\frac{1}{2})^{-5}$	23. $4^{-\frac{1}{2}}$	24. $(\frac{1}{10})^{-1}$
25. 1^0	26. 1^{-1}	27. $1^{\frac{1}{2}}$	28. $1^{-\frac{1}{3}}$
29. $8^{-\frac{2}{3}}$	30. $100^{-1\frac{1}{2}}$	31. $(\frac{1}{16})^{-\frac{3}{4}}$	32. $27^{-\frac{2}{3}}$
33. $32^{-\frac{3}{5}}$	34. $16^{-1\frac{1}{2}}$	35. $25^{-\frac{1}{2}}$	36. $(\frac{1}{2})^3$
37. $(\frac{1}{4})^{-1\frac{1}{2}}$	38. $(\frac{2}{3})^{-2}$	39. $(6\frac{1}{4})^{-1\frac{1}{2}}$	40. $(\frac{8}{27})^{-1\frac{1}{3}}$
41. $x^{\frac{1}{2}} \times x^{\frac{1}{2}}$	42. $x^{\frac{1}{4}} \times x^{-\frac{1}{4}}$	43. $x^2 \div x^3$	44. $x^2 \div x^{-2}$
45. $x^{-\frac{1}{2}} \times x^{-\frac{1}{2}}$	46. $x^{1\frac{1}{2}} \div x^{-\frac{1}{2}}$	47. $(x^2)^{-3}$	48. $(x^4)^{-\frac{1}{2}}$
49. $(8x^6)^{-\frac{2}{3}}$	50. $(9x^4)^{-1\frac{1}{2}}$		

THEORY OF LOGARITHMS

EXAMPLE. *If* $log_{10} 2 = 0\cdot3010$ *and* $log_{10} 3 = 0\cdot4771$, *find the value of:*
(i) $log_{10} 6$; (ii) $log_{10} 9 - log_{10} 8$

(i) Since $6 = 3 \times 2$, $\log 6 = \log 3 + \log 2$
$$= 0\cdot4771 + 0\cdot3010$$
Ans. $= \underline{\underline{0\cdot7781}}$

(ii) Since $9 = 3^2$ and $8 = 2^3$, $\log 9 - \log 8 = \log 3^2 - \log 2^3$
$$2 \log 3 - 3 \log 2 = 2(0\cdot4771) - 3(0\cdot3010)$$
$$= 0\cdot9542 - 0\cdot9030$$
Ans. $= \underline{\underline{0\cdot0512}}$

EXERCISE 107

Simplify without using tables:

1. $\dfrac{\log 9}{\log 3}$ 2. $\dfrac{\log 4}{\log 8}$ 3. $\dfrac{\log 125}{\log 25}$ 4. $\dfrac{\log \sqrt{3}}{\log 3}$

5. $\dfrac{\log 4}{\log \sqrt{2}}$ 6. $\dfrac{\log 9}{\log \sqrt[3]{27}}$ 7. $\dfrac{\log \sqrt{2}}{\log \sqrt[3]{8}}$ 8. $\dfrac{\log 3}{\log \frac{1}{9}}$

9. $\dfrac{\log \frac{1}{25}}{\log 5}$ 10. $\dfrac{\log \sqrt[4]{9}}{\log \sqrt{3}}$

Express the following as simple logarithms:

11. $2 \log 3$ 12. $3 \log 2$ 13. $2 \log 4$ 14. $4 \log 3$
15. $\frac{1}{2} \log 4$ 16. $\frac{1}{3} \log 8$ 17. $\frac{2}{3} \log 27$ 18. $\frac{4}{3} \log 125$
19. $-2 \log 3$ 20. $-\frac{3}{5} \log 32$ 21. $\log 3 + \log 2$ 22. $\log 2 + \log 5$
23. $\log 3 + \log 5$ 24. $\log 8 - \log 2$ 25. $\log 5 - \log 4$
26. $3 \log 3 - 3 \log 2$ 27. $2 \log 5 - 2 \log 2$ 28. $\frac{1}{2} \log 36 + 2 \log 3$
29. $\frac{1}{3} \log 27 + \frac{1}{4} \log 16$ 30. $\frac{4}{3} \log 64 - \frac{3}{2} \log 36$

If $\log 2 = 0\cdot3010$, $\log 3 = 0\cdot4771$ and $\log 5 = 0\cdot6990$, evaluate the following:

31. $2 \log 2$ 32. $4 \log 3$ 33. $3 \log 5$
34. $\log 3 + \log 5$ 35. $\log 3 + \log 10$ 36. $\log 2 + \log 5$
37. $\log 5 + \log 10$ 38. $\log 15$ 39. $\log 27$
40. $\log 16$ 41. $\log 25$ 42. $\log 10$
43. $\log 20$ 44. $\log 30$ 45. $\log 50$
46. $\log 60$ 47. $2 \log 2 + \log 3$ 48. $\log 2 + 2 \log 3$
49. $2 \log 2 + 2 \log 3$ 50. $\log 25 + \log 2$ 51. $2 \log 5 + 2 \log 2$
52. $\log 25 + \log 4$ 53. $\log 100$ 54. $\log 36 - \log 12$
55. $\log 36 - \log 18$ 56. $\log 36 - \log 9$ 57. $\log 90 - \log 9$
58. $\log 15 - \log 3$ 59. $\log 12 + \log 5$ 60. $\log 25 - \log 9$
61. $\log 125 - \log 81$ 62. $\log 48$ 63. $\log 54$
64. $\log 80$ 65. $\log 72$

QUADRATIC EQUATIONS: SUM AND PRODUCT
OF THE ROOTS

If the general quadratic equation $ax^2 + bx + c = 0$ is divided throughout by a, the coefficient of x^2 is reduced to unity. Then $x^2 + \dfrac{b}{a}x + \dfrac{c}{a} = 0.$

Any numerical example will illustrate that

(i) **The sum of the roots** $= -\dfrac{b}{a}$

(ii) **The product of the roots** $= \dfrac{c}{a}$

EXAMPLE. *If α and β are the roots of the equation $2x^2 - 6x + 4 = 0$, find the values of: (i) $\dfrac{1}{\alpha} + \dfrac{1}{\beta}$; (ii) $\alpha^2 + \beta^2$; (iii) $\dfrac{\alpha}{\beta} + \dfrac{\beta}{\alpha}$*

$$\alpha + \beta = \tfrac{6}{2} = 3; \quad \alpha\beta = \tfrac{4}{2} = 2;$$

(i) $\dfrac{1}{\alpha} + \dfrac{1}{\beta} = \dfrac{\beta + \alpha}{\alpha\beta} = \dfrac{3}{2} = 1\tfrac{1}{2}$ *Ans.*

(ii) $(\alpha + \beta)^2 = \alpha^2 + 2\alpha\beta + \beta^2$

$\therefore \alpha^2 + \beta^2 = (\alpha + \beta)^2 - 2\alpha\beta = 3^2 - (2.2) = 9 - 4 = 5$ *Ans.*

(iii) $\dfrac{\alpha}{\beta} + \dfrac{\beta}{\alpha} = \dfrac{\alpha^2 + \beta^2}{\alpha\beta} = \dfrac{5}{2} = 2\tfrac{1}{2}$ *Ans.*

EXERCISE 108

State the sum and product of the roots of the following:

1. $x^2 - 3x - 4 = 0$ 2. $x^2 + 6x - 7 = 0$ 3. $x^2 - 9x + 8 = 0$
4. $x^2 - 2x - 3 = 0$ 5. $x^2 - 16 = 0$ 6. $x^2 + 4x - 5 = 0$
7. $x^2 + 8x + 15 = 0$ 8. $x^2 - 8x + 12 = 0$ 9. $x^2 + x - 42 = 0$
10. $2x^2 + 3x - 2 = 0$ 11. $2x^2 + 3x - 9 = 0$ 12. $2x^2 - 5x - 12 = 0$
13. $3x^2 + 5x - 2 = 0$ 14. $3x^2 + 11x + 6 = 0$ 15. $3x^2 + 8x - 16 = 0$
16. $4x^2 - 1 = 0$ 17. $6x^2 + 13x + 6 = 0$ 18. $8x^2 + 18x - 5 = 0$
19. $5x^2 - 34x + 24 = 0$ 20. $9x^2 - 3x - 20 = 0$ 21. $x^2 - 8x + 16 = 0$

In the following equations, given one root find the other; hence state the value of b or c:

22. $x^2 + bx - 8 = 0$; 4   23. $x^2 + bx + 15 = 0$; -3
24. $x^2 + bx + 20 = 0$; 5   25. $2x^2 + bx + 6 = 0$; $1\tfrac{1}{2}$
26. $2x^2 + bx + 10 = 0$; -2 27. $6x^2 + bx - 6 = 0$; $\tfrac{2}{3}$
28. $x^2 - x + c = 0$; 3   29. $2x^2 + x + c = 0$; -2
30. $3x^2 + 7x + c = 0$; -4 31. $8x^2 + 14x + c = 0$; $\tfrac{3}{4}$

32. If α and β are the roots of $x^2 - 5x + 6 = 0$, find the values of:
(i) $\alpha^2\beta^2$; (ii) $(\alpha + \beta)^2$; (iii) $\alpha^2 + \beta^2$.

33. If α and β are the roots of $2x^2 - 7x = -6$, find the values of:
(i) $\dfrac{1}{\alpha + \beta}$; (ii) $\dfrac{1}{\alpha\beta}$; (iii) $\dfrac{1}{\alpha} + \dfrac{1}{\beta}$.

34. If α and β are the roots of $x^2 + 4x = 5$, find the values of:
(i) $\alpha^2 + \beta^2$; (ii) $\dfrac{1}{\alpha^2\beta^2}$; (iii) $\dfrac{1}{\alpha^2} + \dfrac{1}{\beta^2}$.

35. If α and β are the roots of $x^2 + x = 12$, find the values of:
(i) $\dfrac{\alpha + \beta}{2}$; (ii) $\dfrac{2\alpha\beta}{\alpha + \beta}$; (iii) $\dfrac{\alpha\beta}{2\alpha + 2\beta}$.

36. If α and β are the roots of $2x^2 + x = 10$, find the values of:
(i) $\dfrac{\alpha\beta}{\alpha + \beta}$; (ii) $\dfrac{\alpha}{\beta} + \dfrac{\beta}{\alpha}$; (iii) $\dfrac{1}{\alpha} + \dfrac{1}{\beta}$.

37. If α and β are the roots of $2x^2 - 13x + 20 = 0$, find the value of $(\alpha - \beta)^2$.

38. If α and β are the roots of $3x^2 + 16x = 12$, find the values of:
(i) $(\alpha + 1)(\beta + 1)$; (ii) $(\alpha - 1)(\beta - 1)$.

39. If α and β are the roots of $x^2 - 4x + 4 = 0$, find the value of $\dfrac{\alpha}{\beta + 1} + \dfrac{\beta}{\alpha + 1}$.

40. If α and β are the roots of $x^2 - 3x + 2 = 0$, find the equation whose roots are $\left(\dfrac{1}{\alpha} + \dfrac{1}{\beta}\right)$ and $(\alpha^2 + \beta^2)$.

VARIATION

Two variables, x and y, may be associated in such a way as to obey one of several laws.

    (i) If $y \propto x$, then $y = kx$ (Where k is a constant.)

    (ii) If $y \propto x^2$, then $y = kx^2$ (or $y = kx^n$)

    (iii) If $y \propto \dfrac{1}{x}$, then $y = \dfrac{k}{x}$ $\left(\text{i.e. } y = k\dfrac{1}{x}\right)$

    (iv) If $y \propto \dfrac{1}{x^2}$, then $y = \dfrac{k}{x^2}$ $\left(\text{or } y = \dfrac{k}{x^n}\right)$

    (i) and (ii) show **direct** variation, (iii) and (iv) show **inverse** variation, as does $y = \dfrac{k}{\sqrt{x}}$.

From (i), the value of k may be obtained from the **gradient** of the graph, i.e. $\tan \theta$.

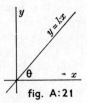

fig. A:21

Reason: $y = kx$ $\therefore$ $k = \dfrac{y}{x} = \tan \theta$

From (ii), the value of k may be obtained if y is plotted against x^2 instead of against x.

Reason: $y = kx^2$ $\therefore$ $k = \dfrac{y}{x^2} = \tan \theta$ [fig. A : 22 (ii)]

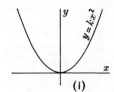

fig. A : 22

(i)   (ii)

QUESTIONS

(i) If $y = \dfrac{k}{x} \left(\text{i.e. } y = k\dfrac{1}{x}\right)$, what must y be plotted against to find k?

(ii) If $y = \dfrac{k}{x^n}$, what must y be plotted against to find k?

NOTE. If a **constant** term is present in the original function its value will be indicated by the intercept of the linear graph on the y-axis.

EXERCISE 109

1. $P(0, 3)$ and $Q(3, 7)$ are two points of the form (x, y) on the graph of $y = mx + c$. Sketch the graph to show these points and use it to find m and c; hence state $f(x)$. Find y when $x = 6$, and x when $y = 8$.

2. In the equation $y = ax + b$ it is found that $y = 20$ when $x = 3$ and $y = 32$ when $x = 6$. Find a, b and state $f(x)$; hence find the value of y when $x = 4$, and x when $y = 45$.

3. If $(3, 2)$; $(4, 5)$ are two points on the graph of $y = kx + c$ find k, c and state $f(x)$; hence find y when $x = 1$, and x when $y = 20$.

4. The velocity v (ft/s) of a moving particle is given by $v = u + at$, where u is the initial velocity (ft/s), a is the acceleration (ft/s²) and t the time (sec). If $v = 60$ when $t = 2$, and $v = 180$ when $t = 6$, find u, a and state $f(t)$; hence find v when $t = 3\frac{1}{2}$, and t when $v = 90$.

5. It is assumed that $°$ C and $°$ F are connected by a formula of the type $C = mF + k$. If $C = 0$ when $F = 32$, and $C = 100$ when $F = 212$ express: (i) C as a function of F; (ii) F as a function of C. For what value are C and F equal?

6. Using a lifting machine, it is found that the effort E (in lb) and the load L (in cwt) are related by the law $E = aL + b$. An effort of 30 lb will raise a load of 4 cwt, and a load of 15 cwt requires an effort of 74 lb. Find the values of a, b and complete the formula.

7. The following table gives the values of two variables x and y, plot the values and draw the "line of best fit". Assuming a relationship $y = mx + c$, use your graph to find probable values for m and c.

x	1	2	3	4	5	6
y	12·75	16	19·25	21·8	25	28·2

8. Plot the graph of the variables x, y and from your graph verify that they obey the law $y = ax + b$. Use your graph to find a and b.

x	$1\frac{1}{2}$	3	$4\frac{1}{2}$	6	9
y	6	7	8	9	11

9. Two variables v and t are related by the formula $v = u + at$. Plot a graph from the following table and use the graph to find u and a.

t	-2	1	4	8	10
v	$-5\frac{1}{2}$	$-3\frac{1}{4}$	-1	2	$3\frac{1}{2}$

10. Plot the graph of P against F from the table. From your graph verify that $P = aF + b$ and find a, b.

F	-1	$-0·5$	1·5	3
P	3·9	3·7	2·9	2·3

11. If $y = \dfrac{k}{x}$, use the given table of values to calculate values for $\dfrac{1}{x}$ and include this line of results in your own table. Plot y against $\dfrac{1}{x}$ using your graph to obtain a value for k. From the table verify that $k = xy$.

x	1	2	3	4	5
y	5	$2\frac{1}{2}$	$1\frac{2}{3}$	$1\frac{1}{4}$	1

12. If $y = \dfrac{k}{x} + c$, provide an additional line to the given table to enable you to plot a straight-line graph and thus obtain the values of k and c.

x	1	2	3	4	5
y	7	5	4·3	4	3·8

13. In Questions 11 and 12 the relationship between x and y is one of **inverse variation.** Using a separate pair of axes for each, plot the graphs of y against x in each case. Note the shape of the curves, each is a **rectangular hyperbola.** No matter how far the axes are extended, they will never touch the curves, though axes and curves approach closer and closer together. In such cases the axes are called **asymptotes.**

14. Use the following table to plot y against x. What type of curve results? What do you conclude about the relationship between x and y?

x	1	2	3	4	6	8
y	72	18	8	$4\frac{1}{2}$	2	$1\frac{1}{8}$

Provide an additional line of values to enable you to construct a graph from which may be obtained the value of k, given that $y = \dfrac{k}{x^2}$.

15. Given the following table and the fact that $y = \dfrac{k}{x^2} + c$, find, graphically, the values of k and c.

x	1	2	3	4	5	6
y	42	15	10	$8\frac{1}{4}$	$7\frac{11}{25}$	7

16. Express the following each as an algebraic equation:

(a) y varies directly as x.   (b) y varies inversely as x.
(c) y varies jointly as x and z.
(d) y varies directly as x and inversely as z.
(e) y is partly constant and partly varies as x.
(f) y varies partly as x^2 and partly as x.
(g) y varies partly as x^2, partly inversely as x and is partly constant.
(h) y varies partly as x^3, partly as x^2 and partly inversely as x.

17. The area A of a figure varies jointly as π and the square of the radius r. Express this as an equation. What is the figure?

18. The electrical resistance R of a wire varies directly as the length l and inversely as the square of the diameter d. Two wires are such that one is twice the length of the other, but the shorter one is twice the thickness of the other. Find the ratio of their resistances in the form $m : 1$.

19. The magnification m of a lens is found to vary jointly as v, the distance of the lens from the image, and the reciprocal of f, the focal length of the lens; the variation of m is also partly constant. Produce an equation to show the relationship.

v	9	15	18	30	42
m	0·5	1·5	2	4	6

Plot the graph of m against v and use it to find: (i) the value of f for the lens used in this set of data; (ii) the value of the constant term; (iii) the value of m when $v = 16$; (iv) the value of v when $m = 5$.

20. The wind pressure to which a vertical wall is subjected varies jointly with the area of the wall and the square of the wind's velocity. If the velocity of the wind is likely to increase by 20%, what percentage change must be made to the area of the wall if the pressure is to remain unaltered?

ARITHMETICAL PROGRESSIONS

If, in a set of values, the items follow each other in succession according to the same law, they provide the separate **terms** of a **series**.

e.g.   (i) 3, 5, 7, 9, . . .   (ii) 2, 4, 8, 16, . . .

An **Arithmetical Progression** is a series in which the difference between each term and its successor is constant. This constant is called the **common difference.**

State the common difference in the following:

(i) 1, 4, 7, . . . (ii) 13, 11, 9, . . .
(iii) $1\frac{1}{2}$, 6, $10\frac{1}{2}$, . . . (iv) a, $a + d$, $a + 2d$, . . .

THE GENERAL, OR 'n'th, TERM OF AN A.P.

If a is the first term and d the common difference:

$$1\text{st term} = a \qquad\qquad 2\text{nd term} = a + d$$
$$3\text{rd term} = (a + d) + d \ = a + 2d$$
$$4\text{th term} = (a + 2d) + d = a + 3d$$
$$\therefore \ \textbf{nth term} = \boldsymbol{a + (n - 1)d}$$

NOTE

(i) For this purpose, n must always be a positive integer.

(ii) n indicates the **position** of a term in a series and not the value of the term.

EXERCISE 110

Find the common difference and the 8th term:

1. 1, 3, 5, . . .   **2.** 30, 27, 24, . . .
3. 5, 12, 19, . . .   **4.** $x, 4x, 7x,$ . . .
5. 3, 1, -1, . . .   **6.** $\frac{1}{4}, \frac{3}{4}, 1\frac{1}{4},$ . . .
7. $x^2, -2x^2, -5x^2,$ . . .   **8.** $15x, 11x, 7x,$ . . .
9. Find the 12th term of 5, 12, 19, . . .
10. Find the 6th term of 9, 4, -1, . . .
11. Find the 20th term of 1, $2\frac{1}{2}$, 4, . . .
12. Find the 18th term of 20, $23\frac{1}{4}$, $26\frac{1}{2}$, . . .

Find the simplest expression for the nth term:

13. 1, 6, 11, . . .   **14.** 7, 2, -3, . . .
15. 4, $4\frac{1}{2}$, 5, . . .   **16.** 8, $5\frac{3}{4}$, $3\frac{1}{2}$, . . .

Find the first three terms in the series:

17. nth term $= 2n + 3$ **18.** nth term $= 3n - 5$

19. nth term $= 5 - \frac{1}{2}n$ **20.** nth term $= \dfrac{1}{n+1}$

Find the common difference and the first term in the following A.P.s:

21. 3rd term 6; 15th term 18   **22.** 5th term 12; 18th term 64
23. 6th term 23; 20th term 2   **24.** 8th term 14; 16th term -12
25. nth term $(6n + 4)$ **26.** nth term $(18 - 3n)$

Find the number of terms in the following A.P.s:

27. 5, 9, 13, . . . 89   **28.** 86, 80, 74, . . . 20
29. $-16, -8, 0,$ . . . 336   **30.** 34, 27, 20, . . . -85

SUMMATION OF AN A.P.

Add together all the integers from 1 to 10, inclusive. Your total should be 55. Here is an interesting alternative method; the terms are set down first forwards and then in reverse order. We will call the sum of the terms S.

$$S_1 = 1 + 2 + 3 + 4 + 5 + 6 + 7 + 8 + 9 + 10$$
$$S_2 = 10 + 9 + 8 + 7 + 6 + 5 + 4 + 3 + 2 + 1$$
$$S_1 + S_2 = 11 + 11 + 11 + \ldots\ldots\ldots + 11 + 11$$
$$\therefore 2S = 11 \times 10 \qquad \therefore S = \frac{11 \times 10}{2} \qquad Ans. = \underline{\underline{55}}$$

Similarly, it may be shown that for an A.P. containing n terms, where a is the first and l the last term:

$$S = \frac{n}{2}(a + l) \text{ ———————— } ①$$

Since n is the position of *any* term, it may also represent l and $\therefore l = a + (n - 1)d$. Substituting for l in ① above:

$$S = \frac{n}{2}[a + \{a + (n - 1)d\}]$$

$$\therefore S = \frac{n}{2}[2a + (n - 1)d] \text{ ———————— } ②$$

EXERCISE 111

Find the sum of the following A.P.s:

1. First term 2, last term 20, 10 terms.
2. First term 3, last term 30, 10 terms.
3. First term 1, last term 45, 12 terms.
4. First term 60, last term 3, 20 terms.
5. First term 80, last term -20, 6 terms.
6. First term 20, last term -50, 15 terms.

Find the last term and the sum of:

7. 3, 8, 13, . . . to 8 terms.   **8.** $-12, -9, -6,$ . . . to 17 terms.
9. 2, 6, 10, . . . to 21 terms.   **10.** 8, $13\frac{1}{2}$, 19, . . . to 14 terms.
11. 10, $6\frac{1}{2}$, 3, . . . to 9 terms.   **12.** 40, 32, 24, . . . to 33 terms.

Find the sum of:

13. 5, 9, 13, . . . to 12 terms.   **14.** 3, 6, 9, . . . to 25 terms.
15. 32, 26, 20, . . . to 10 terms.   **16.** 60, 53, 46, . . . to 15 terms.
17. $8\frac{1}{2}$, 10, $11\frac{1}{2}$, . . . to 23 terms.   **18.** 15, $12\frac{1}{2}$, 10, . . . to 19 terms.

Find the number of terms required to give the sum:

19. $-1\frac{1}{2}, \frac{1}{2}, 2\frac{1}{2},$ . . . $S = 6$ **20.** $-9, -8, -7,$ . . . $S = 10$
21. $-22, -19, -16,$ . . . $S = 8$ **22.** 20, 16, 12, . . . $S = -160$
23. 24, 28, 32, . . . $S = 420$ **24.** 80, 72, 64, . . . $S = -88$

25. The 3rd term of an A.P. is 11, the 10th term is 32; find the sum of the first 15 terms.

26. The 5th term of an A.P. is 18, the 12th term is 46; find the sum of the first 24 terms.

27. The 2nd term of an A.P. is 60, the 14th term is -12; find the sum of the first 20 terms.

28. The 4th term of an A.P. is 0, the 20th term is 80; find the sum of the first 18 terms.

29. The 8th term of an A.P. is 1, the 16th term is 25; find the sum of the first 30 terms.

30. The 2nd term of an A.P. is $13\frac{1}{2}$, the 8th term is $-13\frac{1}{2}$; find the sum of the first 9 terms.

31. The nth term of an A.P. is $4n + 3$; find the sum of the first n terms.

32. The nth term of an A.P. is $5n - 4$; find the sum of the first n terms.

33. A man's salary starts at £500 per annum and rises by £30 a year for 15 years. Find his annual income after completing 15 years and his total income during the first 16 years.

34. Find the sum of all the numbers, divisible by 4, from 100 to 200.

35. Each year a man saves £18 more than he did the previous year. If he saves £117 in the first year, how long will it take him to save a total of £2592?

ARITHMETIC MEAN

If three terms are in A.P., the middle term is the **arithmetic mean** of the first and last term.

E.g., If a, b, c are in A.P. and d is the common difference:

Then $\quad d = b - a = c - b$

$$\therefore\ 2b = a + c \text{ and } b = \frac{a + c}{2}$$

The arithmetic mean of n terms is the sum of the terms divided by n.

EXAMPLE. *Insert 4 arithmetic means between 11 and 46.*

The complete A.P. will contain 6 terms, thus:

$$11,\ p,\ q,\ r,\ s,\ 46$$

$a = 11$ and the 6th term $= a + (n - 1)d$

$\therefore\ 46 = 11 + 5d$ and $d = 7$

$\therefore$ The A.P. is 11, 18, 25, 32, 39, 46

EXERCISE 112

1. Insert 5 arithmetic means between 11 and 29.

2. Insert 6 arithmetic means between 4 and 46.

3. Insert 7 arithmetic means between 2 and 30.

4. Insert 3 arithmetic means between 27 and 3.

5. Insert 7 arithmetic means between 35 and -37.

6. Insert 5 arithmetic means between 46 and 31.

7. Insert 3 arithmetic means between a and b.

8. Insert 4 arithmetic means between $(x + y)$ and $(x - y)$.

GEOMETRICAL PROGRESSIONS

A **Geometrical Progression** is a series in which the quotient obtained by dividing each term into its successor is constant. This constant is called the **common ratio.**

State the common ratio in the following:

(i) 2, 4, 8, . . .   (ii) 18, 6, 2, . . .
(iii) x, $-x^2$, x^3 . . .   (iv) -16, 8, -4, . . .

THE GENERAL, OR 'n'th, TERM OF A G.P.

If a is the first term and r the common ratio:

1st term $= a$ 2nd term $= ar$
3rd term $= (ar)r = ar^2$
4th term $= (ar^2)r = ar^3$

$\therefore$ **nth term** $= ar^{(n-1)}$

EXERCISE 113

Find the common ratio and the 6th term:

1. 3, 6, 12, . . .   **2.** 17, $8\frac{1}{2}$, $4\frac{1}{4}$, . . .
3. 2, -6, 18, . . .   **4.** -64, 16, -4, . . .
5. a^6x, a^4x^3, a^2x^5, . . .   **6.** x, $-\dfrac{1}{x}$, $\dfrac{1}{x^3}$, . . .

7. Find the 7th term of 12, 18, 27, . . .
8. Find the 5th term of 35, 21, $12\frac{3}{5}$, . . .
9. Find the 10th term of 1, 2, 4, . . .
10. Find the 6th term of 240, -180, 135, . . .

Find the simplest expression for the nth term:

11. 8, 16, 32, . . .   **12.** $2\frac{1}{2}$, $6\frac{1}{4}$, $15\frac{5}{8}$, . . .
13. 3, -18, 108, . . .   **14.** 64, -48, 36, . . .

Find the first three terms in the series:

15. nth term $= 2^n$ **16.** nth term $= 3^{(n-1)}$
17. nth term $= \left(\frac{2}{3}\right)^{(2n+1)}$ **18.** nth term $= (2n + 1)^n$

Find the common ratio and the first term in the following G.P.s:

19. 3rd term 18; 6th term 486   **20.** 2nd term 24; 7th term $-\frac{3}{4}$
21. 4th term $13\frac{1}{2}$; 8th term $68\frac{11}{32}$ **22.** 2nd term $-\frac{1}{3}$; 5th term $\frac{8}{81}$
23. 7th term 3; 11th term 243   **24.** 3rd term -36; 6th term $10\frac{2}{3}$
25. nth term $4^n \cdot 5^{(3-n)}$ **26.** nth term $2^{(5-n)} \cdot (-5)^{(n-1)}$

Find the number of terms in the following G.P.s (*Use logarithms*):

27. $\frac{1}{4}, \frac{1}{2}, 1, \ldots 2048$ **28.** 1, 3, 9, . . . 6561

29. 2, 8, 32, . . . 8192   **30.** 8000, 4000, 2000, . . . $15\frac{5}{8}$

31. A sheet of paper is folded on itself, thus doubling its thickness at each fold. If the paper is 0·01 in thick, use tables to find, to 3 sig. fig., the thickness of the paper after 49 folds. Give the answer in miles How many more folds would be required for the paper to reach the sun? Is this a practical proposition? If not, why not?

32. The arc through which a pendulum swings decreases by $\frac{1}{4}$ on each successive swing both forward and backward. If the pendulum first swings through 80°, find the angle at the 6th swing.

SUMMATION OF A G.P.

The following series is a G.P. whose sum is 62.

$$2 + 4 + 8 + 16 + 32$$

Since the first term $(a) = 2$ and the common ratio $(r) = 2$.
The series can be written:

$$S = 2 + 2.2 + 2.2^2 + 2.2^3 + 2.2^4$$

Each side of this equation is now multiplied by the common ratio and the new terms on the R.H.S. are placed to the right of their previous position:

$$\therefore 2S = \quad\quad 2.2 + 2.2^2 + 2.2^3 + 2.2^4 + 2.2^5$$

By subtracting $2S$ from S, most of the terms disappear to leave the terms at the top left and bottom right positions:

$$S - 2S = 2 \quad * \quad * \quad * \quad * \quad - 2.2^5$$
$$\therefore -S = 2 - 64 \quad\quad \therefore S = \underline{\underline{62}}$$

In general, the sum of n terms of a G.P. is given by:

$$S = a + ar + ar^2 + \ldots + ar^{(n-1)}$$
$$rS = \quad ar + ar^2 + ar^3 \ldots + ar^{(n-1)} + ar^n$$
$$S - rS = a \quad * \quad\quad * \quad\quad * \quad - ar^n$$
$$S(1 - r) = a - ar^n$$
$$S(1 - r) = a(1 - r^n) \quad\quad \therefore \boldsymbol{S = \frac{a(1 - r^n)}{1 - r}}$$

If $r > 1$, then both numerator and denominator would be negative; in such cases the formula is more conveniently expressed:

$$\boldsymbol{S = \frac{a(r^n - 1)}{r - 1}}$$

GEOMETRIC MEAN

If three terms are in G.P., the middle term is the **geometric mean** of the first and last term.

E.g., If x, y, z are in G.P. and r is the common ratio:

$$\text{Then} \qquad r = \frac{y}{x} = \frac{z}{y} \text{ and } y^2 = xz$$

$$\therefore y = \pm \sqrt{xz}$$

The geometric mean (mean proportional) of two terms is the square root of their product.

(*We are generally concerned with the +ve root only.*)

EXAMPLE. *Insert 3 geometric means between $7\frac{1}{2}$ and 120.*

The complete G.P. will contain 5 terms, thus:

$$7\tfrac{1}{2}, \quad b, \quad c, \quad d, \quad 120$$

$$a = 7\tfrac{1}{2} \text{ and the 5th term} = ar^{(n-1)}$$

$$\therefore 120 = 7\tfrac{1}{2} \cdot r^4 \qquad \therefore r^4 = \frac{120}{7\frac{1}{2}} = 16$$

$$\therefore r = \pm 2$$

$$\therefore \text{ The G.P. is } 7\tfrac{1}{2}, \pm 15, 30, \pm 60, 120$$

EXERCISE 114

Find the sum of the following G.P.s:

1. 3, 6, 12, . . . to 6 terms.   **2.** 2, 6, 18, . . . to 6 terms.

3. 1, 2, 4 . . . to 10 terms.   **4.** 1, 3, 9, . . . to 10 terms.

5. 800, 400, 200, . . . to 8 terms.   **6.** 64, 48, 36, . . . to 7 terms.

7. 81, 27, 9. . . . to 8 terms.   **8.** $\frac{1}{2}$, $-\frac{1}{4}$, $\frac{1}{8}$, . . . to 8 terms.

9. $\frac{2}{3}$, $\frac{4}{9}$, $\frac{8}{27}$, . . . to 6 terms.   **10.** $39\frac{1}{16}$, $-15\frac{5}{8}$, $6\frac{1}{4}$, . . . to 8 terms.

Use logs to estimate the number of terms required to give the sum:

11. 1, 2, 4, . . . $S = 127$ **12.** 2, 6, 18, . . . $S = 728$

13. $\frac{1}{5}$, -1, 5, . . . $S = 2604\frac{1}{5}$ **14.** -36, 48, -64, . . . $S = 33\frac{1}{3}$

   15. The 2nd term of a G.P. is 10, the 5th term is 80; find the sum of the first 6 terms.

   16. The 3rd term of a G.P. is 36, the 6th term is 972; find the sum of the first 8 terms.

   17. The 2nd term of a G.P. is 10, the 5th term is $156\frac{1}{4}$; find the sum of the first 6 terms.

   18. The 2nd term of a G.P. is 32, the 4th term is 8; find two values of the common ratio and the appropriate sums for the first 6 terms.

19. After a year's trading a firm shows a profit of £1000. In each of the following 9 years the profit shows an increase of 5% on the previous year. Find, with the aid of logs, the total profit during the 10 years. (NOTE. $r = 1.05$.) Give answer to 3 sig. fig.

20. Each arc through which a pendulum swings is $\frac{3}{4}$ the length of the previous arc. If the first arc is 4 ft, find, to the nearest foot, the distance through which the bob moves during 6 swings.

Find the geometric mean of:

21. 4, 9   **22.** 4, 16

23. 8, 18   **24.** 5, 45

25. Insert 2 geometric means between 5 and 135.

26. Insert 3 geometric means between -14 and -224.

27. Insert 2 geometric means between 1 and $-15\frac{5}{8}$.

28. Insert 4 geometric means between $\frac{9}{16}$ and $\frac{2}{27}$.

29. Find the geometric mean of $x - y$ and $x^3 - 3x^2y + 3xy^2 - y^3$ and the common ratio.

30. The arithmetic mean of two numbers is 10 and their geometric mean is 8. Find the two numbers.

PART B: ARITHMETIC

DECIMALS

Evaluate the following:

1. 0.207×100	**2.** $3.07 \div 10$	**3.** $417 \div 100$
4. 3.09×10	**5.** 51.7×100	**6.** $8010 \div 100$
7. $0.613 \div 10$	**8.** 92.14×1000	**9.** 1201×100
10. $0.0073 \div 100$	**11.** 0.0401×100	**12.** 0.005×1000
13. 0.0101×1000	**14.** $0.101 \div 100$	**15.** $0.1 \div 1000$
16. 1.0011×1000	**17.** $1100 \div 1000$	**18.** $1.001 \div 100$
19. 0.01×0	**20.** $1010 \div 10\,000$	**21.** 0.0101×10
22. $0.000\,01 \times 1000$	**23.** 0×1000	**24.** $173\,000 \div 10\,000$

25. $18.09 + 13.23$ $\qquad$ **26.** $170.301 - 98.27$

27. $230.1 - 109.07$ $\qquad$ **28.** $1.05 + 13.1 + 107.009$

29. $1001 + 1.001 + 0.101$ $\qquad$ **30.** $11.01 - 0.909$

31. $57.031 - 9.107$ $\qquad$ **32.** $0.701 - 9.983 + 14.06$

33. $25 + 0.25 + 1.05 + 6.009 + 13.77 - 10.09$

34. $132.07 - 86.103 - 11.31 - 0.509 + 1.201$

35. $20.1 + 0.067 - 19.208 - 14.69 + 100.1 - 0.008$

36. $40.05 - 10.303 + 17.61 + 0.919 - 15.97 - 5.08$

37. 29.3×12	**38.** 2.93×12	**39.** 0.293×12
40. 0.008×8	**41.** 0.8×8	**42.** 0.08×8
43. 12.3×16	**44.** 1.23×1.6	**45.** 1230×0.016
46. 39.5×2.3	**47.** 4.76×7.6	**48.** 0.913×8.1
49. 5.23×17.9	**50.** 64.2×2.34	**51.** 892×0.143
52. 41.7×53.6	**53.** 0.321×2.19	**54.** 0.746×0.549
55. 30.02×13.3	**56.** 49.37×28.1	**57.** 0.0249×33.3
58. 0.0053×716	**59.** 5.103×20.02	**60.** 90.09×10.01
61. $4.6 \times 0.03 \times 17$	**62.** $230 \times 0.12 \times 4.5$	**63.** $0.003 \times 2.6 \times 30$
64. $(10.1)^3$	**65.** $(0.103)^3$	**66.** $(0.001)^3 \times 10^3$
67. $13.11 \div 2.3$	**68.** $0.1311 \div 0.23$	**69.** $1.311 \div 2.3$
70. $1.311 \div 0.023$	**71.** $0.1311 \div 230$	**72.** $0.1311 \div 0.0023$
73. $8.74 \div 2.3$	**74.** $4.08 \div 1.7$	**75.** $3.498 \div 3.3$
76. $5.1308 \div 5.08$	**77.** $0.057 \div 1.9$	**78.** $0.111 \div 37$
79. $398.52 \div 12.3$	**80.** $203.342 \div 3.47$	**81.** $72.422 \div 0.98$
82. $779.332 \div 17.2$	**83.** $0.178\,224 \div 0.047$	**84.** $5817.698 \div 1.006$
85. $10.631\,52 \div 38.52$	**86.** $200.4002 \div 20.02$	**87.** $3.699\,08 \div 12.01$

88. $0.000\,202\,91 \div 0.0394$ $\qquad$ **89.** $0.000\,261\,59 \div 0.0037$

90. $0.020\,9209 \div 2.09$

WRITING MIXED AMOUNTS OF POUNDS AND PENCE*

100 pence (new pence) = £1

2 half-pence (½p) = 1p

The abbreviation for pence (new pence) is p and for pounds we continue to use £ (libra).

The decimal point is used as a "place holder", to separate pounds and pence but the half-penny is expressed as a vulgar fraction: ½p.

EXAMPLES.

Thirty-seven pounds = £37
Fifty pounds = £50
Eighty-five pence = £0·85
Seventy pence = £0·70
5p = £0·05
2½p = £0·02½
Forty-three pounds sixty-two pence = £43·62
Sixty pounds and three pence = £60·03

NOTE. (i) There should always be two digits in the pence column. (£0·6 could give rise to doubt.)

(ii) It is wrong to use the £ symbol *and* the penny abbreviation (p) together, or to use both with a decimal point.

Wrong	Right
£89·63p	£89·63
£36·45p	£36·45

(iii) In writing cheques, receipts etc. it is advisable to use a dash instead of the decimal point. Could you suggest a reason for this?

Words	Figures
Thirty-five pounds =	£35—00
Eighteen pounds 32 =	£18—32
Sixty-eight pence =	£0—68

EXERCISE 2

Express your answers (i) in pence (ii) in £s:

1. 32p × 9
2. 17p × 24
3. 23p × 16
4. 48p × 22
5. 56p × 25
6. 72p × 34
7. 84p × 18
8. 87p × 27
9. 95p × 45
10. £1·18 × 12
11. £12·40 × 18
12. £18·76 × 24
13. £3·84 ÷ 16
14. £5·76 ÷ 18
15. £8·64 ÷ 24
16. £6·75 ÷ 15
17. £17·28 ÷ 36
18. £20·14 ÷ 38
19. £25·62 ÷ 42
20. £40·81 ÷ 53
21. £52·51 ÷ 59
22. £68·82 ÷ 62
23. £587·65 ÷ 73
24. £840·84 ÷ 84

* "Decimal Currency"—H.M.S.O.

INDICES

3×3 may be written as 3^2, which we read as "3 to the power 2" or "3 squared". The small figure 2 at the top right-hand corner is called an **index number**, it tells us the number of 3 s to be multiplied together.

QUESTIONS. What are the index forms for:

(i) $2 \times 2 \times 2$?   (ii) $5 \times 5 \times 5 \times 5$?

(iii) 6×6?   (iv) 7?

FINDING PRIME FACTORS

To find the prime factors of a number we divide by prime numbers, starting with 2. When 2 will no longer divide we try 3, and so on with 5, 7, 11, etc.

EXAMPLE. *Express* 1152 *in prime factors, using index notation.*

$$\begin{aligned}
1152 &= 2 \times 576 \\
&= 2 \times 2 \times 288 \\
&= 2 \times 2 \times 2 \times 144 \\
&= 2 \times 2 \times 2 \times 2 \times 72 \\
&= 2 \times 2 \times 2 \times 2 \times 2 \times 36 \\
&= 2 \times 2 \times 2 \times 2 \times 2 \times 2 \times 18 \\
&= 2 \times 2 \times 2 \times 2 \times 2 \times 2 \times 2 \times 9 \\
&= 2 \times 2 \times 2 \times 2 \times 2 \times 2 \times 2 \times 3 \times 3
\end{aligned}$$

$\therefore\ 1152 = 2^7 \times 3^2$ *Ans.*

EXERCISE 3

Simplify the following, using prime factors and indices:

1. $2 \times 2 \times 2 \times 3 \times 3 \times 4$ 2. $3 \times 4 \times 3 \times 4 \times 25$
3. $2 \times 2 \times 5 \times 5 \times 7 \times 7$ 4. $2 \times 3 \times 5 \times 3 \times 5$
5. $2^2 \times 2 \times 3^2 \times 3$ 6. $2 \times 2^2 \times 2^3 \times 3 \times 27$

Give the prime factors of the following, using index notation where possible:

7. 12	**8.** 16	**9.** 36	**10.** 48	**11.** 15
12. 25	**13.** 75	**14.** 72	**15.** 144	**16.** 900
17. 360	**18.** 216	**19.** 225	**20.** 1323	**21.** 1800
22. 432	**23.** 864	**24.** 5184	**25.** 10 800	**26.** 11 025

SQUARE ROOTS

A **square root** is that factor of a number which has been **multiplied by itself** to provide the number given.

The symbol for a square root is $\sqrt{}$, it is called the **radical sign.**

Since	$81 = 9 \times 9$
Then	$\sqrt{81} = 9$
Using factors we see:	$81 = 3 \times 3 \times 3 \times 3$
	$= 3^2 \times 3^2$

Clearly $\sqrt{3^2}$ is 3, so we take the **square root of each factor**, thus

$$81 = 3^2 \times 3^2$$
$$\sqrt{81} = 3 \times 3$$
$$\therefore \ \sqrt{81} = 9$$

81 is called a **"perfect square"** because its square root is a **whole number.**

EXAMPLE. *Use factors to find* $\sqrt{5184}$

$$
\begin{aligned}
5184 &= 2 \times 2592 \\
&= 2 \times 2 \times 1296 \\
&= 2 \times 2 \times 2 \times 648 \\
&= 2 \times 2 \times 2 \times 2 \times 324 \\
&= 2 \times 2 \times 2 \times 2 \times 2 \times 162 \\
&= 2 \times 2 \times 2 \times 2 \times 2 \times 2 \times 81 \\
&= 2^6 \times 3 \times 27 \\
&= 2^6 \times 3 \times 3 \times 9 \\
&= 2^6 \times 3 \times 3 \times 3 \times 3 \\
&= 2^6 \times 3^4
\end{aligned}
$$

Since	$5184 = (2^3 \times 3^2) \times (2^3 \times 3^2)$
Then	$\sqrt{5184} = 2^3 \times 3^2$
	$= 8 \times 9$
$\therefore$	$\underline{\underline{\sqrt{5184} = 72}} \ Ans.$

NOTE. The answer can be checked by multiplying 72 by itself.

EXERCISE 4

Use factors to find the square roots of the following, *check your results by multiplying*:

1. 576	**2.** 144	**3.** 324	**4.** 1296	**5.** 225
6. 196	**7.** 8100	**8.** 2025	**9.** 3969	**10.** 729
11. 625	**12.** 256	**13.** 1225	**14.** 32 400	**15.** 50 625

H.C.F. AND L.C.M.

The **Highest Common Factor** of two or more numbers is the greatest number which **will divide exactly** INTO **each** of them.

Thus 9 is the H.C.F. of 18, 27 and 45.

The **Lowest Common Multiple** of two or more numbers is the least number which **can be divided exactly** BY **each** of them.

Thus 60 is the L.C.M. of 12, 15 and 20.

EXAMPLE. *Find the H.C.F. and L.C.M. of* 18, 24 *and* 36

$$
\begin{array}{c|c}
2 & 18 \\
\hline
3 & 9 \\
\hline
 & 3
\end{array}
\qquad
\begin{array}{c|c}
2 & 24 \\
\hline
2 & 12 \\
\hline
2 & 6 \\
\hline
 & 3
\end{array}
\qquad
\begin{array}{c|c}
2 & 36 \\
\hline
2 & 18 \\
\hline
3 & 9 \\
\hline
 & 3
\end{array}
$$

Note the alternative method of setting down. The factors appear at the side and along the bottom of the grid (frame).

Thus

$$18 = 2 \times 3 \times 3$$
$$24 = 2 \times 2 \times 2 \times 3$$
$$36 = 2 \times 2 \times 3 \times 3$$

From these factors we can see that

the H.C.F. $= 2 \times 3 = 6$

and   the L.C.M. $= 2 \times 2 \times 2 \times 3 \times 3 = 72$

H.C.F. $= 6$ L.C.M. $= 72$

EXERCISE 5 (*Oral*)

Find the H.C.F. of:

1. 4, 6, 12	**2.** 3, 6, 12	**3.** 4, 12, 16	**4.** 8, 12, 20
5. 12, 15, 18	**6.** 12, 18, 30	**7.** 20, 30, 40	**8.** 48, 64, 16
9. 3, 5, 9	**10.** 10, 15, 30	**11.** 21, 14, 35	**12.** 4, 14, 23

Find the L.C.M. of:

13. 3, 6, 12	**14.** 3, 6, 9	**15.** 2, 4, 16	**16.** 3, 4, 16
17. 3, 4, 15	**18.** 6, 9, 12	**19.** 6, 8, 16	**20.** 3, 5, 12
21. 5, 6, 7	**22.** 14, 21, 42	**23.** 8, 10, 12	**24.** 9, 12, 18

EXERCISE 6

Find the H.C.F. and L.C.M. of the following:

1. 6, 18, 12	**2.** 12, 18, 36	**3.** 14, 21
4. 30, 36	**5.** 8, 27, 36	**6.** 24, 36, 54
7. 45, 75	**8.** 30, 105	**9.** 36, 54, 48
10. 16, 81	**11.** 72, 210, 315	**12.** 225, 105, 42

PROBLEMS

Exercise 7

1. A kitchen wall 69 in long is to be covered with plastic tiles to a height of 48 in. What is the largest size of square tile which can be used without cutting?

2. The floor of a shed is 80 in by 56 in, it is to be paved with square stone slabs. What is the size of the largest slab which can be used without cutting?

3. What is the smallest number by which 405 must be multiplied to produce a perfect square? (HINT. *Factorize first.*)

4. What is the smallest number by which 1575 must be multiplied to produce a perfect square?

5. The front wheels of a cart have a circumference of 20 in, the back wheels have a circumference of 36 in. How far must the cart move in order that both pairs of wheels may make a complete number of revolutions?

6. Apples may be packed in boxes containing 28 lb, 30 lb or 40 lb. What is the least quantity of apples which can be packed, without any over, allowing any one type of box to be used? (*Answer in pounds.*)

7. On three different routes buses leave the terminus at intervals of 4 min, 10 min and 15 min respectively. If three buses leave together, how long will it be before they do so again?

8. What is the smallest sum of money which can be made up from a whole number of 2p or 5p coins?

9. Two cars take part in a road race, one maintains an average speed of 80 mile/h, the other 120 mile/h. What is the least distance they can both travel in an exact number of hours?

10. What is the smallest sum of money which may be divided exactly into shares of 5p or 8p or 9p or 12p each?

11. In an Eastern temple two bells are rung, one at 12-min intervals the other at 16-min intervals. How many times will they ring together in 24 hr?

12. Three bells ring at intervals of 12 s, 15 s, 20 s respectively. How many times will they ring together in 1 hr?

13. A wall is to be covered by hanging strips of material from the ceiling to the floor. The material may be purchased in widths of 36 in, 42 in or 48 in. It is found that if any *one* of these widths of material is used the wall may be covered by a whole number of strips. How long is the wall? (*Answer in yards.*)

14. If a sea mile is 6080 ft, find in sea miles the smallest distance which can be expressed as an exact number of both land and sea miles.

15. A box has internal dimensions of 8 in by 12 in by 20 in. It is exactly filled by equal cubes. What is the least number of cubes?

VULGAR FRACTIONS

Vulgar fractions are those which are "in common use"—they are sometimes called **common fractions.**

QUESTION. What values have the following:

(i) $\frac{0}{4}$; $\frac{0}{2}$; $\frac{0}{1}$; $\frac{0}{2}$; $\frac{0}{5}$?

(ii) $\frac{1}{5}$; $\frac{1}{2}$; $\frac{1}{4}$; $\frac{1}{8}$; $\frac{1}{100}$; $\frac{1}{0}$; $\frac{5}{0}$?

Note the sign for **infinity** ∞.

EXERCISE 8

Find the values of:

1. $\frac{1}{2}$ of £1 (In p)
2. $\frac{1}{10}$ of £1 (In p)
3. $\frac{1}{2}$ of 1 hr (In min)
4. $\frac{1}{4}$ of £2 (In p)
5. $\frac{3}{4}$ of £4 (In £)
6. $\frac{2}{5}$ of £1 (In p)
7. $\frac{1}{2}$ of rt. $\angle$ (In °)
8. $\frac{1}{6}$ of 1 day (In hr)

What fraction is:

9. 25p of £1
10. 75p of £1
11. 20 min of 1 hr
12. 20p of £2
13. £5 of £6
14. 50p of £3
15. 60° of 90°
16. 6 hr of 1 day

Reduce to their lowest terms:

17. $\frac{16}{24}$ 18. $\frac{48}{56}$ 19. $\frac{81}{96}$ 20. $\frac{21}{36}$ 21. $\frac{72}{102}$
22. $\frac{72}{90}$ 23. $\frac{45}{75}$ 24. $\frac{144}{192}$ 25. $\frac{720}{960}$ 26. $\frac{720}{1200}$
27. $\frac{256}{1024}$ 28. $\frac{768}{1024}$ 29. $\frac{576}{720}$ 30. $\frac{1575}{2700}$ 31. $\frac{630}{1470}$
32. $\frac{400}{960}$ 33. $\frac{182}{455}$ 34. $\frac{1540}{2695}$ 35. $\frac{3375}{3600}$ 36. $\frac{3080}{3465}$

Arrange the following in ascending order of size:

37. $\frac{1}{2}$, $\frac{3}{5}$, $\frac{7}{10}$ 38. $\frac{2}{5}$, $\frac{1}{4}$, $\frac{3}{10}$ 39. $\frac{3}{4}$, $\frac{13}{16}$, $\frac{15}{32}$ 40. $\frac{7}{16}$, $\frac{3}{8}$, $\frac{1}{4}$
41. $\frac{4}{5}$, $\frac{11}{15}$, $\frac{17}{20}$ 42. $\frac{1}{2}$, $\frac{10}{21}$, $\frac{3}{7}$ 43. $\frac{2}{9}$, $\frac{1}{3}$, $\frac{5}{18}$ 44. $\frac{5}{6}$, $\frac{29}{36}$, $\frac{7}{9}$
45. $\frac{4}{11}$, $\frac{13}{33}$, $\frac{7}{22}$ 46. $\frac{3}{4}$, $\frac{5}{8}$, $\frac{4}{9}$ 47. $\frac{7}{10}$, $\frac{13}{20}$, $\frac{21}{31}$ 48. $\frac{3}{7}$, $\frac{1}{2}$, $\frac{5}{9}$

Express as mixed or whole numbers:

49. $\frac{16}{5}$ 50. $\frac{22}{7}$ 51. $\frac{42}{21}$ 52. $\frac{39}{15}$
53. $\frac{69}{18}$ 54. $\frac{115}{20}$ 55. $\frac{114}{12}$ 56. $\frac{198}{54}$
57. $\frac{116}{58}$ 55. $\frac{369}{81}$ 59. $\frac{420}{144}$ 60. $\frac{1617}{882}$

Express as improper fractions:

61. $1\frac{2}{3}$ 62. $2\frac{3}{5}$ 63. $3\frac{4}{7}$ 64. $4\frac{5}{9}$
65. $5\frac{6}{11}$ 66. $6\frac{7}{13}$ 67. $7\frac{8}{15}$ 68. $8\frac{9}{17}$
69. $11\frac{23}{25}$ 70. $16\frac{25}{32}$ 71. $23\frac{9}{16}$ 72. $4\frac{45}{64}$

FOUR RULES OF FRACTIONS

ADDITION AND SUBTRACTION OF FRACTIONS

NOTE.

(i) Any improper fractions are written as mixed numbers.

(ii) The whole numbers are dealt with first.

(iii) The fractions are converted so that the numerators appear above a common denominator.

EXAMPLE 1. *Simplify* $2\frac{3}{5} + \frac{19}{15} + 3\frac{11}{20}$

$$2\frac{3}{5} + \frac{19}{15} + 3\frac{11}{20} = 2\frac{3}{5} + 1\frac{4}{15} + 3\frac{11}{20}$$

$$= 6\,\frac{36 + 16 + 33}{60}$$

$$= 6\frac{85}{60} = 7\frac{\overset{5}{\cancel{25}}}{\underset{12}{\cancel{60}}}$$

$$= 7\frac{5}{12} \ Ans.$$

EXAMPLE 2. *Simplify* $4\frac{1}{3} - 2\frac{1}{2}$

$$4\frac{1}{3} - 2\frac{1}{2} = 2\,\frac{2 - 3}{6}$$

$$= 1\,\frac{6 + 2 - 3}{6}\ ^* = 1\,\frac{8 - 3}{6}$$

$$= 1\frac{5}{6} \ Ans.$$

NOTE. * To carry out the necessary subtraction we have taken 1 from the whole number 2 and converted it into $\frac{6}{6}$. The numerator is then increased by 6.

EXERCISE 9

Simplify:

1. $\frac{1}{2} + \frac{1}{3} + \frac{1}{4}$ 2. $1\frac{1}{2} + \frac{5}{3} + 2\frac{3}{4}$ 3. $2\frac{5}{14} + \frac{6}{7} + \frac{34}{28}$

4. $3\frac{3}{4} + \frac{9}{16} + 1\frac{5}{8}$ 5. $3\frac{1}{5} + \frac{9}{4} + \frac{11}{15}$ 6. $3\frac{5}{9} + \frac{7}{3} + \frac{9}{4}$

7. $3\frac{1}{6} - 2\frac{4}{9}$ 8. $\frac{39}{7} - \frac{23}{6}$ 9. $3\frac{1}{8} - \frac{23}{12}$

10. $1\frac{7}{9} - \frac{4}{5}$ 11. $\frac{17}{3} - \frac{43}{16}$ 12. $3\frac{2}{3} - 1\frac{10}{11}$

13. $2\frac{1}{2} + 1\frac{1}{3} - 3\frac{1}{4}$ 14. $3\frac{2}{3} - 2\frac{5}{6} + 1\frac{1}{2}$

15. $3\frac{7}{10} - 1\frac{1}{2} - 1\frac{3}{5}$ 16. $4\frac{1}{2} + 1\frac{1}{8} - 3\frac{3}{4}$

17. $2\frac{1}{2} + 1\frac{5}{6} + 4\frac{5}{12}$ 18. $5\frac{5}{9} - 2\frac{2}{3} - 1\frac{5}{6}$

19. $1\frac{1}{2} + 2\frac{7}{15} - 3\frac{3}{10}$ 20. $4\frac{5}{6} - 6\frac{7}{8} + 3\frac{5}{12}$

21. $3\frac{7}{12} - 1\frac{3}{4} - 1\frac{2}{3}$ 22. $2\frac{5}{7} - 5\frac{3}{4} + 6\frac{11}{14}$

23. $3\frac{1}{6} - 1\frac{9}{30} - 1\frac{9}{20}$ 24. $1\frac{7}{15} - 1\frac{22}{45} + \frac{19}{30}$

25. $6\frac{59}{64} - \frac{37}{16} - \frac{23}{8} - 1\frac{17}{32}$ 26. $1\frac{2}{3} + 1\frac{3}{5} - \frac{9}{5} - 2\frac{7}{10}$

27. $1\frac{23}{12} - \frac{7}{18} - \frac{31}{20}$ 28. $\frac{31}{16} - \frac{25}{12} + \frac{55}{48}$

MULTIPLICATION AND DIVISION OF FRACTIONS

EXAMPLE 1. *Simplify* $2\frac{1}{2} \times 1\frac{1}{5}$.

$$\frac{\overset{1}{\cancel{5}}}{\underset{1}{\cancel{2}}} \times \frac{\overset{3}{\cancel{6}}}{\underset{1}{\cancel{5}}} = \frac{1 \times 3}{1 \times 1} \overset{*}{} = \underset{=}{3} \; Ans.$$

* This step has been included only as an explanation.

NOTE. When multiplying fractions or mixed numbers:

(i) Mixed numbers are expressed as improper fractions, except in easy cases such as $2 \times 1\frac{1}{4}$, etc.

(ii) *Any* numerator may be "paired" with *any* denominator and *each* divided by a common factor. (Cancelled.)

(iii) Cancelling must be carried out with great care. (Accurately and neatly.)

EXAMPLE 2. *Simplify* $3\frac{1}{5} \div 1\frac{7}{25}$

$$3\frac{1}{5} \div 1\frac{7}{25} = \frac{16}{5} \div \frac{32}{25}$$
$$= \frac{\overset{1}{\cancel{16}}}{\underset{1}{\cancel{5}}} \times \frac{\overset{5}{\cancel{25}}}{\underset{2}{\cancel{32}}} = \frac{5}{2}$$
$$= \underset{=}{2\frac{1}{2}} \; Ans.$$

NOTE. When dividing by a fraction or mixed number:

(i) Express mixed numbers as improper fractions.

(ii) **Turn the divisor upside down and multiply.**

(iii) Proceed as for multiplication of fractions, cancelling, etc., with care.

EXERCISE 10

Simplify:

1. $\frac{1}{2} \times \frac{1}{4}$ 2. $\frac{1}{3} \times \frac{3}{4}$ 3. $\frac{2}{3} \times \frac{3}{8}$ 4. $\frac{3}{4} \times \frac{2}{3}$

5. $\frac{4}{5} \times \frac{5}{6}$ 6. $\frac{6}{7} \times \frac{7}{8}$ 7. $\frac{2}{3} \times 1\frac{1}{2}$ 8. $\frac{3}{4} \times 1\frac{1}{3}$

9. $\frac{2}{5} \times 2\frac{1}{2}$ 10. $1\frac{3}{4} \times 2$ 11. $2\frac{2}{3} \times 3$ 12. $3\frac{1}{5} \times 1\frac{1}{4}$

13. $3\frac{3}{4} \times 1\frac{3}{5}$ 14. $3\frac{1}{7} \times 1\frac{3}{11}$ 15. $1\frac{1}{5} \times 3\frac{3}{4}$ 16. $1\frac{5}{7} \times 3\frac{1}{2}$

17. $1\frac{11}{16} \times \frac{8}{9}$ 18. $1\frac{13}{20} \times 2\frac{2}{11}$ 19. $3 \times \frac{7}{15} \times 3\frac{3}{14}$

20. $5\frac{1}{2} \times 2\frac{2}{5} \times 1\frac{4}{11}$ 21. $\frac{14}{39} \times 1\frac{4}{9} \times 2\frac{4}{7}$ 22. $3\frac{5}{6} \times \frac{27}{46} \times 1\frac{5}{9}$

23. $1\frac{3}{13} \times 2\frac{3}{11} \times 0$ 24. $1\frac{14}{17} \times 1\frac{26}{93} \times 1\frac{13}{14}$

25. $\frac{1}{2} \div 2$ 26. $2 \div \frac{1}{2}$ 27. $\frac{2}{3} \div 2$ 28. $3 \div \frac{3}{4}$

29. $\frac{1}{2} \div \frac{1}{8}$ 30. $8 \div \frac{1}{4}$ 31. $\frac{3}{5} \div 3$ 32. $\frac{3}{4} \div \frac{2}{3}$

33. $\frac{3}{4} \div 1\frac{1}{2}$ 34. $0 \div \frac{1}{2}$ 35. $\frac{5}{12} \div \frac{5}{24}$ 36. $2\frac{1}{3} \div 3\frac{1}{9}$

37. $3\frac{1}{5} \div 1\frac{7}{25}$ 38. $2\frac{4}{7} \div 1\frac{5}{7}$ 39. $\frac{7}{8} \div 2\frac{11}{12}$ 40. $1\frac{19}{26} \div 1\frac{2}{13}$

41. $1\frac{1}{13} \div \frac{14}{27}$ 42. $\frac{68}{69} \div \frac{17}{23}$ 43. $1\frac{7}{17} \div 1\frac{25}{51}$ 44. $3\frac{18}{31} \div 1\frac{6}{31}$

USE OF BRACKETS

Consider the expression: $2\frac{1}{2} + 3\frac{1}{3} \times 1\frac{1}{5}$

It is essential to indicate quite clearly what is required, and brackets help us to do this. Here is the above expression as it should have appeared:

$$(a)\ (2\frac{1}{2} + 3\frac{1}{3}) \times 1\frac{1}{5}$$
$$\text{or} \quad (b)\ 2\frac{1}{2} + (3\frac{1}{3} \times 1\frac{1}{5})$$

Clearly they are different expressions and will simplify to different answers.

EXAMPLE 1. *Simplify* $(2\frac{1}{2} + 3\frac{1}{3}) \times 1\frac{1}{5}$

$$(2\frac{1}{2} + 3\frac{1}{3}) \times 1\frac{1}{5} = \left(5\frac{3+2}{6}\right) \times 1\frac{1}{5}$$

$$= 5\frac{5}{6} \times 1\frac{1}{5} = \frac{\overset{7}{\cancel{35}}}{\cancel{6}} \times \frac{\overset{1}{\cancel{6}}}{\cancel{5}} = \frac{7}{1}$$

$$\textit{Ans.} = \underline{\underline{7}}$$

EXAMPLE 2. *Simplify* $2\frac{1}{2} + (3\frac{1}{3} \times 1\frac{1}{5})$

$$2\frac{1}{2} + (3\frac{1}{3} \times 1\frac{1}{5}) = 2\frac{1}{2} + \left(\frac{\overset{2}{\cancel{10}}}{\cancel{3}} \times \frac{\overset{2}{\cancel{6}}}{\cancel{5}}\right)$$

$$= 2\frac{1}{2} + 4 = \underline{\underline{6\frac{1}{2}}} \ \textit{Ans.}$$

NOTE. The brackets serve to enclose the various calculations which are necessary. You should NOT work out these minor calculations on scrap paper and insert the results in the sum as if "by magic".

EXERCISE 11

Simplify:

1. $\frac{1}{2} + (\frac{1}{3} \times \frac{1}{4})$ 2. $(\frac{1}{2} + \frac{1}{3}) \times \frac{1}{4}$

3. $\frac{1}{2} - (\frac{1}{3} \times \frac{1}{4})$ 4. $(\frac{1}{2} - \frac{1}{3}) \times \frac{1}{4}$

5. $(1\frac{1}{2} + \frac{2}{3}) \div \frac{3}{4}$ 6. $(3\frac{1}{4} \div 1\frac{5}{8}) - 1\frac{5}{6}$

7. $(3\frac{1}{7} \times 1\frac{1}{2}) \div 4\frac{5}{7}$ 8. $3\frac{1}{7} \times (1\frac{1}{2} \div 4\frac{5}{7})$

9. $3\frac{3}{7} - 2\frac{9}{14} - \frac{23}{28} + 2\frac{3}{4}$ 10. $2\frac{8}{9} + 3\frac{5}{6} - 4\frac{26}{27} - 1\frac{2}{3}$

11. $(5\frac{4}{9} \times 2\frac{1}{7}) - 5\frac{4}{5}$ 12. $(7\frac{1}{7} \div 2\frac{19}{28}) + 4\frac{1}{4}$

13. $\frac{2}{3} + 2\frac{3}{5} - (1\frac{1}{2} \times \frac{2}{3})$ 14. $(4\frac{3}{5} \div 3\frac{9}{20}) + (3\frac{2}{5} \div 1\frac{21}{30})$

15. $(4\frac{4}{5} + 1\frac{1}{2}) \times (\frac{5}{9} - \frac{10}{27})$ 16. $(3\frac{3}{4} - 2\frac{1}{3}) \div (3\frac{3}{8} + 1\frac{7}{12})$

17. $\frac{1}{2} + [\frac{2}{3} \div (\frac{3}{4} + \frac{5}{8})]$ 18. $[(\frac{7}{8} - \frac{2}{3}) \times 2\frac{2}{5}] - \frac{1}{6}$

19. $\frac{1}{2} + (\frac{2}{3} \times \frac{1}{4}) - (\frac{4}{5} \div \frac{8}{15}) + (4\frac{2}{3} \times 1\frac{1}{14})$

20. $[(5\frac{1}{4} - 2\frac{5}{8}) \div 1\frac{3}{4}] + [2\frac{1}{5} \div (\frac{2}{3} + \frac{4}{5})] - [2 \times (\frac{5}{16} + 1\frac{3}{8} - \frac{3}{16})]$

FRACTIONS: HARDER EXAMPLES

Consider the expression: $3 \times 5 + 2$

It is important to realize that there are only TWO terms present: (i) (3×5), (ii) 2. We must simplify the first term BEFORE we add. Thus the expression now appears as: $15 + 2 = 17$

If we are required to add first, then this part of the expression MUST appear in brackets, thus: $3 \times (5 + 2)$

Alternatively, the multiplication sign may be omitted, thus:

$$3(5 + 2)$$
$$= 3 \times 7 = 21$$

REMEMBER

(i) If brackets are used, their contents must be simplified first.

(ii) Terms must be simplified (i.e., multiply or divide) *before* they can be added or subtracted.

EXAMPLE. *Simplify:* $\dfrac{1\frac{3}{11}(1\frac{1}{2} - \frac{2}{5})}{\frac{4}{5} \times 2\frac{1}{4} + \frac{3}{8}}$

The denominator contains a term which must be simplified—we will insert our own brackets *:

$$\frac{\frac{3}{11}(1\frac{1}{2} - \frac{2}{5})}{*(\frac{4}{5} \times 2\frac{1}{4}) + \frac{3}{8}} = \frac{\frac{3}{11}(1\frac{5-4}{10})}{\left(\frac{4}{5} \times \frac{9}{4}\right) + \frac{3}{8}}$$

$$= \frac{\frac{3}{11} \times 1\frac{1}{10}}{1\frac{4}{5} + \frac{3}{8}} = \frac{\frac{3}{11} \times \frac{11}{10}}{1\frac{32+15}{40}}$$

$$= \frac{3}{10} \div 1\frac{47}{40} = \frac{3}{10} \times \frac{40}{87}$$

$$Ans. = \underline{\underline{\frac{4}{29}}}$$

EXERCISE 12

Simplify:

1. $\dfrac{\frac{1}{2} + \frac{1}{3}}{\frac{1}{2} - \frac{1}{3}}$

2. $\dfrac{\frac{1}{2}(\frac{3}{4} - \frac{1}{3})}{\frac{1}{4}(\frac{2}{3} - \frac{1}{4})}$

3. $\dfrac{\frac{2}{3}(1\frac{1}{4} - \frac{3}{5})}{\frac{3}{4}(2\frac{1}{4} - 1\frac{1}{3})}$

4. $\dfrac{\frac{3}{5}(2\frac{5}{6} + 1\frac{1}{4})}{\frac{4}{9}(3\frac{2}{3} - 1\frac{5}{8})}$

5. $\dfrac{4\frac{1}{2} - 1\frac{5}{8} \times \frac{3}{4}}{1\frac{1}{4} + 1\frac{2}{3} \div \frac{5}{6}}$

6. $\dfrac{3\frac{1}{3} \times \frac{5}{8} - \frac{3}{4}}{1\frac{1}{3} - 2\frac{1}{4} \div 1\frac{4}{5}}$

7. $\dfrac{\frac{3}{4} \text{ of } \frac{5}{9} + \frac{2}{7} \div \frac{3}{14}}{2\frac{1}{4} \div 1\frac{1}{2} - \frac{4}{5} \text{ of } 1\frac{1}{4}}$

8. $\dfrac{1\frac{7}{8} + \frac{2}{3} \text{ of } 2\frac{1}{4} - 1\frac{1}{4}}{1\frac{3}{5} - 4\frac{1}{2} \div 1\frac{4}{5} + 1\frac{1}{3}}$

9. $\dfrac{\frac{7}{12} \div 3\frac{1}{2}}{1\frac{1}{2} + \frac{1}{3}} \times \dfrac{2\frac{9}{20} - \frac{4}{5}}{4\frac{1}{5} \div 2\frac{1}{3}}$

10. $\dfrac{\frac{4}{9} + \frac{3}{4} \text{ of } (1\frac{3}{5} - \frac{2}{3})}{(\frac{5}{8} - \frac{3}{5}) \div (\frac{4}{7} + \frac{1}{2})}$

AREA OF A RECTANGLE

REMEMBER $A = l \times b$

(i) Keep *all* dimensions in the same units.

(ii) Area is measured in square units.

(iii) When calculating areas employ the units in which you require the answer. Thus for an answer in square feet, work in feet; for an answer in square yards, work in yards. This will avoid the need to divide by 144, etc.

(iv) The *acre* is a unit of area, you *cannot* have a field 3 *acres long.*

EXERCISE 13

1. Find the areas of the following rectangles: (i) 6 in by 9 in; (ii) $9\frac{1}{2}$ in by 10 in; (iii) $6\frac{1}{2}$ in by $10\frac{1}{2}$ in; (iv) $3\frac{1}{2}$ ft by $8\frac{1}{2}$ ft; (v) $8\frac{1}{2}$ yd by $1\frac{1}{2}$ yd; (vi) $4\frac{1}{4}$ miles by $\frac{1}{2}$ mile.

2. A room is 9ft high, 25 ft long, 18 ft wide. Find the total surface area. (*Answer in ft².*)

3. A square field has a side of 110 yd. Find the area in yd².

4–7. Find the perimeters and the areas of the following figures: B:1, B:2, B:3, B:4. All dimensions are in inches and angles are right angles.

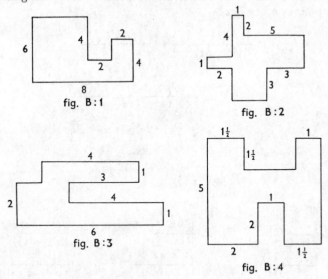

fig. B:1

fig. B:2

fig. B:3

fig. B:4

8. Given that the area of a tennis court is 312 yd² and the width is 12 yd, find the length of the court in yards and also the perimeter.

9. The main runways at London Airport are $1\frac{3}{4}$ miles long, each has a surface area of 308 000 yd². Find the width of each runway.

FURTHER WORK WITH AREAS

EXERCISE 14

1. Find the total area of the four walls of a room: (i) 16 ft long, 10 ft wide, 7ft high; (ii) $12\frac{3}{4}$ ft long, $9\frac{1}{2}$ ft wide, $8\frac{1}{4}$ ft high; (iii) $14\frac{2}{3}$ ft long, $11\frac{5}{12}$ ft wide, $9\frac{1}{6}$ ft high. Give each answer to the nearest square foot.

2. If each roll or "piece" of paper is 36 ft long and $1\frac{3}{4}$ ft wide, find the cost of the paper for the following rooms: (i) 11 ft long, 10 ft wide, $7\frac{1}{2}$ ft high, at 50p per piece; (ii) $14\frac{1}{2}$ ft long, 10 ft wide, 9 ft high, at 65p per piece; (iii) $16\frac{1}{4}$ ft long, $14\frac{1}{4}$ ft wide, $8\frac{3}{4}$ ft high, at 58p per piece.

3. A garden 60 ft long and 50 ft wide is to be fenced to a height of 6 ft using boards $\frac{1}{2}$ ft wide. What is the cost of fencing if the boards cost: (i) 6p per foot length; (ii) $4\frac{1}{2}$p per square foot?

4. A room is 20 ft long, 15 ft wide, 10 ft high. There are two windows, each 6 ft by 4 ft and a door $7\frac{1}{3}$ ft by 3 ft. Find: (i) the cost of papering the walls if each "piece" costs $47\frac{1}{2}$p and measures 36 ft by $1\frac{3}{4}$ ft; (ii) the cost of painting the ceiling at 1p per ft²; (iii) the cost of carpeting the floor at 25p per ft²; (iv) the total cost.

5. A room 18 ft by 14 ft has a carpet in the middle, leaving all round, a border covered with lino. Find the cost of the lino at 14p per ft², if the border is: (i) 1 ft wide; (ii) $1\frac{1}{4}$ ft wide.

6. Carpet at 30p per ft² is laid in a room 14 ft by $12\frac{1}{2}$ ft so as to leave a border $1\frac{1}{2}$ ft wide all round. Find: (i) the cost of carpeting; (ii) cost of staining the border at 2p per ft²; (iii) total cost.

7. A room measures 18 ft by 12 ft, the floor may be covered in one of two ways: (i) strips of carpet $2\frac{1}{4}$ ft wide at 85p per ft length fitted over the whole floor; or (ii) a carpet 12 ft by 9 ft costing £60, surrounded by a border of linoleum at 15p per ft². Which is the cheaper method and by how much?

8. An ornamental garden measures 100 ft by 80 ft. It contains four flower-beds each 16 ft by 8 ft surrounded by grass, and round the edge of the grass a path 6 ft wide. Find the cost of the grass seed if it is spread at $\frac{1}{8}$ lb for every 9 ft² and costs 20p per pound.

9. A rectangular pond 10 yd by $6\frac{2}{3}$ yd is surrounded by a path 2 yd wide. Find the area of the path in yd².

10. If a roll of paper is 36 ft by $1\frac{3}{4}$ ft and costs 75p, find the cost of papering a room 18 ft long, 12 ft wide, 8 ft high, allowing 39 ft² for doors, etc.

VOLUME OF A CUBOID

A solid whose faces are all rectangles is called a **rectangular solid** or **cuboid**; and if each face is a square, the solid is called a **cube**.

REMEMBER $V = l \times b \times h$

(i) Keep *all* dimensions in the same units.

(ii) Volume is measured in cubic units.

(iii) When calculating volumes employ the units in which you require the answer. Thus for an answer in cubic feet, work in feet; for an answer in cubic yards, work in yards. This will avoid the need to divide by 1728, etc.

(iv) An object occupies space (its own volume) whether solid or hollow.

EXERCISE 15

1. Find the volume of the following cuboids: (i) 2 in by 3 in by 4 in; (ii) 5 ft by 7 ft by 9 ft; (iii) 4 yd by 6 yd by 8 yd; (iv) $\frac{5}{12}$ ft by 9 ft by 12 ft; (v) $2\frac{1}{2}$ ft by $3\frac{1}{3}$ ft by 6 ft; (vi) $1\frac{1}{2}$ ft by $2\frac{2}{3}$ ft by $8\frac{3}{4}$ ft.

2. The volume of a cuboid is 600 ft³. Find its height if the base is 5 ft by 12 ft. (ii) Find its height if it is 16 ft long, 10 ft wide. (iii) Find its length if it is 5 ft high, 6 ft wide. (iv) Find its length if it is 2 ft high, 6 ft wide. (v) Find its width if it is $13\frac{1}{3}$ ft long, $4\frac{1}{2}$ ft high. (vi) Find its breadth if it is 25 ft long, $10\frac{2}{3}$ ft high.

3. A tank 5 ft long, 4 ft wide contains water to a depth of $2\frac{1}{2}$ ft. If the water is pumped into a tank 6 ft long, 3 ft wide, what will be the depth of water in the second tank?

4. If the second tank in No. 3 is 3 ft deep, how many more gallons of water will be required to completely fill it? ($1 ft^3 = 6\frac{1}{4} gal.$)

5. A steel block of volume 1 ft³ is in process of being rolled into strip metal. Find the length of strip (*in feet*): (i) when it is $\frac{1}{2}$ ft wide, $\frac{1}{6}$ ft thick; (ii) when it is $\frac{1}{2}$ ft wide, $\frac{1}{12}$ ft thick; (iii) when it is $\frac{1}{4}$ ft wide, $\frac{1}{48}$ ft thick.

6. How many gallons of water are required to fill a tank 40 ft long, 15 ft wide to a depth of 8 ft? ($1 ft^3 = 6\frac{1}{4} gal.$)

7. How long will it take to supply the water for the tank in No. 6 if water is pumped in at the rate of 4 ft³ s?

8. What is the weight of the water in the tank in No. 6 if 1 gal of water weighs 10 lb? (*Answer in tons, to the nearest ton.* 1 *ton* = 2240 *lb*).

SOLIDS OF UNIFORM CROSS-SECTION

A **section** is the shape we see when we cut through an object, the shape depending to some extent on the angle at which the cut is made. For our purposes we shall always assume that our "slices" are taken **at**

right-angles to the length, they are then known as **cross-sections.**
When all cross-sections through a solid are of **the same shape and
size** they are said to be **uniform**, and the solid is then described as
a solid of uniform cross-section.

EXERCISE 16

Find the volumes of the following uniform solids. All angles are
right-angles, dimensions are in inches.

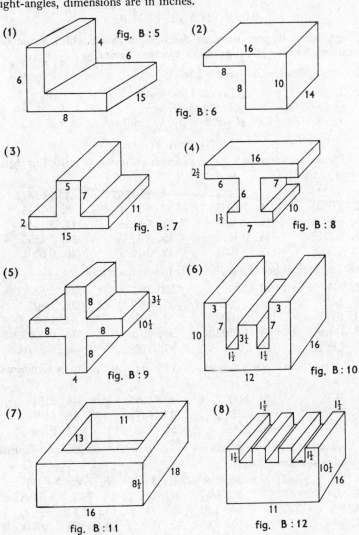

(1) fig. B : 5

(2) fig. B : 6

(3) fig. B : 7

(4) fig. B : 8

(5) fig. B : 9

(6) fig. B : 10

(7) fig. B : 11

(8) fig. B : 12

CORRECT TO A GIVEN DECIMAL PLACE

EXAMPLE. *Consider the number* 25·6375

(i) 25·6   ≃ 26   ∴ 25·6375 = 26 (*Corr. to nearest whole no.*)
(ii) 25·63   ≃ 25·6   ∴ 25·6375 = 25·6 (*Corr. to 1 dec. pl.*)
(iii) 25·637   ≃ 25·64   ∴ 25·6375 = 25·64 (*Corr. to 2 dec. pl.*)
(iv) 25·6375 ≃ 25·638   ∴ 25·6375 = 25·638 (*Corr. to 3 dec. pl.*)

SIGNIFICANT FIGURES

Significant figures may be defined as those which must be retained for any position of the decimal point.

EXAMPLE. *Consider the number* 25·6375

(i) 25·6375 = 30   (*Corr. to 1 sig. fig.*)
(ii) 25·6375 = 26   (*Corr. to 2 sig. fig.*)
(iii) 25·6375 = 25·6 (*Corr. to 3 sig. fig.*) and so on.

EXERCISE 17

Express the following numbers correct to (i) one, (ii) two, (iii) three places of decimals:

1. 0·1234	**2.** 1·3454	**3.** 23·4565	**4.** 12·5673
5. 5·0505	**6.** 7·1053	**7.** 9·5151	**8.** 6·7545
9. 0·9009	**10.** 2·3345	**11.** 4·0636	**12.** 3·1549
13. 5·2753	**14.** 0·5555	**15.** 8·3333	**16.** 7·4556
17. 1·0006	**18.** 3·1001	**19.** 2·1027	**20.** 0·0001

State the number of significant figures in the following:

21. 2	**22.** 20	**23.** 202	**24.** 20·2
25. 2·02	**26.** 0·202	**27.** 0·02	**28.** 10·02
29. 1002	**30.** 0·1002	**31.** 1020	**32.** 1·02
33. 1·0202	**34.** 10·202	**35.** 1020·2	**36.** 0·0010202
37. 19·01	**38.** 109·1	**39.** 0·0009	**40.** 9 000 000

Express the following numbers correct to (i) one, (ii) three significant figures:

41. 1505	**42.** 2115	**43.** 2056	**44.** 3154
45. 7149	**46.** 6358	**47.** 6978	**48.** 6596
49. 31·07	**50.** 591·2	**51.** 0·007 049	**52.** 0·046 98

Evaluate the following, correct to the number of significant figures shown in brackets:

53. 3·7 × 2·3 (2)	**54.** 5·8 × 3·9 (3)	**55.** 7·3 × 4·6 (2)
56. 509 ÷ 14 (1)	**57.** 60·7 ÷ 1·7 (2)	**58.** 7·13 ÷ 0·23 (1)
59. 9·5 × 6·7 (2)	**60.** 1·01 × 3·3 (3)	**61.** 12·4 × 5·9 (3)
62. 74·9 ÷ 2·7 (3)	**63.** 825 ÷ 32 (3)	**64.** 0·863 ÷ 0·048 (3)

APPROXIMATIONS IN DECIMAL COINAGE

$$1{\cdot}4p = 1\tfrac{1}{2}p \quad (Correct\ to\ nearest\ \tfrac{1}{2}p)$$
$$1{\cdot}4p = 1p \quad (Correct\ to\ nearest\ p)$$
$$£2{\cdot}158 = £2{\cdot}16 \quad (Correct\ to\ nearest\ p)$$
$$£2{\cdot}158 = £2 \quad (Correct\ to\ nearest\ £)$$

EXERCISE 18

Express the following in £s, correct to: (i) nearest penny; (ii) nearest £:

1. £7·054	**2.** £2·715	**3.** £3·508	**4.** £6·494
5. 82·7p	**6.** 123·3p	**7.** 9·61p	**8.** 10·87p
9. £4·62	**10.** £4.602	**11.** £4·006	**12.** £4·206
13. 200·7p	**14.** 207·7p	**15.** 270·7p	**16.** 277p
17. £8·154	**18.** £80·541	**19.** £80·045	**20.** £80·005

CHANGING DECIMALS TO VULGAR FRACTIONS

From our knowledge of place value it is clear that:

$$0{\cdot}1 = \tfrac{1}{10}, \ 0{\cdot}01 = \tfrac{1}{100}, \ 0{\cdot}001 = \tfrac{1}{1000}$$

The numerator is formed by the digits in the decimal; the denominator is formed by placing a 1 for the decimal point and a nought for each figure after it.

EXERCISE 19

Change the following to fractions expressed in their lowest terms:

1. 0·5	**2.** 0·25	**3.** 0·75	**4.** 0·125	**5.** 0·375
6. 0·625	**7.** 0·875	**8.** 0·2	**9.** 0·8	**10.** 0·24
11. 0·36	**12.** 0·45	**13.** 1·12	**14.** 9·025	**15.** 17·105
16. 5·95	**17.** 11·104	**18.** 8·31	**19.** 4·18	**20.** 3·05

CHANGING VULGAR FRACTIONS TO DECIMALS

To change a vulgar fraction to a decimal we **divide the numerator by the denominator.**

EXERCISE 20

Change the following to decimals, giving answers to 3 sig. fig. where necessary:

1. $\tfrac{1}{2}$	**2.** $\tfrac{1}{4}$	**3.** $\tfrac{3}{4}$	**4.** $\tfrac{1}{8}$	**5.** $\tfrac{3}{8}$
6. $\tfrac{5}{8}$	**7.** $\tfrac{7}{8}$	**8.** $\tfrac{2}{3}$	**9.** $\tfrac{4}{5}$	**10.** $\tfrac{9}{20}$
11. $\tfrac{13}{40}$	**12.** $\tfrac{27}{40}$	**13.** $\tfrac{59}{80}$	**14.** $\tfrac{73}{80}$	**15.** $1\tfrac{17}{23}$
16. $\tfrac{25}{30}$	**17.** $1\tfrac{14}{25}$	**18.** $3\tfrac{21}{27}$	**19.** $5\tfrac{13}{15}$	**20.** $9\tfrac{13}{17}$

SYSTÈME INTERNATIONAL d'UNITÉS*

The idea of a decimal system of units was conceived by Simon Stevin (1548–1620), who also developed the even more important concept of decimal fractions. Decimal units were also considered in the early days of the French Académie des Sciences founded in 1666, but the adoption of the metric system as a practical measure was part of the general increase in administrative activity in Europe which followed the French Revolution. Advised by the scientists of his day, the statesman Talleyrand aimed at the establishment of an international decimal system of weights and measures. It was based on the metre as the unit of length (it was intended to be one ten-millionth part of the distance from the North Pole to the equator at sea level through Paris, but the circumstances did not permit this aim to be achieved with any great accuracy) and the gramme as the unit of quantity of matter. The gramme was to be the mass of one cubic centimetre of water at $0°$ C.

Although the metric system was primarily devised as a benefit to industry and commerce, physicists soon realized its advantages and it was adopted also in scientific and technical circles. In 1873 the British Association for the Advancement of Science selected the centimetre and the gramme as basic units of length and mass for physical purposes. The United Kingdom is changing to the metric system at a time when a rationalized system of metric units, the Système International d'Unités (SI), is coming into international use. The SI derives all the quantities needed in all technologies from only six basic units. This contrasts with the metric systems currently used, in which additional quantities (for instances "calorie" and "horsepower") are differently defined in different metric countries. Relationships between units are thus greatly simplified in the SI, the introduction of which offers existing metric countries a unique opportunity to harmonize their measuring practices. This opportunity is being seized, and many countries have passed or are preparing legislation to make the SI the only legal system of measurement. It is therefore a logical choice for the United Kingdom. In most respects the use of SI units instead of metric technical units will have little effect on everyday life or trade. The metre and the kilogramme remain the units of linear measure and mass, and the litre, now accepted for most purposes as a special name for the cubic decimetre, will be commonly used as a unit of volume.

The kilogramme is still defined in terms of the international prototype at Sèvres, but the metre is now defined in terms of a number of wavelengths of a particular radiation of light.

* With acknowledgements to British Standards Institution PD 5686 "The Use of SI Units".

THE METRIC SYSTEM

Let us consider the complete set of metric weights and measures:

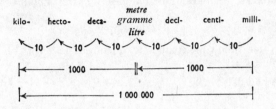

The following example illustrates the simplicity of the system:

1 kg   2 hg   3 dg   4 g   5 dg   6 cg   7 mg

By inserting a decimal point after a unit the answer is given in those units, thus:

1·234 567   kg
12·345 67   hg
123·4567   dag
1234·567   g
etc.

WEIGHT

In the metric system the standard unit of weight is the **gramme**, a very small weight, as you can judge from the fact that a thousand of them (1 kilogramme) approximate to $2\frac{1}{5}$ lb. 1 gramme is the weight of 1 cubic centimetre (1 cm³) of water.

CAPACITY

In the metric system the unit of capacity is the **litre**, which is equal to 1000 cm³. Since 1 cm³ of water weighs 1 g, then 1000 cm³ will weigh 1000 g, or 1 kg. Therefore 1 litre (l) of water weighs 1 kg. A litre is about $1\frac{3}{4}$ pints.

AREA AND VOLUME

AREA.   100 mm² = 1 cm²; 100 cm² = 1 dm²
VOLUME.   1000 mm³ = 1 cm³; 1000 cm³ = 1 dm³
etc.

NOTE

(i) Quantities are usually expressed in one unit only.

(ii) The most common units are: (*a*) kilometres, metres, millimetres; (*b*) kilogrammes, grammes, milligrammes; (*c*) litres, millilitres.

EXERCISE 21

Express the following compound quantities in terms of the unit shown in brackets:

1. 5 kl 4 hl 3 dal 2 l (l)
2. 4 dag 6 g 2 dg 7 cg (g)
3. 8 hm 1 dam 9 m 5 dm (m)
4. 7 km 4 hm (km)
5. 8 kg 6 dag (kg)
6. 4 dal 6 l (l)
7. 5 kg 4 g (kg)
8. 8 dam 2 dm (m)
9. 3 l 5 dl (l)
10. 7 g 9 cg (g)
11. 6 cm 8 mm (m)
12. 4 l 3 cl (l)
13. 2 m 5 cm (cm)
14. 6 dg 7 cg (g)
15. 3 g 9 cg (mg)
16. 5 dm 2 mm (cm)
17. 10 km 8 cm (m)
18. 5 hg 3 dg (kg)
19. 7 mg (g)
20. 3 kg 3 mg (mg)

EXERCISE 22

Add the following, expressing the answer in the *highest* units given:

1.	g	dg	cg	2.	km	hm	dam	3.	l	dl	cl
	4	3	9			4	7		9	3	
	2	0	7		6	3	5		1	0	7
		9	6		8	1	3		2	4	8
	5	1			2	0	9			6	5

4.	kg	hg	dag	5.	m	dm	cm	6.	l	dl	cl
	7	4	9		2	1	9		6	5	8
	3	0	7		4	3	2		2	0	7
		2	8		1	7	5		7	0	4
	5	1	4			8	6		1	3	9

Subtract the second quantity from the first, expressing the answer in the *lowest* units given:

7.	hm	dam	m	8.	dg	cg	mg	9.	hl	dal	l
	9	4	3		3	1	5		5	2	7
	2	5	7		2	2	9		3	6	8

10.	g	dg	cg	11.	m	dm	cm	12.	hg	dag	g
	2	0	4		4	3	1		7	1	5
	1	0	5		2	0	2		3	0	6

MONEY

France: 100 centimes (c.) = 1 franc (fr.)
America: 100 cents (c.)   = 1 dollar ($)

MISCELLANEOUS EXAMPLES

EXERCISE 23

Give answers correct to three significant figures where necessary.

1. Find, in francs, the total cost of 27 articles at 1 fr. 13 c. each.

2. A rectangle measures 3 m 4 dm by 8 dm 7 cm. Find: (i) the perimeter in metres; (ii) the area in square metres.

3. Express: (i) 35 dm in km; (ii) 35 mg in g.

4. How many pieces of tape, each 2 dm 3 cm long, can be cut from a length of 3 m and how many centimetres remain?

5. A pile of 30 metal plates is 0·72 dm high; find the thickness of each plate in millimetres.

6. A quantity of chemical weighing 3·05 kg is divided into packets, each containing 50 g. How many packets are there and how many grams remain?

7. How many litres in: (i) 0·15 hl; (ii) 0·15 cl?

8. (i) How many cm³ are there in 0·3 m³? (ii) How many m² are there in 680 cm²?

9. If 1 kg is approximately equal to 2·205 lb, express 1 lb in kilogrammes.

10. If 1 gallon is approximately equal to 4·546 litres, express 1 litre in pints.

11. A box measures 3·6 dm by 0·04 m by 8·2 cm, find the volume in cubic centimetres.

12. How many articles costing 7 fr. 70 c. each can be purchased for 500 fr., and how much money remains?

13. The area of a rectangle is 3·42 m², the width is 11·7 dm; find the length in centimetres.

14. If 1 cm³ of gold weighs 19·3 g, find the weight of 5·45 cm³ of gold.

15. A can measures 0·8 dm by 9·3 cm by 14·6 mm, find its capacity in litres. (HINT. *Work in cm³*.)

16. A beaker full of water weighs 215 g, when half full of water it weighs 145 g. Find the weight of the empty beaker.

17. A lawn 16·5 m long, 12·4 m wide, has a flower-bed, 1·6 m wide, surrounding it. Find the area of the flower-bed in square metres.

18. A dish has a base 20 cm square, its capacity is 2·52 litres; find its depth in millimetres.

19. 1 kilometre ≃ 0·6214 mile. Express 1 mile in kilometres.

20. 1 cm³ of alcohol weighs 0·78 g. Find the weight of 3½ litres of alcohol: (i) in kilogrammes; (ii) in pounds. (1 *kg* = 2·205 *lb*.)

FURTHER MENSURATION

Area of parallelogram = Base × Height

$$\text{Area of triangle} = \frac{\text{Base} \times \text{Height}}{2}$$

Area of trapezium = Average width × Height

Volume of uniform solid = Area of section × Length

Volume of uniform solid = Area of base × Height

EXERCISE 24

Give answers correct to three figures where necessary.

Nos. 1–20 refer to fig. B : 13.

Find the area of parallelogram $ABCD$, when:

1. $CD = 5$ in, $AX = 3$ in   **2.** $AD = 4$ in, $CZ = 10$ in
3. $BQ = 6.5$ in, $AB = 20$ in   **4.** $BC = 3\frac{3}{4}$ in, $PD = 4.8$ in
5. $CD = 8.05$ in, $BQ = 3.12$ in   **6.** $AB = 5.9$ cm, $AX = 45$ mm
7. $AD = 2.45$ m, $CZ = 18$ dm   **8.** $BQ = 0.05$ dam, $CD = 0.15$ m
9. $CD = 1\frac{1}{2}$ yd, $AX = 3\frac{1}{3}$ yd
10. $BC = 3.12$ ft, $CZ = 5.6$ ft

Find the area of triangle ACD, when:

11. $CD = 8$ in, $AX = 5$ in
12. $AD = 5\frac{1}{2}$ in, $CZ = 9\frac{1}{4}$ in
13. $CZ = 9.05$ cm, $AD = 3.4$ cm
14. $CD = 17$ in, $AX = 10$ in
15. $AD = 1.15$ m, $CZ = 5.8$ m
16. $AX = 4\frac{1}{2}$ yd, $CD = 4\frac{1}{2}$ yd

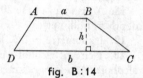

fig. B : 13

Find the area of triangle BCD, when:

17. $BC = 9\frac{1}{2}$ cm, $PD = 0.19$ m
18. $CD = 10\frac{1}{6}$ yd, $BQ = 3\frac{1}{2}$ yd
19. $BC = 4.05$ in, $PD = 13.3$ in
20. $CD = 130$ mm, $BQ = 5$ cm

Nos. 21–25 refer to fig. B : 14.

Find the area of trapezium $ABCD$, when:

21. $a = 6$ in, $b = 18$ in, $h = 10$ in
22. $a = 8$ cm, $b = 2$ dm, $h = 0.7$ dm
23. $a = 16$ in, $b = 21$ in, $h = 8.75$ in
24. $a = 0.1$ m, $b = 1.85$ m, $h = 25$ cm
25. $a = 8\frac{3}{4}$ in, $b = 22\frac{1}{2}$ in, $h = 10.2$ in

fig. B : 14

26. If the area of a parallelogram is 30·06 in² and the base is 5·5 in, find the height.

27. If the area of a parallelogram is 328 cm² and the height is 32 mm, find the base length.

28. If the area of a parallelogram is 216 in² and the base is 20·5 in, find the height.

29. If the area of a triangle is 30 in² and the base is 15 in find the height.

30. If the area of a triangle is 25 cm² and the height is 25 mm, find the base length.

31. If the area of a triangle is 2·4 yd² and the base is $\frac{11}{12}$ yd, find the height.

32. If the area of a trapezium is 35 m² and the average width is 1 dam 5 m, find the distance between the parallel sides.

33. If the area of a trapezium is 460·8 in² and the distance between the parallel sides 48 in, find the average width.

34. If the area of a trapezium is $49\frac{1}{2}$ in², the height $5\frac{1}{2}$ in and one of the parallel sides 8 in, find the length of the other parallel side.

35. If the area of a trapezium is 1350 cm², the height 9 cm and one of the parallel sides 2 m, find the length of the other parallel side.

EXERCISE 25

Give answers correct to three figures where necessary.

1. A swimming-bath is 60 ft long and 30 ft wide, the depth of water increases uniformly from 3ft to 10 ft. If 1 ft³ water equals $6\frac{1}{4}$ gal, find the number of gallons of water in the bath.

2. If 1 ft³ of water weighs $\frac{1}{36}$ ton, find, in tons, the weight of the water in No. 1.

3. A triangular glass prism is 5·6 cm long. The triangle has a base of 3·2 cm and a height of 2·4 cm. If 1 cm³ of glass weighs 2·75 g, find the weight of the prism.

4. Fig. B : 15 represents a shed.

   $AB = 15$ ft
   $DE = 12$ ft
   $CD = 9$ ft
   $EF = 22$ ft

Find the volume.

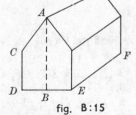

fig. B : 15

5. A tank used for electro-plating is 43 ft long, 7 ft wide and contains 10 000 gal of fluid. Find the depth of the fluid if 1 gal = 0·1605 ft³.

6. Fig. B: 16 represents a lean-to shed. Find the volume.

7. Find the cost of painting the three exposed walls of the shed in No. 6 at 3p per ft².

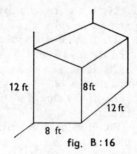

fig. B : 16

8. Fig. B: 17 shows a metal prism where $a = 18$ in, $b = 20$ in, $h = 12$ in, $l = 12$ in. Four dozen of these prisms are lowered into the tank described in No. 5, find, to the nearest inch, the rise in the level of the fluid.

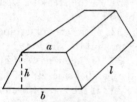

fig. B : 17

9. A railway cutting is to be constructed. It is 40 ft wide at the top 20 ft wide at the bottom and 20 ft deep. If 1 ft³ soil weighs $\frac{1}{20}$ ton, how many tons of soil must be removed per 3000 ft of cutting?

10. An aquarium contains 22 litres of water to a depth of 25 cm. If the tank is 40 cm long, find the width.

11. Six scoops of sand are poured into the tank in No. 10, the water rises 6 cm. Find the volume of sand in one scoop.

12. A swimming-bath is 30 ft wide and the depth increases uniformly from 3 ft to 10 ft. The bath is nearly emptied so that the bottom is just covered with water, the quantity remaining being 52 500 gallons. Find the length of the bath and the number of gallons it contains when full. $(1\,ft^3 = 6\frac{1}{4}\,gal.)$

AVERAGES

EXERCISE 26

Find the average of:

1. 5, 8, 12, 19, 21   **2.** 37, 43, 57, 68, 72, 89
3. 211, 273, 327, 358, 434, 485   **4.** 2·73, 1·07, 3·59, 4·6, 0·96
5. 2·09 ft, 2·162 ft, 2·27 ft, 2·086 ft
6. £2·50, £3·83, £1·98, £4·25
7. 35 g, 13·6 dag, 260 dg, 82 g, 0·76 hg
8. 350 dam, 0·79 km, 20·1 hm, 4620 m
9. $7\frac{3}{4}$, $5\frac{1}{2}$, $3\frac{2}{3}$, $2\frac{1}{4}$, $4\frac{1}{3}$
10. 69p, £0·95, £1·51, 107p, £1·48

PROPORTION: UNITARY METHOD

EXAMPLE. *If it requires* 4 *days for* 10 *men to dig a trench, find how long it will take* 8 *men.*

Estimate: Fewer men will require longer time.

10 men take 4 days

∴ 1 man takes (4 × 10) days

∴ 8 men take $\frac{40}{8}$ days　　　　*Ans.* = 5 days

EXERCISE 27

1. If 7 doz. articles cost £3·50, find the cost of 9 doz.
2. If 130 articles cost £2·86, find the cost of 70 articles.
3. If 9 yd of cloth cost £2·52, find the cost of $5\frac{1}{2}$ yd.
4. If 2 tons of coal cost £19·80, find the cost of $3\frac{1}{4}$ tons.
5. If 1 gal of water weighs 10 lb, find the weight of $\frac{11}{16}$ gal.
6. If 1 km = $\frac{5}{8}$ mile, find $\frac{1}{2}$ mile in metres.
7. If I buy 5 m of material for £1·87$\frac{1}{2}$, how many metres can I buy for £3?
8. A man owes £200 but possesses only £150. How much should he pay to someone to whom he owes £40?
9. 6 men build a wall in 10 days, find how long it will take 4 men.
10. If 40 men complete a job in 21 days, find how long it will take 56 men.
11. A garrison has sufficient food to feed 1000 men for 4 weeks. How long will the food last for 1400 men?
12. 6 pipes empty a bath in 1 hr 20 min. How long will it take if only 5 pipes are used?
13. 25 cm³ of gold weigh 482·5 g. Find the weight of a gold prism 1·5 cm by 7·5 mm by 1·2 cm.
14. A journey by car takes 2 h at 24 mile/h. How long would the journey take at 30 mile/h?
15. A car travels 35 miles in 45 min. How far will it travel in 55 min?
16. 30 men can complete a job in 12 days. How many extra men will be required if the job is to be completed in 10 days?
17. A map uses a scale of 6 cm to 1 km. What distance on the map will represent 1200 m?
18. The scale of a map is 6 cm to 1 km. What area of ground, in acres, is represented by 27 cm² on the map?
19. If a train goes 40 miles in 2 h, how far will it go in 3 h?
20. If a train goes d miles in h h, how far will it go in t h?
21. A train completes a journey in 4 h travelling at 30 mile/h. How long will it take at 40 mile/h?
22. A train travelling at u mile/h takes t h to complete a journey. How long will it take at v mile/h?

RATIO

NOTE

(i) A ratio offered as an answer to a question MUST contain the ratio as a statement in words as well as figures, the order of the figures being in agreement with the written statement.

(ii) The ratio should be in its simplest form: $6 : 4 = 3 : 2$ or $1\frac{1}{2} : 1$.

COMPARISON OF RATIOS

To ascertain which of a number of ratios is the largest, it is necessary to express each ratio in the form $m : 1$.

REPRESENTATIVE FRACTIONS

An interesting application of ratio is to be found on maps, where the scale is often given in the form $1 : m$. Thus:

$$(a) \ 1 : 2\,500\,000; \quad (b) \ 1 : 2\,000\,000$$

When written as a fraction, $(a) \ \dfrac{1}{2\,500\,000}$, is called the **representative fraction** or R.F. of the map. Express (b) as an R.F.

EXERCISE 28

1. A earns £1200 a year, B earns £1500 a year. What is the ratio of A's income to B's income?

2. A saves £450 a year, B saves £550 a year. What is the ratio of B's savings to A's savings?

3. A earns £1200 and saves £450, B earns £1500 and saves £550. Which savings to earnings ratio is larger?

4. A school contains 250 girls, 300 boys. Find the ratio of: (a) number of girls to number of boys; (b) number of girls to total number of pupils.

5. In a single year a 72-in tree grows 5 in and a 14-ft tree grows 1 ft. Which ratio of growth to height is the greater?

6. Income tax is levied at $37\frac{1}{2}$p in the £. Express the tax as a ratio of income. What tax would be paid on £400?

7. A rectangle measures 3 in by 4 in. If the sides are doubled in length, find the ratio of the new area to the original area.

8. A 2-in cube has the dimensions trebled. Find the ratio of the new volume to the original volume.

9. A map of New Zealand has a scale of 1 cm = 138 km. Find the R.F. correct to two significant figures.

10. Give the ratio of the height of Everest 29 002 ft to the height of Mont Blanc 15 781 ft in the form $m : 1$ correct to three significant figures.

RATIO AND PROPORTION

EXAMPLE. *If* 30 *men have sufficient food for* 28 *days, how long will the food last for* 40 *men?*

30 men have food for 28 days

Decrease in ratio
30 : 40 $\}$ ∴ 40 men have food for $\left(\frac{28}{1} \times \frac{30}{40}\right)$ days

Ans. = $\underline{\underline{21 \text{ days}}}$

EXERCISE 29 (1–12 *for Oral Discussion*)

1. Increase 30 in the ratio 6 : 5.
2. Increase $2\frac{1}{4}$ lb in the ratio 9 : 4.
3. Increase £2·40 in the ratio 7 : 6.
4. Increase $3\frac{1}{3}$ ft in the ratio 8 : 5.
5. Increase $1\frac{1}{2}$ days in the ratio 14 : 9.
6. Decrease 40 in the ratio 5 : 8.
7. Decrease $5\frac{5}{8}$ lb in the ratio 5 : 6.
8. Decrease 96p in the ratio 15 : 28.
9. Decrease 33 yd in the ratio 7 : 11.
10. Decrease 9 h 15 min in the ratio 4 : 5.
11. What multiplying factor changes 50 into 25?
12. What multiplying factor changes 2 ft into 4 ft?
13. The price of a commodity changes from 48p to 42p. In what ratio has the price changed?
14. If 7 lb of grass seed cost £1·12, find the cost of 5 lb.
15. The wages bill for 50 men is £825, find the bill for 36 men.
16. If 12 men can be employed for £105, how many can be employed for £131·25?
17. If 15 men can build a wall in 48 hr, how many men will be required to do the work in 30 hr?
18. A car takes 2 hr for a journey at an average speed of x mile/h. How long will it take at an average speed of y mile/h?
19. A solid block of material measures 2 in by 3 in by $4\frac{1}{2}$ in and weighs 27 oz. Find the weight of a block $2\frac{1}{2}$ in by 3 in by 4 in of the same material.
20. Out of 28 lb of apples only $24\frac{1}{2}$ lb are fit to eat. How many pounds are wasted out of 120 lb of the same apples?
21. A 112-lb bag of fuel costs 56p. If 2 lb of the fuel turns out to be slate, what price per bag should be charged for the fuel?
22. With income tax at 36p in the £, a man pays £33 in tax. What will he pay if the tax is increased to 48p in the £?
23. It was estimated that by employing 32 men a piece of work could be completed in 10 days. Extra men were engaged and the work was finished in 8 days. How many extra men were taken on?

COMPOUND PROPORTION

EXAMPLE. *The wages for 6 men for 5 days amount to £75. Find the wages for 8 men for 12 days.*

6 men in 5 days earn £75

Increase in ratio $\left.\begin{array}{l}\\12:5 \text{ days}\end{array}\right\}$ $\therefore$ 6 men in 12 days earn $£(\frac{75}{1} \times \frac{12}{5})$

Increase in ratio $\left.\begin{array}{l}\\8:6 \text{ men}\end{array}\right\}$ $\therefore$ 8 men in 12 days earn $£(\frac{75}{1} \times \frac{12}{5} \times \frac{8}{6})$

NOTE   *Ans.* $= \underline{\underline{£240}}$

(i) Deal with one variable at a time. In the example the first multiplying factor was the ratio 12 days to 5 days; the second multiplying factor was the ratio 8 men to 6 men.

(ii) Each ratio must be arrived at after considering whether the *changing quantity* will produce an *increase* or *decrease* in the *quantity to be found*.

EXERCISE 30

1. In 10 days 8 men earn £192, find how much 10 men earn in 12 days.

2. 12 men earn £576 in 16 days. How much will 20 men earn in 21 days?

3. If 10 men can earn £270 in 9 days, how much will 8 men earn in 12 days?

4. In 20 days 16 men earn £960, find how much 24 men earn in 15 days.

5. 20 men earn £200 in 4 days. How long can 16 men be employed for £120?

6. In 6 days 5 men build a wall 4 ft high. How long will it take 3 men to build a wall 3 ft high?

7. One mile of road can be built by 20 men in 10 days. How long will 30 men take to build $2\frac{1}{4}$ miles of road?

8. Working 9 hr a day, 15 men complete a job in 16 days. How many men, working 8 hr a day, will complete the work in $11\frac{1}{4}$ days?

9. A lawn 20 ft by 15 ft costs £6·50 to turf. Find the cost of turfing a lawn 30 ft by 25 ft.

10. The cost of transporting 40 crates 130 miles is £16·50. Find the transport costs for 32 crates for a distance of 175 miles.

11. Electricity charges are reduced from 1·25p per unit to 1·125p per unit. At the same time a householder finds his consumption has risen from 480 units to 520 units. In what ratio does the cost increase or decrease?

12. In 30 days 60 men complete $\frac{3}{4}$ of a job. How many extra men are required to complete the job in a total of 36 days?

PROPORTIONAL PARTS

EXAMPLE. *Divide £126 so that A has 3 times as much as B, who has twice as much as C.*

Let   C's share be 1 unit
Then   B's share is 2 units
and   A's share is 6 units
∴ Total number of units = 9
∴ Value of 1 unit   = £$\frac{126}{9}$
∴ C's share   = £14
B's share   = £14 × 2
A's share   = £14 × 6

Ans. = *A*'s share £84; *B*'s share £28;

C's share £14

Check: £84 + £28 + £14 = £126

EXERCISE 31

1. Divide 18 in the ratio 1 : 2 : 3

2. Divide 27 ft in the ratio 1 : 3 : 5

3. Divide £48 in the ratio 3 : 4 : 5

4. Divide 112 lb in the ratio 5 : 10 : 13

5. Divide $1\frac{1}{2}$ hr in the ratio 7 : 10 : 13

6. Divide 13 in in the ratio $\frac{1}{2}$: $\frac{1}{3}$: $\frac{1}{4}$

7. Divide £15 in the ratio 4 : 2 : 1 : $\frac{1}{2}$

8. In what ratio are £1·50; £3; £0·87$\frac{1}{2}$?

9. In what ratio are $3\frac{1}{3}$ ft; $2\frac{1}{2}$ ft; $1\frac{5}{6}$ ft?

10. In what ratio are $1\frac{1}{2}$ tons; $\frac{4}{5}$ tons; $1\frac{3}{5}$ tons?

11. *A, B* and *C* share £1800 in the ratio $1\frac{1}{4}$: $3\frac{1}{2}$: $4\frac{1}{4}$. What does each receive?

12. *A, B* and *C* invest £350, £400, £850 respectively in a business. If the profit is £1500, what should each receive?

13. A manufacturing concern owes £380, £640, £460, £520 respectively to four suppliers *A, B, C* and *D*. If £1500 is available, how much should each supplier receive?

14. *A, B* and *C* each invest the same sum in a business. *A* invests his money for three times as long as *B*, who has invested his money only half as long as *C*. How should a profit of £120 be shared?

15. In his will a man leaves half his fortune to his wife. One-eighth of the remainder is given to charity and the children, *A, B* and *C*, share the rest in the ratio 2 : $2\frac{1}{4}$: $2\frac{3}{4}$ respectively. If *A* receives £466·25, what is the total fortune?

STRAIGHT-LINE GRAPHS

Graphs which represent a pair of **continuous variables** have an advantage over those forms illustrating **categorical data** because of the mathematical relationship between the two variables. For every value of the **independent variable** the value of the **dependent variable** will be **in proportion**. This makes it possible to read off intermediate values of the variables. The process is called **interpolation**, and some degree of approximation is occasionally necessary.

REMEMBER

(i) Give the graph a title.

(ii) Let the horizontal axis represent the independent variable and the vertical axis represent the dependent variable.

(iii) Choose convenient scales for each axis and mark the scale at regular intervals.

(iv) Label each axis.

(v) Show interpolation clearly.

EXERCISE 32

Ready Reckoner Graphs

1. Construct a vertical axis to represent 0–50 pence. Construct a horizontal axis to represent 0–5 kg. Draw graph lines to represent: (i) 25p per kg; (ii) 10p per kg; (iii) 6p per kg. In each case: (a) How many kg can be bought for 25p? (b) What is the cost of 1·75 kg?

2. Using a horizontal axis for 0–40 hours and a vertical axis for 0–£20, construct a wages ready reckoner suitable for the following rates of pay: (i) 20p per hour; (ii) 35p per hour; (iii) 50p per hour. In each case: (a) What pay is received for 36 hr work? (b) How many hours are worked for a wage of £3?

Conversion Graphs

3. Given that 1 km = $\frac{5}{8}$ mile (8 km = 5 miles), construct a conversion graph, miles/kilometres. From your graph estimate: (a) 1 mile in km; (b) 5 km in miles.

4. Given that 1 l = $1\frac{3}{4}$ pt (4 l = 7 pt); construct a conversion graph, pints/litres. From your graph estimate: (a) 1 pt in l; (b) 3·5 l in pt.

5. Given that 60 mile/h = 88 ft/s, construct a conversion graph, ft/s against mile/h. From your graph estimate: (a) 50 mile/h in ft/s; (b) 60 ft/s in mile/h.

6. Given that 1 kg = $2\frac{1}{5}$ lb, construct a conversion graph (0–10 kg), kg/lb. From your graph estimate: (a) 5 lb in kg; (b) $7\frac{1}{2}$ kg in lb.

7. In an experiment to compare the two temperature scales a Fahrenheit and Celsius thermometer were placed in water which was being heated. Readings were taken simultaneously, and the following results were obtained:

Temperature, °C	0°	20°	40°	60°	80°	100°
Temperature, °F	30°	65°	114°	144°	172°	214°

Let the horizontal axis represent °C and use a suitable scale for each axis. Allowing for experimental error, draw the "line of best fit" and from the graph estimate: (*a*) the temperature in °F which corresponds to 8°C; (*b*) the temperature in °C which corresponds to 160°F.

8. During a journey a cyclist noted the distance travelled and the time elapsed since he started. The results are shown in the table:

Time elapsed (*T*)	30 min	1h	1h 36 min	2h	2h 20 min	3h	3h 30 min
Distance travelled (*D*)	7 miles	14 miles	21 miles	26 miles	32 miles	40 miles	48 miles

Draw the distance/time graph and from the line of best fit estimate: (*a*) the distance travelled in 45 min; (*b*) the time taken to travel 35 miles.

NOTE. *By convention we represent "time" on the horizontal axis and in doing so we regard time as the independent variable.*

9. In an experiment to investigate the elasticity of a length of rubber cord (Hooke's Law) the following observations were made:

Load in g (*L*)	0	50	100	200	350	525
Length of cord in cm (*l*)	20	20·4	21	22·1	23·7	25·1

Allowing for experimental error, construct a graph to represent the data. From the graph estimate: (*a*) the length of the cord under a load of 300 g; (*b*) the load carried when the cord is 21·5 cm long.

10. The following is a record of a baby's weight and the corresponding average daily consumption of food in calories:

Weight in lb (*W*)	6	8	10	12	14	16
Food in calories (*C*)	304	410	490	590	705	807

Represent this data by a suitable straight-line graph and from it estimate:

(*a*) the average intake of food per day (in cal) for a weight of 11 lb;
(*b*) the weight of the baby for a food intake of 350 cal per day.

PERCENTAGE

RULES

(i) **Fractions or decimals to percentages.**

Multiply by $\frac{100}{1}$

    (*a*) $\frac{5}{8} = (\frac{5}{8} \times \frac{100}{1})\% = 62\frac{1}{2}\%$
    (*b*) $0{\cdot}375 = (0{\cdot}375 \times 100)\% = 37{\cdot}5\%$

(ii) **Percentages to fractions or decimals.**

Express as a fraction whose denominator is 100

    (*a*) $60\% = \frac{60}{100} = \frac{3}{5}$
    (*b*) $43\% = \frac{43}{100} = 0{\cdot}43$

NOTE

(i) Any suitable method may be employed for the multiplication of money.

(ii) If money is expressed as a decimal of £, five decimal places may be required to provide an answer correct to the nearest penny.

EXERCISE 33

1. Express £3·25 as a percentage of £5.

2. Express 1·6 g as a percentage of 4 g.

3. Express 2·7 m as a percentage of 3 m.

4. Express 3·6 l as a percentage of 24 l.

5. Express 385 m as a percentage of 5 km.

6. Find $7\frac{1}{2}\%$ of £800   **7.** Find 38% of 60 m.

8. Find 45% of 1·16 kg.   **9.** Find 28% of £5.

10. Find 35% of £4.

11. Find 56% of £1.

12. Find correct to the nearest penny, 70% of £16·07.

13. Find, correct to the nearest penny, 162% of £2·82.

14. Find, correct to 3 sig. fig., 15% of 9·27 m.

15. Find, correct to 3 sig. fig., $38\frac{1}{2}\%$ of 3·35 kg.

16. A tank of capacity 80 litres is 60% full. How many more litres, are required to fill it?

17. If 22% of a number is 16, what is 33% of the number?

18. An article priced at £2·75 is reduced by 80p. What is the percentage reduction, correct to three significant figures?

19. A running track is 3·5 dm short of the correct length of 400 m. What is the percentage error, correct to three significant figures?

20. Tinsmiths' solder is composed of 49% tin, 2·5% antimony by weight, the rest is lead. In 1 kg of this alloy, what weight is lead? (*Ans. in g.*)

EXAMPLE 1. *Increase £30 by 15%*

$$£\frac{\cancel{30}}{1} \times \frac{\overset{23}{\cancel{115}}}{\underset{2}{\cancel{100}}} = £\frac{69}{2} \qquad Ans. = \underline{\underline{£34 \cdot 50}}$$

The ratio of the **new value to the old** (115 : 100) can be expressed as a fraction $\frac{115}{100}$ and used as **a multiplying factor.** This method (using ratio) is quicker than finding 15% of £30 and adding it to the £30.

EXAMPLE 2. *Decrease £30 by 15%*

$$£\frac{\cancel{30}}{1} \times \frac{\overset{17}{\cancel{85}}}{\underset{2}{\cancel{100}}} = £\frac{51}{2} \qquad Ans. = \underline{\underline{£25 \cdot 50}}$$

EXERCISE 34 (1–20 *Oral*)

Find the required *multiplying factor* to increase a number by:

1. 6%	**2.** 10%	**3.** 25%	**4.** 18%	**5.** $12\frac{1}{2}$%
6. 30%	**7.** 50%	**8.** 100%	**9.** 120%	**10.** 150%

Find the required *multiplying factor* to decrease a number by:

11. 6%	**12.** 10%	**13.** 25%	**14.** 18%	**15.** $12\frac{1}{2}$%
16. 30%	**17.** 50%	**18.** 75%	**19.** 1%	**20.** 100%

21. Increase 160 by 20% **22.** Increase 240 by 25%
23. Increase 80 by 40% **24.** Increase 128 by 75%
25. Decrease 60 by 10% **26.** Decrease 120 by 15%
27. Decrease 96 by 25% **28.** Decrease 60 by 15%
29. Increase 84 by 15% **30.** Decrease 150 by 19%

31. 25% of a number is 35. Find the number.

32. $12\frac{1}{2}$% of a sum of money is £$1 \cdot 37\frac{1}{2}$. Find the sum of money.

33. A man's rent is £180 p.a. If this is 18% of his income, find his annual income.

34. A man saves 10% of his income. Find his income if he spends £900 a year.

35. A man's weight increases from 100 kg to 109 kg. What is the percentage increase in weight, correct to three figures?

36. A rectangle 8 in by 4 in is decreased to 7 in by $3\frac{1}{2}$ in. Find the percentage decrease in area, correct to three figures.

37. The length of a rectangle is increased by 5% and the breadth is decreased by 3%. What is the percentage change in area?

38. When 16% filled, a tank contains 5·6 litres. What will it contain when 35% filled?

39. By what percentage must 550 be changed to give 374?

40. A suit and overcoat together cost £32·50. The coat cost $37\frac{1}{2}\%$ less than the suit. What did each cost separately?

41. What is the error per cent, correct to three figures, in taking 1 kg as $2\frac{1}{5}$ lb instead of 2·205 lb?

42. The difference between two measurements is 2·6 cm. The larger is 105%, the smaller is 92% of the correct value. Find the correct measurement in metres.

43. In a local fruit crop from two farms, 56% of the fruit came from one farm, and this was 12 tonnes more than the crop from the second farm. What was the total crop?

44. An article costs £1·12$\frac{1}{2}$, if the price is increased by 4%, find the new cost.

45. An article costs £3·30, if the price is decreased by 8%, find the new cost correct to 1p.

PROFIT AND LOSS

NOTE

1. Profit or loss per cent is a method of expressing the **profit or loss as a percentage of the cost price.**

2. The number of articles does not affect the profit or loss per cent. Thus the percentage may be based on the C.P., S.P. and profit on **one article** or the **total** C.P., S.P. and profit on 30 articles.

EXERCISE 35 (*Oral*)

Find the profit or loss per cent in the following:

1. C.P. £10, profit £5	**2.** C.P. £5, loss £2
3. C.P. £15, profit £3	**4.** C.P. £20, loss £4
5. C.P. £18, profit £3	**6.** C.P. £9, loss £3
7. C.P. £50, loss £2	**8.** C.P. £75, profit £15
9. C.P. 90p, profit 6p	**10.** C.P. 15p, loss 3p
11. C.P. £1·08, S.P. £1·44	**12.** C.P. £1·20, S.P. £0·90
13. C.P. £5, S.P. £5·50	**14.** C.P. £1·70, S.P. £0·85

Given the C.P., what multiplying factor will give the S.P. in the following:

15. Profit 50%	**16.** Profit 10%	**17.** Profit 6%
18. Profit $12\frac{1}{2}\%$	**19.** Loss 3%	**20.** Loss 30%
21. Loss $33\frac{1}{3}\%$	**22.** Loss $66\frac{2}{3}\%$	

Given the S.P., what multiplying factor will give the C.P. in the following:

23. Loss 3%	**24.** Profit $2\frac{1}{2}\%$	**25.** Profit 4%
26. Loss 8%	**27.** Loss 15%	**28.** Profit 45%
29. Loss $2\frac{1}{2}\%$	**30.** Profit $12\frac{1}{2}\%$	

Find the S.P. in the following:

31. C.P. £10, profit 20%   **32.** C.P. £5, loss 20%
33. C.P. £30, loss 10%   **34.** C.P. 50p, profit 5%
35. C.P. £12, loss 15%

Find the C.P. in the following:

36. S.P. £6, loss 4%   **37.** S.P. £45, loss 10%
38. S.P. £28, gain 12%   **39.** S.P. £53, gain 6%
40. S.P. £18, loss 28%

PROFIT AND LOSS: PROBLEMS

Exercise 36

1. An article purchased for £51 is sold for £68, find the profit per cent.

2. An article purchased for £63 is sold for £56, find the loss per cent.

3. An article purchased for £30 is sold at a profit of 15%, find the selling price.

4. An article purchased for £5 is sold at a loss of 12%, find the selling price.

5. An article is sold for £159 at a profit of 6%, find the cost price.

6. An article is sold for £108 at a loss of 20%, find the cost price.

7. Buttons are bought for 10p per dozen and sold at 1p each, find the profit per cent.

8. Some articles costing 30p each are sold by auction at £3 per dozen, find the loss per cent.

9. Fuel is purchased wholesale at £7·50 per tonne, find the selling price to give a profit of $16\frac{2}{3}$%.

10. 2400 m² of damaged carpeting, worth £4360 new, are disposed of at a loss of 40%. Find the selling price per m².

11. Wine is sold in bulk for £5720 per 10 kl. If this gives the wholesaler a profit of 10%, find the cost per litre to the wholesaler.

12. A merchant disposes of 120 m of shop-soiled material for £22·50 and thus suffers a loss of $37\frac{1}{2}$%. Find the cost price per metre.

13. From a man's income of £1000 business expenses of £100 are deducted and the remainder is the family income. If his income is increased by 5% and business expenses increase by 14%, find the percentage change in the family income.

14. From a man's income 10% is deducted for business expenses, the remainder is the family income. If his income is increased by 15% and business expenses increase by 12%, find the percentage change in the family income.

15. An article sold for £21 gives a profit of 5%, find the profit per cent if the selling price had been £25.

16. An article sold for £1·62½ gives a loss of 35%, find the percentage change if the selling price had been £2·62½.

17. Selling an article for £3·50 gives a profit of 5%, find the selling price required to give a profit of 8%.

18. Selling an article for £8·90 gives a loss of 11%, find the selling price required to give a profit of 1%.

19. An article is manufactured for £16. The manufacturer sells it to the wholesaler at a profit of 15%, the wholesaler sells it to the retailer at a profit of 25%, the retailer sells it to the public for £25·30. What profit per cent does the retailer make?

20. The catalogue price of an article is £8, a retailer purchases from a wholesaler at a discount of 5% of the catalogue price, the wholesaler is allowed 12½% discount from the catalogue price when purchasing from the manufacturer. What profit per cent does the wholesaler make?

PERCENTAGE: MISCELLANEOUS EXAMPLES

EXERCISE 37

1. Express ⅚ as: (i) a decimal; (ii) a percentage, to two figures in each case.

2. Express 0·28 as: (i) a fraction in its lowest terms; (ii) a percentage.

3. Express 27·5% as: (i) a fraction in its lowest terms; (ii) a decimal.

4. Express 6·18 tonnes as a percentage of 9·27 tonnes.

5. Find, correct to 1p, the value of 6% of £0·55.

6. If 34% of a number is 110, find 51% of the number.

7. One inch is taken as 2½ cm instead of 2·54 cm. What is the percentage error correct to two figures?

8. Increase 3·28 m by 15%.

9. Decrease 6·31 l by 20%.

10. The length of a rectangle is halved and the breadth trebled, find the percentage change in area.

11. The three dimensions of a cube are increased by 20%, find the percentage increase in volume.

12. Selling an article for 36p gives a profit of 12½%. If the cost price is increased by 8p, find, to the nearest penny, the new selling price required to continue a profit of 12½%.

13. An article bought for 50p has the price increased by 20%. If the article had been sold previously at 62½p, what is the percentage decrease in profit if this selling price is maintained?

14. The buying price of an article is raised by 8% and the selling price by 6%. If the profit per cent was originally 20%, what is it now?

15. The charge for one unit of electricity is increased from 6p to 6·25p. If a householder reduces his consumption by 4%, by what percentage will his electricity bill change?

THE CIRCLE: CIRCUMFERENCE

REMEMBER. $C = \pi d = 2\pi r$; $\pi = 3\frac{1}{7}$ or $3\cdot142$ (*Correct to 4 sig. fig.*)

EXAMPLE. *Find the circumference of a circle of radius 7 cm.* ($\pi = 3\frac{1}{7}$.)

$\qquad C = 2\pi r = (\frac{2}{1} \times \frac{22}{7} \times \frac{7}{1})$ cm. $\therefore$ *Circumference* = <u>44 cm.</u>

EXERCISE 38

Find the circumference ($\pi = 3\frac{1}{7}$):

1. Radius = 14 cm   **2.** Radius = 28 cm   **3.** Radius = 3·5 cm
4. Diam. = 14 cm   **5.** Diam. = 21 cm   **6.** Diam. = 1·19 m
7. Radius = 35 m   **8.** Diam. = 3·5 m   **9.** Radius = 1·75 m

Find the circumference ($\pi = 3\cdot142$):

10. Radius = 5 cm   **11.** Radius = 10 cm   **12.** Diam. = 2 dm
13. Diam. = 10 in   **14.** Radius = 2·5 m   **15.** Diam. = 5 cm
16. Radius = 3 mm   **17.** Diam. = 4·2 dm   **18.** Diam. = 5·32 m

Find the diameter ($\pi = 3\frac{1}{7}$):

19. Circ. = 11 cm   **20.** Circ. = 5·5 cm   **21.** Circ. = 8 cm
22. Circ. = 22 m   **23.** Circ. = 440 m   **24.** Circ. = 60 dm

Find the radius ($\pi = 3\cdot142$):

25. Circ. = 12·6 cm   **26.** Circ. = 30 cm   **27.** Circ. = 86·3 cm

For the remainder of the exercise use whichever value of π seems the more convenient, but give answers in decimal form correct to three figures where necessary.

28. A cycle wheel is 63 cm in diameter, how many complete revolutions must it make for the cycle to travel 396 m?

29. The minute hand of a clock is 1·75 cm long, how far has the tip moved in 1 hr?

30. A piece of thread is wrapped fifty times round a reel 1·75 cm in diameter, find the length of thread in metres.

31. The minute hand of Big Ben is 11 ft long, how far does the tip move in 10 min?

32. A cyclist is travelling at 29·7 km/h if the wheels of his machine are 63 cm in diameter, how many revolutions does each make in 1 min?

33. A protractor is 10 cm in diameter, find its perimeter. (*Think carefully.*)

34. To give the correct measurements, distances round a running-track are measured 1 ft from the inside edge. Find the length of the inside edge of a circular track if the running distance is 1320 ft.

35. A piece of wire bent to form a rectangle 3·5 cm by 5·5 cm is now bent to form a semicircle complete with diameter. Find the radius of the semicircle so formed. ($\pi = 3\frac{1}{7}$.)

THE CIRCLE: AREA

REMEMBER. $A = \pi r^2$

EXAMPLE 1. *Find the area of a circle, radius $3\frac{1}{2}$ cm.* $(\pi = 3\frac{1}{7})$

$$A = \pi r^2$$
$$= (3\tfrac{1}{7} \times 3\tfrac{1}{2} \times 3\tfrac{1}{2}) \text{ cm}^2$$
$$= (\tfrac{22}{7} \times \tfrac{7}{2} \times \tfrac{7}{2}) \text{ cm}^2$$
$$= \tfrac{77}{2} \text{ cm}^2$$
$$\therefore Area = 38\tfrac{1}{2} \text{ cm}^2$$

EXAMPLE 2. *Find the area of a circle, diameter 9 cm.* $(\pi = 3\cdot142)$

$$A = \pi r^2$$
$$= (3\cdot142 \times 4\cdot5 \times 4\cdot5) \text{ cm}^2$$
$$= 63\cdot6255 \text{ cm}^2$$
$$\therefore Area = 63\cdot6 \text{ cm}^2 \ (Corr. \ to \ 3 \ sig. \ fig.)$$

EXERCISE 39

Find the area of the circles $(\pi = 3\frac{1}{7})$:

1. Radius = 7 cm   **2.** Diam. = 7 m   **3.** Radius = $10\frac{1}{2}$ m
4. Radius = 21 m   **5.** Diam. = 5·6 in   **6.** Diam. = 630 m

Find the areas of the circles $(\pi = 3\cdot142)$:

7. Radius = 3 in   **8.** Diam. = 12 m   **9.** Radius = 2·4 ft
10. Diam. = 600 miles

For the remainder of the exercise use whichever value of π seems the more convenient, but give answers in decimal form correct to three figures where necessary.

11. Find the area of a semicircle, radius 4 cm.

12. A running-track is in the form of a rectangle, 100 m by 50 m, with a semicircle on each of the shorter sides. Find the distance round the inside edge and also the area enclosed by the track.

13. Fig. B : 18 shows a rectangle, 8 yd by 3 yd, with four quadrants (quarter circles) radius $1\frac{1}{2}$ yd removed, one from each corner. Find the area remaining.

fig. B:18

14. A path 4 m wide surrounds a circular pond of radius 54 m. Find the area of the path. (*Use the "area by subtraction" method.*)

15. The circumference of a circular plate is 88 cm, find its area in square centimetres. (*Find radius first.*)

THE CIRCULAR CYLINDER

Area of Curved Surface = $2\pi rh$ or πdh

Total Surface Area = $2\pi rh + 2\pi r^2$

$$V = \pi r^2 h$$

EXAMPLE. *Find the total surface area and the volume of a cylinder, radius 5 cm, height 6 cm.*

Total Surface Area:

Curved surface	$= \pi dh$
	$= (3 \cdot 142 \times 10 \times 6)$ cm²
	$= 188 \cdot 52$ cm²
Area of ends	$= 2\pi r^2$
	$= (2 \times 3 \cdot 142 \times 25)$ cm²
	$= 157 \cdot 1$ cm²
Total $= 345 \cdot 62$ cm²

$\therefore$ *Total Surface Area* $= \underline{\underline{346 \text{ cm}^2}}$ (*Corr. to 3 sig. fig.*)

Volume:

$$V = \pi r^2 h$$
$$= (3 \cdot 142 \times 25 \times 6) \text{ cm}^3$$
$$= 471 \cdot 3 \text{ cm}^3$$

$\therefore$ *Volume* $= \underline{\underline{471 \text{ cm}^3}}$ (*Corr. to 3 sig. fig.*)

EXERCISE 40

Find the total surface area and the volume of the following cylinders:

1. Radius 4 cm, height 5 cm   **2.** Diameter 6 cm, height 4 cm
3. Diameter 10 cm, height 5 cm   **4.** Radius 2 cm, height 10 cm
5. Radius 6 cm, height 8 cm   **6.** Diameter 8 cm, height 10 cm
7. Diameter 10 cm, height 8 cm   **8.** Radius 1 ft, height 1 ft
9. Radius 2·5 cm, height 6 cm   **10.** Radius 3·5 m, height 4·5 m

11. A garden roller is 50 cm in diameter and 70 cm wide. Find the area rolled in 36 revolutions. (*Answer in square metres.*)

12. Find the weight of water, in kilograms, contained in a butt whose internal dimensions are 1·4 m high, 75 cm diameter. (1 *cm³ water weighs* 1 *g.*)

13. Find the radius of a cyclinder whose curved surface is 231 in² and height 7 in.

14. Find the height of a cylinder whose curved surface is 1936 cm² and radius 7 cm.

15. Find the radius of a cylinder whose volume is 44 ft³ and height $3\frac{1}{2}$ ft.

SQUARE ROOTS

EXERCISE 41

Find the square roots of:

1. 529	**2.** 1024	**3.** 1444	**4.** 1681
5. 3025	**6.** 4096	**7.** 5776	**8.** 6724
9. 7921	**10.** 9409	**11.** 12 544	**12.** 17 956
13. 25 281	**14.** 34 969	**15.** 49 284	**16.** 343 396

As in division much trouble can be caused by the careless omission of the cipher (0).

Examine the following examples carefully.

EXAMPLE 1.

$\quad$ Find $\sqrt{41209}$

```
              2 0 3 ·
           )4'12'09·
            4
40          12
            00
403         1209
            1209
   Ans. = 203
```

EXAMPLE 2.

$\quad$ Find $\sqrt{16·4025}$

```
              4 · 0 5
           )'16·40'25'
            16
80          40
            00
805         4025
            4025
   Ans. = 4·05
```

NOTE. The working of such examples can be somewhat reduced but by setting them down as shown, we avoid possible errors.

EXERCISE 42

Find the square roots of:

1. 10 201	**2.** 11 449	**3.** 11 881	**4.** 14 400
5. 25 600	**6.** 40 401	**7.** 48 400	**8.** 41 616
9. 4 080 400	**10.** 9 054 081		

THE SQUARE ROOT OF A DECIMAL

To find square roots of decimals we employ the same methods as before, remembering to **mark off pairs of digits to the left and right of the decimal point.** Bring down **two digits** at the appropriate stages. (See EXAMPLE 2 above.)

EXERCISE 43

Find the square roots of:

1. 6·25   2. 11·56   3. 23·04   4. 51·84
5. 25·6036   6. 1·1664   7. 0·5329   8. 0·7056
9. 0·362 404   10. 0·185 761   11. 0·0064   12. 0·01
13. 0·000 001   14. 0·000 121   15. 0·001 089   16. 0·000 104 04

Find the square roots correct to three figures:

17. 893·204   18. 7074·9   19. 0·935 71   20. 20·0013
21. 737 569   22. 4231   23. 3·248   24. 68·57
25. $\frac{16}{9}$ 26. $\frac{25}{9}$ 27. $\frac{16}{25}$ 28. $\frac{36}{25}$
29. $\frac{25}{36}$ 30. $\frac{49}{64}$ 31. $\frac{16}{49}$ 32. $\frac{25}{64}$
33. $1\frac{7}{9}$ 34. $2\frac{7}{9}$ 35. $1\frac{11}{25}$ 36. $2\frac{1}{4}$
37. $3\frac{5}{8}$ 38. $4\frac{1}{4}$ 39. $5\frac{5}{9}$ 40. $7\frac{4}{11}$

PROBLEMS ON SQUARE ROOTS

EXERCISE 44

Find the radii of the following circles, correct to two figures:

1. Area = 20 in²   2. Area = 45 cm²
3. Area = 100 cm²   4. Area = 78 ft²
5. Area = 168 yd²   6. Area = 3582 m²

Find the radii of the following cylinders, correct to two figures:

7. Volume = 9316 ft³, height 6 ft
8. Volume = 4812 cm³, height 40 cm
9. Volume = 83 746 ft³, height 96 ft
10. Volume = 10 000 m³, height 10 m

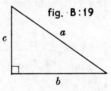

fig. B:19

Fig. B:19 shows a right-angled triangle whose sides are a units, b units and c units of length. By Pythagoras' Theorem find the length of the third side, correct to two figures, given:

11. $b = 3$ in, $c = 4$ in   12. $b = 15$ cm, $c = 20$ cm
13. $b = 12$ cm, $c = 16$ cm   14. $b = 5$ cm, $c = 12$ cm
15. $b = 5$ ft, $c = 12$ ft   16. $b = 25$ cm, $c = 60$ cm
17. $a = 20$ cm, $b = 16$ cm   18. $a = 39$ cm, $c = 15$ cm
19. $a = 91$ cm, $b = 84$ cm   20. $a = 15$ cm, $c = 44$ cm
21. $b = 4·8$ cm, $c = 6·3$ cm   22. $a = 5·3$ cm, $b = 2·1$ cm
23. $a = 3·2$ m, $c = 1·8$ m   24. $b = 6·7$ m, $c = 5·4$ m

Find the length of a side of a square whose area is:

25. 5000 m²   26. 300 cm²
27. 10 000 yd²   28. $6\frac{1}{4}$ ares
29. 3000 km²   30. 1 mile²

USING TABLES OF SQUARES

MEAN DIFFERENCES

	0	1	2	3	4	5	6	7	8	9	1	2	3	4	5	6	7	8	9
35	1225	1232	1239	1246	1253	1260	1267	1274	1282	1289	1	1	2	3	4	4	5	6	6

EXAMPLE. *To find* $(35 \cdot 47)^2$

We look for 35 in the left-hand column, under 4 in the centre columns we find 1253 and under 7 in the difference columns we find the figure **5**. By adding the 5 to 1253 we obtain the answer **1258**, and comparing this with the value obtained by multiplication (1258·1209) we can see that our answer is correct to four figures.

THE POSITION OF THE DECIMAL POINT

The decimal points are omitted from the tables and must be placed in each answer by careful inspection.

The actual digits obtained from $(3 \cdot 547)^2$ will be exactly the same as those obtained from $(3547)^2$, but the decimal point will be in a different position.

EXAMPLE 1. *Find* $(3 \cdot 547)^2$

Since 3·547 lies between 3 and 4, then $(3 \cdot 547)^2$ will lie between 3^2 and 4^2 (i.e., between 9 and 16).

$$Ans. = \underline{\underline{12 \cdot 58}}$$

EXAMPLE 2. *Find* $(354 \cdot 7)^2$

Since 354·7 lies between 300 and 400, then $(354 \cdot 7)^2$ will lie between $(300)^2$ and $(400)^2$ (i.e., between 90 000 and 160 000).

$$Ans. = \underline{\underline{125\ 800}}$$

QUESTION. What is the square of: (*a*) 0·3547; (*b*) 3547?

EXERCISE 45

Use four-figure tables to find the squares of:

1. 18	**2.** 34	**3.** 45	**4.** 5·1	**5.** 55
6. 60	**7.** 6·3	**8.** 72	**9.** 7·9	**10.** 86
11. 97	**12.** 10·2	**13.** 191	**14.** 28·3	**15.** 3·64
16. 426	**17.** 51·9	**18.** 0·647	**19.** 78·1	**20.** 832
21. 8·75	**22.** 91·3	**23.** 0·913	**24.** 9·13	**25.** 913
26. 10·32	**27.** 15·55	**28.** 2·009	**29.** 253·7	**30.** 3·642
31. 49·28	**32.** 0·5341	**33.** 612·4	**34.** 7·358	**35.** 84·67
36. 0·073	**37.** 3·2597	**38.** 1·001	**39.** 0·099	**40.** 91·354

USING TABLES OF SQUARE ROOTS

Consider these examples:

(i) 6·356 lies between 4 and 9   ∴ $\sqrt{6\cdot356}$ must lie between
     2 and 3   i.e., 2·??

(ii) 63·56 lies between 49 and 64   ∴ $\sqrt{63\cdot56}$ must lie between
     7 and 8   i.e., 7·??

(iii) 635·6 lies between 400 and 900   ∴ $\sqrt{635\cdot6}$ must lie between
     20 and 30   i.e., 2?·?

(iv) 6356 lies between 4900 and 6400   ∴ $\sqrt{6356}$ must lie between
     70 and 80   i.e., 7?·?

Although the digits remain constant, the position of the decimal point produces square roots which start with either 2 or 7. For this reason **two sets** of tables are necessary, and to use them it is essential to estimate the first figure of the root in order to select the correct line in the tables.

Here is an extract from the square-root tables. *Always use a ruler or other straight-edge to identify the required line of figures.*

	0	1	2	3	4	5	6	7	8	9	1	2	3	4	5	6	7	8	9
63	2510	2512	2514	2516	2518	2520	2522	2524	2526	2528	0	0	1	1	1	1	1	2	2
	7937	7944	7950	7956	7962	7969	7975	7981	7987	7994	1	1	2	3	3	4	4	5	6

As with other tables, the use of the figures in the difference columns may produce *slight* inaccuracies. The decimal point must be placed in each answer by careful inspection, a useful aid is the first stage of finding square roots by arithmetic (i.e., marking off pairs of digits).

(i) $\dfrac{2\cdot\ ?}{\sqrt{6\cdot35'60'}}$ (ii) $\dfrac{7\cdot\ ?}{\sqrt{'63'56'}}$

(iii) $\dfrac{2\ ?\cdot}{\sqrt{6'35\cdot60'}}$ (iv) $\dfrac{7\ ?\cdot}{\sqrt{'63'56'\cdot}}$

From the table extract verify these results:

(i) 2·521   (ii) 7·973   (iii) 25·21   (iv) 79·73

EXERCISE 46

Find, as accurately as four-figure tables will allow, the square roots of the following:

1. 10·35	**2.** 5·132	**3.** 17·49	**4.** 26·83	**5.** 3·142
6. 37·41	**7.** 8·256	**8.** 50·27	**9.** 1·739	**10.** 63·21
11. 67·54	**12.** 6·754	**13.** 675·4	**14.** 0·6754	**15.** 6754
16. 112·7	**17.** 21·39	**18.** 367·2	**19.** 417·8	**20.** 638·2
21. 6·382	**22.** 0·059	**23.** 0·4513	**24.** 9832	**25.** 0·098 32
26. 27315	**27.** 38·414	**28.** 0·009 47	**29.** 873·12	**30.** 5714·83

EXERCISE 47

Give answers correct to three figures.

Find the length of side of the following squares:

1. Area = 125 yd²   **2.** Area = 153 m²
3. Area = 9680 yd²   **4.** Area = 13·1 km²
5. Area = 98·25 ft²   **6.** Area = 5 miles²

Find the length of the diagonal of the following squares:

7. Area = 47·76 in²   **8.** Area = 0·9375 yd²
9. Area = 324·9 ft²
10. Area = 0·0074 m², answer in cm
11. Area = 873·27 km², answer in km
12. Area = 3·24 km², answer in m

13. A ladder 32 ft long rests against a vertical wall at a point 28 ft above the ground. Find the distance between the foot of the ladder and the base of the wall.

14. *AB* is a tangent touching a circle, centre *O*, at *B*. If *AB* is 12·4 cm and *AO* is 18·5 cm, find the diameter of the circle.

15. Find the radius of a circle which can be inscribed in a square of area 234·7 ft² so that the circle just touches the four sides of the square.

16. Find the area of a square which can be inscribed in a circle of radius 8·93 in so that the four corners of the square just touch the circle.

17. An aircraft flies 124 miles due North, then 98 miles due West, then 45 miles due South. Find: (i) the greatest distance from the starting-point reached during the flight; (ii) the distance from the starting-point to the final position.

18. A room is 18 ft long, 12 ft wide and 7½ ft high. Find the distance from a corner of the ceiling to the opposite corner of the floor.

19. The diagonals of a rhombus are 9·6 cm and 15·4 cm. Calculate: (i) the length of side; (ii) the area.

20. The length of a rectangle is five times its breadth. If the area is 173·2 in², find the perimeter.

LOGARITHMS

When a number is expressed in the form 2^5 the figure 2 is called the **base** and the index figure 5 is called the **logarithm.**

Although any base number may be employed, "common logarithms" employ 10 as the base.

If we convert ordinary numbers into powers of 10 we can multiply these numbers by adding the logarithms, because they will have the same base.   e.g., (i) $1000 \times 100 = 10^3 \times 10^2 = 10^5$
                    (ii) $27 \times 81 = 3^3 \times 3^4 = 3^7$

What is (i) the base, (ii) the logarithm in the second example?

LOGARITHMS OF NUMBERS GREATER THAN 1

Multiples of 10 are readily expressed as powers of the base number 10, thus: $10 = 10^1$; $100 = 10^2$; $1000 = 10^3$; $10\,000 = 10^4$, etc.

Because $67 \cdot 59$ lies between 10 and 100, the power of 10 must lie between 10^1 and 10^2. $\therefore$ $67 \cdot 59 = 10^{1 \cdot SOMETHING}$

Because $983 \cdot 2$ lies between 100 and 1000, the power of 10 must lie between 10^2 and 10^3. $\therefore$ $983 \cdot 2 = 10^{2 \cdot SOMETHING}$

Why does $7593 = 10^{3 \cdot SOMETHING}$?

NOTE

(i) A logarithm is composed of two parts: (*a*) a whole number called the **characteristic**; (*b*) a decimal portion called the **mantissa.**

(ii) The value of the characteristic is one less than the number of digits before the decimal point in the original number. (See above examples.)

(iii) "Log$_{10}$ 1000 = 3" means "the logarithm of 1000 to the base 10 is 3".

EXERCISE 48

Read off the characteristics of the logarithms of:

1. 1234	**2.** 5678	**3.** 91 011	**4.** 121 314	**5.** 1004
6. 100 00	**7.** 100	**8.** 40	**9.** 500	**10.** 7070
11. 23·4	**12.** 893·2	**13.** 89·32	**14.** 8·932	**15.** 8932
16. 893 20	**17.** 1·751	**18.** 107·2	**19.** 7·004	**20.** 41·21
21. 359·245	**22.** 3·592 45	**23.** 35 924·5	**24.** 3592·45	

FINDING THE MANTISSA FROM TABLES

EXAMPLE. *Find the logarithm of* $372 \cdot 8$

(i) The characteristic of $372 \cdot 8$ is 2.

(ii) From the log tables we select the line of figures against 37. (The first two digits of $372 \cdot 8$.)

MEAN DIFFERENCES

	0	1	2	3	4	5	6	7	8	9	1	2	3	4	5	6	7	8	9
37	5682	5694	5705	5717	5729	5740	5752	5763	5775	5786	1	2	3	5	6	7	8	9	10

line 37 under 2 in the main columns gives $0 \cdot 5705$
line 37 under 8 in the difference columns gives $0 \cdot 0009$
by addition, the mantissa $= 0 \cdot 5714$

The results are suitably tabulated thus:
This means $372 \cdot 8 = 10^{2 \cdot 5714}$

NO.	LOG
372·8	2·5714

What are the logarithms of:

(i) 3728   (ii) 3·728   (iii) 37·28   (iv) 372 80?

EXERCISE 49

Find the logarithms of the following from four-figure tables. *Tabulate your work.*

1. 12·78	**2.** 523	**3.** 9·371	**4.** 4973	**5.** 1000
6. 5·135	**7.** 9324	**8.** 100·1	**9.** 10·01	**10.** 1·001
11. 43 120	**12.** 73·4	**13.** 7·34	**14.** 2354	**15.** 235·4
16. 1·000	**17.** 2·000	**18.** 30	**19.** 400	**20.** 5000
21. 8	**22.** 91 919	**23.** 53 713	**24.** 70 710	**25.** 63·255
26. 3·01	**27.** 50·02	**28.** 7009	**29.** 80 010	**30.** 10

THE TABLE OF ANTILOGARITHMS

To find the number whose logarithm is 2·0947 we use the first two digits of the mantissa (0·09) to obtain the appropriate line of numbers in the **table of antilogarithms.** Under 4 in the main columns we find the number 1242 and under 7 in the difference columns we find the number 2; by addition we now have the number 1244. The characteristic 2 is now employed to position the decimal point:

$$\text{thus} \qquad 10^{2 \cdot 0947} = 124 \cdot 4$$

NOTE

(i) To position the decimal point in the antilog let it first be placed (mentally) between the first two digits of the number. The characteristic now tells us the number of places to move the decimal point.

At present our characteristics are **positive** *and the decimal point moves to the* **right.** *Later we shall employ negative characteristics. Which way will the decimal point move then?*

(ii) When using the difference columns you may be unable to find the exact digit required. In such cases select the digit nearest in value.

EXERCISE 50

From four-figure antilog tables find the numbers whose logarithms are as given. *Tabulate your work.*

1. 2·573	**2.** 1·347	**3.** 0·851	**4.** 0·6423	**5.** 1·5345
6. 1·0941	**7.** 2·0941	**8.** 0·0941	**9.** 2·7324	**10.** 3·7324
11. 0·8793	**12.** 1·0000	**13.** 3·5758	**14.** 1·0759	**15.** 2·0051
16. 4·9326	**17.** 2·7171	**18.** 3·052	**19.** 2·9998	**20.** 4·6398

USING LOGARITHMS TO MULTIPLY AND DIVIDE

EXAMPLE. *From four-figure tables evaluate* 60·01 × 713·8

NO.	LOG
60·01	1·7783
713·8	2·8536
EXPRESSION	4·6319
ANTILOG	42850

Ans. = 42850

EXERCISE 51

Using four-figure tables, evaluate the following:

1. $5 \cdot 73 \times 9 \cdot 82$	**2.** $4 \cdot 71 \times 5 \cdot 36$	**3.** $29 \cdot 8 \times 4 \cdot 17$
4. $9 \cdot 42 \div 2 \cdot 16$	**5.** $7 \cdot 45 \div 4 \cdot 36$	**6.** $63 \cdot 2 \div 8 \cdot 73$
7. $1 \cdot 753 \times 23 \cdot 17$	**8.** $17\,657 \times 131 \cdot 4$	**9.** $1 \cdot 521 \times 3 \cdot 76$
10. $129 \cdot 4 \div 33 \cdot 47$	**11.** $58 \cdot 43 \div 1 \cdot 57$	**12.** $19 \cdot 52 \div 5 \cdot 314$
13. 2329×1173	**14.** $452 \cdot 1 \times 452 \cdot 1$	**15.** $2137 \times 1 \cdot 5214$
16. 5280×1760	**17.** $2897 \div 119 \cdot 8$	**18.** $317 \cdot 4 \div 9 \cdot 52$
19. $101 \div 1 \cdot 002$	**20.** $74 \cdot 93 \div 42 \cdot 17$	**21.** $15\,930 \div 5438$
22. $51\,315 \div 13\,123$	**23.** $48 \times 67 \times 94$	**24.** $201 \cdot 5 \div 114 \cdot 32$
25. $73 \times 73 \times 73$	**26.** $9783 \cdot 4 \div 98 \cdot 537$	**27.** $1 \cdot 23 \times 12 \cdot 3 \times 123$
28. $\dfrac{4 \cdot 241 \times 7 \cdot 362}{5 \cdot 213}$	**29.** $\dfrac{83 \cdot 27 \times 43 \cdot 96}{65 \cdot 16}$	**30.** $\dfrac{672 \cdot 4 \times 923 \cdot 1}{4798}$

USING LOGARITHMS TO OBTAIN POWERS
AND ROOTS

EXAMPLE. *From four-figure tables evaluate* $(1 \cdot 374)^4$

NO.	LOG
$1 \cdot 374$	$0 \cdot 1380$
$(1 \cdot 374)^4$	$0 \cdot 5520$
ANTILOG	$3 \cdot 565$

Ans. $= 3 \cdot 565$

NOTE. If **exact answers** are required, standard methods of multiplication, etc., must be employed.

EXERCISE 52

Using four-figure tables, evaluate the following:

1. $(12 \cdot 4)^3$	**2.** $(1 \cdot 24)^4$	**3.** $(124)^2$	**4.** $\sqrt[3]{124}$
5. $\sqrt{12 \cdot 4}$	**6.** $\sqrt{1 \cdot 24}$	**7.** $(58 \cdot 4)^2$	**8.** $\sqrt[4]{5840}$
9. $(3 \cdot 527)^4$	**10.** $\sqrt[3]{352 \cdot 7}$	**11.** $\sqrt{3527}$	**12.** $(21 \cdot 53)^3$
13. $(91 \cdot 46)^2$	**14.** $\sqrt[4]{71 \cdot 38}$	**15.** $\sqrt[3]{112 \cdot 4}$	**16.** $(7 \cdot 592)^5$
17. $\dfrac{(5 \cdot 314)^3}{\sqrt[3]{81 \cdot 52}}$	**18.** $\dfrac{(12 \cdot 57)^2}{(8 \cdot 931)^2}$	**19.** $\dfrac{\sqrt{1984}}{\sqrt[3]{2415}}$	**20.** $\dfrac{\sqrt[3]{7436}}{(4 \cdot 257)^2}$

FRACTIONAL INDICES

$$16^{\frac{1}{2}} = \sqrt{16} \quad = ? \qquad\qquad 8^{\frac{1}{3}} = \sqrt[3]{8} \quad = ?$$

To evaluate $16^{\frac{3}{4}}$ we would need to multiply log 16 by $\frac{3}{4}$, i.e., we divide the log by 4 (giving $\sqrt[4]{16}$) and multiply the result by 3 [giving $(\sqrt[4]{16})^3$]. Thus $16^{\frac{3}{4}} = (\sqrt[4]{16})^3 = (2)^3 = 8$

Similarly $8^{\frac{2}{3}} = (\sqrt[3]{8})^2 = (2)^2 = 4$

NOTE. **If an index number is a fraction, the denominator indicates a root and the numerator indicates a power. Find the root first.**

NEGATIVE INDICES

Expressions of the form $x^3 \div x^5$ may be simplified by two methods:

(i) Subtracting indices   (ii) Cancelling terms

$$x^3 \div x^5 = x^{-2} \qquad\qquad \frac{x^3}{x^5} = \frac{\overset{1}{\cancel{x}}.\overset{1}{\cancel{x}}.\overset{1}{\cancel{x}}}{x.x.\underset{1}{\cancel{x}}.\underset{1}{\cancel{x}}.\underset{1}{\cancel{x}}} = \frac{1}{x^2}$$

Since the results must be equivalent $x^{-2} = \dfrac{1}{x^2}$

NOTE. **If a term contains a negative index number, the negative sign is removed by employing the reciprocal.**

EXERCISE 53

Express the following in the simplest form:

1. $4^{\frac{1}{2}}$	**2.** 2^{-1}	**3.** 3^{-2}	**4.** $27^{\frac{1}{3}}$	**5.** $9^{-\frac{1}{2}}$
6. 1^{-2}	**7.** $1^{-\frac{1}{2}}$	**8.** $27^{\frac{2}{3}}$	**9.** $4^{-\frac{1}{2}}$	**10.** $32^{\frac{3}{5}}$
11. $81^{\frac{1}{4}}$	**12.** $36^{-\frac{1}{2}}$	**13.** $81^{\frac{3}{4}}$	**14.** $125^{\frac{2}{3}}$	**15.** $1000^{-\frac{1}{3}}$
16. $\dfrac{1}{8^{-\frac{2}{3}}}$	**17.** $9^{1\frac{1}{4}} \div 9^{\frac{3}{4}}$	**18.** $8^{-1\frac{2}{3}}$	**19.** $16^{\frac{1}{2}} \times 2^{-2}$	**20.** $\dfrac{2^{-3}}{16^{-\frac{3}{4}}}$

HARDER EXAMPLES

EXAMPLE. *From four-figure tables evaluate* $\dfrac{16\cdot79 \times (137\cdot4)^3}{\sqrt{(43\cdot25)} \times 7521}$ *correct to three figures.*

NO.		LOG
$16\cdot79$		$1\cdot2251$
$(137\cdot4)^3$	$2\cdot1380$	$6\cdot4140$
NUMERATOR		$7\cdot6391$
$\sqrt{43\cdot25}$	$1\cdot6360$	$0\cdot8180$
7521		$3\cdot8763$
DENOMINATOR		$4\cdot6943$
EXPRESSION		$2\cdot9448$
ANTILOG		$880\cdot6$

NOTE

(i) If a log requires work to be done on it the log is placed in the second column. The final result is placed in the third column.

(ii) The log of the denominator is subtracted from the log of the numerator to give the log of the expression.

Ans. $= \underline{\underline{881}}$ *(Corr. to 3 fig.)*

EXERCISE 54

Using tables, evaluate the following correct to three significant figures:

1. $\dfrac{1\cdot579 \times 23\cdot24}{5\cdot32 \times 2\cdot48}$ **2.** $\dfrac{3\cdot75 \times 15\cdot7}{3\cdot513 \times 2\cdot414}$

3. $\dfrac{(57\cdot75)^2}{(9\cdot36)^3}$ **4.** $\dfrac{\sqrt{9447}}{\sqrt[3]{8375}}$

5. $\dfrac{439\cdot2 \times 27\cdot31}{35\cdot59 \times 27\cdot31}$

6. $\dfrac{137\cdot6 \times 73\cdot24}{98\cdot62 \times 44\cdot35}$

7. $\dfrac{\sqrt{823\cdot4} \times 4\cdot71}{\sqrt[3]{7362} \times 3\cdot83}$

8. $\dfrac{(7\cdot34)^3 \times 5\cdot24}{5\cdot24 \times (7\cdot34)^2}$

9. $\dfrac{57\cdot37 \times (6\cdot217)^3}{(7\cdot983)^2 \times 43\cdot24}$

10. $\dfrac{137\cdot4 \times \sqrt{101\cdot9}}{89\cdot42 \times (3\cdot147)^2}$

11. $(57\cdot36)^{1\frac{3}{8}} - (33\cdot47)^{1\frac{3}{8}}$

12. $(32\cdot51)^{1\frac{1}{2}} + (27\cdot36)^{\frac{3}{4}}$

13. $\dfrac{37\cdot35 + 42\cdot59}{103\cdot8 - 57\cdot24}$

14. $\dfrac{\sqrt{143\cdot7} + \sqrt[3]{1437} + 10}{(5\cdot21)^2 - (1\cdot25)^3}$

15. If $I = \dfrac{P \times R \times T}{100}$, find I when $P = 1675$, $R = 5\cdot05$ and $T = 7\cdot125$.

16. Find, in hectares, the area of a rectangular plot of ground 3759 m by 2438 m.

17. If $V = \pi r^2 h$, find V when $\pi = 3\cdot142$, $r = 27\cdot4$ and $h = 572$.

18. Find, in feet, the circumference of a circle whose diameter is 1·545 ft. ($\pi = 3\cdot142$.)

19. If $A = 2\pi r^2 + 2\pi r h$, find A when $\pi = 3\cdot142$, $r = 3\cdot43$, $h = 16\cdot57$.

20. Find the area of a circle of radius 4·735 in.

21. If $r = \sqrt{\dfrac{A}{\pi}}$, find the value of r when $A = 734\cdot8$ and $\pi = 3\cdot142$.

22. Find, in inches, the diameter of a circle whose circumference is 64·416 in.

23. If $V = \pi r^2 h$, find r when $V = 306\cdot3$, $h = 17\cdot58$ and $\pi = 3\cdot142$.

24. Find the area of a circle whose circumference is 26·52 ft.

25. If $A = \sqrt{[S(S - a)(S - b)(S - c)]}$ and $S = \frac{1}{2}(a + b + c)$, find A when $a = 132$, $b = 167$, $c = 81$.

26. Find the area of a parallelogram of base 16·44 in and perpendicular height 12·42 in.

27. Find the area of an isosceles triangle ABC in which $AB = AC = 5\cdot28$ in and $BC = 3\cdot56$ in.

28. Find the volume of a circular cylinder of radius 7·54 in, height 22·5 in and the weight of water it contains when full if 1 ft³ of water weighs 62·3 lb.

29. Find the area of a trapezium $ABCD$ if $AB = 10\cdot3$ in, $CD = 22\cdot4$ in and the perpendicular distance AB to CD is 17·6 in.

30. Find, in square decimetres, the total surface area of a circular cylinder of radius 12·48 cm and height 12·48 cm.

LOGARITHMS OF NUMBERS LESS THAN 1

$$10\,000 = 10^4; \quad 1000 = 10^3; \quad 100 = 10^2; \quad 10 = 10^1$$

As we divide the numbers on the L.H.S. by 10 the power of 10 decreases by 1 with each step.

Continuing the table we obtain:

$$1 = 10^0; \quad 0.1 = 10^{-1}; \quad 0.01 = 10^{-2}; \quad 0.001 = 10^{-3}$$

It should now be clear that for numbers less than 1 the characteristic of the logarithm is **negative.**

NOTE

(i) The mantissas are obtained as for numbers greater than 1, by employing the first four significant figures. (In the table of logarithms.)

(ii) The characteristic of a number less than 1 is obtained by counting the number of decimal places from the point to the first significant figure. Such a characteristic is negative.

(iii) While the characteristic of a logarithm may be either positive or negative, the mantissa portion is always positive.

(iv) If the minus sign is placed before the characteristic the whole logarithm may be thought of wrongly as negative. To avoid this confusion the sign (now termed a **bar**) is placed above the characteristic only.

In such logarithms the characteristics would be referred to as "bar 1", "bar 2", etc.

THE LOGARITHM OF 1

Remembering that a **logarithm** is another name for a **power**, it is interesting to consider 1 as a power of different bases.

By subtracting indices: By division:

$$10^1 \div 10^1 = 10^0 \qquad 10 \div 10 = 1 \qquad \therefore \; \underline{10^0 = 1}$$

$$2^3 \div 2^3 = 2^0 \qquad 8 \div 8 = 1 \qquad \therefore \; \underline{2^0 = 1}$$

$$7^2 \div 7^2 = 7^0 \qquad 49 \div 49 = 1 \qquad \therefore \; \underline{7^0 = 1}$$

$$x^5 \div x^5 = x^0 \qquad x^5 \div x^5 = 1 \qquad \therefore \; \underline{x^0 = 1}$$

CONCLUSION. **Any term (base) to the power nought is equal to 1.**

EXERCISE 55

From four-figure tables find the logarithms of the following:

1. 5·013   **2.** 0·3471   **3.** 0·5542 **4.** 0·0047 **5.** 131·2
6. 0·000 85 **7.** 0·7234   **8.** 0·0932 **9.** 23·63 **10.** 0·000 04
11. 8·0001   **12.** 0·02473   **13.** 0·0001 **14.** 0·0095 **15.** 2571
16. 0·4398   **17.** 0·000 052 47 **18.** 1·0000 **19.** 23   **20.** 0·005 298

ANTILOGARITHMS

As in previous work the decimal point starts between the first two digits of the antilog. If the characteristic is **positive** the point moves to the **right** the number of places indicated by the value of the characteristic. If the characteristic is **negative** the point moves the appropriate number of places to the **left**.

EXERCISE 56

From four-figure tables find the numbers whose logarithms are as given. *Tabulate your work.*

1. 0·5354	**2.** $\bar{1}$·5354	**3.** 2·5354	**4.** $\bar{2}$·8751	**5.** 3·8751
6. 0·0000	**7.** $\bar{3}$·0000	**8.** 4·0000	**9.** $\bar{1}$·4783	**10.** $\bar{2}$·0057
11. 2·1749	**12.** $\bar{3}$·1042	**13.** $\bar{2}$·9103	**14.** 1·3010	**15.** $\bar{1}$·3010
16. $\bar{4}$·0059	**17.** $\bar{2}$·7318	**18.** 3·2001	**19.** $\bar{3}$·9891	**20.** $\bar{1}$·2198

USING LOGARITHMS WITH NEGATIVE CHARACTERISTICS

EXAMPLE 1. *Using four-figure tables evaluate:* $0·0549 \times 0·002\,473$

NO.	LOG
0·0549	$\bar{2}$·7396
0·002 473	(+1)$\bar{3}$·3932
EXPRESSION	$\bar{4}$·1328
ANTILOG	0·000 1358

Explanation:
(i) Simplify lower line characteristic:
  $+1 - 3 = -2$ (i.e., $\bar{2}$)
(ii) Add upper and lower characteristics:
  $(-2)+(-2) = -2 - 2 = -4$ (i.e., $\bar{4}$)

Ans. = 0·000 1358

NOTE (Addition of logs)

Since the mantissas are positive, anything "carried" from the addition of the mantissas will be positive; hence the (+1) shown on the lower line.

EXAMPLE 2. *Using four-figure tables, evaluate:* $0·2061 \div 0·002\,739$

NO.	LOG
0·2061	$\bar{1}$·3141
0·002 739	(+1)$\bar{3}$·4376
EXPRESSION	1·8765
ANTILOG	75·25

Explanation:
(i) Simplify lower line characteristic:
  $+1 - 3 = -2$ (i.e., $\bar{2}$)
(ii) Subtract lower charact. from upper:
  $(-1) - (-2) = -1 + 2 = +1$

Ans. = 75·25

NOTE (Subtraction of logs)

Since the mantissas are positive, anything "borrowed" (to make the subtraction of the mantissas possible) must also be positive; hence (+1) is "paid back" to the lower-line characteristic.

When the lower-line characteristic has been simplified we may apply the rule "change the sign of the lower line and add".

EXERCISE 57

The following are **logarithms**. Deal with them in accordance with the signs. **Do not antilog.** *Tabulate your work.*

1. $\bar{1}\cdot0523 + 1\cdot9241$ 2. $\bar{1}\cdot1437 + 1\cdot8742$ 3. $\bar{2}\cdot5731 + \bar{1}\cdot6422$
4. $3\cdot4732 + \bar{2}\cdot6471$ 5. $3\cdot1357 - \bar{1}\cdot0563$ 6. $\bar{1}\cdot2371 - 1\cdot3435$
7. $2\cdot3483 - 3\cdot4579$ 8. $\bar{1}\cdot2539 - \bar{3}\cdot3714$ 9. $\bar{2}\cdot1541 + 0\cdot9729$
10. $2\cdot5737 + \bar{1}\cdot6452$ 11. $\bar{2}\cdot1541 - 0\cdot9729$ 12. $2\cdot5737 - \bar{1}\cdot6452$
13. $\bar{1}\cdot5927 - \bar{2}\cdot9439$ 14. $\bar{5}\cdot4923 + 2\cdot9431$ 15. $\bar{1}\cdot5927 + \bar{2}\cdot9439$
16. $\bar{5}\cdot4923 - 2\cdot9431$ 17. $\bar{2}\cdot5371 - \bar{3}\cdot7524$ 18. $1\cdot1951 + 3\cdot8595$

FURTHER WORK WITH NEGATIVE CHARACTERISTICS

EXAMPLE 1. *Using four-figure tables evaluate* $\sqrt[3]{0\cdot1357}$

NO.	LOG
0·1357	$\bar{1}$·1326
0·1357	$\bar{3}$·²1326*
$\sqrt[3]{0\cdot1357}$	$\bar{1}$·7108
ANTILOG	0·5139

Ans. = 0·5139

Explanation:

$$\bar{1} + 0\cdot1326 \equiv \bar{3} + 2\cdot1326$$
$$(\bar{3} + 2\cdot1326) \div 3 = \bar{1} + 0\cdot7108$$

By this method a negative characteristic is made into an exact multiple of the divisor (i.e., the required root). Thus we avoid "carrying" a negative number over to the mantissa.

* NOTE. With experience this stage can be carried out mentally.

EXAMPLE 2. *Using four-figure tables, evaluate* $(0\cdot08375)^{-1\frac{2}{3}}$

Simplifying:

$$(0\cdot08375)^{-1\frac{2}{3}} = \frac{1}{(0\cdot083\,75)^{1\frac{2}{3}}} = \frac{1}{(0\cdot083\,75)^{\frac{5}{3}}} = \frac{1}{(\sqrt[3]{0\cdot083\,75})^{5}}$$

NO.	LOG	
1·0000		0·0000
0·083 75	$\bar{2}$·9230	
$\sqrt[3]{0\cdot083\,75}$	$\bar{1}$·6410	
$(\sqrt[3]{0\cdot083\,75})^{5}$		(+1)$\bar{2}$·2050
EXPRESSION		$\bar{1}$·7950
ANTILOG		62·37

Explanation:
(i) $\bar{2}\cdot9230 \equiv \bar{3} + 1\cdot9230$
   $\therefore \bar{2}\cdot9230 \div 3 = \bar{1} + 0\cdot6410$
(ii) After subtracting lower mantissa:
   Lower character. $= +1 - 2 = -1$
(iii) Subtract lower characteristic from upper: $0 - (-1) = 0 + 1 = 1$

Ans. = 62·37

EXERCISE 58

The following are operations with **logarithms. Do not antilog.**

1. $\bar{1}\cdot3571 \times 3$ 2. $\bar{2}\cdot5936 \times 2$ 3. $\bar{3}\cdot4575 \times 3$ 4. $\bar{2}\cdot1734 \div 2$
5. $\bar{4}\cdot3793 \div 2$ 6. $\bar{3}\cdot5841 \div 3$ 7. $\bar{6}\cdot2594 \div 3$ 8. $\bar{1}\cdot2572 \div 2$
9. $\bar{3}\cdot4986 \div 2$ 10. $\bar{5}\cdot2736 \div 2$ 11. $\bar{1}\cdot4293 \div 3$ 12. $\bar{2}\cdot3427 \div 3$
13. $\bar{3}\cdot1716 \div 3$ 14. $\bar{1}\cdot9324 \times 3$ 15. $\bar{3}\cdot2656 \div 4$ 16. $\bar{2}\cdot9876 \times 4$
17. $\bar{8}\cdot4524 \div 4$ 18. $\bar{1}\cdot2898 \div 4$ 19. $0\cdot5321 \div 3$ 20. $0\cdot9384 \times 4$
21. $\bar{1}\cdot1131 \div 3$ 22. $\bar{1}\cdot9387 \times 4$ 23. $\bar{4}\cdot5945 \div 4$ 24. $\bar{5}\cdot2497 \div 4$

EXERCISE 59

Using four-figure tables, evaluate the following:

1. $23 \cdot 49 \times 0 \cdot 5736$
2. $0 \cdot 037 \times 0 \cdot 0042$
3. $2 \cdot 045 \times 0 \cdot 0902$
4. $3 \cdot 591 \div 4 \cdot 973$
5. $0 \cdot 942 \div 1 \cdot 597$
6. $2 \cdot 38 \div 0 \cdot 7652$
7. $498 \cdot 6 \times 0 \cdot 572$
8. $0 \cdot 009\ 376 \div 0 \cdot 074$
9. $0 \cdot 4593 \div 1 \cdot 717$
10. $0 \cdot 1593 \times 0 \cdot 0424$
11. $4 \cdot 128 \times 0 \cdot 005\ 16$
12. $3 \cdot 295 \div 0 \cdot 0686$
13. $(0 \cdot 1245)^3$
14. $(0 \cdot 003\ 18)^2$
15. $(0 \cdot 0408)^4$
16. $\sqrt{6 \cdot 846}$
17. $\sqrt[3]{0 \cdot 005\ 47}$
18. $\sqrt{0 \cdot 000\ 923}$
19. $\sqrt[3]{7 \cdot 231}$
20. $\sqrt[3]{0 \cdot 6823}$
21. $\sqrt{0 \cdot 7595}$
22. $\sqrt[4]{0 \cdot 0067}$
23. $(0 \cdot 5374)^{-\frac{2}{3}}$
24. $(0 \cdot 0632)^{-1\frac{2}{3}}$

MISCELLANEOUS EXAMPLES

EXERCISE 60

Using tables, evaluate the following correct to three significant figures:

1. $(5 \cdot 731)^3$
2. $\dfrac{1 \cdot 707 \times 0 \cdot 512}{43 \cdot 21 \times 0 \cdot 069}$
3. $\sqrt{0 \cdot 009872}$

4. $\dfrac{32 \cdot 49 \times 13 \cdot 06}{4782 \times 0 \cdot 436}$
5. $\dfrac{(2 \cdot 084)^2}{(1 \cdot 325)^3}$
6. $(0 \cdot 004\ 14)^{1\frac{1}{2}}$

7. $\dfrac{1\frac{3}{5} + \sqrt{1 \cdot 27}}{(3 \cdot 142)^2 - 5\frac{2}{3}}$
8. $(56 \cdot 48)^{-\frac{1}{2}}$
9. $\dfrac{\sqrt[3]{42 \cdot 46} \times 0 \cdot 9872}{5896 \times (0 \cdot 0534)^2}$

10. $(0 \cdot 076\ 82)^{-2\frac{1}{2}}$
11. $\dfrac{(7 \cdot 84)^{1\frac{1}{2}} \times (7 \cdot 84)^{\frac{3}{4}}}{(7 \cdot 84)^{2\frac{1}{4}} \times (7 \cdot 84)^{1\frac{1}{2}}}$
12. $\dfrac{1}{(0 \cdot 0856)^{-1\frac{1}{2}}}$

13. $\dfrac{(9 \cdot 474)^3}{\sqrt[3]{4749}}$
14. $\dfrac{1}{\sqrt{0 \cdot 524}} + \dfrac{1}{(0 \cdot 524)^2}$
15. $\dfrac{(5 \cdot 34)^3 - \sqrt[3]{1256}}{\sqrt[2]{72 \cdot 64} + (4 \cdot 07)^2}$

16. The radius of a circle is $9 \cdot 804$ dm. Find the area in square metres.

17. The internal dimensions of a rectangular box are $26 \cdot 4$ cm by $5 \cdot 87$ dm by $1 \cdot 26$ m. Find the internal volume in cubic decimetres.

18. Find, in cubic metres, the volume of a circular cylinder of radius $84 \cdot 2$ cm and height $23 \cdot 4$ dm.

19. Find the length of ladder if the foot is $1 \cdot 436$ m from the base of a vertical wall and the top rests against the wall $6 \cdot 58$ m above the ground.

20. The cross-section of a trough is an isosceles trapezium 94 cm wide at the top and 30 cm wide at the bottom, the sides being at $45°$ to the top. If the trough is $18 \cdot 6$ dm long, find the volume in cubic metres.

21. Town B is due North of A and town C is due East of A. If AB is $89 \cdot 43$ hm and BC is $9 \cdot 26$ km, find AC in metres.

22. Find, in square metres, the total surface area of a circular cylinder of radius $0 \cdot 0758$ m and height $0 \cdot 852$ m.

23. The base of a glass prism is an equilateral triangle of side 24·2 mm. If the prism is 5·44 dm long, find the volume in cubic decimetres.

24. Find, in square centimetres, the total surface area of the prism in Question 23.

25. A square field of side 98·56 m is to be sown with grass seed at the rate of 65 g per metre². If the seed costs 4·6 New Francs per kilogram, find the total cost of the seed.

GENERAL REVISION
Exercise 61

1. Simplify: (i) $3\frac{1}{2} - 1\frac{1}{4} (2\frac{1}{3} \div 1\frac{3}{4})$; (ii) $(2\frac{5}{6} + 1\frac{3}{4}) \div (4\frac{1}{4} - 3\frac{1}{3})$.

2. Divide £132·65 by 35.

3. Find the H.C.F. and L.C.M. of 36, 54, 48.

4. Evaluate without the use of tables $\sqrt{1371\cdot9616}$.

5. Arrange the following in descending order of size: $\frac{21}{31}, \frac{7}{10}, \frac{13}{20}$

6. Find the cost of papering the walls of a room 15 ft long, 12 ft wide, $8\frac{1}{2}$ ft high, allowing 81 ft² for doors, windows, etc., if the paper costs $52\frac{1}{2}$p per roll, 36 ft long, $1\frac{3}{4}$ ft wide.

7. Find the cost of 4 tonnes at £10·15 per tonne.

8. Evaluate the following, correct to three figures:

     (i) $(38\cdot47)^3$ (ii) $\sqrt[3]{0\cdot5858}$ (iii) $\dfrac{12\cdot05 \times 0\cdot007\,32}{7\cdot68 \times 0\cdot4724}$

9. How many pieces of wire each 3·78 dm long can be cut from a roll of such wire 85·08 m in length?

10. A swimming-bath is 85 ft long and 32 ft wide. The depth of the water increases uniformly from $2\frac{3}{4}$ ft to $8\frac{1}{2}$ ft. Find, correct to three figures, the number of gallons the bath contains if 1 ft³ = $6\frac{1}{4}$ gal.

11. Find, in pounds, the average of $4\frac{5}{12}$ lb, $3\frac{1}{8}$ lb, 31 lb, $9\frac{3}{8}$ lb. Answer correct to three figures.

12. If 1 km = $\frac{5}{8}$ mile, find $\frac{5}{22}$ mile in metres, correct to three figures.

13. Working 8 hr a day, 15 men complete a job in 18 days. How many men, working 10 hr a day, will complete the work in 12 days?

14. A, B and C invest £350, £400, £850 respectively in a business. If the profit is £2400, what should each receive?

15. Find 45% of $1\frac{4}{5}$ tonnes. Answer in kilogrammes.

16. If 16% of a sum of money is 32p, find the sum.

17. A rectangle 9·5 m by 4·2 m is reduced to 8·15 m by 3·45 m. Find the percentage decrease in area, correct to three figures.

18. An article is sold for £147 at a loss of 16%. Find the cost price.

19. An article sold for £24 gives a profit of 8%. Find the profit per cent if the selling price had been £25.

20. A protractor is 12·5 cm in diameter. Find its perimeter, correct to three figures.

STANDARD FORM

A number is in **standard form** when only one digit appears before the decimal point. To express any number in standard form we have to multiply or divide by an appropriate power of 10. Thus:

$$250\ 000 = 2{\cdot}5 \times 10^5; \ 0{\cdot}000\ 025 = 2{\cdot}5 \times 10^{-5}$$

Standard form is a particularly convenient form of expressing very large and very small numbers.

E.g., The Earth is 93 million miles from the Sun, i.e. $(9{\cdot}3 \times 10^7)$ miles.

EXERCISE 62

Express in standard form, the numbers given in the following:

1. The Moon is 240 000 miles from the Earth.

2. The speed of light is 186 000 miles/s.

3. Light travels approximately six million million miles in a year, this is a unit called the **light year.**

4. The star Sirius is 8 light years from Earth. Give the distance in miles.

5. In August 1961 a Russian spaceman, Gherman Titov, travelled seventeen times round the Earth in 25 h; a distance of 435 000 miles at 18 000 mile/h.

6. The planet Neptune is 2800 million miles from the Sun.

7. The circumference of the Earth at the Equator is 25 000 miles.

8. The mass of the Earth is estimated to be
   6 000 000 000 000 000 000 000 tons

9. The Earth's orbit is 580 million miles.

10. The speed of the Earth travelling round the Sun is 66 000 mile/h.

11. The attraction between Earth and Sun is equal to a pull of 3 600 000 000 000 tons.

12. One **joule** is equal to 10 000 000 **ergs.**

13. One **horse-power** is the force required to raise 33 000 lb one foot in one minute.

14. A **micron** is a millionth part of a metre.

15. A **millivolt** is a thousandth part of a volt.

16. One **megohm** is equal to a million ohms.

17. There are 800 000 birch seeds to a pound.

18. There are 1500 pea seeds to a pound.

19. One **mil** is a thousandth part of an inch.

20. To turn cubic millimetres into cubic inches multiply by 0·000 061.

21. To turn yards into kilometres multiply by 0·000 914 4.

22. One **watt-hour** equals 3600 joules. Express this in ergs.

23. To turn grains into kilogrammes multiply by 0·000 064 8.

24. To turn kilogrammes into tons multiply by 0·000 984.

25. One short ton equals 2000 lb. One pound equals 7000 grains. Express one short ton in grains.

EXERCISE 63

Give answers correct to three significant figures.

1. If $x^4 = 8{\cdot}73 \times 10^{-9}$, find x.   **2.** If $x^5 = 0{\cdot}000\ 357$, find x.

3. If $x^{\frac{3}{2}} = 6{\cdot}8$, find x.   **4.** If $x^{\frac{1}{4}} = 0{\cdot}8$, find x.

5. If $x^{\frac{1}{3}} = 4{\cdot}2 \times 10^{-2}$, find x.   **6.** $x^{\frac{2}{3}} = 1{\cdot}4 \times 10^{-3}$, find x.

7. If $2^x = 8$, find x.   **8.** If $5^x = 3284$, find x.

9. If $10^x = 25{\cdot}65$, find x.

(Answer Nos. 10–20 without the use of tables.)

10. If $10^{3x} = 1000$, find x.   **11.** Evaluate $\log 2 + \log 50$.

12. Evaluate $\log 40 + \log 0{\cdot}25$ **13.** Evaluate $\log 50 - \log 0{\cdot}5$.

14. Evaluate $\log 75 - \log \frac{3}{4}$.   **15.** If $3^x = 9^{x-2}$, find x.

16. If $\log x = \log 5 + \log 3$, find x.

17. If $\log x - \log 4 = \log 3$, find x.

18. If $\log x + \log 3 = \log 15$, find x.

19. Evaluate $\log_{10} 25 + \log_{10} 24 - \log_{10} 60$.

20. Evaluate $\log_{10} 80 + \log_{10} 60 - \log_{10} 48$.

21. Solve the equation $x \log_{10} 5{\cdot}473 = \log_{10} 3{\cdot}586$.

22. Evaluate $\dfrac{xy}{z^2}$ when $x = 1{\cdot}4 \times 10^4, y = 2{\cdot}3 \times 10^{-5}, z = 1{\cdot}8 \times 10^{-4}$. Give the answer in standard form.

23. Evaluate $(\log_{10} 2{\cdot}424) \times (\log_{10} 5{\cdot}151)$. Give the answer in standard form.

24. If $x(\log_{10} 6{\cdot}84 + \log_{10} 2{\cdot}56) = \log_{10} 9{\cdot}82 - \log_{10} 1{\cdot}42$, find x.

HARDER MENSURATION—THE PYRAMID

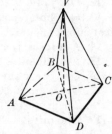

fig. **B: 20**

(i) The base of a pyramid is a **polygon** and may have any number of sides.

(ii) The top of the solid is called the **vertex**.

(iii) The edges from the vertex to each corner of the base are called **slant edges**.

(iv) If the slant edges are all equal the solid is called a **right pyramid**.

(v) In a right pyramid the line joining the vertex to the centre of the base is perpendicular to the base and is called the perpendicular height.

(vi) Each face of a pyramid is a triangle and the number of faces will depend upon the number of sides of the base. These features control the total surface area of the solid.

$$\textbf{Volume of a pyramid} = \frac{\textbf{Area of base} \times \textbf{Height}}{3}$$

EXERCISE 64

Give answers correct to three significant figures.

Find the volume of each of the following right pyramids:

1. Height 6 in; area of base 4 in².
2. Height 10 cm; area of base 18·4 cm².
3. Height 16·4 in; area of base 35·09 in².
4. Height 0·8 dm; base 18 cm square.
5. Height 9·6 in; rectangular base, 10·2 in by 8·4 in.
6. Height 6·2 ft; rectangular base, 4·5 ft by 5·3 ft.
7. Height 4·2 in; triangular base, sides 3 in, 4 in, 5 in.
8. Height 5·8 in; triangular base, sides 5 in, 7 in, 8 in.
9. Slant edges 10 in; rectangular base, 6 in by 8 in.
10. Slant edges 12·8 cm; rectangular base, 10·6 cm by 8·4 cm.

Find the total surface area of the following right pyramids:

11. Height 8 in; base 10 in square.
12. Height 10 cm; base 8 cm square.
13. Height 8 in; rectangular base, 8 in by 10 in.
14. Height 12 ft; rectangular base, 9 ft by 15 ft.
15. Slant edge 16 cm; base equilateral triangle of side 8 cm.
16. Slant edge 14·6 in; rectangular base, 7·6 in by 9·5 in.

17. The volume of a right pyramid is 136 in³ and the base is a square of side 6 in. Find the height and the length of the slant edges.

18. The volume of a right pyramid is 152 in³. If the height is 9 in and the base is a square, find the length of a side of the base and of the slant edges.

19. The base of a pyramid is a rectangle 7·6 cm by 9·2 cm and the slant edges are each 12·4 cm. Find: (i) the volume of the pyramid, and (ii) the total surface area.

20. A pyramid on a square base has a height of 8 in and slant edges of 10 in. Find: (i) the volume of the pyramid, and (ii) the total surface area.

HARDER MENSURATION—THE CONE

A **circular cone** is one in which the cross-section, taken through the solid at right-angles to the axis, is always a circle.

Since a cone may be regarded as a pyramid, the formula for the volume is the same, an appropriate calculation being necessary to find the area of the base.

$$\text{Volume of cone} = \frac{\text{Area of base} \times \text{Height}}{3}$$

$$\therefore \text{Volume} = \frac{\pi r^2 h}{3}$$

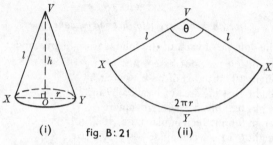

(i)   fig. B:21   (ii)

Area of curved surface of cone $= \pi r l$
Total surface area of cone $= \pi r^2 + \pi r l$

(i) $\pi r^2 + \pi r l = \pi r(r + l)$

Converting an expression into factor form may sometimes simplify the calculation.

(ii) If the angle of the sector is given the area may be obtained by using the ratio $\dfrac{\theta°}{360°}$, thus: Area of sector $= \pi l^2 \times \dfrac{\theta}{360}$

EXERCISE 65

Give answers correct to three significant figures; use log $\pi = 0\cdot4971$.

Find the volume of each of the following circular cones:

1. Height 5 in; base radius 5 in.
2. Height 7·4 cm; base diameter 6·8 cm.
3. Height 3·6 in; base circumference 7·2 in.
4. Height 12 in; slant height 13 in.
5. Height 4·5 dm; slant height 6·3 dm.
6. Slant height 10·4 in; base radius 4·2 in.
7. Slant height 12·5 in; base diameter 16·2 in.
8. Area of base 35 cm²; slant height 10 cm.
9. Area of curved surface 120 in²; base radius 5·2 in.

Find the area of the curved surface of each of the following circular cones:

10. Slant height 7 in; base radius 3 in.
11. Slant height 12 cm; base diameter 0·8 dm.
12. Height 12 in; slant height 13 in.
13. Height 1·6 dm; slant height 24 cm.
14. Height 9 in; base circumference 18 in.
15. Slant height 8 in; base circumference 12 in.
16. Find the volume and the total surface area of a cone, height 7·5 in, base radius 2·4 in.
17. Find the volume and total surface area of a cone, slant height 9·6 cm, area of base 27·4 cm².

HARDER MENSURATION—THE SPHERE

Volume of sphere $= \frac{4}{3}\pi r^3$
Area of surface of sphere $= 4\pi r^2$

Exercise 66

Give answers correct to three significant figures; use log $\pi = 0.4971$.

1. Find the volume of a sphere, radius 3·05 cm.

2. Find the volume of a sphere, diameter 11·7 in.

3. Find the radius of a sphere, volume 9·8 in³.

4. Find the radius of a sphere, volume 129·8 in³.

5. Find the surface area of a sphere, diameter 15 cm.

6. Find the radius of a sphere, surface area 25 cm².

7. Find the radius of a sphere, surface area 129·8 cm².

8. Find the volume of a sphere, surface area 64 in².

9. Find the volume of a sphere, surface area 2·5 in².

10. Find the surface area of a sphere, volume 20 cm³.

11 Find the surface area of a sphere, volume 14 in³.

12. Ten lead spheres each of radius 0·3 in are melted down to form one sphere. Find the radius of this sphere.

13. A lead sphere of radius 5·5 cm is melted down to form 6 equal spheres. Find the radius of each.

14. Four lead spheres, each of surface area 24 cm², are melted down to form one sphere. Find the surface area of this sphere.

15. A hemispherical solid has a total surface area of 86 in². Find the radius.

16. A hemispherical solid has a total surface area of 32 cm². Find the volume.

17. An open hemispherical bowl has an internal surface area of 60 cm². A solid metal sphere of surface area 30 cm² is placed in the bowl, which is then filled to the brim with water. Find the volume of water in the bowl.

HARDER MENSURATION—THE ANNULUS

The area between two concentric circles is called an **annulus**.

From fig. B : 22, if the larger circle has a radius of R units, its area will be πR^2 units².

If the smaller circle has a radius of r units, its area will be πr^2 units².

$\therefore$ The shaded area $= \pi R^2 - \pi r^2$

$$= \pi(R^2 - r^2)$$

Area of annulus $= (\pi R + r)(R - r)$ **units².**

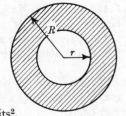

fig. B : 22

VOLUME OF MATERIAL IN A PIPE

From the general formula: Volume = Area of section × Length, the volume of material used in making a pipe of circular section will be given by:

$$\textbf{Volume} = \pi l(R + r)(R - r)$$

EXERCISE 67

Give answers correct to three significant figures: $\pi = 3.142$, *log* $\pi = 0.4971$, $1\ ft^3 = 6\frac{1}{4}\ gal$.

Find the area of the annulus between two concentric circles of radii:

1. 1·5 cm, 0·5 cm
2. 1·1 cm, 0·9 cm
3. 5·4 ft, 4·6 ft
4. 55 yd, 45 yd
5. 16·7 m, 3·3 m
6. 25·9 cm, 4·1 cm

Find the volume of material required to make a pipe of the following dimensions:

7. Ext. rad. 6 in; int. rad. 4 in; length 20 in.
8. Ext. rad. 17 cm; int. rad. 13 cm; length 1 m. (*Answer in dm³.*)
9. Ext. rad. 3·5 cm; int. rad. 3·25 cm; length 30 cm.
10. Ext. diam. 1 ft; int. diam. $\frac{5}{6}$ ft; length 16 ft.
11. Int. diam. 8·5 in; thickness of pipe 0·25 in; length 240 in.
12. Ext. diam. 3 ft; thickness of pipe $\frac{1}{4}$ ft; length 4 ft.
13. Water flows through a pipe, radius $\frac{1}{24}$ ft, at 8 ft/s. Find the volume of water flowing in 1 min.
14. Water flows through a pipe, diameter $\frac{1}{8}$ ft, at 14 ft/s. Find the time taken to fill a 40-gal cask.
15. Water flows through a pipe, diameter $\frac{1}{16}$ ft, at 6 ft/s. Find, in minutes, the time taken to fill a rectangular tank 2 ft by 3 ft by 4 ft.
16. Water flows through a pipe, diameter $\frac{1}{24}$ ft, at 5 ft/s. Find the weight of water collected in a container after 5 min. (1 gal of water weighs 10 lb.)
17. The water flowing from a pipe of $\frac{1}{8}$ ft diameter will fill a 10-gal drum in 2 min. Calculate the speed in ft/s at which the water is passing through the pipe.
18. The water flowing from a pipe of $\frac{1}{16}$ ft diameter fills a pint bottle in 2·5 s. Calculate the rate of flow of the water in ft/s.
19. A rectangular tank 2$\frac{1}{2}$ ft by 3 ft by 4$\frac{1}{3}$ ft is filled with water from a pipe, diameter $\frac{1}{6}$ ft in 5 min 12s. Find the rate of flow of the water in ft/s.
20. The water flowing from a pipe, diameter $\frac{1}{24}$ ft, is collected in a container weighing 2$\frac{3}{16}$ lb. After 6 min the weight of the container and contents is 37$\frac{7}{16}$ lb. Calculate the rate at which the water is flowing in ft/s. (1 gal of water weighs 10 lb.)

21. A circular cylinder of radius 3 in has the same volume as a sphere of the same radius. Find the length of the circular cylinder.

22. Water is flowing from a pipe of diameter $\frac{1}{2}$ ft at the rate of 10 ft/s. If the water contains $\frac{1}{4}$ oz of sediment per gallon, find the number of tons of sediment passing from the pipe in four days. [1 ft^3 = 6$\frac{1}{4}$ gal; 1 ton = (2240 × 16) oz.]

23. A rectangular tank of base 54 in by 39 in contains water. A solid metal pyramid with a square base of side 16 in and slant edge 28 in is lowered into the tank. Calculate the rise in the level of the water.

24. A cylinder of external diameter 6·5 in has an internal depth of 14 in. If the cylinder holds 1$\frac{1}{2}$ gal of water when full, calculate the thickness of the cylinder wall. (1 gal = 277 in^3.)

25. An advertising display model of an ice-cream cone consists of an inverted cone of base radius $\frac{1}{2}$ ft surmounted by a hemisphere of the same radius. If the complete volume of the model is 0·851 ft^3, calculate the over-all height.

26. A sphere has a volume of 16 cm^3. Find the length of a circular cylinder which has the same radius and volume.

27. A rectangular tank of base 3$\frac{1}{3}$ ft by 2$\frac{1}{2}$ ft contains water which just covers the vertex of a solid metal pyramid whose base is square. When the pyramid is removed the level of the water falls by $\frac{1}{2}$ ft. If the height of the pyramid is equal to the length of side of the base, calculate the height.

28. A test-tube consists of a cylindrical portion 12 cm long joined to a hemisphere of radius 1·3 cm. Find the external surface area of the test-tube.

29. The cross-section of a trough is a trapezium whose upper width is 2 ft and lower width 1$\frac{2}{3}$ ft, the trough being 8 ft long and $\frac{3}{4}$ ft deep The trough is supplied by a pipe of diameter $\frac{1}{16}$ ft through which water flows at 6 ft/s. Find, in minutes, the time taken to fill the trough.

30. A flat rectangular roof is 52 ft by 3 ft. The rain falling on the roof is carried to a cylindrical tank of radius 1$\frac{2}{3}$ ft. Calculate the increase in the depth of water in the tank for a rainfall of $\frac{1}{40}$ ft.

31. A planetarium is constructed by building a cylinder with vertical walls 10 ft high surmounted by a hemispherical roof. If the area of the floor is 800 ft^2, find the volume of the building.

32. A solid cylinder of radius $\frac{1}{3}$ ft is 5 ft long. A hole of square section, side $\frac{1}{8}$ ft, is made through the whole length of the cylinder. Find: (i) the volume, in cubic feet, of the material remaining, and (ii) the entire surface area in square feet.

AVERAGES AND MIXTURES

EXAMPLE. *In what ratio must tea at 8p per $\frac{1}{4}$ lb be mixed with tea at 11p per $\frac{1}{4}$ lb to produce a mixture giving a profit of 20% when sold at 42p per lb?*

To give a profit of 20%, cost price $= \frac{100}{120}$% of S.P.

$\therefore$ C.P. $= (42 \times \frac{100}{120}) = (\frac{42}{1} \times \frac{5}{6})$p per lb

$\therefore$ C.P. of mixture $= 35$p per lb

In such a mixture:

1 lb of tea at 32p will produce a *profit* of 3p
1 lb of tea at 44p will produce a *loss* of 9p.

If the mixture were sold at *cost price* there would be neither a profit nor a loss; therefore possible profits or losses must be so arranged to balance each other, thus:

3 units of tea at 32p give a profit of 9p.
1 unit of tea at 44p gives a loss of 9p.

Ans. Tea at 8p per $\frac{1}{4}$ lb is mixed with tea at

11p per $\frac{1}{4}$ lb in the ratio 3 : 1.

EXERCISE 68

1. A tobacconist mixes 6 oz of tobacco costing 25p per oz with 10 oz of tobacco costing 28p per oz. Calculate the cost of the mixture per oz, correct to the nearest penny.

2. In 15 innings a batsman scores an average of 33·4 runs. If he scores 59 runs in the following innings, find his new average. How many runs must he score in the 17th innings to obtain an average of 38 runs?

3. In an examination a student obtains an average mark of 47 on three papers and an average of 42 on two other papers. Find the average for the five papers.

4. By selling a certain blend of tea at $10\frac{1}{2}$p per $\frac{1}{4}$ lb a grocer would make a profit of 40% on the cost price. He mixes 8 lb of this tea with 12 lb of another blend, the cost price of which is 15p per $\frac{1}{4}$ lb. Find the cost price per $\frac{1}{4}$ lb of the mixture and the selling price per $\frac{1}{4}$ lb if he makes a profit of 25% on his total outlay.

5. In what ratio must a chemical at 22p per kg be mixed with a chemical at 31p per kg to produce a mixture worth 29p per kg?

6. What weight of chemical at 54p per kg must be mixed with 7 kg of a chemical at 73p per kg to produce a mixture worth 68p per kg?

7. In six subjects a student secures an average mark of 46, and in four of them the average is 45. In the remaining two subjects one mark was 40, calculate the other.

8. A form of 30 children sold £2·60 worth of tickets for a school concert, but not all the children sold tickets. If all the children had sold tickets at the same average rate the form would have collected £3. How many children sold tickets?

9. In one week 12 skilled men and 3 apprentices earn a total of £224·25. If the average wage for an apprentice is £4·75, find the average wage for a skilled man.

10. A wine merchant buys 135 l of wine at 95p per litre and mixes this with 240 l of wine at 70p per litre. Find, correct to the nearest penny, the price per litre at which he must sell the mixture to produce a profit of 28% on his outlay.

11. In 15 innings a batsman made the following scores: 23, 47, 12, 38, 56, 19, 6, 84, 72, 30, 42, 20, 16, 21, 45. After 5 more innings the batsman's average score had increased by 4·6 runs. Find the average score for these last 5 innings.

12. A student takes five subjects in an examination, and in three of them obtains the following marks: 27, 48, 65. What minimum average mark must be scored in the remaining two subjects to secure an over-all average of 50?

13. Over a period of seven years a business made an average profit of £584 per annum. In the first two years there were losses of £150, £119. In the next three years there were profits of £132, £480, £875. What was the average profit in the next two years?

14. A certain substance contains 2% by weight of water, while a second substance contains 18% by weight of water. In what proportions by weight should the two substances be mixed to give a mixture containing 12% by weight of water?

15. In what ratio must tea at 60p per kg be mixed with tea at £1 per kg to produce a mixture giving a profit of 25% when sold at £1·10 per kg?

16. What weight of tea at 62p per kg must be mixed with 15 kg of tea at £1·02 per kg to produce a mixture giving a profit of 25% when sold at 107½p per kg?

17. German silver is composed of 60% copper, 20% zinc, 20% nickel. Admiralty gun-metal is composed of 88% copper, 10% tin, 2% zinc. If gun-metal is mixed with German silver in the ratio 5 : 2, find the percentages of tin and of nickel contained in the mixture.

18. Two types of solder are made up as follows:

Type A   64% tin, 1% antimony, 35% lead.
Type B   50% tin, 3% antimony, 47% lead.

In what ratio should they be mixed to form: (i) solder containing 2% antimony; (ii) solder containing 42% lead? What percentage of mixture (ii) is tin?

AVERAGE SPEED

(i) Distance travelled is given by the area under a speed/time graph.

(ii) **Average speed** $= \dfrac{\textbf{Total distance travelled}}{\textbf{Total time taken}}$

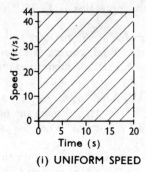

(i) UNIFORM SPEED

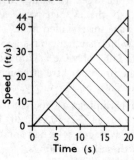

(ii) VARIABLE SPEED
(UNIFORM ACCELERATION)

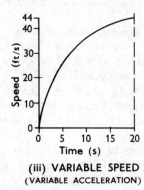

(iii) VARIABLE SPEED
(VARIABLE ACCELERATION)

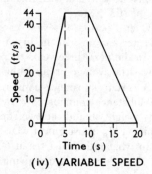

(iv) VARIABLE SPEED

fig. B:23

EXERCISE 69

Give answers correct to three significant figures where necessary.

1. Express the following in feet per second: (i) 30 mile/h; (ii) 40 mile/h; (iii) 45 mile/h; (iv) 25 mile/h.

2. Express the following in miles per hour: (i) 22 ft/s; (ii) 132 ft/s; (iii) 17·6 ft/s; (iv) $73\frac{1}{3}$ ft/s.

3. Express 18 km/h in metres/min.

4. Express 15 metres/s in km/h.

5. An athlete runs a mile in 4 min. What is the average speed in: (i) mile/h; (ii) ft/s?

6. An athlete runs 100 metres in 9·6 s. What is the average speed in km/h?

7. Taking the mean radius of the Earth's orbit as 93×10^6 miles and a year as 365 days, find the speed of the Earth in orbit in mile/h.

8. A cyclist rides for $2\frac{1}{2}$ h at 10 mile/h and a further $1\frac{1}{2}$ h at 12 mile/h. Find his average speed.

9. A motorist travels for 75 miles at 30 mile/h and for a further 50 miles at 40 mile/h. Find his average speed.

10. A man walks for 15 min at 5 mile/h and for 3 miles at 4 mile/h. Find his average speed in ft/s.

11. A train travels 15 miles at 20 mile/h, 20 miles at 25 mile/h and 25 miles at 30 mile/h. Find the average speed.

12. If a car travels 32 miles in 1 h 20 min, find the average speed. If the average speed for the first 12 miles is 18 mile/h, find the average speed for the remaining 20 miles.

13. An aircraft flies 20 miles in a S.E. direction and then 20 miles in a N.E. direction taking a total time of 6 min. Find: (i) the average speed of flight: (ii) the effective speed in an Easterly direction.

14. A cyclist rides for 36 min at 10 mile/h. After taking a short rest he travels a further 5 miles at 15 mile/h, completing the whole journey at an average speed of 11 mile/h. How long did he rest?

15. The engine of an express train blows its whistle as it passes through a level crossing. An observer standing on a bridge 3850 ft from the crossing observes the "steam" from the whistle, but does not hear the sound until the engine is level with a village station 308 ft beyond the crossing. If sound travels at 1100 ft/s, find the speed of the train in mile/h. (5280 ft = 1 mile)

16. A body moves at a uniform speed of 30 ft/s. Show this graphically, and from your graph obtain the distance travelled in 6 s. Verify by calculation.

17. A body starts from rest and accelerates uniformly to a speed of 60 ft/s in 8 s. Show this graphically, and from your graph obtain the distance travelled in: (i) 4 s; (ii) 6 s; (iii) 8 s. Verify your results by calculation.

18. A body moving at 40 mile/h is retarded uniformly and comes to rest in 5 s. Show this graphically, and from your graph obtain the distance travelled in (i) 2 s, (ii) 5 s after retardation commences.

19. A body starts from rest and accelerates uniformly to reach a speed of 30 ft/s in 3 s. It maintains this speed for 5 s, then retards uniformly to come to rest in 2 s. Find graphically the distance travelled, and hence calculate the average speed.

20. A body moves with an initial velocity of 40 mile/h and accelerates uniformly to 60 mile/h in 12 s, then it is uniformly retarded to rest. If the total time taken is 16 s, find graphically: (i) the distance travelled; (ii) the time taken to reach 45 mile/h; and calculate (iii) the average speed in ft/s.

RELATIVE SPEED

If we compare the speed of one body with that of another we obtain their **relative speed.**

(i) If two bodies are moving in opposite directions their relative speed is equal to the sum of their individual speeds.

(ii) If two bodies are moving in the same direction their relative speed is equal to the difference between their individual speeds.

EXAMPLE. *A train 60 yd long travelling at 40 mile/h overtakes a second train moving at 30 mile/h in the same direction and passes it completely in $29\frac{1}{4}$ s. Find the length of the second train.*

fig. B: 24

Train A gains on train B at $(40 - 30)$ mile/h $= 10$ mile/h
For A to completely pass B it must travel $(x + 60)$ yd
$\therefore A$ travels $(x + 60)$ yd in $29\frac{1}{4}$ s at 10 mile/h

$$10 \text{ mile/h} = \frac{10 \times 1760}{60 \times 60} \text{ yd/s.}$$

$\therefore$ In $29\frac{1}{4}$ s, at 10 mile/h, distance $= \left(\frac{10 \times 1760}{60 \times 60} \times \frac{117}{4}\right)$ yd

$$= 143 \text{ yd}$$
$$\therefore x + 60 = 143$$
$$x = 83$$

Ans. Second train is 83 yd long.

EXERCISE 70

1. A train 140 yd long takes 15 s to pass completely a station platform 80 yd long. Find the speed of the train and the time taken for the driver to pass the platform completely.

2. A train 160 yd long takes 30 s to pass completely over a viaduct 170 yd long. Find the speed of the train.

3. A train 110 yd long travelling at 35 mile/h overtakes a second train moving at 25 mile/h in the same direction and passes it completely in 30 s. Find the length of the second train and the time it will take to pass completely the driver of the first train.

4. A train 88 m long travelling at 54 km/h overtakes a second train moving at 42 km/h in the same direction and passes it completely in 45 s. Find the length of the second train and the time taken for its driver to pass completely the first train.

5. A train 100 m long travelling at 72 km/h passes a second train moving at 48 km/h in the opposite direction. If the trains pass each other completely in 6 s, find the length of the second train.

6. A train travelling at 48 km/h passes a second train, 104 m long, moving at 56 km/h in the opposite direction. If the trains pass each other completely in 4½ s, find the length of the first train and the time taken for its driver to pass the second train completely.

7. A train 125 yd long travelling at 43 mile/h overtakes a second train, 139 yd long, moving at 28 mile/h in the same direction. Find the time taken for the trains to pass each other completely and the time taken for the first train to pass completely the tail lamp of the second train.

8. A train 88 yd long travelling at 35 mile/h passes a second train 132 yd long, moving at 25 mile/h in the opposite direction. Find the time taken for the trains to pass each other completely and the time taken for the second train to pass completely the tail-lamp of the first train.

9. A train 92 m long travelling at 48 km/h completely overtakes a second train 84 m long and moving in the same direction, in 1 min 12 s. Find the speed of the second train.

10. A train 115 m long travelling at 50 km/h completely passes a second train, 105 m long and moving in the opposite direction, in 7½ s. Find the speed of the second train.

11. A sprinter running at 15 mile/h overtakes a road-walker moving at 6 mile/h in the same direction. If the latter received a start of ¼ mile, how long will it take the sprinter to draw level with him?

12. A police car travelling at 104 km/h pursues a suspect travelling in a second car at 80 km/h. If the suspect had a 12-minute start, how far must the police travel in order to catch him?

13. A train passes a signal in 5 s and completely through a station 264 yd long in 17 s. Find the speed of the train and its length.

14. A train passes completely through one station, 114 yd long, in 9 s and completely through another, 180 yd long, in 12 s. Find the speed of the train and its length.

15. A motorist travelling at 45 mile/h overtakes a cyclist moving in the same direction at 15 mile/h. Fifteen minutes later he passes a second cyclist moving in the opposite direction at 15 mile/h. How much time must now elapse for the two cyclists to meet?

GRAPHICAL SOLUTION OF SPEED AND DISTANCE PROBLEMS

EXAMPLE. *Two towns, A and B, are* 40 *miles apart. A car leaves A at* 9.00 A.M. *travelling towards B at an average speed of* 30 *mile/h.* 30 *min later a car leaves B travelling towards A at an average speed of* 45 *mile/h. When the cars meet the drivers talk for* 10 *min, then each returns to his starting-point, both arriving back at* 10.20 A.M. *Find, graphically:* (i) *the time at which the cars meet:* (ii) *the distance from A when the cars meet:* (iii) *the average speed at which each car travels on the return journey.* Answers:
(i) 9.50 A.M.
(ii) 25 miles
(iii) Car returns to *A* at 75 mile/h.

Car returns to *B* at 45 mile/h.

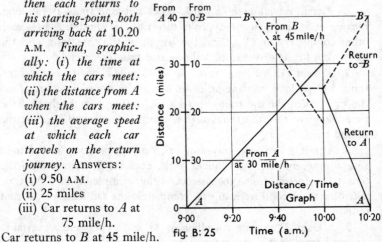

fig. B: 25

After considering the speed of each car, convenient times and distances are selected from which each distance/time graph may be drawn. (Taking into account the 30 min delay for the car starting from *B*.) The point of intersection of the two lines gives the answers to Questions (i) and (ii). The answers to Question (iii) are obtained after considering the respective distance each car has to travel and the time taken. (20 min in each case.)

EXERCISE 71

1. A motorist is travelling at an average speed of 30 mile/h, a cyclist at 14 mile/h and a pedestrian at 5 mile/h. If all three start from the same point at the same time and move in the same direction, represent the information graphically, using the same pair of axes. From your graph find: (i) the distance between the cyclist and the pedestrian after 30 min, and (ii) the time taken for the motorist and the cyclist to be 12 miles apart. Check your results by calculation.

2. A motorist begins a journey of 30 km at an average speed of 30 km/h. After 15 km the car develops engine trouble and the motorist stops to carry out repairs. Eventually he continues his journey at an average speed of 20 km/h and arrives at his destination after a total time of $1\frac{1}{2}$ hours. Represent this information graphically and find the time spent on repairs.

3. Two towns, A and B, are 25 miles apart. A motorist leaves A at 10.00 A.M. travelling towards B at 30 mile/h. After 30 min he stops and waits for a friend who left B at 10.15 A.M. travelling towards A at 20 mile/h. On being joined by his friend, the first motorist returns to A, leading the way for the second motorist. The journey is completed at 11.00 A.M. From a graph, determine: (i) the time the first motorist had to wait; (ii) his average speed when returning to A.

4. A motorist leaves town A travelling towards town B, 30 km away. After travelling for 25 km at an average speed of 60 km/h he takes a 5-min rest and returns to A at the same average speed. If a car left B at 6.00 P.M. (the same time as the car left A) and travelled at an average speed of 30 km/h towards A, find from a graph: (i) the times at which the cars would pass each other; (ii) the distances of these points from A; (iii) the times at which the cars arrived at A.

5. Using the same axes as Question 4, find from the graph the speed of a car leaving C at 6.00 P.M. travelling towards A along the route through B, passing the car from A at 6.25 P.M. and the car from B at a point 10 km from A. Find, also, the time at which this car would arrive at A and the distance from C to A.

6. From town B a motorist approaches town A, 30 miles away, at an average speed of 20 mile/h. A motor-cyclist leaves town A 15 min later, approaching town B at 40 mile/h. From a graph find: (i) the time taken for the motor-cyclist to reach the motorist; (ii) the distance from B when the two pass each other; (iii) the time at which they will each be the same distance from their respective starting-points if the motorist started at 1.00 P.M.

7. A and B are two towns 30 km apart. At 12.00 noon a motor-cyclist leaves B travelling towards A at 36 km/h; at the same time a motorist and cyclist leave A travelling towards B, the motorist at 24 km/h and the cyclist at 12 km/h. From a graph determine: (i) the time at which the motor-cyclist passed (a) the motorist, (b) the cyclist; (ii) the increase in the distance separating the motorist and the cyclist during this period of time.

8. A cyclist travelling at an average speed of 10 mile/h leaves point A at the same time as a pedestrian travelling in the same direction at 5 mile/h. After 45 min the cyclist stops for a rest and the pedestrian is given a lift by a passing motorist travelling at an average speed of 30 mile/h. As the motorist passes the resting cyclist, the latter begins the return journey to A, arriving there $37\frac{1}{2}$ min later at the same time as the motorist and his passenger complete their journey to point B. Represent this information graphically and from your graph determine: (i) the length of time the cyclist was resting; (ii) the speed at which the cyclist returned to A; (iii) the distance from A to B.

SIMILAR FIGURES: REVISION
EXERCISE 72

Find the ratio of the two similar rectangles, given:

   1. Two shorter sides of 4 cm and 6 cm respectively.

   2. Two longer sides of 28 cm and 35 cm respectively.

   3. Two perimeters of 72 in and 180 in respectively.

Find the ratio of the areas of two similar triangles, given:

   4. Bases of 3 in and 4 in respectively.

   5. Perpendicular heights of 6 m and 8 m respectively.

   6. Perimeters of 9 cm and 36 cm respectively.

Find the ratio of the areas of two circles, given:

   7. Radii of 2 m and 5 m respectively.

   8. Diameters of 5 cm and 8 cm respectively.

   9. Circumferences of 27 in and 36 in respectively.

   10. Find the ratio of (i) the surface areas, (ii) the volumes of two spheres of radii 16 cm and 24 cm respectively.

   11. A cube of edge 6 in weighs 10 lb. Find the weight of a cube, edge 8 in, of the same material.

   12. A cone of height 15 cm has a volume of 500 cm³. Find the volume of a similar cone of height 12 cm.

   13. A sphere of circumference 35 cm has a weight of 250 g. Find the weight of a sphere of the same material, circumference 49 cm.

   14. A cylindrical container whose girth is 28 in holds 16 pints. Find, in pints, the capacity of a similar container of girth 21 in.

   15. It costs £5 to paper a wall 10 ft high. Find the cost of papering a similar wall 8 ft high.

   16. Two solid spheres have surface areas in the ratio 9 : 16. If the smaller weighs $1\frac{11}{16}$ lb, find the weight of the larger, in pounds.

   17. Two similar cylindrical containers have volumes in the ratio 8 : 27. If the larger has a surface area of $2\frac{1}{4}$ ft², find the surface area of the smaller.

   18. Two solid cubes of the same material have weights of 54 g and 250 g respectively. If the smaller has an edge of 6 cm, find the length of edge of the larger cube.

   19. Two solid spheres of the same material have weights of 24 lb and 81 lb respectively. If the larger has a diameter of $13\frac{1}{2}$ in, find the radius of the smaller sphere.

   20. On a map of scale $\dfrac{1}{20\,000}$ an area of forest land is represented by an area of $7\frac{1}{2}$ in². What area would be required to show the forest on a map of scale $\dfrac{1}{50\,000}$?

THE RECTANGLE AND CUBOID
MISCELLANEOUS EXAMPLES

EXERCISE 73

1. Calculate the number of tiles, 10 cm by 15 cm required to tile the walls of a room 2 m by 2·4 m to a height of 1·5 m.

2. Calculate the number of whole tiles, 12 cm by 15 cm, which can be fitted to the walls of a bathroom 2·16 m by 2·4 m to a height of 1·6 m and find the area left untiled.

3. What length of linoleum, 2 m wide, is required to cover a floor 4·95 m by 3·2 m? Answer to the nearest metre.

4. Find the weight of a steel plate 6 ft by 8 ft if the plate is $\frac{1}{96}$ ft thick and steel weighs 490 lb per ft³.

5. How many billiard balls of radius 1·65 cm can be packed into a box 26·4 cm square and 3·5 cm deep? (Internal dimensions.)

6. Find the weight of a rectangular block of aluminium, $\frac{1}{2}$ ft by $\frac{2}{3}$ ft by $\frac{3}{4}$ ft, if 1 ft³ of the metal weighs 168 lb.

7. How many cubes of edge 5 cm can be fitted into a box whose internal dimensions are 45 cm by 50 cm by 60 cm?

8. How many cuboids, 6 cm by 5 cm by 4 cm, can be fitted into a box whose internal dimensions are 36 cm by 30 cm by 24 cm?

9. A gold medallion of diameter 28 mm is $3\frac{1}{2}$ mm thick and weighs 41·6 g. Find the **density** of the gold. (I.e., find, in grams, the weight of 1 cm³ of the metal.)

10. How many whole tiles, 10 cm by 12 cm, can be fitted into an area 1·75 m by 2·4 m, and what area remains uncovered?

11. How many dusters, 18 in by 20 in, can be cut from a piece of cloth 38 in wide by 180 in long?

12. How many cuboids, 3 in by 4 in by 5 in, can be fitted into a box of internal measurements 45 in by 48 in by 72 in?

13. How many whole flagstones, $2\frac{1}{4}$ ft by $3\frac{1}{3}$ ft, can be fitted into an area 90 ft by 81 ft?

14. How many rolls of paper, 12 m long, 50 cm wide, are required to paper the walls of a room 5 m long, 4 m wide and 3 m high?

15. How many crates, 1·2 m wide, 2·1 m long, may be placed on the floor of a closed van, 4 m long and 2 m wide?

16. How many whole tiles, 6 in square, may be fitted into an area 100 in by 57 in, and what area remains uncovered?

17. A closed wooden box of external dimensions 64 cm by 100 cm by 144 cm is made of timber 2 cm thick. If the wood weighs 0·5 kg per dm³, find the weight of the box.

18. How many cubes of edge 5 cm may be fitted into a box whose internal dimensions are 18 cm by 27 cm by 36 cm, what is the volume, in dm³ left unoccupied?

INCREASE AND DECREASE PER CENT
EXERCISE 74

Give answers correct to three figures, where necessary.

1. If a car is sold for £600, a dealer loses £250. What must he sell the car for to gain 8%?

2. If a car is sold for £506, a dealer gains 15%. What is the profit or loss per cent if the car is sold for £400?

3. If a houseowner sells a house for £5200, he gains £500. What is the selling price if he loses 6%?

4. A tree, 40 ft high, increases its height by 5% of its height each year. How tall is the tree after three years?

5. If an article is sold for £36, a shopkeeper loses 4%. What is the selling price if there is a profit of 6%?

6. A house is sold for £4850 at a loss of 3%. What is the profit or loss per cent if the house is sold for £4950?

7. An article sold for 7% above the cost price gives a profit of £1·05. Find the selling price if the article is sold at a profit of 10%.

8. An article costs £1·25 to manufacture. The manufacturer catalogues the article to give a profit of 20%, but allows the wholesaler a reduction of 15% on the catalogue price. What does the wholesaler pay for the article?

9. A manufacturer sells an article to a wholesaler at a profit of 40%. The wholesaler sells to the retailer at a profit of 25%. The retailer sells to the public for £3·85 and makes a profit of 10%. Find the cost of manufacture.

10. A salesman increased his earnings in 1960 by 8% of his salary in 1959. In 1961 his 1960 salary was increased by 25%, and his 1962 salary was 112% of his 1961 salary. By what percentage did his salary in 1962 exceed that of 1959?

11. When the cost of fuel rose by 5%, a householder decreased his consumption by 5%. Find the percentage increase or decrease in the householder's expenditure on fuel.

12. When the cost of tobacco rose by 20%, a man reduced his consumption by 15%. Find the percentage change in his expenditure.

13. When the cost of petrol rose by $12\frac{1}{2}$%, a motorist decreased his weekly consumption by 20% so that his weekly expenditure was now £1·80. Find the reduction in his expenditure per week.

14. If the import duty on a certain article is increased by 25%, by what per cent must the quantity imported be reduced in order that an importer shall not pay an increased Customs fee?

15. A tradesman marks goods 25% above cost price, but gives a discount of 6p in the pound for a cash sale. If he receives £2·35 cash, what profit does he make?

SIMPLE INTEREST

$$S.I. = \frac{P \times R \times T}{100} \quad \text{and} \quad \textbf{Amount} = P + S.I.$$

EXERCISE 75

Find the simple interest on:

1. £450 at 5% for 3 yr
2. £750 at 6% for 4 yr
3. £200 at $3\frac{1}{2}$% for 5 yr
4. £300 at $4\frac{1}{2}$% for 6 yr
5. £250 at $3\frac{1}{2}$% for 5 yr
6. £350 at $4\frac{1}{2}$% for 6 yr
7. £450 at $5\frac{1}{2}$% for 3 yr
8. £750 at $6\frac{1}{2}$% for 3 yr
9. £440 at $5\frac{1}{4}$% for $3\frac{3}{4}$ yr
10. £765 at $6\frac{2}{3}$% for 4 yr 8 mth
11. £384 at $4\frac{1}{4}$% for 6 yr 8 mth
12. £432 at $2\frac{3}{4}$% for 20 mth
13. £528 at $3\frac{1}{3}$% for 33 mth
14. £1095 at 4·2% for 100 days

Find, correct to 1p, the simple interest on:

15. £612 at 4% for 3 yr
16. £436 at 3% for 3 yr
17. £239 at 3% for 7 yr
18. £352 at $3\frac{1}{4}$% for $3\frac{1}{2}$ yr
19. £216·92 at $2\frac{1}{2}$% for $4\frac{1}{2}$ yr
20. £438·46 at 4% for 5 yr
21. £242·70 at $3\frac{1}{3}$% from 28th February to 5th October.
22. £584 at 5% from 1st January to 11th April.
23. £428 at $2\frac{1}{4}$% from 14th March to 31st December.

INVERSE PROBLEMS

EXERCISE 76

Find the principal required to produce a simple interest of:

1. £36 after 2 yr at 3%
2. £45 after 3 yr at 5%
3. £72 after 4 yr at 6%
4. £67·50 after 5 yr at $4\frac{1}{2}$%

Find the rate per cent per annum, simple interest, at which:

5. £300 will amount to £325 in 2 yr
6. £240 will amount to £320 in 5 yr
7. £270 will amount to £405 in $7\frac{1}{2}$ yr
8. £310 will amount to £356·50 in $4\frac{1}{2}$ yr

Find the time in which:

9. £250 at 4% produces £20 interest.
10. £360 at $2\frac{1}{2}$% produces £36 interest.
11. £325 at 4% produces £65 interest.
12. £336 at 5% produces £28 interest.

Find the principal which amounts to:

13. £750 after 10 yr at $2\frac{1}{2}$%
14. £525 after 6 yr at $3\frac{1}{3}$%
15. £499 after $5\frac{1}{2}$ yr at $4\frac{1}{2}$%
16. £689·25 after $4\frac{1}{4}$ yr at $3\frac{1}{2}$%

COMPOUND INTEREST

EXAMPLE. *Find the compound interest on £225·27½ for 3 yr at 4%.*

$$£225·27\tfrac{1}{2} = £225·275$$

Princ. for 1st yr.	£225·27500	—*Start mult. by 4*
Int. at 4%	9·01100	*Move 2 pl. to rt.*
Princ. for 2nd yr.	234·28600	—*Start mult. by 4*
Int. at 4%	9·37144	*Move 2 pl. to rt.*
Princ. for 3rd yr.	243·65744 *	—*Start mult. by 4*
Int. at 4%	9·74630	*Move 2 pl. to rt.*
Amount after 3 yr.	253·40374	
Deduct 1st princ.	225·27500	
Comp. interest	£28·12874	

$$£28·12874 = £28·13 \text{ (Corr. to 2 dec. pl.)}$$

$$Ans. = £28·13 \text{ (Corr. to 1p.)}$$

EXERCISE 77

Give answers correct to the nearest penny.

Find the compound interest, payable yearly, on the following:

1. £400 for 2 yr at 3%
2. £500 for 2 yr at 2%
3. £350 for 3 yr at 3%
4. £450 for 3 yr at 2%
5. £256·42½ for 2 yr at 3%
6. £312·57½ for 2 yr at 4%
7. £415 for 1½ yr at 3%
8. £150·87½ for 1½ yr at 3%
9. £182 for 2 yr at 2½%
10. £275·50 for 2 yr at 2¾%
11. £315·77½ for 3 yr at 3½%
12. £422·62½ for 2½ yr at 4½%
13. £185·67½ for 1½ yr at 2½% p.a., payable half-yearly.
14. £268·7625 for 1 yr at 4% p.a., payable quarterly.
15. £342·916 67 for 2½ yr at 4% p.a., payable half-yearly.
16. £467·666 67 for 1¼ yr at 5% p.a., payable quarterly.
17. £568·7375 for 2½ yr at 4½% p.a., payable half-yearly.
18. A tree increases its height by 4% each year. What will be the height of the tree in 3 years' time, if it is 23½ ft now?
19. An insect larva increases its volume by 12% each day. If it has a volume of 2 mm³ now, what will be its volume after 5 days' growth?
20. A baby's weight increases by 10% each month. If the baby weighed 7½ lb at birth, find its weight after 6 mth.
21. In successive laps of a race track a car increased its average speed by 3½%. If the average speed for the first lap was 87·4 mile/h, find the average speed for the third lap.

REPAYMENTS. *A man borrows £600 at 5% compound interest (payable yearly) and repays £150 at the end of each year. How much has still to be repaid at the beginning of the third year?*

DEPRECIATION. *A machine purchased for £985, depreciates in value by 6% each year. Find, to the nearest penny, the value of the machine at the end of four years.*

REPAYMENTS

Princ. 1st yr.	£600·0
Int. at 5%	30·0
Debt 1st yr.	630·0
Repay £150	150·0
Princ. 2nd yr.	480·0
Int. at 5%	24·0
Debt 2nd yr.	504·0
Repay £150	150·0
Princ. 3rd yr.	£354·0

Ans. = £354

DEPRECIATION

Value 1st yr.	£985·0
Deprec. 6%	59·10
Value 2nd yr.	925·90
Deprec. 6%	55·5540
Value 3rd yr.	870·3460
Deprec. 6%	52·22076
Value 4th yr.	818·12524
Deprec. 6%	49·08751
Value after 4 yr.	£769·03773

Ans. £769·04 (Corr. to 1p.)

EXERCISE 78

Give answers correct to the nearest penny.

Calculate the difference between the simple interest and compound interest (each payable yearly) on:

1. £100 for 3 yr at 2%
2. £200 for 2 yr at 5%
3. £150 for 4 yr at 3%
4. £250 for 3 yr at 4%
5. £300 for 2 yr at $2\frac{1}{2}$%
6. £400 for 2 yr at $4\frac{1}{2}$%
7. £450 for 3 yr at $3\frac{1}{2}$%
8. £550 for 4 yr at 6%

9. A man borrows £500 at 4% compound interest (payable yearly) and repays £125 at the end of each year. How much is still to be repaid at the beginning of the third year?

10. A man borrows £236·75 at 5% compound interest (payable yearly) and repays £75 at the end of each year. How much does the man owe at the beginning of the third year?

11. A car purchased for £650, depreciates in value by 8% each year. Find, to the nearest penny, the value of the car at the end of the fourth year.

12. A material is found to shrink by 2% of its length after washing and by 3% after dyeing. What is the new length of material, originally 16·75 ft, after twice being washed and then dyed once.

13. It is observed that the average speed of a long distance runner decreases by 3% of itself for each interval of 2 miles after the half-way mark. What is the total decrease per cent at a point 10 miles past the half-way mark?

RATES, TAXES, INSURANCE, ETC.

RATES

A **Local Authority** is responsible for certain **local services** (Schools, Hospitals, Highways, etc.). To meet the cost of such services **Local Rates** are **levied** on all property within the control of the authority.

Each property is **assessed** according to its size and amenities and is given a **Rateable Value**. (This must not be confused with the market value, which is a matter to be decided between the buyer and seller.)

A rate of **1p in the** £ would mean that the owner (or sometimes the occupier) of a house would pay, each year, 1p in rates for every £1 rateable value.

EXAMPLES

    1. If the rates are 25p in the £:

      (*a*) On a house of rateable value £100 the rates are
$$25p \times 100 = \underline{£25 \text{ p.a.}}$$

      (*b*) On a district total rateable value £750 000 the rates are
$$25p \times 750\ 000 = \underline{£187\ 500 \text{ p.a.}}$$

      (*c*) If a man pays £30 rates, the rateable value of his property is
$$£\frac{30 \times 100}{25} = \underline{£120}$$

    2. If a Local Authority requires to spend £200 000 in a particular year, and the total rateable value of the district is £800 000:

    The local rate would be $\left(\frac{200\ 000}{800\ 000} \times \frac{100}{1} \right)$p in the £

$$= \underline{25p \text{ in the } £}$$

    3. If a Local Authority collects a total of £84 000 from rates at 25p in the £, the total rateable value of the district

$$= \frac{£84\ 000 \times 100}{25}$$

$$= \underline{£336\ 000}$$

INCOME TAX

The **Government** is also responsible for certain expenses concerning the **nation** (Education, National Defence, Health Service, etc.). To balance **Revenue** and **Expenditure** it is necessary to impose certain **taxes.** (Customs Duty, Vehicle and Radio Licences, Purchase Tax, Income Tax, etc.)

Income tax varies with the needs of the national economy, but it is based on a "sliding scale" according to an individual's income and responsibilities. For example, **allowances** are made for children and vary according to age (under 11 yr £115; 11–15 yr, £140; over 16 yr but still at school, £165).

When these and various other allowances have been deducted from a man's annual income the remaining **Taxable Income** might be taxed as follows:

> the first £100 is taxed at 20p in the £
> the next £200 is taxed at 30p in the £
> the remainder is taxed at 40p in the £.

By the "Pay-As-You-Earn" Scheme (**P.A.Y.E.**), a person claims the allowances to be set against **earned income.** He is then given a **tax-code number**, issued by the local Income Tax Office. The appropriate tax is deducted from his earnings by his employer, who then passes it to the Income Tax Department.

EXAMPLE 1. *A man earns £1800 in a year* (**Gross Income**). *He is allowed £1000 free of tax and pays tax on the rest as shown above. Calculate the total income tax paid.*

$$\text{Taxable Income} = £1800 - £1000 = £800$$
$$\text{Tax on } £100 \text{ at 20p in } £ = £20$$
$$\text{Tax on } £200 \text{ at 30p in } £ = £60$$
$$\text{Tax on } £500 \text{ at 40p in } £ = £200$$
$$\text{Tax paid on } £800 = £280$$

Net Income $= £1800 - £280 = £1520$

EXAMPLE 2. *A man invests £350 for 4 years at 4% p.a. C.I. Tax is deducted from the interest at 20p in the £. Calculate the actual interest received from the investment.*

A tax of 20p in the £ leaves 80p in the £.
∴ Every £1 of interest is actually worth 80p ($£\frac{4}{5}$).
∴ 4% interest is actually worth $\frac{4}{5}$ of 4%
∴ Actual interest $= 3\frac{1}{5}\%$

A normal calculation of compound interest is now carried out using an interest of $3\frac{1}{5}\%$ p.a. and not 4% p.a.

BANKRUPTCY

If a business man is unable to pay his debts he may be declared **bankrupt** and the business equipment may be sold by a responsible official to secure all available **assets**. If the **liabilities** (debts) cannot be paid in full the **creditors** (to whom money is owing) will receive only a proportion of the money owing to them.

EXAMPLE. *If the liabilities are £1200 and the assets £960, each creditor will receive $\frac{4}{5}$ (i.e., $\frac{960}{1200}$) of the money due to him. The bankrupt is paying at 80p in the £.*

GENERAL INSURANCE

All kinds of "possibilities" may be **insured against**. (Risk of fire, flood, burst pipes, etc.) A **policy** (agreement) is drawn up between the **Insurer** (the Insurance Co.) and the **Insured** (the individual), who pays an annual **premium** to cover the possible event. The greater the risk, the greater the premium.

EXAMPLE. *The contents of a house may be covered by a "Comprehensive Policy" (fire, flood, etc.) for 25p per cent per annum.*

Thus, if the contents are considered to be worth £1000, the premium would be $\frac{1000}{100} \times 25p$.

$$= \underline{£2\cdot50 \text{ p.a.}}$$

LIFE ASSURANCE

This is based on "probability", and an Insurance Company can generally give a quite accurate assessment of the **expectation of life** for an individual, given the facts relating to present age, physique and health, nature of employment. The premium is based upon such facts.

EXAMPLE. *For a premium of perhaps £10 p.a., a man might assure his dependants of, say, £500 on his death. For a higher premium the* **assured sum** *might be increased by profits or* **bonuses** *accrued during the years.*

HIRE PURCHASE

Hire Purchase makes possible the purchase of goods without possessing the full purchase price. A cash deposit of perhaps 20% of the purchase price is first paid and the balance, plus interest on the loan (i.e., the sum outstanding after the deposit is paid), is paid in equal sums over 1 year, 2 years or even 3 years, depending upon the nature of the goods.

In attempting to reduce **inflationary** trends in the nation's economy (too many goods being bought on **credit**), the **Chancellor of the Exchequer** may, from time to time, change the percentage of deposit required, and also the repayment period, on certain commodities. Similarly, changes may be made in the various forms of taxation. (Purchase Tax exerts a considerable influence on individual purchasing power.)

EXAMPLE. *A TV set may be purchased for cash at £70·35 or (a) £3 dep., 39 weekly payments of £1·85 or (b) £14·07 dep. and 24 monthly payments of £2·70.*

> By cash the set costs £70·35
> By method (a)   ,,   £73·20
> By method (b)   ,,   £78·87

What is the percentage deposit required in (b) (i.e., % dep. of the cash price)?

Method (*a*) is really called **deferred terms** and not hire purchase, but note the fairly short repayment period (9 mth.) and the higher level of repayments.

MISCELLANEOUS

EXERCISE 79

1. If the rateable value of a town is £3,696,000, what is the income obtained from a rate of: (*a*) 1p in the £; (*b*) 10p in the £; (*c*) 75p in the £?

2. If the estimated income from a 1p rate in the County of Middlesex was £1,328,000, find the rateable value of the county. What was the income from a rate of 30p in the £?

3. A man's income is £1960 p.a., of which £950 is tax free. Find the net income if tax is deducted as follows: first £100 at 20p in the £; next £200 at 30p in the £; the remainder at 40p in the £.

4. Find the premium required to insure the contents of a house at 28p per cent if the contents are valued at £1250.

5. If 40 000 copies of a book are sold at 75p each, find the author's gross income if he receives a **royalty** as follows: 5% on the first 10 000 copies, $7\frac{1}{2}$% on the next 20 000 copies and 10% on any further sales. If income tax is deducted at 40p in the £, find the author's net income.

6. A washing machine may be purchased for £63 cash, or on H.P. terms: deposit £12·60 and 36 monthly payments of £1·85. Find: (*a*) the percentage deposit required, and (*b*) the extra cost by H.P. terms.

7. In Question 6 the extra cost by H.P. is accounted for by the interest charged on the loan (i.e., the sum outstanding after the deposit is paid). Express the interest as a fraction of the loan and calculate the effective *percentage* interest.

8. £450 is invested at 4% p.a. C.I. If the interest is subject to an income tax deduction of 37½p in the £, find the amount after 4 years.

9. A bankrupt's assets are £2400 and the liabilities £3000, what does he pay in the £ to his creditors? If a creditor were owed £150, what would he receive?

10. In a town of rateable value £8,619,600 the income from the rates is £5,171,760. Find: (*a*) the local rate; (*b*) the rates on a house, rateable value £108; (*c*) the rateable value of a house on which the rates were £72.

11. Calculate the value of the contents of a house if the Fire Insurance premium was £2·64 when charged at the rate of 22p per cent.

12. £386·75 is invested at 4½% p.a. C.I. If the interest is subject to an income tax deduction of 30p in the £, find the interest received after 4 years.

13. A man purchases a £6000 house by making a deposit of 20% of the purchase price and borrowing the rest from a Building Society. If his repayments over 25 years are £33·50 per month, calculate the price actually paid for the house.

14. If a bankrupt pays 60p in the £, and his liabilities are £3600, find his assets. If a creditor received £144, what was the actual debt?

15. A bankrupt owed one of his creditors £20, but could pay him only £13·50. If the total assets were £1917, find the total liability.

16. A man bought a house for £4520, the rateable value was £100 and the rates 60p in the £. If he charges a weekly rent of £5·50, inclusive of rates, what is the profit per cent, on his outlay, at the end of one year?

17. The rateable value of a house is £80 and the rates are 55p in the £. What rates are due on the house? Out of this amount paid in rates, £10·56 goes towards Education. Find the rate in the £ for the cost of Education.

18. The rateable value of a house is £108 and the rates are 60p in the £. If the property is reassessed at £128 and the rates are reduced to 50p in the £, calculate the increase or decrease in the rates of the house.

19. A man's income for one year was £1350. Tax was not paid on ⅔ of the income, nor on the next £450. On the remaining taxable income, tax was charged as follows: 20p in the £ on the first £100; 30p in the £ on the next £200; 40p in the £ on the remainder. Calculate the total tax paid.

20. The income from the rates of a town is £762,120 and a 1p in the £ rate produces £13,140. Find the rateable value of the town and the rate in the £. If the education rate is 18p in the £, find the annual cost of education in the town.

STOCKS AND SHARES

Large organizations, such as Governments or Corporations, may borrow money by an "Issue of Stock"; similarly, money may be invested in business companies by the purchase of "Shares".

Stocks are issued in £100 units, and the **nominal value** of all stock is £100, whereas the nominal value of shares varies (e.g., 25p, £1, £5). Once issued, stocks are frequently bought and sold in fractional quantities, but in shares transactions only whole numbers of shares are transferable.

It is important to distinguish between **nominal value** and **cash value**; if a company is doing well public confidence in that organization increases and the price of the shares rises:

e.g.,   25p shares stand at 30p   (A **premium** of 5p)
          £100 stock stands at £100   (At **par**.)
          £1 shares stand at 75p   (A **discount** of 25p)

Interest on investments is received in the form of a **dividend**, which is calculated as a given percentage of the nominal value per share. "5% £1 shares at £1·25" means that a dividend of 5% of £1 (i.e., 5p) will be paid on each share held, although each share will cost £1·25 to buy.

The **yield** of an investment is obtained by expressing the dividend as a percentage of the cash value of a share, i.e., Yield $= \dfrac{\text{Income}}{\text{Outlay}} \times \dfrac{100}{1}\%$

(4% in the above case.)

If stocks or share prices are given in the form "£1 share at 2" or "stock at 120" they mean £2 and £120 respectively. Buying and selling by the public is done through a **Stockbroker**, who charges **brokerage**; he, in turn, deals with a **Jobber**, who also charges for his services, and hence a share price might be quoted as "52p to 54p", meaning that a seller receives 52p and a buyer pays 54p per share. (Plus brokerage in each case.)

Exercise 80

Find the cost of and the dividend from:

1. 100 5% £1 shares at £1·25.
2. 200 4% 50p shares at 62½p.
3. £300 of 5% stock at 120.
4. £450 of 6% stock at 98.

Find the quantity of stock or numbers of shares which can be purchased and the dividend obtained by investing:

5. £154 in 5% £1 shares at 87½p.
6. £342 in 6% 50p shares at 45p.
7. £570 in 4½% stock at 95.
8. £754 in 4% stock at 104.

Find the proceeds from selling:

9. 240 3% £1 shares at par.

10. 300 4½% 50p shares at a premium of 4p.

11. £200 of 5% stock at a discount of £5.

12. £450 of 6% stock at 99 to 99½.

Find the price of the following and the dividend obtained:

13. 180 3½% £1 shares cost £144.

14. 320 4¼% 25p shares cost £89·60.

15. £350 of 5½% stock cost £393·75.

16. £540 of 3¼% stock cost £564·84.

Find the yield on the following investments:

17. 25p shares paying 4% stand at 20p.

18. £5 shares paying 5% stand at £6·25.

19. 3½% stock stands at 98.

20. 4¾% stock stands at 114.

Find the cost of buying and the proceeds from selling:

21. 140 shares at £1·10 to £1·12½, brokerage 1p per share.

22. 386 shares at 28p to 30p, brokerage ½p per share.

23. £750 of stock at 95 to 95½, brokerage ½%.

24. £2400 of stock at 83¾ to 84¼, brokerage ⅛%.

25. A man receives a dividend of £28 from some 5% 50p shares for which he paid £448. Find: (i) the number of shares; (ii) the purchase price per share; (iii) the yield on the investment.

26. A quantity of 4½% stock costing £864 produces a dividend of £36. Find: (i) the quantity of stock; (ii) the purchase price of £100 stock; (iii) the yield on the investment.

27. 2500 £1 shares are bought for £2000 and produce a dividend of £93·75. Find: (i) the premium or discount at which the shares stand; (ii) the rate % of the dividend; (iii) the yield on the investment.

28. £5600 of stock is bought at par and produces a dividend of £308, find the rate % of the dividend. If the stock is sold at a premium of 12½ and the proceeds are reinvested in 5% stock at a discount of 10, find the change in income.

29. £840 is invested in £5 shares. If a 4¼% dividend produces £51, find the price of a share and the yield on the investment.

30. £3250 is invested in stock. If a 5½% dividend produces £143, find the price of £100 stock and the yield on the investment.

31. £429 is invested in 4% 25p shares at 22p. The shares are sold when the price has risen to 24p and the proceeds are reinvested in 5% stock at 96. Find the change in income.

32. £2456 is invested in 5¼% stock at 105. The stock is sold at par and the proceeds are reinvested in 4⅕% stock at a discount of 20. Find the change in income.

33. A company makes a profit of £87,460 on a year's trading. From this, a dividend of 5% is first paid on 200 000 £1 Preference Shares, then £22,500 is set aside in a reserve fund. The remaining profits are distributed as a dividend to the holders of 916 000 £1 Ordinary Shares. Find: (i) the rate % of the Ordinary Share dividend; (ii) the dividend received by a man holding 1280 Ordinary Shares for which he had paid £1·12½ each; (iii) the yield on this man's investment after Income Tax at 37½p in the £ had been deducted from his dividend.

34. A company makes a year's profit, of which one-sixth is first distributed as a 6½% dividend on a number of £5 Preference Shares, then a fifth of the remainder is set aside to a reserve fund and the rest is distributed to the holders of 195 000 £1 Ordinary Shares as a 4¼% dividend after Income Tax has been deducted at 25p in the £. Find: (i) the pre-tax rate % of the Ordinary Share dividend; (ii) the total sum paid as untaxed dividend to the Ordinary Share holders; (iii) the total untaxed profit for the year; (iv) the amount set aside as reserve; (v) the number of Preference Shares issued.

BRITISH AND FOREIGN UNITS
EXERCISE 81

1. If 2·5 litres cost 29 fr 70 cent, find the cost per pint, given that £1 = 13·5 francs and 1 litre = 1·76 pints.

2. If 6 litres cost 9 lire, find the cost per gallon, given that £1 = 17·5 lire and 1 gallon = 4·55 litres.

3. Given that 1 kilometre = 0·6214 mile, express 40 feet per second in km/h, correct to 3 sig. fig. (5280 ft = 1 mile.)

4. If 1 litre of water weighs 1 kilogramme, prove that 1 cubic foot of water weighs approximately 1000 ounces, given that 1 litre = 61·03 cubic inches and 1 kilogramme = 2·205 lb. (16 oz = 1 lb.)

5. Express 17½p per square foot in francs per square metre, correct to 3 sig. fig., given that £1 = 12·09 francs and 3 ft = 0·9144 metre.

6. If 4 kilogrammes cost 22·4 marks, find the cost per pound, to the nearest penny, given that £1 = 11·2 marks and 1 pound = 0·454 kilogramme.

7. Given that 1 mile = 1·609 kilometres, express 45 km/h in feet per second correct to 3 sig. fig. (5280 ft = 1 mile.)

8. If 3·45 metres cost 34·5 guilders, find the cost per yard, given that £1 = 10·16 guilders and 1 yd = 0·9144 m.

9. If 1 ounce costs 31·25p, find, in francs, correct to 3 sig. fig., the cost of 50 grammes, given that 1 kilogramme = 2·2 pounds and £1 = 12·5 francs. (16 oz = 1 lb.)

10. Express an atmospheric pressure of 15 pounds per square inch in kilogrammes per square centimetre, correct to 3 sig. fig., given that 1 pound = 0·4536 kilogramme and 1 inch = 2·54 centimetres.

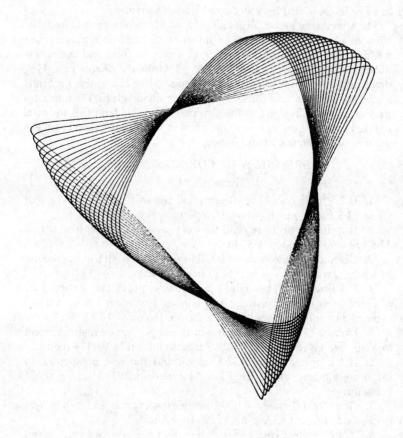

PART C: GEOMETRY

AXIOMS

PRACTICAL

Draw a straight line XY. Mark a point P about an inch below XY. Draw as many straight lines as you can through P and parallel to XY. How many such lines can you draw?

Your answer (which should be **one only**) provides the basis for a well-known **axiom** or self-evident truth.

Playfair's Axiom.

Only one straight line can be drawn through a given point parallel to a given straight line.

ANGLES AT A POINT

AXIOM 1

If a straight line stands on another straight line, the sum of the two adjacent angles is equal to two right-angles.

fig. C:1

DEFINITION. Adjacent angles are the two angles which lie one on each side of a common arm. See fig. C : 1. $\angle P$ and $\angle Q$ are adjacent.

Fig. C : 2 shows how to use a protractor. From the diagram name pairs of adjacent angles.

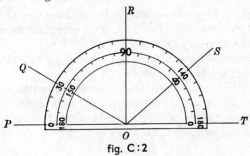

fig. C:2

Read off the size of the following angles:

(i) *POQ*	(ii) *POR*	(iii) *POS*	(iv) *QOR*	(v) *QOS*
(vi) *QOT*	(vii) *ROS*	(viii) *ROT*	(ix) *SOT*	(x) *POT*

What can you say about the sum of the following adjacent angles?

(i) *POQ* and *QOT* (ii) *POR* and *ROT*
(iii) *POS* and *SOT* (iv) *TOS* and *SOP*

THE CONVERSE

A number of geometrical propositions have a **converse**; that is to say the given facts and the conclusion are placed in reverse order. Thus the converse of Axiom 1 is stated as follows:

If the sum of two adjacent angles is equal to two right-angles, the exterior arms of the angles form a straight line.

Note the difference between the proposition and its converse, each must be thoroughly known and understood.

THE COROLLARY

In certain cases the satisfactory proof of one fact in geometry will lead quite naturally to certain other simple facts easily deduced from the first but nevertheless dependent on it. Such additional facts are called **corollaries**. The corollary to Axiom 1 is as follows:

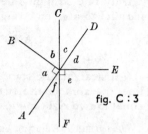

fig. C : 3

If any number of straight lines meet at a point, the sum of all the angles so formed is equal to four right-angles. (See fig. C : 3.)

VOCABULARY

1. An angle of 180° may sometimes be termed a **straight angle.**
2. Two angles whose sum is 90° are called **complementary angles.** One is the complement of the other.
3. Two angles whose sum is 180° are called **supplementary angles.** One is the supplement of the other.
4. If two straight lines intersect (i.e. cross each other) the opposite angles so formed are called **vertically opposite angles.**
5. An angle less than one right-angle is called an **acute angle.**
6. An angle between one and two right-angles is called an **obtuse angle.**
7. An angle between two and four right-angles is called a **reflex angle.**

NOTE
  (i) In diagrams a right-angle is indicated thus ⌐ .
  (ii) When two adjacent angles on a straight line are equal they must each be a right-angle (rt. ∠).

EXERCISE 1 (*Oral*)

Refer to fig. C : 3. Given that AD is a straight line and $\angle c + \angle d = 90°$, answer the following, giving reasons—the lines meet at point O:

1. What is the sum of $\angle b + \angle c$?
2. Give the name for $\angle AOB$ and $\angle BOD$.
3. Give the name for $\angle c$ and $\angle d$.
4. Name the vertically opposite angles.
5. Name the straight angles.
6. Why does $\angle AOC$ equal $\angle FOD$?
7. Name the supplementary angles.
8. Name the complementary angles.
9. What is the name given to $\angle BOE\,(b + c + d)$?
10. What is the name given to $\angle BOE\,(a + f + e)$?

VERTICALLY OPPOSITE ANGLES

AXIOM 2

If two straight lines intersect, the vertically opposite angles are equal. (See fig. C : 4.)

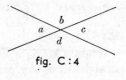

fig. C : 4

Although we are regarding this as an axiom, it provides an excellent example of the setting down of a simple formal proof:

Given: Two intersecting straight lines as in fig. C : 4.
To prove: $\angle a = \angle c$ and $\angle b = \angle d$.
Proof: $a + b = 180°$ (adj. $\angle$s on st. line)
              $b + c = 180°$ (adj. $\angle$s on st. line)
$\therefore\ a + b = b + c$
$\therefore\ \angle a = \angle c$
Similarly $\angle b = \angle d$ Q.E.D.

NOTE. The formal setting down of a proof should follow the same pattern at all times—the headings are important –diagrams should be fairly large and clearly lettered.

A proof may be concluded with the letters Q.E.D., an abbreviation for *Quod erat demonstrandum*, which means "that which was to be proved".

QUESTIONS

1. What have you learnt about vertically opposite angles?
2. Name the supplementary angles in fig. C : 4.
3. What are the steps in the formal proof to show $\angle b = \angle d$ in fig. C : 4?
4. Reconsider Q. 6, Exercise 1. Give the steps for a formal proof.

PARALLEL STRAIGHT LINES
VOCABULARY

A straight line which cuts two or more straight lines is called a **transversal**.

In fig. C : 5, *PT* is a transversal.

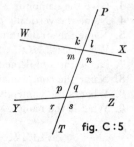

The angles *k*, *l*, *r*, *s* are called **exterior angles**.

The angles *m* and *n* are called **interior angles**. Name two more interior angles.

Angles *m* and *p* are called interior angles on the same side of the transversal; name two more such angles.

fig. C : 5

Angles *m* and *q* are called **alternate angles**; they are sometimes described as Z angles, can you suggest why? Name another pair of alternate angles.

Angles *k* and *p* are a pair of **corresponding angles**, so also are angles *n* and *s*. Name two more pairs of corresponding angles.

AXIOM 3

When a straight line cuts two other straight lines, these two straight lines are parallel IF

    (i) a pair of alternate angles are equal; or

    (ii) a pair of corresponding angles are equal; or

    (iii) a pair of interior angles on the same side of the transversal are together equal to two right-angles (supplementary).

CONVERSE

IF a straight line cuts two parallel straight lines, then

    (i) the alternate angles are equal;

    (ii) the corresponding angles are equal;

    (iii) the interior angles on the same side of the transversal are supplementary.

AXIOM 4

Straight lines which are parallel to the same straight line are parallel to one another.

In diagrams parallel lines are indicated thus:

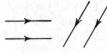

fig. C : 6

EXERCISE 2

Calculate the unknown angles in the following figures, give reasons:

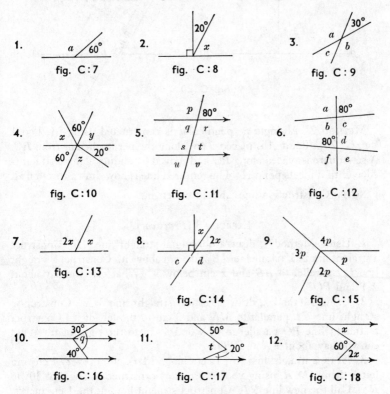

1.   fig. C:7

2.   fig. C:8

3.   fig. C:9

4.   fig. C:10

5.   fig. C:11

6.   fig. C:12

7.   fig. C:13

8.   fig. C:14

9.   fig. C:15

10.   fig. C:16

11.   fig. C:17

12.   fig. C:18

CONSTRUCTION 1

CONSTRUCTION OF PARALLEL LINES

PROBLEM. *To construct a straight line parallel to a straight line PQ and 5 cm from it.*

METHOD 1. A straight-edge is placed along *PQ* (fig. C:19) and two perpendiculars are constructed with the aid of a set-square. Points *A* and *B* are measured 5 cm along the perpendiculars from *PQ*.

The required line passes through *A* and *B*.

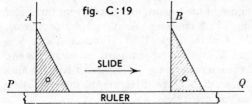

fig. C:19

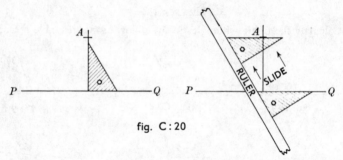

fig. C : 20

METHOD 2. A single perpendicular is constructed using Method 1 (fig. C : 20). Point A is placed 5 cm along the perpendicular from PQ. A set-square is placed along PQ and moved by sliding along a straight-edge until it meets point A. The required line is now drawn through A.

NOTE. Construction lines should be left in.

EXERCISE 3 (*Practical*)

1. Using Method 1, draw a horizontal straight line AB. Construct a straight line XY parallel to AB and 5 cm above it. Construct a straight line PQ parallel to AB and 2 cm below it. What can you say about XY and PQ?

2. Using Method 2, draw a vertical straight line MN. Construct a straight line KL parallel to MN and 3 cm to the left of it. Construct a straight line PQ parallel to MN and 4 cm to the right of it. What can you say about KL and PQ?

3. Draw a straight line RS, 10 cm long. Mark points every 1 cm and letter them. At R, using your protractor, construct an angle of 30° to RS. Call the new line RT. Construct straight lines at the 1 cm marks, all parallel to RT. What have these lines in common? Do you agree with Axiom 4?

4. Construct a square of side 4 cm. At a distance of 2 cm draw a line parallel to each side, outside the square. Make the new lines long enough to intersect. What is the shape of the new figure?

Figures which are the same SHAPE (they may differ in size) are called similar figures.

Can you construct a figure inside the original, 2 cm from the sides?

5. Draw a straight line XY, 8 cm long. At X, using a protractor, construct an angle of 30°. Call the angle YXT. Produce XT to S, a distance of 10 cm. Divide XS into five equal parts by measurement. Join S and Y. At the 2 cm intervals along XS construct straight lines parallel to SY, cutting XY at A, B, C, D. Measure the intervals along XY. What conclusion do you reach?

CONSTRUCTION 2

The division of a straight line into a given number of equal parts.

Let XY be the given straight line. We are required to divide XY into five equal parts.

Construction:

1. Through X draw XS at a convenient angle to XY.
2. Using compasses step off along XS five equal intervals XA, AB, BC, CD, DE.
3. Join EY. Through D, C, B, A draw lines parallel to EY cutting XY at d, c, b, a.

Then Xa, ab, bc, cd, dY are the required parts.

fig. C:21

EXERCISE 4 (*Practical*)

By construction:

1. Divide a 8-cm straight line into 3 equal parts.
2. Divide a 8-cm straight line into 5 equal parts.
3. Divide a 7-cm straight line into 4 equal parts.
4. Divide a 7-cm straight line into 5 equal parts.
5. Divide a 10-cm straight line into 7 equal parts.

RECTILINEAR FIGURES

A figure bounded by any number of straight sides is called a **rectilinear figure**. The corners are called **vertices** (*Sing.* **vertex**).

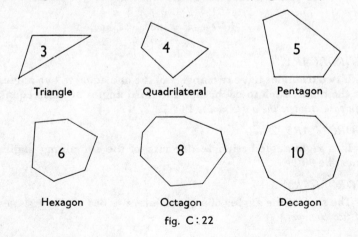

| Triangle | Quadrilateral | Pentagon |
| Hexagon | Octagon | Decagon |

fig. C:22

A figure with more than four sides may be described as a **polygon**.

A **regular polygon** is a polygon in which all the sides are equal and all the angles are equal, e.g., a regular pentagon, a regular octagon, etc.

THE TRIANGLE

NOTE

1. To construct a triangle, the sum of the two shortest sides must be greater than the longest side.
2. Every triangle must contain at least two acute angles.

THEOREM 1

1. **If one side of a triangle is produced, the exterior angle is equal to the sum of the two interior opposite angles.**

2. **The sum of the angles of a triangle is equal to two right-angles.**

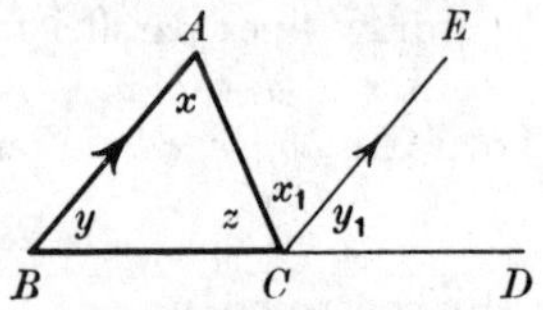

NOTE. $\angle ACD$ *is the exterior angle.* $\angle x$ *and* $\angle y$ *are interior and opposite.*

fig. C: 23

Given: ABC is a triangle with BC produced to D.

To prove: 1. $\angle ACD = x + y$
2. $x + y + z = 180°$

Construction: Through C draw $CE \| BA$.

Proof: 1. $x_1 = x$ (alt. $\angle$s)
$y_1 = y$ (corr. $\angle$s)
$\therefore\ x_1 + y_1 = x + y$
$\therefore\ \angle ACD = x + y$

2. Add z to each side of the equation:
Then $\angle ACD + z = x + y + z$
But $\angle ACD + z = 180°$ (a st. line)
$\therefore\ x + y + z = 180°$ Q.E.D.

COROLLARY 1

If two triangles have two angles of the one equal to two angles of the other, each to each, then the third angles are also equal. (*In each triangle the angle sum is* 180°.)

COROLLARY 2

In a right-angled triangle the sum of the remaining angles must be 90°.

COROLLARY 3

The sum of the angles of a quadrilateral is 360°. (*The angle sum of two triangles.*)

TRIANGLE PROPERTIES

Equilateral—3 equal sides. Isosceles—2 equal sides.
Scalene—no equal sides. Right-angled—contains a rt. ∠.
Obtuse-angled—one angle greater Acute-angled—3 acute angles.
  than 90°.

EXERCISE 5

Find the numbered angles in the following:

1.
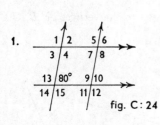
fig. C:24

2.
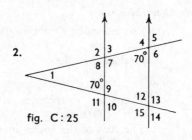
fig. C:25

3.
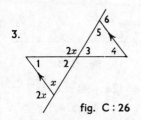
fig. C:26

4.
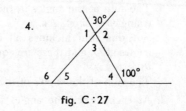
fig. C:27

5. Prove that if one side of a triangle is produced, the exterior angle is greater than either of the interior opposite angles.

6. From fig. C:28 prove the sum of angles 1, 2, 3 is equal to 4 right-angles.

7. From fig. C:29 prove the triangle is right-angled.

8. From fig. C:30, prove ∠a = ∠b and ∠c = ∠d.

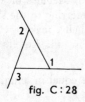
fig. C:28

fig. C:29

fig. C:30

THEOREM 2

1. In a polygon of *n* sides, the sum of the interior angles is equal to (2*n* — 4) right-angles.

2. If the sides of a convex polygon are produced in order, the sum of the exterior angles so formed is equal to four right-angles.

Given: $ABCD \ldots$ is a polygon of n sides.

To prove: 1. $a + b + c + \ldots$
$$= (2n - 4) \text{ rt. } \angle \text{s.}$$
       2. $p + q + r + \ldots = 4 \text{ rt. } \angle \text{s.}$

Construction:

1. Join any point O within the polygon to the vertices A, B, C, $D, \ldots$

2. Produce the sides in order.

Proof:

1. There are n triangles AOB, BOC, COD, etc.
   The sum of the angles in each triangle is 2 rt. $\angle$s.

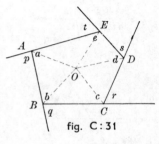

fig. C : 31

∴ The sum of the angles in the n triangles is $2n$ rt. $\angle$s.
But this sum includes all the angles at O, and these are equal to 4 rt. $\angle$s.

∴ The sum of the angles of the polygon $= (2n - 4)$ rt. $\angle$s.

2. At each vertex the angles form 2 rt. $\angle$s. Thus $a + p = 2$ rt. $\angle$s; $b + q = 2$ rt. $\angle$s, etc. Since there are n vertices, the sum of these angles $= 2n$ rt. $\angle$s.

∴ $(a + b + c + \ldots) + (p + q + r + \ldots) = 2n$ rt. $\angle$s.

But $(a + b + c + \ldots) + \quad 4 \text{ rt. } \angle \text{s} \quad = 2n$ rt. $\angle$s.

∴ $(p + q + r + \ldots) = 4$ rt. $\angle$s.

                                        Q.E.D.

NOTE. All theorems have to be read many times before they are clearly understood, then it is necessary to LEARN the proof. Do not attempt to learn by heart; instead make numerous sketches on rough paper and go through the proof from memory, referring to the sketch as: "This angle plus this angle is . . ., etc." Finally, make a careful diagram suitably lettered and write out the proof in full. Letters must follow each other round a diagram.

REMEMBER. In a regular polygon the sides are equal in length and the angles are equal.

CALCULATION

Find the interior angle of a regular pentagon.

METHOD 1. Sum of interior $\angle$s $= (2n - 4)$ rt. $\angle$s.

For 5 sides $= (10 - 4)$ rt. $\angle$s.

$= 6$ rt. $\angle$s.

But the 5 interior $\angle$s will be equal:

$\therefore$ 1 interior $\angle = \dfrac{540°}{5} = 108°$ $\therefore$ *Reqd. angle* $= \underline{\underline{108°}}$

METHOD 2. Sum of exterior $\angle$s $= 360°$

But the 5 exterior $\angle$s will be equal:

$\therefore$ 1 exterior $\angle = \dfrac{360°}{5} = 72°$

$\therefore$ 1 interior $\angle = 108°$ (supp. to $72°$) $\therefore$ *Reqd. angle* $= \underline{\underline{108°}}$

EXERCISE 6

1. Calculate the sum of the interior angles of: (i) a hexagon; (ii) an octagon.

2. What is the sum of the exterior angles of: (i) a quadrilateral; (ii) a decagon?

3. Find the interior angle of a regular polygon of: (i) 9 sides; (ii) 7 sides.

4. Find the exterior angle of a regular polygon of: (i) 12 sides; (ii) 20 sides.

5. By adapting Method 2 above, calculate the number of sides in a regular polygon whose exterior angle is: (i) $45°$; (ii) $40°$; (iii) $36°$; (iv) $30°$.

6. How many sides has a regular polygon whose interior angle is: (i) $60°$; (ii) $156°$; (iii) $160°$?

7. Construct a regular pentagon of side 5 cm.

8. Construct a regular hexagon of side 4 cm.

9. Construct a regular octagon of side 3 cm.

10. Four angles of a pentagon are $90°$, $100°$, $110°$, $120°$. Find the other angle.

11. Five angles of a hexagon are $80°$, $95°$, $105°$, $138°$, $152°$. Find the other angle.

12. Calculate the number of sides in a regular polygon in which the interior angle is twice the exterior angle.

13. Calculate the number of sides in a regular polygon in which the exterior angle is twice the interior angle.

14. *ABCDEF* is an irregular polygon whose opposite sides are parallel. Prove that the pairs of opposite angles are equal.

CONSTRUCTION 3

To construct an angle equal to a given angle.

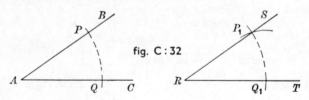

fig. C : 32

Given: Angle *BAC*.

To construct: Angle *SRT* equal to angle *BAC*.

Construction: With centre *A* and any radius draw an arc of a circle cutting *AB*, *AC* at *P*, *Q*.

With centre *R* and the same radius draw an arc of a circle cutting *RT* at Q_1.

With centre Q_1 and radius equal to *PQ*, draw an arc of a circle cutting arc P_1Q_1 at P_1.

Draw *RS* passing through P_1. The required angle is $\angle SRT$.

NOTE. Construction lines should be left in.

CONSTRUCTION 4

To bisect a given angle.

Given: Angle *PQR*.

To construct: A line bisecting $\angle PQR$.

Construction: With centre *Q* and any radius, draw an arc cutting *PQ*, *RQ* at *X*, *Y*.

With centres *X*, *Y* and a sufficient radius draw arcs of equal radius intersecting at *S*.

Then *QS* bisects $\angle PQR$.

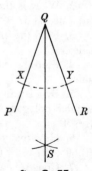

fig. C : 33

CONSTRUCTION 5

To draw the perpendicular bisector of a given straight line.

Given: Straight line *PQ*.

To construct: A straight line bisecting *PQ* at right angles.

Construction: With centres *P*, *Q* and a sufficient radius draw arcs of equal radius intersecting at *R*, *S*.

Then *RS* is the required line bisecting *PQ* at *T*.

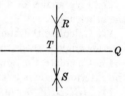

fig. C : 34

CONSTRUCTION 6

To draw a straight line perpendicular to a given straight line from a given point in it.

Given: A straight line *PQ* containing a point *X*. (Fig. C : 35.)

To construct: A straight line from *X* perpendicular to *PQ*.

Construction: With centre *X* and any radius, draw an arc of a circle cutting *PQ* at *R*, *S*.

With centres *R*, *S* and a sufficient radius draw arcs of equal radius intersecting at *T*.

Then *TX* is perpendicular to *PQ*.

NOTE the abbreviation for perpendicular: ⊥.

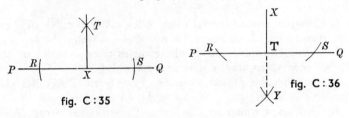

fig. C : 35   fig. C : 36

CONSTRUCTION 7

To draw a straight line perpendicular to a given straight line from a given point outside it.

Given: A straight line *PQ* and a point *X* outside *PQ*. (Fig. C : 36.)

To construct: A straight line from *X* perpendicular to *PQ*.

Construction: With centre *X* and any radius, draw an arc of a circle cutting *PQ* at *R*, *S*.

With centres *R*, *S* and a sufficient radius draw arcs of equal radius intersecting at *Y*.

Join *XY* cutting *PQ* at *T*.

Then *TX* is ⊥ to *PQ*.

FOR CLASS DISCUSSION

1. How can we construct an angle of 90°?

2. Can we now obtain an angle of 45°? How should this be done?

3. Can we readily obtain an angle of 60°? Does the equilateral triangle possess any useful properties? Might this triangle be called equiangular as well as equilateral? (**Equiangular** actually refers to figures in which corresponding angles are equal.)

4. How might we obtain an angle of 30°?

5. Suggest methods for constructing an angle of 120°, 150°, 135°. Which is the quickest method in each case? Does it help to consider the supplement of each angle?

VOCABULARY

A **quadrilateral** is a plane figure bounded by four straight sides. Opposite vertices are joined by a **diagonal**.

A **parallelogram** is a quadrilateral with opposite sides parallel.

A **rhombus** is a parallelogram with adjacent sides equal.

A **rectangle** is a parallelogram, one angle of which is a right-angle.

A **square** is a rectangle in which the adjacent sides are equal.

A **trapezium** is a quadrilateral having only one pair of opposite sides parallel.

An **isosceles trapezium** is a trapezium with its non-parallel sides equal.

EXERCISE 7

1. Construct a quadrilateral whose sides, taken in order are 2 cm, 4 cm, 6 cm, 8 cm. Measure the interior angles in order.

2. Construct a quadrilateral whose angles taken in order are 90°, 65°, 140°, 65°. Measure the sides in order.

3. Construct an equilateral triangle of side 6 cm. Bisect each side. What do you observe about the perpendicular bisectors?

4. Construct a square of side 6 cm. Construct the perpendicular bisector on each side. Draw the two diagonals. What do you observe about the bisectors and the diagonals?

5. Construct an isosceles triangle of base 8 cm, and base angles 70°. Bisect each angle of the triangle. How do the bisectors meet? Can you construct a circle which just touches the three sides of the triangle?

6. Construct a square of side 8 cm. Bisect each angle. Draw the diagonals. What conclusion do you reach?

7. Construct a rhombus of side 8 cm. Bisect each angle. Draw the diagonals. What conclusion do you reach? Measure the "halves" of each diagonal; measure the angles where they intersect. Have you reached any further conclusions? Check your conclusions with Question 6.

8. Construct a parallelogram of adjacent sides 4 cm and 6 cm. Repeat the constructions of Question 7. Do you reach the same conclusions?

9. Construct a rectangle of adjacent sides 4 cm and 6 cm. Repeat the constructions of Question 7. Do you reach the same conclusions?

10. Construct a quadrilateral $ABCD$ in which AB is 4 cm, $\angle A$ is 100°, $\angle B$ is 60°, AD is 3 cm, $\angle D$ is 80°. What name is given to such a figure? Give a reason. Measure $\angle C$. Measure sides CD, BC. Do the diagonals: (i) bisect the angles at the vertices; (ii) bisect each other?

11. If a parallelogram contains one right-angle, prove that the remaining angles are right-angles.

12. If a quadrilateral contains four right-angles, prove that the figure is a parallelogram.

CONGRUENT TRIANGLES 1

The word **congruent** is used to describe figures which are **alike in every respect**. This means not only the **same shape** but also the **same size**.

We shall see that it is possible to prove a pair of triangles congruent by matching the properties of one triangle with the corresponding properties of the other.

How many properties does a triangle possess?

Consider fig. C : 37, we can see that there are six features which concern us: 3 sides, 3 angles.

fig. C : 37

NOTE

1. A figure may be lettered clockwise or anticlockwise, but the letters MUST follow round the figure.

2. The sides carry the small letters of the opposite vertices.

3. The angle enclosed by two sides of a triangle is called the **included angle**. In fig. C : 37 $\angle A$ is the included angle of sides b and c.

AXIOM 5

If two triangles have two sides and the included angle of the one equal to two sides and the included angle of the other, the triangles are congruent. (Fig. C : 38.)

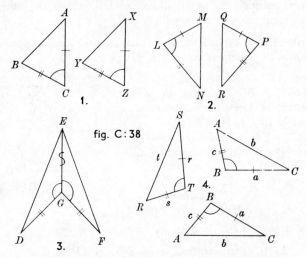

fig. C : 38

NOTE. Abbreviations: $\triangle ABC$ means triangle ABC; $\equiv$ means congruent; S.A.S. means side, angle, side (i.e., 2 sides and inc. $\angle$).

CONGRUENT TRIANGLES 2

AXIOM 6

If two triangles have two angles of the one equal to two angles of the other, each to each, and also one side of the one equal to the corresponding side of the other, the triangles are congruent.

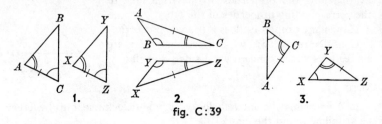

fig. C:39

NOTE. To establish congruency the pairs of sides or angles must **correspond.**

Note the method for indicating equal sides and angles.

THE USE OF CONGRUENCY

If triangles are congruent **all corresponding angles and sides are equal.** To ensure that two sides or two angles **correspond** we may need mentally to superimpose one triangle on the other and "juggle" them round until they fit.

RIDER. *ABCD is a quadrilateral in which AB is equal and parallel to CD. Prove BC equal and parallel to AD. (Fig. C : 40.)*

Given: $AB = CD$, $AB\|CD$.
To prove: 1. $BC = AD$. 2. $BC\|AD$.
Construction: Draw the diagonal AC.
Proof: 1. In $\triangle$s ABC, CDA.

$$AB\|CD \text{ (given)}$$
AC is a transversal across AB, CD
Then $x_1 = x_2$ (alt. $\angle$s)
            $AB = CD$ (given)
            AC is common
$\therefore$ $\triangle ABC \equiv \triangle CDA$ (S.A.S.)

In particular: $BC = AD$

2. AC is a transversal across BC and AD

Alt. $y_1 =$ Alt. y_2 ($\triangle ABC \equiv \triangle CDA$)
$\therefore$ $BC\|AD$ Q.E.D.

THE SOLUTION OF RIDERS

The following points are intended to help you:

(i) Draw a good diagram, try to let it conform to the given data, i.e., make equal sides *look* equal, though to do so by measurement is unnecessary.

(ii) Mark the given equalities on your diagram; if these lead to others through your knowledge of axioms or earlier theorems mark these facts also, e.g., angles equal due to parallel lines.

(iii) Do not assume equalities because they *appear* to be so on your diagram; instead, see if you can *prove* them to be so, providing the resulting facts will be of use in the main problem.

(iv) If, after considering *all* the given facts, you are still unable to establish the proof, work backwards, i.e., assume that what you have to prove is true and then examine *all* the facts that result from this assumption. This should lead you to see those facts which have to be proved.

EXERCISE 8 (*Proof by S.A.S.*)

1. *LMN, PQR* are two triangles. By suitably marking your diagrams, decide which of the following pairs of triangles are congruent (S.A.S.): (i) $l = p, n = r, \angle M = \angle Q$; (ii) $l = r, n = p, \angle N = \angle R$; (iii) $l = q, m = p, \angle N = \angle R$; (iv) $n = q, l = r, \angle M = \angle P$; (v) $m = q, l = p, \angle L = \angle R$.

2. *P, Q, R, S* are respectively the mid-points of *AB, BC, CD, DA* the sides of a square. Prove that $PQ = QR = RS = SP$.

3. Prove that the angle bisectors of an equilateral triangle are equal in length.

4. In fig. C : 41 the circle, centre *O*, passes through *P, Q, R*. If $\angle POQ$ is equal to $\angle ROQ$, prove $PQ = RQ$.

5. In fig. C : 42 the circle, centre *O*, passes through *A, B*. If *OC* bisects $\angle AOB$, prove $AX = BX$ and *OX* is $\perp$ to *AB*.

6. In fig. C : 43 *ABCD* and *BEFG* are squares. Prove $EC = AG$.

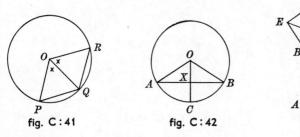

fig. C : 41   fig. C : 42

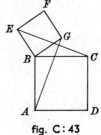

fig. C : 43

7. *RS* is the perpendicular bisector of a straight line *AB* and cuts *AB* at *X*. Prove *AR* = *BR*.

8. *ABCD* is a quadrilateral in which *AB* = *AD* and the diagonal *AC* bisects ∠*BAD*. If *BC* is 9 cm find *CD*. Give reasons.

9. *ABC* is an isosceles triangle in which *AB* = *BC*. Prove that *BD*, the bisector of ∠*ABC*, bisects *AC* and is at right-angles to it.

10. Prove that the angles opposite the equal sides of an isosceles triangle are also equal. (HINT. *Construct the bisector of the third angle.*)

EXERCISE 9 (*Class Discussion*)

Which of the following pairs of triangles are congruent either for A.S.A. or A.A.S.? State which.

In △s *ABC*, *XYZ*:

1. ∠*A* = ∠*X*, ∠*B* = ∠*Y*, *c* = *z*
2. ∠*C* = ∠*Z*, ∠*B* = ∠*Y*, *b* = *y*
3. ∠*A* = ∠*X*, ∠*C* = ∠*Z*, *b* = *z*
4. ∠*B* = ∠*Z*, ∠*C* = ∠*Y*, *a* = *x*
5. ∠*B* = ∠*Z*, ∠*C* = ∠*Y*, *c* = *y*
6. ∠*A* = ∠*X*, ∠*B* = ∠*Z*, *b* = *z*

EXERCISE 10

Nos. 1–7 Proof by A.S.A. or A.A.S.

1. *DA* is the bisector of ∠*BAC*. *BD* and *CD* are the perpendiculars from *D* to *AB* and *AC* respectively. Prove *AB* = *AC* and *BD* = *CD*.

2. In fig. C : 44 ∠*ABC* = ∠*DCB*, *DB* and *AC* bisect ∠*ABC* and ∠*DCB* respectively. Prove *AB* = *DC* and *AC* = *DB*.

3. In fig. C : 45 *O* is the centre of the circle and tangents *TR* and *TS* are at right-angles to *OR* and *OS* respectively. If *OT* bisects ∠*T* prove *RT* = *ST*.

4. In fig. C : 46 ∠*B* = ∠*C*. *BY* and *CX* bisects ∠*B* and ∠*C* respectively. Prove that *BY* = *CX*.

5. *BD* is the bisector of ∠*B* in △*ABC*. If *BD* meets *AC* at *D* perpendicularly, prove that △*ABC* is isosceles.

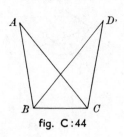

fig. C : 44

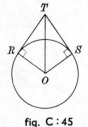

fig. C : 45

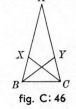

fig. C : 46

6. *O* is the mid-point of straight line *BC*. *AB* and *DC* are perpendiculars to *BC*. If *AOD* is a straight line prove *AB* = *DC*.

7. *BX*, the perpendicular to straight line *DAXCE*, bisects ∠*ABC* and ∠*DBE*. Prove that *DA* = *EC*.

Revision of S.A.S.

8. *ABCDE* is a regular pentagon. (i) Prove that *AC* = *BE*. (ii) Prove that △*ABD* is isosceles.

9. *ABCDEF* is a regular hexagon. Prove that *AC* = *FD*. Hence prove that *ACDF* is a rectangle.

10. *P*, *Q*, *R* are the mid-points of *AB*, *BC*, *CA* respectively, the sides of equilateral △*ABC*. Prove that △*PQR* is also equilateral. Hence prove that *BPRQ* is a parallelogram.

THEOREM 3

If two sides of a triangle are equal, the angles opposite to these sides are equal.

Given: *ABC* is a triangle in which *AB* = *AC*.
To prove: ∠*B* = ∠*C*.
Construction: Let *DA* be the bisector of ∠*A*.

Proof: In △s *ABD*, *ACD*
$$AB = AC \text{ (given)}$$
$$x_1 = x_2 \text{ (constr.)}$$
$$AD \text{ is common.}$$
∴ △*ABD* ≡ △*ACD* (S.A.S.)

In particular: ∠*B* = ∠*C*   Q.E.D.

fig. C : 47

CONVERSE

If two angles of a triangle are equal, the sides opposite to these angles are equal.

Given: *ABC* is a triangle in which ∠*B* = ∠*C*.
To prove: *AB* = *AC*.
Construction: Let *DA* be the bisector of ∠*A*.

Proof: In △s *ABD*, *ACD*
$$\angle B = \angle C \text{ (given)}$$
$$x_1 = x_2 \text{ (constr.)}$$
$$AD \text{ is common}$$
∴ △*ABD* = △*ACD* (A.A.S.)

In particular: *AB* = *AC*   Q.E.D.

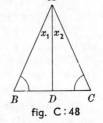
fig. C : 48

NOTE. ∠*A* is called the **vertical angle.** ∠*B* and ∠*C* are called the **base angles.**

CONGRUENCY 3

AXIOM 7

If two triangles have the three sides of the one equal to the three sides of the other, the triangles are congruent.

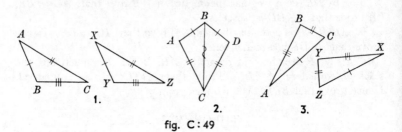

fig. C : 49

EXERCISE 11

1. Show the construction necessary to bisect an angle. Prove your construction.

2. Show how to construct the perpendicular bisector of a given straight line. Prove your construction. (HINT. *See fig.* C : 34. $\triangle RPS \equiv \triangle RQS$, $\angle PRS = \angle QRS$. *Now use* $\triangle PRT$ *and* $\triangle QRT$. *Prove* $PT = QT$ *and then* $\angle RTP = \angle RTQ = 90°$.)

3. Show how to draw a straight line perpendicular to a given straight line from a given point in it. Prove your construction.

4. Show how to draw a straight line perpendicular to a given straight line from a given point outside it. Prove your construction. (HINT. *Proceed as for Question 2*.)

5. *ABCD* is a square, prove that *AC* bisects the figure.

6. *ABC* is a triangle in which *D* is the mid-point of *BC*. If *AB = AC*, prove that *AD* is $\perp$ to *BC*.

7. *ABCD* is a quadrilateral in which *AB = CD*. If the diagonals are equal, prove $\angle ABC = \angle DCB$.

8. Prove that the diagonals of a square bisect each other at right-angles. (HINT. *Proceed as for Question 2, but remember to prove each diagonal bisects the other*.)

9. Two circles of equal radius and centres *O* and *Q* intersect at *A* and *B*. *OQ* cuts *AB* at *D*. Prove that *OQ* is the perpendicular bisector of *AB*.

10. *A*, *B* and *C* are three points on a circle such that *AB = BC*. If *O* is the centre of the circle, prove: (i) $\angle AOB = \angle COB$; (ii) *OB* bisects $\angle ABC$; (iii) $\triangle$s *AOB* and *COB* are both isosceles triangles.

CONGRUENCY 4

AXIOM 8

Two right-angled triangles are congruent if the hypotenuse and a side of one are equal to the hypotenuse and a side of the other.

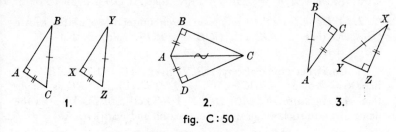

fig. C:50

MEASUREMENT OF LENGTH

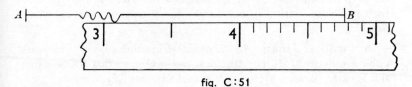

fig. C:51

The diagram, fig. C:51, shows a ruler, the $\frac{1}{10}$-in scale much enlarged, placed alongside line AB, whose length we are to measure. We read off the units as shown in the stages below, whole inches first, then whole tenths, then an intelligent guess at the decimal portion of the next tenth.

STAGE			TOTAL
1.	4 whole inches	$= 4\cdot0$	4·0 in
2.	7 whole tenths	$= 0\cdot7$	4·7 in
3.	$\frac{7}{10}$ of a tenth $= \frac{7}{100} = 0\cdot07$		4·77 in

The final stage depends on your ability to imagine a tenth of an inch divided into a further ten equal parts.

"Just over a tenth" would be 0·01 or 0·02
"Just under a tenth" would be 0·08 or 0·09

NOTE. The centimetre scale may be employed in a similar fashion.

QUESTIONS

 (i) What would "half a tenth" be?
 (ii) What would "a third of a tenth" be?
 (iii) What would "three-quarters of a tenth" be?
 (iv) What would "two-thirds of a tenth" be?
 (v) How close are you attempting to approximate in the last three questions?

MISCELLANEOUS EXAMPLES

EXERCISE 12

1. Draw a circle of radius 4 cm, centre O. Draw any diameter AOB. C and D are two points on the circumference such that $AC = BD =$ 4 cm. Prove $\triangle AOC \equiv \triangle BOD$. Prove that $\triangle COD$ is equilateral.

2. A, B and C are three points on the circumference of a circle centre O such that $\angle AOC = \angle COB = \angle BOA$. Prove that $AC = CB = BA$.

3. Construct a quadrilateral $ABCD$ given: $AD = 1\cdot7$ in, $AB = 2\cdot2$ in, $DC = 2\cdot6$ in; $\angle ABC = 70°$, $\angle BAD = 110°$. If BC is longer than AD, measure BC, AC, BD, $\angle ADC$ and $\angle BCD$. What is the name given to such a figure? Give a reason and justify it.

4. Construct an isosceles triangle in which the base angle is 70° and the perpendicular bisector of the base is 3 in. Measure the base and the equal sides. (HINT. $\perp$ *bisector also bisects vert.* $\angle$; *construct rt.* $\angle$ *first.*)

5. Construct a square $ABCD$ of side 1 in using ruler and compass only. On diagonal AC construct a square $ACFE$ so that B lies within this square. Measure AC. Prove that AF bisects CE.

6. A, B and C are three points on the circumference of a circle centre O, such that $AB = BC = CA$. Prove that $\angle AOB = \angle BOC = \angle COA$.

7. $ABCDEF$ is an irregular polygon. What name is given to such a figure? By joining AC, AD, AE prove that the sum of the interior angles of the polygon is equal to 8 rt. $\angle$s.

8. Fig. C : 52 shows a stellate (star-shaped figure) constructed by producing the sides of a regular hexagon $ABCDEF$. Calculate $\angle Z$ and prove that points A and B trisect XY.

9. Construct a stellate by producing the sides of a regular pentagon. Calculate the size of the acute angle at a point of the star.

10. In fig. C : 53, $PQ\|RS$. Calculate the value of x.

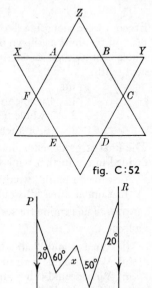

fig. C : 52

fig. C : 53

CONSTRUCTION 8

A more accurate construction for parallel straight lines.

Given: Straight line AB and X a point outside AB.

To construct: A straight line through X and parallel to AB.

Construction: With centre X
describe an arc QP cutting AB
at P. With centre P and same
radius describe an arc cutting
AB at R. With centre R and
same radius describe an arc
cutting arc QP at Q. The required line passes through X and Q.

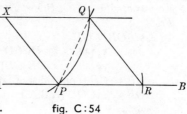

fig. **C:54**

Proof:

By construction Isos. $\triangle XPQ \equiv$ Isos. $\triangle RQP$ (S.S.S.)

$\therefore$ Alt. $\angle XQP =$ Alt. $\angle RPQ$

$\therefore XQ \| RP$

THE PARALLELOGRAM 1

DEFINITION. (*To remind you.*)

A parallelogram is a quadrilateral whose opposite sides are parallel.

THEOREM 4

(i) The opposite sides of a parallelogram are equal.

(ii) The opposite angles of a parallelogram are equal.

(iii) Each diagonal bisects the parallelogram.

Given: ABCD is a ∥gram.

To prove:

(i) $AB = CD$, $AD = BC$

(ii) $\angle A = \angle C$, $\angle D = \angle B$

(iii) BD and AC each bisect the ∥gram.

Construction: Join BD.

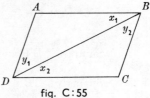

fig. **C:55**

Proof: In $\triangle$s ABD, CDB

$$x_1 = x_2 \text{ (alt. } \angle s, \ AB\|CD)$$
$$y_1 = y_2 \text{ (alt. } \angle s, \ AD\|BC)$$
$$BD \text{ is common}$$
$$\therefore \ \triangle ABD \equiv \triangle CDB \text{ (A.S.A.)}$$

In particular:

(i) $AB = CD$, $AD = BC$

(ii) $\angle A = \angle C$

(iii) BD bisects the ∥gram because the two triangles are congruent.

Similarly, by joining AC it may be proved that $\angle D = \angle B$, and that AC bisects the ∥gram.   Q.E.D.

THEOREM 5

The diagonals of a parallelogram bisect each other.

Given: $ABCD$ is a ‖gram in which the diagonals AC and DB intersect at O.

To prove: $AO = CO$, $BO = DO$.

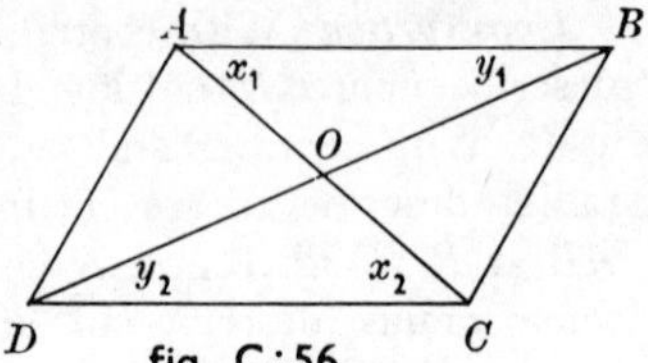

Proof:

In △s AOB, COD

$$x_1 = x_2 \text{ (alt. } \angle\text{s, } AB\|CD)$$
$$y_1 = y_2 \text{ (alt. } \angle\text{s, } AB\|CD)$$
$$AB = CD \text{ (opp. sides of ‖gram)}$$
$$\therefore \quad \triangle AOB \equiv \triangle COD \text{ (A.S.A.)}$$

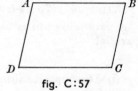

fig. C:56

In particular:

$$AO = CO$$
$$BO = DO$$

Q.E.D.

ALTERNATIVE PROOFS

The opposite angles of a parallelogram are equal.

ALTERNATIVE 1

Given: $ABCD$ is a ‖gram.

To prove: $\angle A = \angle C$, $\angle B = \angle D$.

fig. C:57

Proof: $\angle A + \angle D = 180°$ (on the same side of transversal AD and between parallels AB, CD)

$\angle D + \angle C = 180°$ (on the same side of transversal CD and between parallels AD, BC)

$\therefore \angle A + \angle D = \angle D + \angle C$

Subtract $\angle D$ from both sides:

$$\angle A = \angle C$$

Similarly $\angle B = \angle D$ Q.E.D.

ALTERNATIVE 2

Given: $ABCD$ is a ‖gram.

To prove: $\angle A = \angle C$, $\angle B = \angle D$.

Construction: Produce DC to E.

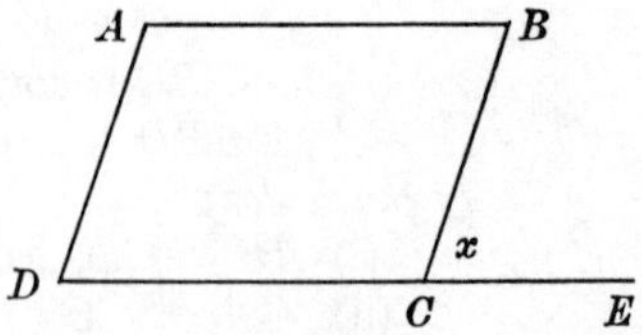

Proof:

$\angle x = \angle B$ (alt. $\angle$s, $AB\|DE$)

$\angle x = \angle D$ (corr. $\angle$s, $AD\|BC$)

$\therefore \angle B = \angle D$

fig. C:58

Similarly, by producing CD to F, we can prove $\angle A = \angle C$.

Q.E.D.

THE PARALLELOGRAM 2

THE CONVERSE THEOREMS

THEOREM 6

A quadrilateral is a parallelogram IF

   (i) the opposite sides are equal;
   (ii) the opposite angles are equal;
  (iii) the diagonals bisect each other;
  (iv) one pair of opposite sides are equal and parallel.

PROOF (i)

Given: $ABCD$ is a quadrilateral in which $AB = CD$, $AD = BC$.
To prove: $ABCD$ is a ∥gram.
Construction: Join BD.
Proof: In △s BAD, DCB

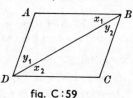

$$AB = CD \text{ (given)}$$
$$AD = BC \text{ (given)}$$
$$BD \text{ is common}$$
$$\therefore \triangle BAD \equiv \triangle DCB \text{ (S.S.S.)}$$

fig. C:59

In particular: $x_1 = x_2$
These are alternate ∠s, ∴ $AB \parallel CD$
Similarly $y_1 = y_2$
These are alternate ∠s, ∴ $AD \parallel BC$
      ∴ $ABCD$ is a ∥gram   Q.E.D.

PROOF (ii)

Given: $ABCD$ is a quadrilateral in which $\angle A = \angle C$, $\angle B = \angle D$.

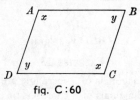

fig. C:60

To prove: $ABCD$ is a ∥gram.
Proof: The sum of the ∠s in a quadrilateral $= 360°$
$$\therefore x + x + y + y = 360°$$
$$\therefore x + y = 180°$$
But any pair of ∠s, x, y are on the same side of a transversal.
∴ The transversal cuts parallel lines.
       ∴ $AB \parallel CD$, $AD \parallel BC$
       ∴ $ABCD$ is a ∥gram   Q.E.D.

PROOF (iii)

Given: $ABCD$ is a quadrilateral in which $AO = CO$, $BO = DO$.

To prove: $ABCD$ is a ||gram.

Proof: In △s AOB, COD

$$AO = CO \text{ (given)}$$
$$BO = DO \text{ (given)}$$
$$x_1 = x_2 \text{ (vert. opp.)}$$
$$\therefore \ \triangle AOB \equiv \triangle COD \text{ (S.A.S.)}$$

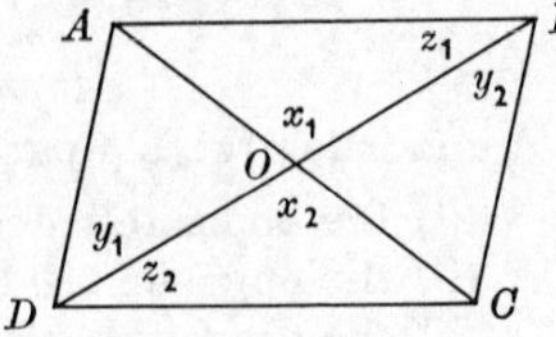

fig. C:61

In particular: $z_1 = z_2$

   But these are alternate ∠s,   ∴ $AB \| CD$

   Similarly in   △s AOD, COB

   It may be proved that $y_1 = y_2$,   ∴ $AD \| BC$

                $\therefore$ $ABCD$ is a ||gram   Q.E.D.

NOTE. Each of these proofs arrives at the definition of the parallelogram and satisfies the requirement that "opposite sides are parallel".

PROOF (iv)

Given: $ABCD$ is a quadrilateral in which $AB = CD$, $AB \| CD$.

To prove: $ABCD$ is a ||gram.

Construction: Join BD.

Proof:

   To satisfy the definition of a ||gram we have to prove $AD \| BC$.

   In △s ABD, CDB

$$AB = CD \text{ (given)}$$
$$x_1 = x_2 \text{ (alt. } \angle s, AB \| CD)$$
$$BD \text{ is common}$$
$$\therefore \ \triangle ABD \equiv \triangle CDB \text{ (S.A.S.)}$$

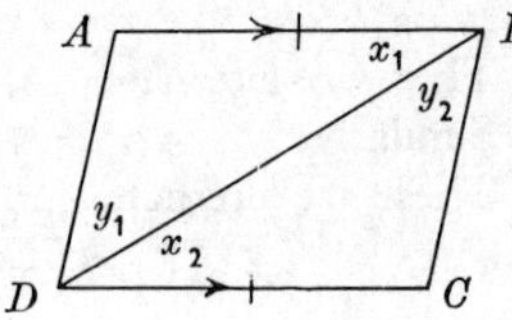

fig. C:62

In particular: $y_1 = y_2$

But these are alternate ∠s,   ∴ $AD \| BC$

                $\therefore$ $ABCD$ is a ||gram   Q.E.D.

EXERCISE 13

1. Prove that the diagonals of a rectangle are equal.

2. Prove that the diagonals of a rhombus bisect each other at right-angles.

3. Prove that the diagonals of a square are equal and bisect each other at right-angles.

4. P, Q, R and S are the mid-points respectively of the sides AB, BC, CD and DA of a parallelogram $ABCD$. Prove that $PQRS$ is a parallelogram.

5. In fig. C:63 $PQRS$, $PQYZ$ are parallelograms. Prove that $SRYZ$ is a parallelogram.

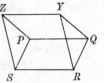

fig. C:63

6. If the diagonals of a parallelogram are equal, prove that the parallelogram is a rectangle.

7. Two circles of equal radius and centres O and P intersect at Q and R. Prove that $OQPR$ is: (i) a parallelogram; (ii) a rhombus.

8. AC and BD are two diameters of a circle. Prove that $ABCD$ is a parallelogram.

9. ABC is an isosceles triangle. AC is produced to D so that $AC = DC$, BC is produced to E so that $BC = EC$. Prove that $ABDE$ is a parallelogram.

10. Triangle ABC is right-angled at A. $AFGB$ is a square outside the triangle and so is $ACDE$. Prove: (i) GAD is a straight line; (ii) $BCEF$ is an isosceles trapezium.

11. $ABCDE$ is a regular polygon. Prove that $\triangle BED$ is isosceles. Calculate the angles of this triangle.

12. $ABCDEF$ is a regular polygon. Prove that $BCEF$ is a rectangle. Calculate the angles of: (i) $\triangle ABF$; (ii) $\triangle BEF$.

AREA OF PARALLELOGRAMS
THEOREM 7

The area of a parallelogram is equal to the area of a rectangle on the same base and between the same parallels.

Given: $ABCD$ is a rectangle on the same base CD as ∥gram $EFCD$. The figures are between the same parallels AF and CD.

To prove: $ABCD = EFCD$.

Proof:

In △s AED, BFC

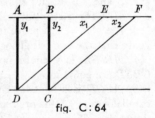

$$y_1 = y_2 \text{ (corr. } \angle\text{s, } AD \| BC)$$
$$x_1 = x_2 \text{ (corr. } \angle\text{s, } ED \| FC)$$
$$AD = CB \text{ (opp. sides of rect.)}$$
$$\therefore \triangle AED \equiv \triangle BFC \text{ (A.A.S.)}$$

fiq. C : 64

Each of these equal triangles is subtracted, in turn, from the complete figure $AFCD$. Thus:

$$AFCD - \triangle BFC = AFCD - \triangle AED$$
$$\therefore ABCD = EFCD$$

The rectangle and parallelogram are equivalent, i.e., their areas are equal.   Q.E.D.

NOTE

(i) The phrase "between the same parallels" means that **a pair of opposite sides of each figure must actually lie on the parallels.**

(ii) The area of a rectangle is given by the product of two adjacent sides, from fig. C : 64   Area of rectangle $= CD \times BC$

$$\therefore \text{ Area of } \| \text{gram} = CD \times BC$$

COROLLARY 1

Parallelograms on the same base and between the same parallels (i.e., the same altitude) are equal in area. (Fig. 65.)

COROLLARY 2

Parallelograms on equal bases and of the same altitude are equal in area. (Fig. 66.)

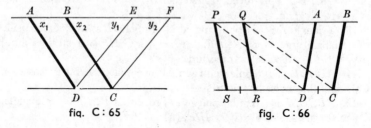

fig. C: 65   fig. C: 66

EXERCISE 14

1. *ABCD* is a parallelogram with *AB* produced to *E* so that *BECD* is a parallelogram with side *BD*. Prove that parallelograms *ABCD* and *BECD* are equivalent.

2. *PQRS* is a parallelogram in which *X* is the mid-point of *PQ*. *PQ* is produced to *Y* so that *XYRS* is a parallelogram. Prove that *PQRS* and *XYRS* are equal in area.

3. *ABCD* is a rectangle. *BA* is produced to *R*, *AB* is produced to *S*. If *RA* = *AB* = *BS*, prove that *RACD* and *SBDC* are parallelograms. Prove that the two parallelograms are equal in area.

4. *ABCDE* is a regular polygon. Prove that the figures *ABDE* and *CBED* are equal in area.

5. *ABCD* is a square. *BC* is produced to *Y* so that *ACYD* is a parallelogram. *AD* is produced to *X* so that *BDXC* is a parallelogram. Prove that *ACYD* and *BDXC* are equal in area.

6. In fig. C : 67 *ACEG* is a rectangle. If *HD* is parallel to *AC* and *BF* is parallel to *CE*, prove that the shaded areas are equal in area.

7. *ABCDEF* is a regular polygon. Prove that the figures *ABDE* and *BCEF* are equal in area.

8. *ABCD* is a rectangle in which *P*, *Q*, *R* and *S* are the mid-points respectively of sides *AB*, *BC*, *CD* and *DA*. Prove that the area of *PQRS* is half the area of *ABCD*. (HINT. *Draw PR and SQ intersecting at O. Consider △POS and △PAS. Repeat with rest of figure.*)

fig. C: 67

CONSTRUCTION 9

Construct a rectangle equal in area to a given parallelogram.

Given: Parallelogram *ABCD*.

To construct: A rectangle equal in area
to ‖gram *ABCD*.

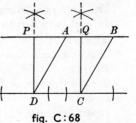

Construction:

(i) Erect a ⊥ at *D* to cut *BA* produced at *P*.

(ii) Erect a ⊥ at *C* to cut *BA* at *Q*.

Then *PQCD* is the required rectangle.

fig. C:68

From Theorem 7, a parallelogram is equal in area to a rectangle on the same base and between the same parallels.

VOCABULARY

Perpendicular means at right-angles to a given line or plane.

Vertical means at right-angles to a **horizontal** plane.

We have already seen that the **altitude of a parallelogram** is the perpendicular distance between a pair of parallel sides with one of those sides as a base.

The **altitude of a triangle** is given by the perpendicular from a vertex to the opposite side as base. From fig. C : 69 we see that a triangle has three altitudes each employing a different side as base. These altitudes are **concurrent**, i.e., they meet at one point *O*, called the **orthocentre**.

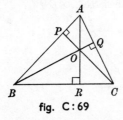

fig. C:69

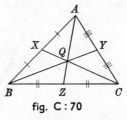

fig. C:70

A **median** is the line joining the mid-point of a side of a triangle to the opposite vertex. From fig. C : 70 we see that the three medians are concurrent at point *Q*, called the **centroid.**

NOTE

(i) The centroid lies on each median at a point one-third of the distance from the side to the opposite vertex.

(ii) The centroid is the point at which a triangular **lamina** (thin sheet) will rest in **equilibrium** (balance). It is sometimes called the **centre of gravity.** (C. of G.)

TRIANGLES

THEOREM 8

The area of a triangle is half that of a rectangle on the same base and between the same parallels.

Given: $ABCD$ is a rectangle with triangle BCE on the same base BC. $AE\|BC$.

To prove: $\triangle BCE = \frac{1}{2}ABCD$.

Construction: Draw $BF\|CE$. Then $BCEF$ is a $\|$gram.

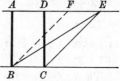

fig. C:71

Proof: Diagonal BE bisects $\|$gram $BCEF$

$$\therefore \quad \triangle BCE = \frac{1}{2}BCEF$$

But $\quad BCEF = ABCD$ (same base, same $\|$s)

$$\therefore \quad \triangle BCE = \frac{1}{2}ABCD \qquad \text{Q.E.D.}$$

COROLLARY 1

The area of a triangle is given by half the product of the base and the altitude. Area $= \dfrac{\text{Base} \times \perp \text{Ht.}}{2}$

COROLLARY 2

The area of a triangle is half that of a parallelogram on the same base and between the same parallels.

THEOREM 9

Triangles on the same base and of the same altitude are equal in area.

Given: $\triangle ABC$, $\triangle DBC$ on the same base BC. $AD\|BC$.

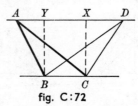

fig. C:72

To prove: $\triangle ABC$, $\triangle DBC$ equal in area.

Construction: Erect $\perp$s at B and C, then $BCXY$ is a rectangle.

Proof: $\quad \triangle ABC = \frac{1}{2}BCXY$ ($\triangle = \frac{1}{2}$ rect. on same base, etc.)

and $\quad \triangle DBC = \frac{1}{2}BCXY$ ($\triangle = \frac{1}{2}$ rect. on same base, etc.)

$$\therefore \quad \triangle ABC = \triangle DBC \qquad \text{Q.E.D.}$$

NOTE. $\triangle ABC$ is equal in area to $\triangle DBC$, NOT congruent.

COROLLARY

Triangles on equal bases and of the same altitude are equal in area.

THEOREM 10

(CONVERSE OF THEOREM 9.) **Triangles which are equal in area and are on the same side of the same base are of the same altitude (between the same parallels).**

Given: △s *ABC*, *ABD* are on the same base *AB* and are equal in area.

To prove: △s *ABC*, *ABD* equal in altitude.
Construction: Draw *CX*, the altitude of △*ABC* and *DY*, the altitude of △*ABD*.
Proof:
The area of △*ABC* is given by $\frac{1}{2}AB \cdot CX$
The area of △*ABD* is given by $\frac{1}{2}AB \cdot DY$
But these areas are equal (given).

$$\therefore \ \frac{1}{2}AB \cdot CX = \frac{1}{2}AB \cdot DY$$

Dividing both sides by $\frac{1}{2}AB$:

$$CX = DY \qquad\qquad \text{Q.E.D.}$$

fig. C:73

NOTE. Since the altitudes are equal and the triangles are on the same base, they must lie between the same parallels.

COROLLARY

Triangles which are equal in area and are on equal bases are of the same altitude.

CONSTRUCTION 10

Construct a triangle equal in area to a given quadrilateral.

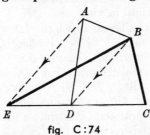

fig. C:74

Given: ABCD is a quadrilateral.
To construct: A triangle equal in area to *ABCD*.
Construction: Join *BD*. Through *A* draw *AE*, parallel to *BD*, meeting *CD* produced at *E*. Join *BE*.
EBC is the required triangle.
Proof: △s *ABD*, *EDB* are equal in area. (Same base *BD*, *AE*∥*BD*.)
  Add to each of these the △ *BDC*.
  Then $ABD + BDC = EDB + BDC$
  ∴ Quad. *ABCD* = △*EBC*

PART C: GEOMETRY

1. Prove that parallelograms on the same base and of the same altitude are equal in area. (Theorem 7, Corollary 1, fig. C : 65.)

2. Prove that parallelograms on equal bases and of the same altitude are equal in area. (Theorem 7, Corollary 2, fig. C : 66.) (HINT. ‖gram *PQCD* is common.)

3. Prove that the area of a triangle is given by half the product of the base and the altitude.

4. Prove that the area of a triangle is half that of a parallelogram on the same base and between the same parallels.

5. Prove that triangles on equal bases and of the same altitude are equal in area.

6. Prove that triangles which are equal in area and are on equal bases are of the same altitude.

7. *BX* is a median in △*ABC*. Prove that △s *ABX, CBX* are equal in area.

8. *ABCD* is a parallelogram with *X* any point on *AB*. Prove △*DXC* = △*AXD* + △*BXC*.

9. *ABCD* is a trapezium whose diagonals *AC* and *BD* intersect at *X*. Prove that △s *AXD, BXC* are equal in area. (*AB*‖*CD*.)

10. *CD* and *BE* are two medians in △*ABC*. Prove that △s *ABE, ACD* are equal in area.

11. *PQRS* is a parallelogram whose diagonals intersect at *O*. Prove that the four triangles so produced are equal in area.

12. *ABCD* is a rhombus whose diagonals intersect at *X*. Prove that △s *ABX, CDX* are equal in area.

13. *ABCD* is a quadrilateral in which *AD* = 7 cm, *DC* = 8 cm, *CB* = 9 cm, ∠*D* = 90°, ∠*C* = 70°. Produce *CD* to *E* to construct △*BCE* equal in area to *ABCD*. Measure *BE*.

14. *ABCD* is a quadrilateral in which *AD* = 8 cm, *AB* = 6 cm, *DC* = 10 cm, ∠*A* = 120°, ∠*D* = 80°. Produce *DC* to *E* to construct △*ADE* equal in area to *ABCD*. Measure *BE*.

15. *ABCD* is a quadrilateral. Construct a triangle equal in area to *ABCD*. Prove your construction.

16. *ABCD* is a quadrilateral in which ∠*A* = 70°, *AD* = 8·6 cm, *CD* = 6 cm, ∠*D* = 110°, ∠*C* = 70°. Construct △*BCX* equal in area to *ABCD*. Prove your construction by finding the area of *ABCD* and △*BCX*.

17. *ABCD* is a regular pentagon with *AE* as base. By producing *EA* to *X* and *AE* to *Y*, construct △*CXY* equal in area to *ABCDE*. Prove your construction.

18. Prove that, if two sides of a triangle are equal, the angles opposite these sides are also equal.

AXIOM 9

The sum of any two sides of a triangle is greater than the third side.

For revision purposes, consider fig. C : 75.

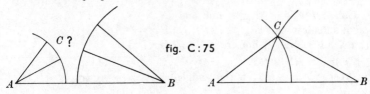

fig. C : 75

THE INEQUALITY THEOREMS

THEOREM 11

If two sides of a triangle are unequal, the greater side has the greater angle opposite to it.

Given: $\triangle PQR$ with $PR > PQ$.
To prove: $\angle Q > \angle R$.
Construction: On PR mark point S so that $PS = PQ$. Join QS.
Proof: $\quad$ $\triangle PQS$ is isosceles because
$$PS = PQ \text{ (const.)}$$
$$\therefore x_1 = x_2$$
Also $\quad x_2 = y_1 + y_2 \text{ (ext. } \angle \text{ of } \triangle QRS)$
$$\therefore x_2 > y_2$$
$$\therefore x_1 > y_2$$
$$\therefore x_1 + y_1 > y_2$$
$$\therefore \angle Q > \angle R \qquad\qquad\qquad \text{Q.E.D.}$$

fig. C : 76

THEOREM 12

(CONVERSE OF THEOREM 11.) **If two angles of a triangle are unequal, the greater angle has the greater side opposite to it.**

Given: $\triangle PQR$ with $\angle Q > \angle R$.
To prove: $PR > PQ$.
Proof: There are three possible conditions: (i) $PR = PQ$;
(ii) $PR < PQ$; (iii) $PR > PQ$.
$\quad$ (i) If $PR = PQ$, then $\angle Q = \angle R$.
But this is untrue. (Contrary to the data.)
$\quad$ (ii) If $PR < PQ$, then $\angle Q < \angle R$.
But this is untrue. (Contrary to the data.)
$$\therefore PR > PQ \qquad\qquad\qquad \text{Q.E.D.}$$

fig. C : 77

THEOREM 13

The perpendicular is the shortest line which can be drawn from a given point to a given straight line.

Given: PQ is a straight line and R a point outside it. RS is $\perp$ to PQ at S. RT is any other straight line from R to PQ.

To prove: $RS < RT$.

Proof: Since $\triangle RST$ is right-angled at $\angle S$, $\angle T <$ a rt. $\angle$.

$$\therefore \angle T < \angle S \qquad \therefore RS < RT$$

But RT is *any* other straight line from R to PQ.

$\therefore RS$ is the shortest distance from R to PQ.   Q.E.D.

fig. C:78

THE EQUAL INTERCEPT THEOREMS

THEOREM 14

The straight line drawn through the middle point of one side of a triangle parallel to another side bisects the third side.

Given: ABC is a triangle in which $AP = BP$, $PQ \| BC$.

To prove: $AQ = CQ$.

Construction: Through C, draw $CR \| AB$ to meet PQ produced at R.

Proof: $BCRP$ is a $\|$gram. $(PR \| BC, BP \| CR)$.

$$\therefore BP = CR$$
$$\therefore AP = CR \ (P \text{ mid-pt. } AB)$$

In $\triangle$s APQ, CRQ

$$AP = CR \text{ (proved)}$$
$$x_1 = x_2 \text{ (alt. } \angle\text{s, } AB \| CR)$$
$$y_1 = y_2 \text{ (vert. opp.)}$$
$$\therefore \triangle APQ \equiv \triangle CRQ \text{ (A.A.S.)}$$

In particular: $AQ = CQ$ Q.E.D.

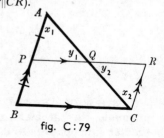

fig. C:79

THEOREM 15

(CONVERSE OF THEOREM 14.) **The straight line joining the middle points of two sides of a triangle is parallel to the third side, and equal to half of it.**

Given: ABC is a triangle in which P and Q are the mid-points of AB and AC respectively.

To prove: (i) $PQ \| BC$; (ii) $PQ = \frac{1}{2}BC$.

Construction: Through C, draw $CR \| AB$ to meet PQ produced at R.

Proof:

In $\triangle$s APQ, CRQ

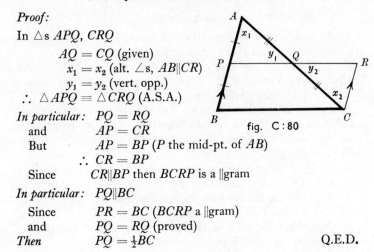

$\qquad AQ = CQ$ (given)

$\qquad x_1 = x_2$ (alt. $\angle$s, $AB \| CR$)

$\qquad y_1 = y_2$ (vert. opp.)

$\therefore\ \triangle APQ \equiv \triangle CRQ$ (A.S.A.)

In particular: $\quad PQ = RQ$

and $\qquad\qquad AP = CR$

But $\qquad\qquad AP = BP$ (P the mid-pt. of AB)

$\qquad\qquad \therefore\ CR = BP$

Since $\qquad\quad CR \| BP$ then $BCRP$ is a $\|$gram

In particular: $\quad PQ \| BC$

Since $\qquad\quad PR = BC$ ($BCRP$ a $\|$gram)

and $\qquad\qquad PQ = RQ$ (proved)

Then $\qquad\qquad PQ = \frac{1}{2}BC$ $\qquad\qquad\qquad\qquad$ Q.E.D.

THEOREM 16

If there are three or more parallel straight lines, and they make equal intercepts on one transversal, then the intercepts on any other transversal will be equal.

Given: Three parallel straight lines cut one transversal at A, B, C; and another transversal at P, Q, R. $AB = BC$.

To prove: $PQ = QR$.

Construction: Draw AX and $BY \|$ to PR and cutting BQ, CR at X, Y respectively.

Proof:

In $\triangle$s ABX, BCY

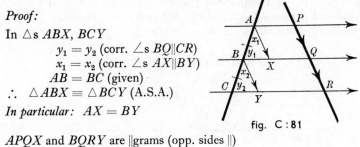

$\qquad y_1 = y_2$ (corr. $\angle$s $BQ \| CR$)

$\qquad x_1 = x_2$ (corr. $\angle$s $AX \| BY$)

$\qquad AB = BC$ (given)

$\therefore\ \triangle ABX \equiv \triangle BCY$ (A.S.A.)

In particular: $AX = BY$

$APQX$ and $BQRY$ are $\|$grams (opp. sides $\|$)

$\qquad\qquad \therefore\ AX = PQ$

and $\qquad\qquad BY = QR$

Since $\qquad\quad AX = BY$ (proved)

Then $\qquad\qquad PQ = QR$ $\qquad\qquad\qquad\qquad$ Q.E.D.

Exercise 16

1. *X*, *Y*, *Z* are the mid-points of the sides of an equilateral triangle *ABC*. Prove that △*XYZ* is equilateral.

2. *X*, *Y* are the mid-points of *AB*, *AC* respectively, two sides of △*ABC*. Prove that △s *BXY*, *CYX* are equal in area. Prove also that △s *ABY*, *ACX* are equal in area.

3. *PQR* is an isosceles triangle. *A* and *B* are the mid-points of *PQ* and *RQ* respectively. Prove that △s *PAR*, *RBP* are equal in area. Prove also that, if *AR* and *BP* intersect at *O*, △s *AOP*, *BOR* are equal in area.

4. *X*, *Y*, *Z* are the mid-points of sides *AB*, *BC*, *CA* in △*ABC*. Prove that *AXYZ* is a parallelogram.

5. *X*, *Y*, *Z* are the mid-points of sides *AB*, *BC*, *CA* in △*ABC*. Prove that △s *BXY*, *YZC* are congruent.

6. Prove that the lines joining the mid-points of the sides of a triangle divide the triangle into four congruent triangles.

7. *P* and *Q* are the mid-points of sides *XY* and *XZ* respectively of △*XYZ*. Prove that the area of △*XPQ* = $\frac{1}{4}$ area △*XYZ*.

8. Divide a straight line *AB* 8 cm long into three equal parts and prove your construction.

9. *Z* is the mid-point of side *QR* in △*PQR*. If *XZ* and *YZ* are drawn parallel to *PR* and *PQ* respectively, meeting these sides at *X* and *Y*, prove that *XY*∥*QR* and is equal to $\frac{1}{2}QR$.

10. *ABCD* is any quadrilateral. *P*, *Q*, *R* and *S* are the mid-points of the sides. Prove that *PQRS* is a parallelogram.

11. *PQR* is an equilateral triangle. The angle bisectors meet at *O*. *X*, *Y*, *Z* are the mid-points of *OP*, *OQ*, *OR* respectively. Prove that △*XYZ* is equilateral. Prove also that △*XYZ* = $\frac{1}{4}$ area △*PQR*.

12. *PQRS* is a parallelogram. The diagonal *RP* is produced to *X* so that *RP* = *PX*. (i) Prove that *QP* produced bisects *XS* and *SP* produced bisects *XQ*. (ii) If *PR*, *SQ* insersect at *O*, prove that *PO* is one-third of the median *XO* in △*XSQ*.

PROPORTIONAL DIVISION

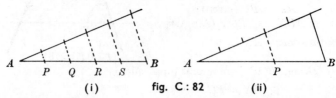

(i)   fig. C : 82   (ii)

Fig. C : 82 (i) shows the usual construction to divide *AB* into 5 equal parts, but suppose we require to divide *AB* into two parts in the ratio 3 : 2. By addition we see that 5 parts are still involved, but we take them in two groups, one of 3 parts the other 2 parts. Thus, in fig. C : 82 (ii) $AP : BP = 3 : 2$ or $\dfrac{AP}{BP} = \dfrac{3}{2}$.

NOTE

(i) Ensure that the literal ratio is in the same order as the numerical ratio.

(ii) Since the point P occurs on AB, we say that AB has been divided **internally.** At a later stage we shall consider the case in which AB is divided **externally** and P occurs on AB produced.

CONSTRUCTION 11

Divide a given straight line internally in a given ratio.

Given: AB is a straight line

To construct: A point E such that $\dfrac{AE}{BE} = \dfrac{x}{y}$.

Construction:

Draw line AQ at a convenient angle to AB.

Along AQ mark $AD = x$ and $DC = y$.

Join BC and draw $DE \| BC$ cutting AB at E.

Then E divides AB in the ratio $x : y$.

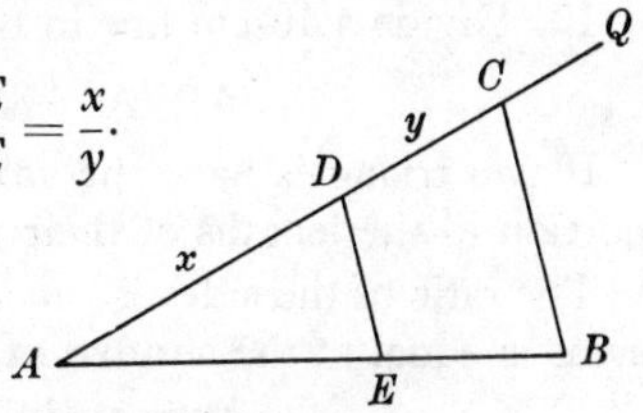

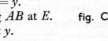

fig. C : 83

THEOREM 17

A straight line parallel to one side of a triangle divides the other two sides proportionally.

Given: ABC is a triangle and $RS \| BC$.

To prove: $\dfrac{AR}{BR} = \dfrac{AS}{CS}$.

Proof: Divide AR into x equal parts and BR into y equal parts of the same kind.

      Through the points of division draw lines parallel to BC.

fig. C : 84

      From the Equal Intercept Theorem (THEOREM 16):

               AS will be divided into x equal parts

and CS will be divided into y equal parts

$\therefore \dfrac{AR}{BR} = \dfrac{x}{y} = \dfrac{AS}{CS}$ $\left\{\begin{array}{l}\text{Where } x \text{ and } y \\ \text{are integers.}\end{array}\right.$

Similarly we may prove $\dfrac{AR}{AB} = \dfrac{AS}{AC}$ and $\dfrac{BR}{AB} = \dfrac{CS}{AC}$.   Q.E.D.

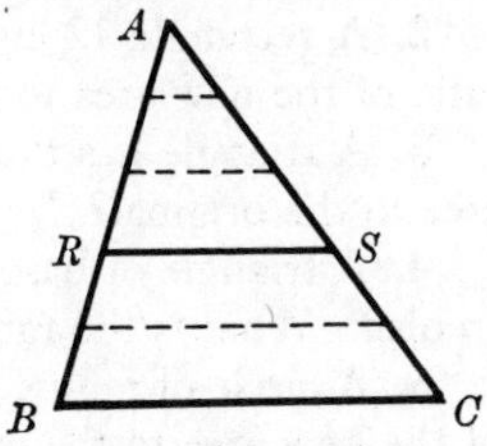

COROLLARY

If a number of transversals lie across several parallel straight lines the corresponding intercepts on the transversals are proportional.

EXERCISE 17 (*Practical*)

1. Divide a 9-cm line into 4 equal parts.
2. Divide a 12-cm line into 5 equal parts.
3. Divide a 10-cm line into 3 equal parts.
4. Divide a 9-cm line into 5 equal parts.
5. Divide a 9-cm line in the ratio 2 : 1.
6. Divide a 9-cm line in the ratio 1 : 2.
7. Divide a 12-cm line in the ratio 2 : 2.
8. Divide a 12-cm line in the ratio 3 : 1.
9. Divide a 12-cm line in the ratio 2 : 3.
10. Divide a 10-cm line in the ratio 3 : 3.
11. Divide a 10-cm line in the ratio 4 : 3.
12. Divide a 10-cm line in the ratio 5 : 2.

AREA AND PROPORTION

If two triangles have the **same altitude** their areas will be in proportion to the lengths of their bases.

The ratio of the sides is called the **linear ratio**, and the ratio of the areas is equal to the **square of the linear ratio**:

$$\textbf{Area ratio} = \textbf{(Linear ratio)}^2$$
$$\textbf{Volumetric ratio} = \textbf{(Linear ratio)}^3$$

EXERCISE 18

1. A square of side 4 cm has the sides halved. What is the ratio of the new area to the original?

2. A rectangle 12 cm by 6 cm has the sides doubled. What is the ratio of the new area to the original?

3. A triangle has the altitude halved. What is the ratio of the new area to the original?

4. A triangle of base 8 cm and altitude 12 cm has both dimensions trebled. What is the ratio of the new area to the original?

5. A circle of radius 6 cm has the radius doubled. What is the ratio of the new area to the original?

6. A square has its sides increased by 50%. What is the ratio of the new area to the original?

7. A cuboid, 2 cm by 4 cm by 6 cm, has its dimensions decreased by 25%. What is the ratio of the new volume to the original? What is the new volume? (HINT. *Use vol. ratio as a mult. fact. for orig. vol.*)

8. A quadrilateral of area 50 cm² has its sides decreased by 20%. What is the new area?

9. A cube of volume 64 cm³ has its sides increased by 50%. What is the new volume?

10. A photograph of area 4 in² is enlarged and now has an area of 9 in². One side of the original measured $1\frac{1}{2}$ in, what will the corresponding side of the enlargement measure?

SIMILARITY

Figures which are **the same shape** are called **similar figures.**

Consider fig. C : 85. *PQRS* has been constructed within *ABCD* in such a way that all the corresponding sides are parallel.

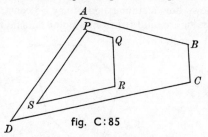

fig. C : 85

From this it follows that **corresponding angles are equal.**

If the phrase "same shape" meant only that corresponding angles were equal (i.e., the figures were equiangular), then *ABCD* and *PQRS* would be similar, but they are NOT the same shape.

DEFINITION

If figures are equiangular and their corresponding sides are proportional they are similar.

CONSTRUCTION OF SIMILAR POLYGONS

The following examples show a number of interesting methods for constructing similar figures, yet they are all based on the division of a line in a given ratio.

In fig. C : 86 the diagonals intersect at *O*. *P*, *Q*, *R*, *S* are the mid-points of *OA*, *OB*, *OC*, *OD*.

The ratio of *PQRS* to *ABCD* is 1 : 2. That is, the new figure is half-scale. We refer only to the linear ratio when giving a scale.

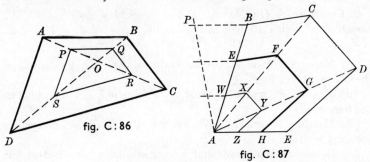

fig. C : 86

fig. C : 87

In fig. C : 87 one vertex *A* has been joined to the opposite vertices *C* and *D*. Side *AB* has been trisected by the usual construction employing line *AP*. *EF* has been drawn parallel to *BC*, and so with each corresponding side in turn. Similarly, the smallest figure has been constructed, starting with $WX\|BC$ and on round the rest of the sides.

Any required scale may be achieved by suitably dividing the first side, but careful thought must be given to the matter before such constructions are begun.

Fig. C : 88 shows how a similar figure may be constructed: (i) from a point inside the figure, and (ii) from a point outside the figure.

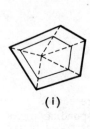
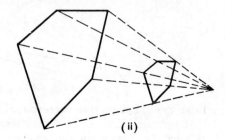

(i)

(ii)

fig. C:88

EXERCISE 19 (*Practical*)

1. Construct a regular pentagon of 5-cm side. Within the polygon construct a similar figure of 2-cm side.

2. Construct a regular hexagon of 3-cm side. Outside the polygon construct a similar figure of 6-cm side.

3. Construct a regular octagon of 4-cm side. Within the polygon construct a similar figure on a scale of 3 : 4.

4. Construct a quadrilateral of sides 3 cm, 4·5 cm, 6 cm, 6 cm. Outside the quadrilateral construct a similar figure on a scale of 3 : 2.

5. Construct any polygon and construct a similar figure from a point outside the polygon.

Triangles

6. Construct $\triangle ABC$ having $\angle A = 50°$, $\angle B = 60°$, $\angle C = 70°$. Construct a similar triangle XYZ having $\angle X = \angle A$, $\angle Y = \angle B$, $\angle Z = \angle C$.

(i) Measure AB and XY. What is the value of $\dfrac{AB}{XY}$ as a decimal in the form $m : 1$?

(ii) Measure BC and YZ. What is the value of $\dfrac{BC}{YZ}$ as a decimal in the form $m : 1$?

(iii) Measure AC and XZ. What is the value of $\dfrac{AC}{XZ}$ as a decimal in the form $m : 1$?

(iv) If your measurements and calculations have been carried out carefully you should find close agreement between the three ratios $\dfrac{AB}{XY}$, $\dfrac{BC}{YZ}$, $\dfrac{AC}{XZ}$.

SIMILAR TRIANGLES
THEOREM 18

If two triangles are equiangular, their corresponding sides are proportional.

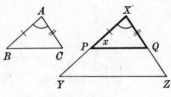

fig. C : 89

Given: $\triangle$s ABC, XYZ are equiangular:

$$\angle A = \angle X, \ \angle B = \angle Y, \ \angle C = \angle Z$$

To prove: $\dfrac{AB}{XY} = \dfrac{BC}{YZ} = \dfrac{AC}{XZ}$

Construction: Along XY mark point P so that $XP = AB$, similarly $XQ = AC$.

Proof:
In $\triangle$s ABC, XPQ
$$AB = XP \text{ (const.)}$$
$$AC = XQ \text{ (const.)}$$
$$\angle A = \angle X \text{ (given)}$$
$$\therefore \ \triangle ABC \equiv \triangle XPQ \text{ (S.A.S.)}$$

In particular: $\angle B = \angle x$

But $\angle B = \angle Y$ (given)

$\therefore \ \angle x = \angle Y$

Since these are corresponding $\angle$s $PQ\|YZ$.

$$\therefore \ \frac{XP}{XY} = \frac{XQ}{XZ} \ (\| \text{ divides sides proportionally, Theorem 17})$$

Since $XP = AB$ and $XQ = AC$ (const.)

Then $\dfrac{AB}{XY} = \dfrac{AC}{XZ}$

Similarly by construction of points on XY and YZ equal to AB and BC respectively we may prove:

$$\frac{AB}{XY} = \frac{BC}{YZ} \qquad\qquad \text{Q.E.D.}$$

NOTE. From the proof we should now understand that **equiangular TRIANGLES are always similar.** Our practical work has shown that this is not necessarily true for other figures.

PYTHAGORAS' THEOREM

THEOREM 19

In a right-angled triangle the square on the hypotenuse is equal to the sum of the squares on the other two sides.

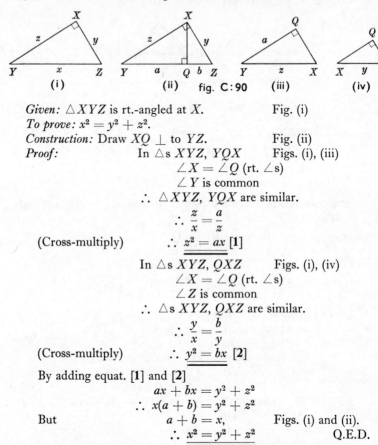

(i)   (ii)   fig. C:90   (iii)   (iv)

Given: $\triangle XYZ$ is rt.-angled at X.   Fig. (i)
To prove: $x^2 = y^2 + z^2$.
Construction: Draw $XQ \perp$ to YZ.   Fig. (ii)
Proof: In $\triangle$s XYZ, YQX Figs. (i), (iii)
          $\angle X = \angle Q$ (rt. $\angle$s)
          $\angle Y$ is common
          $\therefore$ $\triangle XYZ$, YQX are similar.

          $\therefore$ $\dfrac{z}{x} = \dfrac{a}{z}$

(Cross-multiply) $\therefore$ $z^2 = ax$ **[1]**

          In $\triangle$s XYZ, QXZ Figs. (i), (iv)
          $\angle X = \angle Q$ (rt. $\angle$s)
          $\angle Z$ is common
          $\therefore$ $\triangle$s XYZ, QXZ are similar.

          $\therefore$ $\dfrac{y}{x} = \dfrac{b}{y}$

(Cross-multiply) $\therefore$ $y^2 = bx$ **[2]**

By adding equat. **[1]** and **[2]**
          $ax + bx = y^2 + z^2$
          $\therefore$ $x(a + b) = y^2 + z^2$
But $a + b = x,$ Figs. (i) and (ii).
          $\therefore$ $x^2 = y^2 + z^2$ Q.E.D.

APPLICATIONS OF PYTHAGORAS' THEOREM

EXAMPLE. *In fig. C*:91, *b* = 13 *ft.*, *a* = 5 *ft. Find c.*
          $b^2 = a^2 + c^2$
$\therefore$ $13^2 = 5^2 + c^2$
          $169 = 25 + c^2$
$\therefore$ $c^2 = 144$ $\therefore$ $c = 12$ ft.

fig. C:91

EXERCISE 20

Questions 1–8 refer to fig. C : 92.

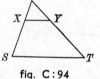

fig. C : 92

1. $b = 8$, $c = 6$, find a.   **2.** $b = 10$, $c = 24$, find a.
3. $b = 24$, $c = 7$, find a.   **4.** $b = 15$, $c = 8$, find a.
5. $b = 12$, $c = 16$, find a.   **6.** $a = 15$, $b = 12$, find c.
7. $a = 34$, $c = 30$, find b.   **8.** $a = 39$, $b = 36$, find c.

NOTE. The two triangles 3, 4, 5 and 5, 12, 13 are worth remembering.

9. The diagonals of a rhombus are 10 cm and 24 cm. Find the side of the rhombus.

10. A ladder 26 ft long is resting against a wall with the foot of the ladder 10 ft from the wall. How high up the wall does the ladder reach?

fig. C : 93

11. In fig. C : 93 PQ is 4 cm, BC is 8 cm, AP is 6 cm, QC is 4 cm. Find BP and AQ. What is $\triangle ABC$?

12. In fig. C : 94 RS is divided in the ratio 1 : 2. $RT = 4\frac{1}{2}$ in, $ST = 4$ in. Find RY, TY, XY.

13. $\triangle ABC$ is right-angled at B and BC is 6 m long. CA is produced to X, BA is produced to Y to form $\triangle AXY$ right-angled at Y with XY 2 m long. If $XC = 17$ m, $BY = 15$ m, find AY and AC.

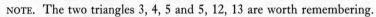

fig. C : 94

14. Find the area of an equilateral triangle of side 8 cm, correct to two significant figures.

15. The diagonal of a square is 10 cm. Find the area.

16. A flag-pole 36 dm high is secured by a wire fastened to the level ground some distance from the pole. If the wire forms an angle of 67° 23′ with the ground, find the distance from the pole. Find also the length of the wire.

17. AD and PS are two transversals cutting four parallel lines at $ABCD$ and $PQRS$ respectively. $AB = 2$ in, $CD = 5$ in, $PS = 12$ in, $RS = 6$ in. Find BC, PQ, QR.

18. An isosceles triangle has a vertical angle of 80° and a base of 12 cm. Find the area.

19. $\triangle ABC$ is right-angled at A. The perpendicular from A to BC divides BC in the ratio 1 : 4, meeting BC at X. XY is drawn parallel to AB and CY is 5 m long. Find the length of AY and the area of $\triangle ABC$. (HINT. $\dfrac{XY}{CY} = \dfrac{AY}{XY}$. *Find XY, then AB*.)

20. From a boat 600 ft from the base of a cliff the angle of elevation to the top of the cliff is 16° 42′. On top of the cliff is a radio mast, and the angle of elevation from the boat to the top of the mast is 23° 26′. Find the height of the mast.

GENERAL REVISION
Exercise 21

1. ABC is a triangle in which $\angle A = 40°$ and $\angle B = 60°$. Find the angle at which the internal bisector of $\angle B$ meets the bisector of the external angle at C when BC is produced.

2. XOY is the $\perp$ bisector of straight line AB, cutting AB at O. Prove that any point P on XY is equidistant from A and B.

3. If three parallel straight lines make equal intercepts on one transversal, prove that the intercepts on another transversal are also equal.

4. Construct an equilateral triangle ABC of side 6 cm. Divide AB internally in the ratio 2 : 1 at point X. Construct the circumcircle of $\triangle AXC$.

5. Prove that the diagonals of a parallelogram bisect each other.

6. The diagonals of a parallelogram $ABCD$ intersect at E. XBY is drawn through B parallel to AC. AG and CF are parallel to EB and meet XY at G and F respectively. Prove $EF = AB$.

7. A and B are two points on the circumference of a circle centre O such that $\angle AOB = 52°$. C is a point on the circumference such that $OC\|AB$. Calculate the size of $\angle BOC$ and $\angle OCB$.

8. YQ is the bisector of $\angle XYZ$. Prove that the $\perp$ distances from XY and ZY to any point P on YQ are equal.

9. Prove that the area of a parallelogram is equal to the area of a rectangle on the same base and between the same parallels.

10. $ABCD$ is a parallelogram in which E, F, G, H are the mid-points of the sides AB, BC, CD, DA respectively. Prove that the area of $\triangle HEF$ is half that of $ABFH$. What fraction is the area of $\triangle EBF$ of the figure $AEFCD$?

11. $ABCD$ is a rhombus in which $\angle ABC$ is 60°. X and Y are points on BD such that $\angle BCX = \angle XCY = \angle YCD$. Prove that $\triangle XCY$ is isosceles.

12. Prove that the straight line joining the middle points of two sides of a triangle is parallel to the third side, and equal to half of it.

13. $ABCD$ is a trapezium in which $AB = 3$ cm, $CD = 5$ cm, $AD = 4$ cm, $\angle BAD = 90°$ and $AB\|CD$. On CD produced, construct $\triangle BCE$ equal in area to $ABCD$. Measure CE. Prove your construction.

14. Prove that the sum of the angles of a triangle is equal to two right-angles.

15. In $\triangle ABC$, $\angle ABC$ is a rt. $\angle$ and $BD \perp AC$. AE cuts BD at F and BC at E. If $BE = EF = FB$, prove that AE bisects $\angle BAC$.

16. Prove that the opposite angles of a parallelogram are equal.

17. $ABCD$ is a parallelogram in which the diagonals intersect at O. P and Q are two points on BD such that $OP = OQ$. Prove that $APCQ$ is a parallelogram.

18. *ABC* is an isosceles triangle, right-angled at *B*. If *BC* is 4 in long, calculate the length of the median *BD*.

19. Calculate: (i) the size of the interior angle of a regular polygon of 15 sides; (ii) the number of sides in a regular polygon if an exterior angle is 20°.

20. *ABCD* is a parallelogram. *AE* is the bisector of $\angle DAB$ and meets *BC* produced at *E*. Prove that $\triangle ABE$ is isosceles.

21. In $\triangle ABC$, *AD* is the bisector of $\angle BAC$ meeting *BC* at *D*. *DE* is drawn parallel to *AB* and meets *AC* at *E*. Prove that $\triangle ADE$ is isosceles.

22. In $\triangle XYZ$, $XY = 3$ in, $YZ = 6$ in, $XZ = 8$ in. *YP*, the bisector of $\angle XYZ$, meets *XZ* at *P*. *PQ* is drawn parallel to *XY*, meeting *YZ* at *Q*. Calculate the length of *PQ*, *QZ* and *PZ*.

LOCI

When a point moves in such a way that it *always* obeys certain conditions, the path traced out by the point is called a **locus**.

EXERCISE 22 (*Practical*)

1. Draw a horizontal line, *AB*, 8 cm long. Construct the locus of a point which is always 3 cm from the line *AB*.

2. Using the line in Question 1, construct the locus of a point which is always: (i) 2 cm from *A*; (ii) 5 cm from *B*.

3. Draw a pair of parallel lines *AB* and *XY* about 5 cm apart. Construct the locus of a point, *P*, which is always equidistant from *AB* and *XY*.

4. Draw a circle, centre *O*, of radius 5 cm. Draw several radii and find the mid-point on each. Construct the locus of the mid-points of the radii.

5. Using the circle in Question 4, draw several other circles of 2 cm diameter such that these circles just touch the outside of the larger circle. Construct the locus of the centres of the smaller circles.

6. Plot the locus of a football which always remains at the same distance from: (*a*) the opposing goal-lines; (*b*) the two touch-lines.

7. A goalkeeper moves so that he is always equidistant from his two goal-posts. Plot the locus, both in front of and behind the goal-line.

8. A ship approaches the entrance to a harbour. Plot the locus of the ship if it moves so as to be equidistant from the two sides of the harbour entrance.

9. A sheep is free to roam in a rectangular pen which is hedged on two adjacent sides. Plot the locus of the sheep, if it moves so as to be equidistant from the two hedges at all times.

10. *BC* is the base of an isosceles triangle *ABC*. Plot the locus of the vertex A for a series of triangles, constructed on both sides of BC, such that $AB = AC$.

11. ABC is an isosceles triangle in which $AB = AC$. Plot the locus of a point such that it is always equidistant from AB and AC.

12. XYZ is any triangle. Plot the loci of P_1, P_2, P_3 such that P_1 is equidistant from X and Y, P_2 is equidistant from X and Z, P_3 is equidistant from Y and Z.

THEOREM 20

The locus of a point which is equidistant from two fixed points is the perpendicular bisector of the straight line joining the two points.

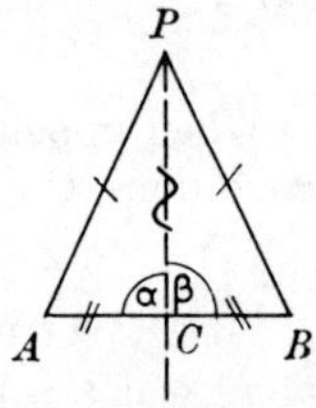

Given: A and B are two fixed points.

$\quad\quad$ P is any point such that $PA = PB$.

To prove: The locus of P is the perpendicular bisector of AB.

Construction: Bisect AB at C. Join PC.

Proof: $\quad\quad\quad\quad$ In $\triangle$s APC, BPC

$\quad\quad\quad\quad\quad\quad$ $AP = BP$ $\quad\quad$ (given)

$\quad\quad\quad\quad\quad\quad$ $AC = BC$ $\quad\quad$ (const.)

$\quad\quad\quad\quad\quad\quad$ PC is common

$\quad\quad\quad\quad$ $\therefore$ $\triangle APC \equiv \triangle BPC$ $\quad$ (S.S.S.)

In particular: $\quad\quad\quad$ $\angle \alpha = \angle \beta$

fig. C:95

But these are adj. $\angle$s on st. line AB

$\quad\quad\quad\quad$ $\therefore$ $\angle \alpha = \angle \beta = 90°$

$\quad$ $\therefore$ PC is the $\perp$ bisector of AB.

$\quad$ $\therefore$ P lies on the perpendicular bisector of AB. $\quad\quad$ Q.E.D.

CONVERSE

Any point on the perpendicular bisector of a straight line is equidistant from the ends of the straight line.

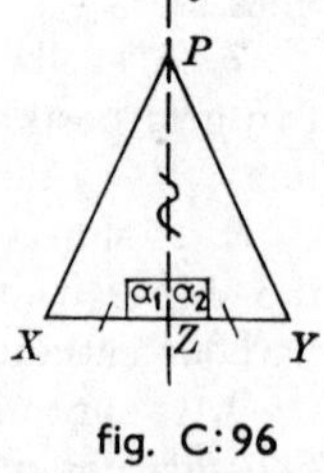

Given: QZ is the $\perp$ bisector of XY.

$\quad\quad\quad$ P is any point on QZ.

To prove: $PX = PY$

Proof: $\quad\quad$ In $\triangle$s PXZ, PYZ

$\quad\quad\quad\quad\quad\quad$ $XZ = YZ$ $\quad\quad$ (given)

$\quad\quad\quad\quad\quad\quad$ PZ is common

$\quad\quad\quad\quad\quad\quad$ $\angle \alpha_1 = \angle \alpha_2$ $\quad$ ($QZ \perp$ to XY)

$\quad\quad\quad\quad$ $\therefore$ $\triangle PXZ \equiv \triangle PYZ$ (S.A.S.)

In particular: $\quad\quad$ $PX = PY$

fig. C:96

$\quad$ $\therefore$ Any point on the perpendicular bisector of XY is equidistant from X and Y. $\quad\quad\quad\quad\quad\quad\quad\quad\quad\quad\quad$ Q.E.D.

THEOREM 21

The locus of a point which is equidistant from two intersecting straight lines is the pair of straight lines which bisect the angles between the given lines.

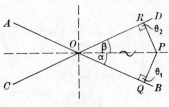

fig. C:97

Given: Straight lines AB and CD intersect at O. The perpendiculars from any point P to AB and CD are equal (i.e., $PQ = PR$).

To prove: P lies on one of the lines bisecting $\angle AOD$ and $\angle BOD$.

Construction: Join PO.

Proof: In $\triangle$s PQO, PRO

$\qquad\qquad\qquad\qquad PQ = PR$ (given)

$\qquad\qquad\qquad\qquad \angle\theta_1 = \angle\theta_2$ (rt. $\angle$s)

$\qquad\qquad\qquad\qquad PO$ is common   (hypot.)

$\qquad\therefore \triangle PQO \equiv \triangle PRO$ (R.H.S.)

In particular: $\angle\alpha = \angle\beta$

$\therefore$ P lies on the line bisecting $\angle BOD$.

Similarly: If P were placed equidistant from AO and DO, it could be shown to lie on the bisector of $\angle AOD$.   Q.E.D.

CONVERSE

Any point on the bisector of an angle is equidistant from the straight lines forming the angle.

Given: P is any point on the bisector of $\angle BOD$.

To prove: The perpendiculars from P to OB and OD are equal.

Construction: Draw $PQ \perp$ to OB and $PR \perp$ to OD.

Proof: In $\triangle$s PQO, PRO

$\qquad\qquad\qquad\qquad \angle\alpha = \angle\beta$ (given)

$\qquad\qquad\qquad\qquad \angle\theta_1 = \angle\theta_2$ (const.)

$\qquad\qquad\qquad\qquad OP$ is common

$\qquad\therefore \triangle PQO \equiv \triangle PRO$ (A.A.S.)

In particular: $PQ = PR$

Similarly: Any point on the bisector of $\angle AOD$ will be equidistant from AB and CD.   Q.E.D.

CONSTRUCTION 12

To construct the circumscribed circle of a triangle.

Given: $\triangle RST$.

To construct: A circle passing through the
vertices R, S and T of $\triangle RST$.

Construction:

    1. Construct AB the $\perp$ bisector of ST.

    2. Construct XY the $\perp$ bisector of RT.

    3. O, the intersection of AB and XY,
is the centre of the required circle and
the radius is OS.

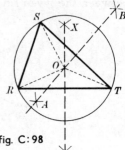

Proof:

    AB is the locus of a point equidistant
from S and T. $\therefore OS = OT$

fig. C:98

    XY is the locus of a point equidistant from R and T.

$$\therefore OR = OT$$
$$\therefore OR = OS = OT$$

$\therefore$ A circle, centre O and radius OS, will pass through R, S and T.

                                                    Q.E.F.

CONSTRUCTION 13

To construct the inscribed circle of a triangle.

Given: $\triangle ABC$.

To construct: A circle within $\triangle ABC$
to touch AB, BC and CA.

Construction:

1. Construct AP the bisector of $\angle A$.
2. Construct CQ the bisector of $\angle C$.
3. Construct XR the $\perp$ from X to AC,
    where X is the intersection of AP
    and CQ.
4. X is the centre of the required circle
    and the radius is XR.

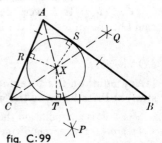

fig. C:99

Proof:

    AP is the locus of a point equidistant from AC and AB.

$$\therefore XR = XS$$

    CQ is the locus of a point equidistant from AC and BC.

$$\therefore XR = XT$$
$$\therefore XR = XS = XT$$

    $\therefore$ A circle, centre X and radius XR, will just touch AB, BC
and CA.   Q.E.F.

NOTE

(i) O is called the **circumcentre** of the triangle. (Fig. C : 98.)

(ii) If a circle passes through any number of points, the points are described as **concyclic**.

(iii) X is called the **incentre** of the triangle. (Fig. C : 99.)

(iv) AB, BC, CA are **tangents** to the circle at the points S, T, R respectively. Each tangent is at right-angles with a radius at the point of contact with the circle.

(v) A new abbreviation has been used, Q.E.F., which means *Quod erat faciendum*—"which was the thing to be made or done."

EXERCISE 23

1. Draw two lines as shown in fig. C : 100, each 10 cm long. Construct the locus of a point which is always 2 cm from AB. Construct the locus of a second point which is always 5 cm from YZ. Find the position of a point which is 2 cm from AB and 5 cm from YZ.

fig. C : 100

2. From fig. C : 100 construct the locus of a point which is equidistant from the line AB and the line YZ. Prove your construction.

3. From fig. C : 100, find a point which is equidistant from AB and YZ and 2 cm from AB.

4. From fig. C : 100, find a point which is 5 cm from both AB and YZ.

5. From fig. C : 100, construct the locus of a point which is equidistant from A and B. Prove your construction.

6. From fig. C : 100, find a point which is equidistant from A and B and is 4 cm from the line AB. Prove your construction.

7. From fig. C : 100, find a point which is equidistant from A and B and is 4 cm from the line YZ.

8. From fig. C : 100, find a point which is equidistant from A and B, and also equidistant from Y and Z though not necessarily the same distance as from A or B.

9. Show that the point found in Question 8 is a common centre for two concentric circles, one passing through A and B, the other through Y and Z.

10. $ABCD$ is a rhombus of side 5 cm and $\angle ADC = 55°$. Construct the inscribed circle and prove your construction.

11. XYZ is an equilateral triangle. Construct the incircle and the circumcircle. Prove that the incircle and circumcircle are concurrent.

12. AB is a straight line of fixed length, $\angle ACB$ is 90°. Plot the locus of C. (On both sides of AB.)

13. In $\triangle PQR$, QR is of fixed position and length 5 cm. PQ is 2 cm long and PR is of no fixed length. Plot the locus of P.

14. *KLMN* is a parallelogram with adjacent sides of 5 cm and 8 cm and ∠*KNM* = 60°. The diagonals intersect at *O*. Construct the incircle in each of the triangles so formed. (*KOL, LOM*, etc.)

15. ∠*BAC* is 40°. *AD* is the locus of a point equidistant from *AB* and *AC*. At any point *O*, on *AD*, construct a circle to touch *AB* and *AC* at *P* and *Q* respectively. Prove that the tangents *AP* and *AQ* are equal in length.

16. Plot the locus of the centre of a circle *O*, of fixed radius *OP*, such that the circle always passes through a fixed point *P*.

17. The base *BC* of △*ABC* is of fixed position and length. If the altitude of the triangle is constant, plot the locus of *A*.

18. *ABC* is an isosceles triangle whose base *BC* is of fixed position and length. If *AD* is the median on *BC*, prove that the locus of *P*, the mid-point of *AC*, is a straight line parallel to *AD*. Prove also that the ⊥ from *P* to *BC* is ½*AD*.

19. A piece of thread *AB*, 4 in long, is wound on a piece of wood whose section is a square of side ½ in. If *B* is secured at a corner of the wood, plot the locus of *A* as the thread unwinds, remaining taut in the process. (*Graph paper, minimum* 7 *in by* 8 *in, will be helpful; compasses essential.*)

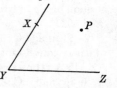

20. In fig. C : 101, *P* is equidistant from *X* and *Z* and also equidistant from *XY* and *ZY*. If *PR* is the ⊥ to *YZ* and *PQ* the ⊥ to *YX* (produced if necessary), prove that *RZ* = *QX*.

fig. C:101

21. In fig. C : 102 (not drawn to scale), ∠*QPR* = 40°, *RP* = *RQ*, *SR* = *SQ*. *OT* the radius of the smaller circle (inscribed) is 1 in. Construct an accurate drawing, full size, and measure *XR*, the radius of the larger circle (circumscribed). Give a brief description of your construction. (*Foolscap paper required.*)

22. Plot the locus of a point *P* which is always equidistant from a fixed vertical line *AB* and a fixed point *Q*, situated 2 cm away from *AB*.

23. *AB* is a fixed line of length 8 cm. State the locus of the vertex *C* if △*ABC* always has an area of 16 cm². Give a reason for your statement.

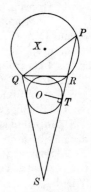

fig. C:102

24. Fig. C : 103 shows a circle centre *O*, diameter *AB* = 4 in (diag. not to scale). *P* is the mid-point of chord *AQ*. Plot the locus of *P*, for various positions of *Q* on both sides of *AB*. Show that the locus is a circle. Find the centre and measure the radius.

fig. C:103

THEOREM 22

(*Pythagoras' Theorem—an alternative proof to* Theorem 19.)

In a right-angled triangle, the square on the hypotenuse is equal to the sum of the squares on the other two sides.

Given: $\triangle ABC$ is right-angled at B. $ABGF$, $BCKH$, $ACDE$ are squares on AB, BC, AC respectively.

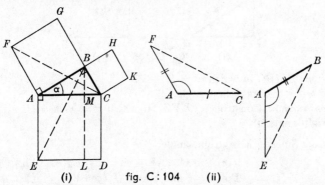

(i)   fig. C : 104   (ii)

To prove: The area of $ACDE = ABGF + BCKH$.

Construction: Join FC, BE. Draw $BML \| AE$ cutting AC and ED at M and L respectively.

Proof:

	In $\triangle$s FAC, BAE	[fig. C : 104 (ii)]
	$AC = AE$	(sides of square $ACDE$)
	$AF = AB$	(sides of square $ABGF$)
	$\angle FAC = \angle BAE$	(each is $90° + \alpha$)
$\therefore$	$\triangle FAC \equiv \triangle BAE$	(S.A.S.)

Since $\angle$s ABG, ABC are right-angles:

GBC is a st. line $\| AF$

$\therefore$ Area $\triangle FAC = \frac{1}{2}ABGF$ (same base AF, $AF \| CG$)

And area $\triangle BAE = \frac{1}{2}AELM$ (same base AE, $AE \| BL$)

But $\triangle FAC \equiv \triangle BAE$ (proved)

$\therefore$ Area $ABGF =$ area $AELM$

Similarly, by joining BD, AK it may be proved that:

Area $BCKH =$ area $CDLM$

$\therefore ABGF + BCKH = AELM + CDLM$

$\therefore$ Area $ACDE = ABGF + BCKH$ Q.E.D.

Algebraically, $AB^2 = AM \cdot AE$ i.e., $AB^2 = AM \cdot AC$

And $BC^2 = CM \cdot CD$ i.e., $BC^2 = CM \cdot AC$

$\therefore AB^2 + BC^2 = (AM \cdot AC) + (CM \cdot AC) = AC(AM + CM)$

$\therefore AB^2 + BC^2 = AC^2$

THEOREM 23

(*Converse of Pythagoras' Theorem.*)

If the square on one side of a triangle is equal to the sum of the squares on the other two sides, then the angle contained by these sides is a right-angle.

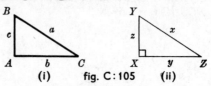

(i)   fig. C:105   (ii)

Given: $\triangle ABC$ in which $a^2 = b^2 + c^2$.

To prove: $\angle A$ is a right-angle.

Construction: Construct $\triangle XYZ$, in which $y = b$, $z = c$ and $\angle X$ is a right-angle.

Proof: Since $\angle X$ is a right-angle:

$$x^2 = y^2 + z^2 \quad \text{(Pythagoras' Th.)}$$

But $y = b$ and $z = c$ (const.)

$$\therefore \ x^2 = b^2 + c^2$$

But $a^2 = b^2 + c^2$ (given)

$$\therefore \ a^2 = x^2$$
$$\therefore \ a = x$$

In $\triangle$s ABC, XYZ

$$a = x \quad \text{(proved)}$$
$$b = y \quad \text{(const.)}$$
$$c = z \quad \text{(const.)}$$
$$\therefore \ \triangle ABC \equiv \triangle XYZ \quad \text{(S.S.S.)}$$

In particular: $\angle A = \angle X$

But $\angle X$ is a rt. $\angle$ (const.)

$$\therefore \ \angle A \text{ is a right-angle.} \qquad \text{Q.E.D.}$$

EXERCISE 24

1. From fig. C : 104, prove that $BCKH$ is equal in area to $CDLM$.

2. $ABCD$ is a square. Prove that the area of $ABCD = \frac{1}{2}BD^2$.

3. ABC is an isosceles triangle, right-angled at A. AD is $\perp$ to BC. Prove that $AD^2 = \frac{1}{4}BC^2$.

4. ABC is an acute-angled triangle in which $AD \perp BC$. Prove that $AB^2 - BD^2 = AC^2 - CD^2$.

5. Repeat Question 4 for $\triangle ABC$, obtuse-angled at B.

6. From fig. C : 104, the area of $AELM = AM \cdot AE$. Prove that $AB^2 = AM \cdot AC$.

7. From fig. C : 104, prove that $BC^2 = MC \cdot AC$.

8. $\triangle ABC$ is right-angled at B and $BD \perp AC$. Prove that $BC^2 = AC \cdot CD$.

9. ABC is an equilateral triangle in which $AD \perp BC$. Prove that $AD^2 = \frac{3}{4}AB^2$.

10. $ABCD$ is a rhombus. Prove that

$$AC^2 + BD^2 = AB^2 + BC^2 + CD^2 + DA^2$$

THE CIRCLE: VOCABULARY

A straight line which joins any two points on the circumference of a circle is called a **chord.** When a chord passes through the centre of a circle it is called a **diameter.**

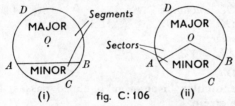

fig. C:106

A part of the circumference of a circle is called an **arc**; ACB is a **minor arc** and ADB is a **major arc** (fig. C : 106).

The area enclosed by an arc of a circle and a chord is called a **segment.** If the segment is smaller than a semicircle it is called a **minor segment.**

The area enclosed by an arc of a circle and two radii is called a **sector.** If the sector is greater than a semicircle it is called a **major sector.**

PROPERTIES OF CHORDS
THEOREM 24

A straight line from the centre of a circle to the mid-point of a chord, which is not a diameter, is at right-angles to the chord.

Given: AB is a chord of a circle, centre O.

       C is the mid-point of AB.

To prove: $OC \perp AB$.

Construction: Join AO, BO.

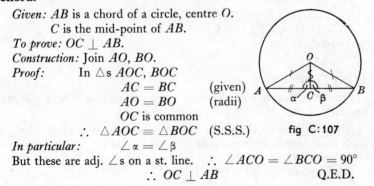

Proof: In $\triangle$s AOC, BOC

                $AC = BC$ (given)

                $AO = BO$ (radii)

                OC is common

    $\therefore \triangle AOC \equiv \triangle BOC$ (S.S.S.)   **fig C:107**

In particular: $\angle \alpha = \angle \beta$

But these are adj. $\angle$s on a st. line. $\therefore \angle ACO = \angle BCO = 90°$

            $\therefore OC \perp AB$ Q.E.D.

CONVERSE

The perpendicular to a chord from the centre of a circle bisects the chord.

Given: XY is a chord of a circle, centre O.
$$OZ \perp XY$$
To prove: $XZ = YZ$.
Construction: Join XO, YO.
Proof:

In $\triangle$s XOZ, YOZ
$\qquad XO = YO$ $\qquad$ (radii)
$\qquad \angle\alpha_1 = \angle\alpha_2 = 90°$ $\quad$ ($OZ \perp XY$)
$\qquad OZ$ is common
$\therefore \triangle XOZ \equiv \triangle YOZ$ $\qquad$ (R.H.S.)
In particular:
$\qquad XZ = YZ$ $\qquad\qquad\qquad\qquad\qquad$ Q.E.D.

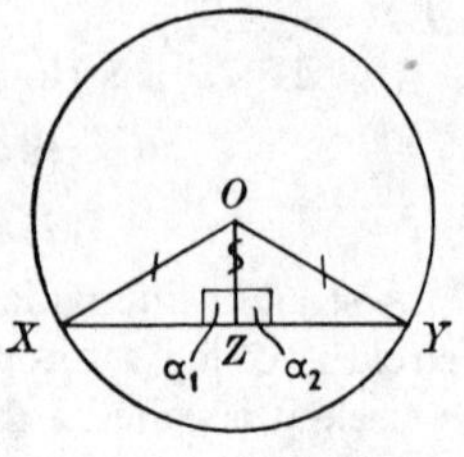

COROLLARY

The perpendicular bisector of a chord passes through the centre of the circle.

THEOREM 25

Equal chords of a circle are equidistant from the centre.

Given: AB and XY are equal chords of a circle, centre O. OC and OZ are $\perp$ to AB and XY respectively.
To prove: $OC = OZ$.
Construction: Join BO, YO.
Proof:

$\qquad AB = XY$ $\qquad$ (given)
$\therefore BC = YZ$ $\qquad$ (equal chords bisected
$\qquad\qquad\qquad\qquad$ by $\perp$ from O)

In $\triangle$s BCO, YZO
$\qquad BC = YZ$ $\qquad$ (proved)
$\qquad BO = YO$ $\qquad$ (radii)
$\qquad \angle\alpha_1 = \angle\alpha_2 = 90°$ $\quad$ (given)
$\therefore \triangle BCO \equiv \triangle YZO$ $\qquad$ (R.H.S.)
In particular:
$\qquad OC = OZ$ $\qquad\qquad\qquad\qquad\qquad$ Q.E.D.

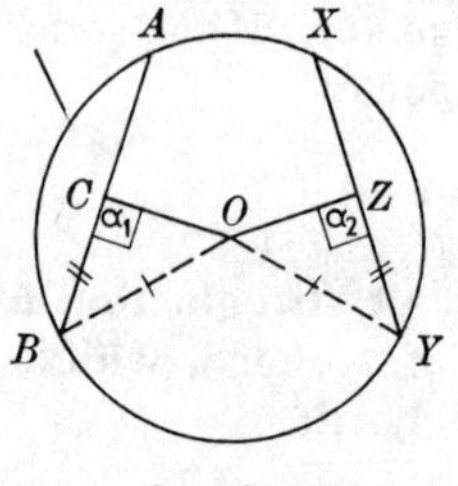

fig. C:108

fig. C:109

CONVERSE

Chords which are equidistant from the centre of a circle are equal.

Given: AB and XY are two chords in a circle, centre O. The $\perp$s from these chords to O are equal (i.e., $OC = OZ$).
To prove: $AB = XY$.
Construction: Join BO, YO.

Proof:

In △s *BCO, YZO*

$$OC = OZ \quad \text{(given)}$$
$$BO = YO \quad \text{(radii)}$$
$$\angle \alpha_1 = \angle \alpha_2 = 90° \quad \text{(given)}$$
$$\therefore \quad \triangle BCO \equiv \triangle YZO \quad \text{(R.H.S.)}$$

In particular:

$$BC = YZ$$

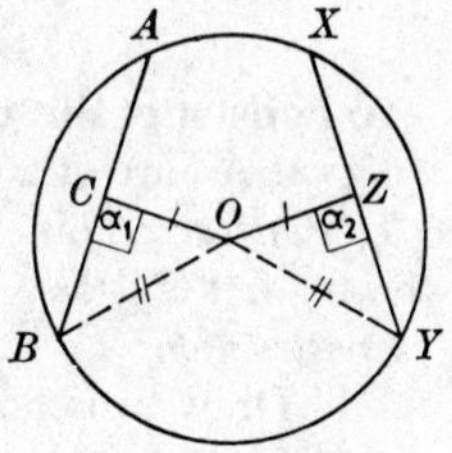

But each chord is bisected by the ⊥ from the centre.

$$\therefore \quad AB = YZ \qquad \qquad \text{Q.E.D.}$$

fig. C:110

THEOREM 26

There is only one circle which passes through three given points not in the same straight line.

Given: *A, B* and *C* are three points, not in the same straight line.

To prove: One circle only can be drawn to pass through *A, B* and *C*.

Construction: Join *AB, BC*. Construct the ⊥ bisectors of *AB* and *BC*, let them intersect at *O*. Then *O* is the centre of the only circle which will pass through *A, B* and *C*.

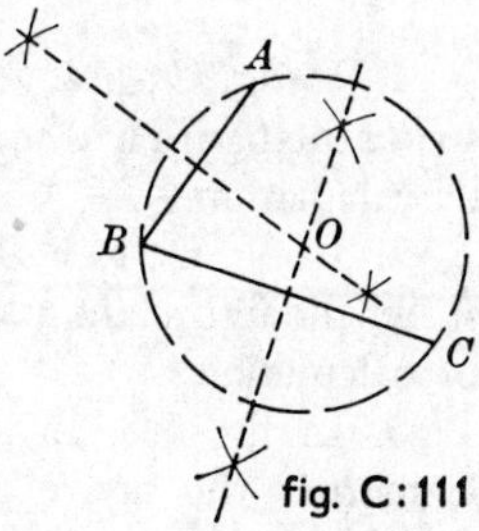

Proof: The ⊥ bisector of *AB* is the locus of points equidistant from *A* and *B*.

$$\therefore \quad AO = BO$$

fig. C:111

The ⊥ bisector of *BC* is the locus of points equidistant from *B* and *C*. $\qquad \therefore \quad BO = CO$

$$\therefore \quad AO = BO = CO$$

Since the ⊥ bisectors of *AB* and *BC* are straight lines, they can intersect at one point only. Therefore *O* is the only point which is equidistant from *A, B* and *C*.

Only one circle can pass through *A, B* and *C*, and that circle is centred at *O*. $\qquad \qquad$ Q.E.D.

COROLLARY

If three points are not in a straight line, the three points must lie on a circle.

REMEMBER

(i) When dealing with properties of chords the case for congruency must be **R.H.S.**

(ii) Points through which a circle passes are called **concyclic** points.

(iii) A polygon whose vertices lie on a circle is said to be **cyclic.**

CONSTRUCTION 14

To construct the centre of a circle, given an arc of that circle.

Given: An arc of a circle, PQR.

To construct: Point O, the centre of a circle whose arc is PQR.

Construction:

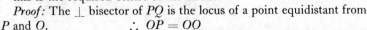

   1. Draw two chords PQ and QR.

   2. Construct the $\perp$ bisector of each chord.

   3. Let the $\perp$ bisectors intersect at O, this is the required centre of the circle.

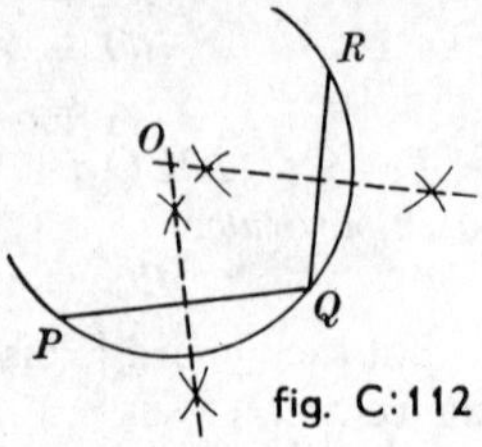

fig. C:112

Proof: The $\perp$ bisector of PQ is the locus of a point equidistant from P and Q. $\therefore\ OP = OQ$

     The $\perp$ bisector of QR is the locus of a point equidistant from Q and R. $\therefore\ OQ = OR$

$$\therefore\ OP = OQ = OR.$$ Q.E.F.

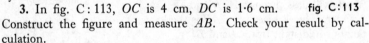

EXERCISE 25

   1. In fig. C : 113, AB is 8 cm, OD is 3 cm. Construct the figure and measure OA. Check your result by calculation.

   2. In fig. C : 113, AB is 9 cm, OD is 6 cm. Construct the figure and measure OA. Check your result by calculation.

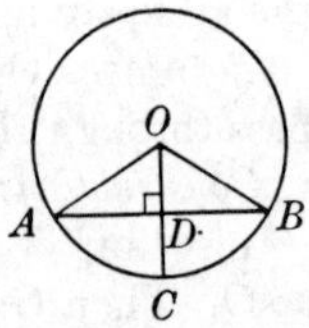

   3. In fig. C : 113, OC is 4 cm, DC is 1·6 cm. Construct the figure and measure AB. Check your result by calculation.

fig. C:113

   4. In fig. C : 113, OA is 0·625 in, OD is 0·375 in. Construct the figure and measure AB. Check your result by calculation.

   5. In fig. C : 113, OA is 1·2 in, AB is 1·2 in. Calculate the length of DC, correct to two decimal places.

   6. AB is the chord of a circle and P is the mid-point of AB. Plot the locus of P.

   7. AC and XY are equal chords of a circle, centre O. Prove that $\angle AOC = \angle XOY$.

   8. AB and CD are equal chords intersecting at X. Prove $AX = CX$. (Fig. C : 114.)

   9. Accepting the proof of Question 8 proceed to prove that $AC\|BD$. (Fig. C : 114.)

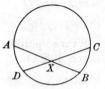

   10. Two circles, centres P and Q, intersect at R and S. Prove that PQ is the $\perp$ bisector of RS.

   11. AB and BC are two chords of a circle, centre O. If OB bisects $\angle ABC$, prove $AB = BC$.

fig. C:114

   12. XY is the common chord of a series of circles. Plot the locus of the centres of these circles and describe it.

13. Two circles, centres P and Q, intersect at R and S. Prove $\angle PRQ = \angle PSQ$.

14. $ABCDE$ is a regular pentagon. Construct such a figure of side 1 in. Construct the circumcircle and prove that points A, B, C, D, E are **concyclic** (i.e., $ABCDE$ is **cyclic**).

15. Two **concentric** circles, centre O, have radii of 1 in. and 1·5 in. respectively. A straight line $AXYB$ cuts the outer circle at A and B and the inner circle at X and Y. Prove $AX = BY$.

16. From Question 15, if C is the mid-point of AB and $OC = 0·8$ in., calculate the length of AX, to three significant figures.

17. Two circles, centres P and Q respectively, intersect at A and B. If AB and PQ bisect each other, prove that the two circles are of equal radius.

18. A circle, centre O, contains two chords which intersect at X. If the chords form equal angles with OX, prove that the chords are equal.

19. AB and CD are two diameters of a circle. P, Q, R, S are the mid-points of AC, CB, BD, DA respectively. Prove that $PQRS$ is a parallelogram and that $ACBD$ is a rectangle.

ANGLE PROPERTIES OF THE CIRCLE

EXERCISE 26 (*Practical*)

1. AB and AC are two chords of a circle centre O. $\angle BAC = 60°$. Measure $\angle BOC$.

2. Draw a circle of radius 2·5 cm. Select a minor arc AB with point P moving anywhere along the minor arc between A and B. Using at least four different positions for P, measure $\angle APB$ for each position.

3. From Question 2, employing the major arc AB and a point Q moving along the arc, measure $\angle AQB$ for at least four different positions of Q. What conclusion may be reached?

4. Draw a circle of radius 2·5 cm. Draw a diameter AB. For various positions of P (on both sides of AB) as P moves along the circumference, measure $\angle APB$.

5. Draw a circle of radius 4 cm. Select any four points A, B, C, D in order, on the circumference. Construct the quadrilateral $ABCD$. Measure $\angle ABC$ and $\angle ADC$, find their sum. Measure $\angle BAD$ and $\angle BCD$, find their sum.

6. Draw a circle of radius 5 cm. Select any four points W, X, Y, Z in order on the circumference. Construct the quadrilateral $WXYZ$ and produce YZ to A. Measure $\angle AZW$ and $\angle WXY$. Produce XY to B and measure $\angle BYZ$, $\angle XWZ$.

7. In any circle construct quadrilateral $WXYZ$. Produce ZY to C and measure $\angle CYX$, $\angle XWZ$. Produce YX to D and measure $\angle DXW$, $\angle YZW$. What conclusion may be reached?

THEOREM 27

The angle which an arc of a circle subtends at the centre is double that which it subtends at any point on the remaining part of the circumference.

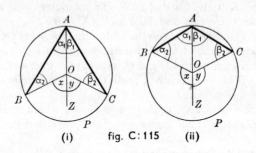

fig. C:115

(i)   (ii)

Given: BPC is an arc of a circle, centre O. A is any point on the remaining part of the circumference.

Arc BPC subtends $\angle BAC$ at the circumference and $\angle BOC$ at the centre.

In fig. C : 115 (ii) arc BPC is greater than a semicircle and $\angle BOC$ is reflex. $(\angle BOC > 180°.)$

To prove: $\angle BOC = 2\angle BAC$

Construction: Join AO and produce to Z.

Proof: $\qquad\qquad \triangle AOB$ is isosceles $\qquad (AO = BO$, radii)

$\qquad\qquad\qquad\quad \therefore\ \angle \alpha_1 = \angle \alpha_2$

$\qquad\qquad\qquad\quad$ But $\angle x = \angle \alpha_1 + \angle \alpha_2$ (ext. $\angle$ = 2 int. opp. $\angle$s)

$\qquad\qquad\qquad\quad \therefore\ \angle x = 2\angle \alpha_1$

Similarly $\qquad\qquad\qquad \angle y = 2\angle \beta_1$

By addition: $\ \angle x + \angle y = 2(\angle \alpha_1 + \angle \beta_1)$

$\qquad\qquad\quad \therefore\ \angle BOC = 2\angle BAC \qquad\qquad\qquad\qquad$ Q.E.D.

NOTE

If O lies outside $\angle BAC$ (fig. C : 116)

Then $\qquad\qquad \angle BOC = \angle x - \angle y$

And $\qquad\qquad \angle x - \angle y = 2(\alpha_1 - \beta_1)$

$\qquad\qquad \therefore\ \angle BOC = 2\angle BAC$

$\qquad\qquad\qquad\qquad$ Q.E.D.

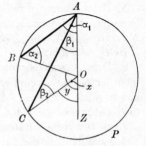

fig. C:116

THEOREM 28

Angles in the same segment of a circle are equal.

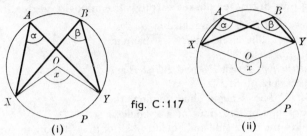

fig. C:117

(i)   (ii)

Given: $\angle XAY$ and $\angle XBY$ are angles in the same segment of a circle, centre O. Each angle is subtended by arc XPY.

To prove: $\angle \alpha = \angle \beta$.

Construction: Join XO, YO.

Proof: Arc XPY subtends $\angle x$ at the centre and $\angle \alpha$, $\angle \beta$ at the circumference. $\therefore$ $\angle x = 2\angle \alpha$ ($\angle$ at centre $= 2 \angle$ at circum.)

Similarly $\angle x = 2\angle \beta$ $\therefore$ $2\angle \alpha = 2\angle \beta$

$\therefore$ $\angle \alpha = \angle \beta$ Q.E.D.

COROLLARY 1

The angle in a segment greater than a semicircle is less than a right-angle. [Fig. C : 117 (i).]

If $XABY$ is a major arc, XPY is a minor arc.

$\therefore$ $\angle x < 180°$ But $\angle x = 2\alpha$ $\therefore$ $\angle \alpha < 90°$

COROLLARY 2

The angle in a segment less than a semicircle is greater than a right-angle. [Fig. C : 117 (ii).]

If $XABY$ is a minor arc, XPY is a major arc.

$\therefore$ $\angle x > 180°$ But $\angle x = 2\alpha$ $\therefore$ $\angle \alpha > 90°$

THEOREM 29

The angle in a semicircle is a right-angle.

Given: XY is a diameter of a circle, centre O. $\angle XAY$ is an angle in a semicircle.

To prove: $\angle \alpha = 90°$.

Proof: Arc XPY subtends $\angle x$ at the centre and $\angle \alpha$ at the circumference.

$\therefore$ $\angle x = 2\angle \alpha$ ($\angle$ at centre $= 2 \angle$ at circum.)

But $\angle x = 180°$ (XOY is a str. line)

$\therefore$ $2\angle \alpha = 180°$

$\therefore$ $\angle \alpha = 90°$ Q.E.D.

fig. C: 118

THEOREM 30

The circle described on the hypotenuse of a right-angled triangle as diameter passes through the opposite vertex.

Given: $\triangle ABC$ is right-angled at A.

To prove: The circle on BC as diameter passes through A.

Construction: Bisect AB and BC at P and Q respectively. Join PQ.

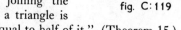
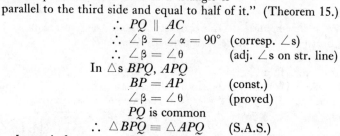

fig. C:119

Proof: "The straight line joining the middle points of two sides of a triangle is parallel to the third side and equal to half of it." (Theorem 15.)

$$\therefore PQ \parallel AC$$
$$\therefore \angle\beta = \angle\alpha = 90° \quad \text{(corresp. } \angle\text{s)}$$
$$\therefore \angle\beta = \angle\theta \quad \text{(adj. } \angle\text{s on str. line)}$$

In $\triangle$s BPQ, APQ

$$BP = AP \quad \text{(const.)}$$
$$\angle\beta = \angle\theta \quad \text{(proved)}$$
$$PQ \text{ is common}$$
$$\therefore \triangle BPQ \equiv \triangle APQ \quad \text{(S.A.S.)}$$

In particular: $\quad BQ = AQ$

But $\quad BQ = CQ \quad$ (const.)

$$\therefore AQ = BQ = CQ$$

$\therefore Q$ is the centre of a circle passing through A, B and C.

$\therefore$ The circle on BC as diameter passes through A.   Q.E.D.

CONSTRUCTION 15

To draw a straight line perpendicular to a given straight line from a given point outside it. (*An alternative to* Const. 7.)

Given: A straight line PQ and a point X outside PQ.

To construct: A straight line from X perpendicular to PQ.

Construction:

1. On PQ select any point R and join RX.
2. Bisect RX at T.
3. With T as centre, construct a circle on RX as diameter. Let the circle cut PQ at S.
4. Join XS. XS is the required straight line from X perpendicular to PQ.

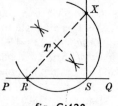

Proof: Since RX is the diameter of a circle, centre T.

$\angle RSX$ is the angle in a semicircle.

fig. C:120

$$\therefore \angle RSX = 90°$$

      Q.E.F.

CYCLIC QUADRILATERALS
THEOREM 31

The opposite angles of a cyclic quadrilateral are supplementary.

Given: $XAYB$ is a cyclic quadrilateral.

To prove: (i) $\angle A + \angle B = 180°$

(ii) $\angle X + \angle Y = 180°$

Construction: Let O be the centre of the circle.

Join XO, YO

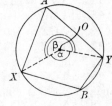

Proof: Angle at centre = twice angle at circumference.

$$\therefore \angle \alpha = 2\angle A$$

And $\qquad \angle \beta = 2\angle B$

But $\quad \angle \alpha + \angle \beta = 360°$

$$\therefore 2\angle A + 2\angle B = 360°$$

$$\therefore \angle A + \angle B = 180°$$

fig. C:121

Similarly, by joining AO, BO it may be proved that

$$\angle X + \angle Y = 180° \qquad \text{Q.E.D.}$$

THEOREM 32

If one side of a cyclic quadrilateral is produced, the exterior angle so formed is equal to the interior opposite angle of the quadrilateral.

Given: $ABCD$ is a cyclic quadrilateral with CD produced to Q.

To prove: $\qquad \angle \alpha_1 = \angle \alpha_2$

Proof: $\quad \angle \alpha_1 + \angle \beta = 180°$

$\qquad\qquad\qquad$ (adj. $\angle$s on str. line)

$$\angle \alpha_2 + \angle \beta = 180°$$

$\qquad\qquad\qquad$ (opp. $\angle$s of cyc. quad.)

$$\therefore \angle \alpha_1 + \angle \beta = \angle \alpha_2 + \angle \beta$$

$$\therefore \angle \alpha_1 = \angle \alpha_2$$

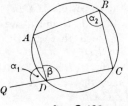

fig. C:122

Q.E.D.

EXERCISE 27

With reference to fig. C:123. Give reasons for your answers.

1. $\angle AOD = 100°$. Find $\angle ABD$.
2. $\angle EOD = 60°$. Find $\angle ACD$.
3. $\angle AOE = 80°$. Find $\angle OAE$.
4. $\angle ODE = 25°$. Find $\angle ABD$.
5. $\angle ACD = 45°$. Find $\angle ABD$.
6. $\angle BAC = 12°$. Find $\angle CDB$.
7. $\angle CBD = 36°$. Find $\angle DAC$.
8. $\angle OAE = 28°$. Find $\angle ACD$.
9. $\angle ACB = 54°$. Find $\angle AOB$.
10. $\angle BOC = 110°$. Find $\angle BDC$.

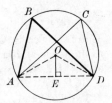

fig. C:123

258   PART C: GEOMETRY

With reference to fig. C : 124. Give reasons for your answers.

11. $\angle XYZ = 100°$. Find $\angle XWZ$.

12. $\angle YXW = 125°$. Find $\angle YZW$.

13. $\angle YOW = 130°$. Find $\angle YZW$.

14. $\angle XYZ = 112°$. Find $\angle PWX$.

15. $\angle YOW = 148°$. Find $\angle WXY$.

16. $\angle XOZ = 180°$. Find $\angle XWZ$.

17. $\angle YOZ = 120°$ and XOZ is a straight line. Find $\angle YXO$.

18. $\angle YOZ = 140°$, XOZ is a straight line and $XY = XW$. Find $\angle QXY$.

fig. C: 124

19. $XY = WX$, $\angle WOX = 132°$. Find $\angle YWX$.

20. $\angle Q\dot{X}Y = 90°$. Find $\angle YOW$.

21. PQ and SR are two chords of a circle. They are produced to meet at X. $QR\|PS$ and $\angle PQR = 120°$. Find $\angle QXR$.

22. $ABCD$ is a cyclic quadrilateral and O is the centre of the circle. If $\angle BOD = 160°$, find $\angle BAD$ and $\angle BCD$.

23. A, B and C are three points on the minor arc of the circumference of a circle, centre O. AB is produced to D and $\angle CBD = 80°$. Calculate the value of $\angle AOC$ in the minor sector.

24. A, B, C, D and E are concyclic points taken in order such that $AB = AE$, $AB\|CE$, BD is a diameter and $\angle BAE = 100°$. Calculate the angles of $\triangle CDE$.

25. A, B and C are three points on a minor arc of a circle centre O, such that $\angle AOC = 90°$. If AB is produced to D, calculate $\angle CBD$.

26. A, B and C are three points on a minor arc of a circle centre O, such that $\angle AOB = \angle OBC = 50°$. Calculate $\angle ABO$, $\angle ACB$ and $\angle CAB$.

27. A, B and C are three points on a minor arc of a circle centre O, such that $\angle AOB = 100°$ and AC bisects $\angle OAB$. Calculate $\angle ABO$, $\angle ACB$ and $\angle OBC$.

28. A, B, C and D are concyclic points forming parallel chords AB and CD. AC and BD intersect at X and $\angle BAC = 40°$. Calculate $\angle AXD$.

29. AOC is the diameter of a circle centre O. (See fig. C : 125.) BD is a tangent to the circle at B and meets AC produced at D. If $\angle BDC = 40°$, calculate $\angle CBD$, $\angle BCD$ and $\angle BAO$.

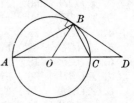

fig. C:125

30. $ABCD$ is a cyclic quadrilateral in which $AB = AC$, $AD = CD$ and $\angle ABC = 80°$. Calculate $\angle BAD$.

EXERCISE 28

1. Prove that the angle which an arc of a circle subtends at the centre is double that which it subtends at any point on the remaining part of the circumference, using the case in which the centre of the circle lies outside the angle formed at the circumference.

2. ABC is an equilateral triangle inscribed in a circle. P is a point on the minor arc BC and BC, AP intersect at X. Prove

$$\angle BXP = \angle ACP.$$

3. XY is a chord. Prove that XY subtends equal angles in the major segment.

4. Prove that the angle in a semicircle is a right-angle.

5. Two circles of equal radius are each drawn to pass through the other's centre. O and P are the respective centres and AP, OC are diameters such that $AOPC$ is a straight line. If the circles intersect at B and D, prove $AB = BC$.

6. Prove that the opposite angles of a cyclic quadrilateral are supplementary.

7. Two circles intersect at X and Y. AXB is a straight line cutting one circle at A and the other at B. Similarly with straight line AYC. Prove $\angle AXC = \angle AYB$.

8. From Question 7, show that $\triangle$s AYB, AXC are equiangular, and hence prove that $\dfrac{AX}{AY} = \dfrac{CX}{BY} = \dfrac{AC}{AB}$.

9. $ABCD$ is a cyclic quadrilateral in which DB bisects $\angle ABC$. Prove that $AD = CD$.

10. If the opposite angles of a cyclic quadrilateral are equal, prove that the other two vertices are joined by a diameter.

11. Two circles intersect at X and Y. $ABXY$ is a cyclic quadrilateral in one circle and $CDXY$ is a cyclic quadrilateral in the other circle. If BXD, AYC are straight lines, prove that $AB\|CD$.

12. $ABCD$ is a cyclic quadrilateral in which $AB\|CD$. Prove that $\angle ABC = \angle BAD$.

13. ABC is a triangle, right-angled at A and $AC > AB$. BC is bisected at D and a circle drawn to pass through A, B, D cutting AC at E. Prove that BE is a diameter of the circle and $\angle AEB = 2\angle EBD$.

14. Two circles intersect at A and B. AC is a diameter of one circle and AD a diameter of the other. Prove that CBD is a straight line.

15. AP and AB are two chords of a circle, centre O. OA bisects $\angle PAB$ and AB is produced to C so that $OB = BC$. Prove that $\angle POC$ (remote from A) is equal to $7\angle OCB$.

THEOREM 33
(*Converse of Theorem 28.*)

If the line joining two points subtends equal angles at two other points on the same side of it, the four points lie on a circle.

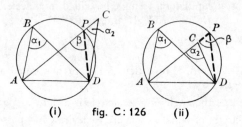

(i)   fig. C : 126   (ii)

Given: $\angle ABD = \angle ACD$ (i.e., $\angle \alpha_1 = \angle \alpha_2$) and points B and C lie on the same side of AD.

To prove: Points A, B, C, D lie on a circle.

Construction: Construct a circle to pass through points A, B and D. Let the circle cut AC (or AC produced) at P, then points A, B, P, D are concyclic.

Proof: $\angle \alpha_1 = \angle \beta$ ($\angle$s in same segment)
                    But $\angle \alpha_1 = \angle \alpha_2$ (given)
                    $\therefore \ \angle \alpha_2 = \angle \beta$

From fig. (i) $\angle \alpha_2 < \angle \beta$ $\left\{ \begin{array}{l} \text{ext. } \angle \text{ of } \triangle \text{ greater than} \\ \text{either int. } \angle \end{array} \right\}$
From fig. (ii) $\angle \alpha_2 > \angle \beta$

$\therefore \ \alpha_2$ can equal β only when P and C coincide.

$\therefore$ Points A, B, C, D must be concyclic.   Q.E.D.

THEOREM 34
(*Converse of Theorem 31.*)

If the opposite angles of a quadrilateral are supplementary the quadrilateral is cyclic.

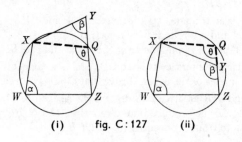

(i)   fig. C : 127   (ii)

Given: $WXYZ$ is a quadrilateral in which
$\quad\quad \angle XWZ + \angle XYZ = 180°$ (i.e., $\angle \alpha + \angle \beta = 180°$)
To prove: Points W, X, Y, Z lie on a circle.
Construction: Construct a circle to pass through points WXZ.
Let the circle cut ZY (or ZY produced) at Q, then points W, X, Q, Z are concyclic.

Proof: $\quad\quad\quad \angle \alpha + \angle \theta = 180°$ (opp. $\angle$s of cyc. quad.)
$\quad\quad$ But $\angle \alpha + \angle \beta = 180°$ (given)
$\quad\quad \therefore \; \angle \alpha + \angle \beta = \angle \alpha + \angle \theta$
$\quad\quad\quad\quad \therefore \; \angle \beta = \angle \theta$

From fig. (i) $\quad\quad \angle \beta < \angle \theta$ $\quad \left\{\begin{array}{l}\text{ext. } \angle \text{ of } \triangle \text{ greater than} \\ \text{ either int. } \angle\end{array}\right\}$
From fig. (ii) $\quad\quad \angle \beta > \angle \theta$

$\therefore \; \beta$ can equal θ only when Q and Y coincide.
$\therefore$ Points W, X, Y, Z must be concyclic. $\quad\quad\quad$ Q.E.D.

COROLLARY

If an exterior angle of a quadrilateral is equal to the interior opposite angle, the quadrilateral is cyclic.

Exercise 29

1. XY is the diameter of a circle. A and B are two points on opposite sides of XY. Prove $\angle AXY = \angle ABY$.

2. XYZ is a triangle in which XQ and YP are the perpendiculars to ZY and ZX respectively. Prove $\angle ZXQ = \angle ZYP$.

3. PQR is a triangle in which PX and QY are the perpendiculars to QR and PR respectively. Prove that P, Q, X, Y lie on a circle of which PQ is a diameter. Prove also that $\angle PQY = \angle PXY$.

4. XY and AB are two chords of a circle intersecting at Q. Prove that triangles AXQ and YBQ are similar.

5. PQ and RS are two chords of a circle, centre O. If the two chords intersect at O, prove that $PRQS$ is a rectangle.

6. B and D are two points on the same side of line AC. BC and AD intersect at X and $\triangle AXC$ is isosceles. If $BC = AD$, prove that A, B, D, C are concyclic points.

7. P and Q are two points on the same side of line AB. AQ and BP intersect at X. If $\angle BXQ = 70°$ and $\angle PAX = 45°$, calculate $\angle APX$. If $\angle QBX = 45°$, prove that $APQB$ is a cyclic quadrilateral.

8. $PQRS$ is a parallelogram in which $\angle P = 90°$. Prove that the figure is a cyclic quadrilateral.

9. AC and BD are two equal lines which bisect each other at right angles. Prove that $ABCD$ is a cyclic quadrilateral.

10. XYZ is an equilateral triangle in which $AY \perp XZ$. B is a point outside XY such that $\angle BYX = 60°$ and $\angle BXY = 30°$. Prove that points A, X, B, Y are concyclic.

11. *ABCD* is a quadrilateral in which *AD∥BC* and ∠*A* = ∠*D*. Prove that *ABCD* is cyclic. (NOTE. **ABCD is called an isosceles trapezium.**)

12. A regular convex polygon has an exterior angle of 60°. Prove that any four adjacent vertices, taken in order, are concyclic points.

AXIOM 10

In equal circles (or in the same circle) if two arcs subtend equal angles at the centre, the arcs are equal.

COROLLARY

In equal circles (or in the same circle) if two arcs subtend equal angles at the circumference, the arcs are equal.

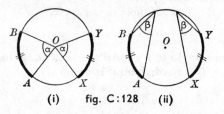

(i)   fig. C:128   (ii)

CONVERSE

In equal circles (or in the same circle) if two arcs are equal, they subtend equal angles at the centre and equal angles at the circumference.

QUESTION

What is the relationship between α and β? (From fig. C:128.) Give a reason.

AXIOM 11

In equal circles (or in the same circle) if two chords are equal, they cut off equal arcs.

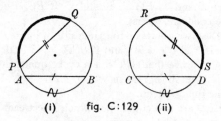

(i)   fig. C:129   (ii)

CONVERSE

In equal circles (or in the same circle) if two arcs are equal, the chords of the arcs are equal.

LENGTH OF AN ARC

EXERCISE 30

Give answers correct to three significant figures

With reference to fig. C : 130, find the length of arc AB, given:

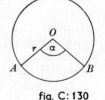

fig. C : 130

1. $\alpha = 90°$, $r = 3\frac{1}{2}$ cm   **2.** $\alpha = 45°$, $r = 7$ cm
3. $\alpha = 22\frac{1}{2}°$, $r = 8$ cm   **4.** $\alpha = 120°$, $r = 1\frac{1}{2}$ cm
5. $\alpha = 60°$, $r = 3$ cm   **6.** $\alpha = 40°$, $r = 4\frac{1}{2}$ cm
7. $\alpha = 60°$, $r = 1\cdot75$ cm   **8.** $\alpha = 70°$, $r = 6$ cm
9. $\alpha = 80°$, $r = 4\cdot5$ cm   **10.** $\alpha = 100°$, $r = 3$ m

AREAS OF SECTORS AND SEGMENTS

EXERCISE 31

Give answers to three significant figures

With reference to fig. C : 131, find the area of sector AOB, given:

1. $\alpha = 90°$, $r = 2$ cm   **2.** $\alpha = 45°$, $r = 4$ cm
3. $\alpha = 22\frac{1}{2}°$, $r = 4$ cm   **4.** $\alpha = 120°$, $r = 3$ cm
5. $\alpha = 80°$, $r = 1\cdot5$ cm   **6.** $\alpha = 50°$, $r = 6$ cm

fig. C:131

Find the area of the minor segment AB, given:

7. $\alpha = 70°$, $r = 6$ cm   **8.** $\alpha = 140°$, $r = 3$ cm
9. $\alpha = 100°$, $r = 6$ cm   **10.** $\alpha = 48°$, $r = 3\cdot5$ cm

11. If $r = 5$ cm and $AB = 6$ cm, find the area of: (i) sector AOB; (ii) minor segment AB.

12. If $\alpha = 90°$ and $AB = 12$ cm, find the area of: (i) sector AOB; (ii) minor segment AB.

TANGENTS TO CIRCLES

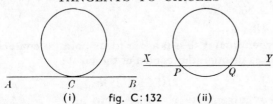

(i)   fig. C:132   (ii)

DEFINITIONS

(i) A **tangent** is a straight line which, however far it may be produced, has only **one point of contact** with a circle. AB is a tangent and C is the point of contact. [Fig. C : 132 (i).]

(ii) A straight line which, if produced, cuts a circle at **two points** is called a **secant**. XY is a secant cutting the circle at P and Q. [Fig. C : 132 (ii).]

NOTE. Previous references to tangents have treated the propositions contained in Theorem 35 as axiomatic.

THEOREM 35

The tangent at any point of a circle and the radius through the point are perpendicular to each other.

Given: B is any point on a circle, centre O. PQ is a tangent to the circle at B.

To prove: $OB \perp PBQ$ (i.e., $\angle \beta = 90°$).

Construction: If OB is not $\perp PBQ$, it will be possible to draw line $OA \perp PAQ$.

Proof: If $\alpha = 90°$

     Then $\beta < 90°$ (sum of $\angle$s in

                                 $\triangle OAB = 180°$)

     Then $OA < OB$ (longer side opp.

                                      greater $\angle$)

     But OB is a radius   (given)

$\therefore$ If $OA < OB$, A must lie inside the circle.

$\therefore$ If BA is produced, it will cut the circle at a second point. This is impossible by definition of tangent PQ.

$\therefore$ It is wrong to assume that β is not a rt. $\angle$.

                         $\therefore OB \perp PBQ$ Q.E.D.

fig. C:133

COROLLARY 1

A straight line perpendicular to the end of a radius of a circle is a tangent to the circle.

COROLLARY 2

At any given point on a circle, only one tangent to the circle can be drawn.

(For every point on a circle, there is only one radius which passes through the point. Therefore, there can be only one tangent to the circle at that point.)

COROLLARY 3

The perpendicular to a tangent at its point of contact with a circle passes through the centre of the circle.

THEOREM 36

The two tangents to a circle from an external point are equal.

Given: PQ and PR are tangents from a point P to a circle, centre O.

To prove: $PQ = PR$.

Construction: Join PO, QO, RO.

Proof:

     $\angle \alpha_1 = 90°$ (tangent $PQ \perp$ rad. QO)

     $\angle \alpha_2 = 90°$ (tangent $PR \perp$ rad. RO)

$\therefore \angle \alpha_1 = \angle \alpha_2 = 90°$

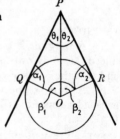

fig. C:134

$$\text{In } \triangle\text{s } PQO, PRO \qquad (\text{rt. } \angle\text{d } \triangle\text{s})$$
$$\angle\alpha_1 = \angle\alpha_2 \qquad (\text{proved})$$
$$QO = RO \qquad (\text{radii})$$

hypotenuse PO is common
$$\therefore \triangle PQO \equiv \triangle PRO \qquad (\text{R.H.S.})$$

In particular: $\qquad PQ = PR \qquad\qquad$ Q.E.D.

COROLLARY 1. (From fig. C : 134, $\theta_1 = \theta_2$.)

The two tangents to a circle from an external point are equally inclined to the line joining the point to the centre.

COROLLARY 2. (From fig. C : 134, $\beta_1 = \beta_2$.)

The two tangents to a circle from an external point subtend equal angles at the centre of the circle.

COROLLARY 3

If two tangents to a circle meet at a point, the line joining that point and the centre of the circle is the perpendicular bisector of the chord of contact of the two tangents.

THEOREM 37

If two circles touch, the point of contact lies on the straight line through the centres.

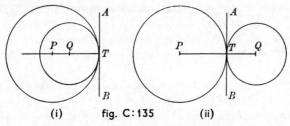

(i)   fig. C: 135   (ii)

Given: Two circles, centres P and Q, touch at T.

To prove: Points P, Q, T are in the same straight line (i.e., **collinear**).

Construction: Draw ATB the common tangent at T. Join PT, QT.

Proof: Since AB is a common tangent at T,
$$\angle ATP = 90° \text{ and } \angle ATQ = 90° \text{ (tangent } \perp \text{ radius)}$$

From fig. (i), $\angle ATP$ and $\angle ATQ$ are coincident.

$\therefore$ Q lies on PT and PQT is a str. line.

From fig. (ii), $\angle ATP$ and $\angle ATQ$ are adjacent.

$\therefore$ PTQ is a str. line. $\therefore$ Points P, Q, T are collinear.

COROLLARY 1 $\qquad\qquad\qquad\qquad\qquad\qquad$ Q.E.D.

If two circles touch internally, the distance between their centres is equal to the difference of their radii. [Fig. C : 135 (i).]

COROLLARY 2

If two circles touch externally, the distance between their centres is equal to the sum of their radii. [Fig. C : 135 (ii).]

EXERCISE 32

1. *PA* and *PB* are tangents from *P* to a circle, centre *O*. Prove that *PO* bisects $\angle APB$.

2. *AX* and *AY* are tangents from *A* to a circle, centre *O*. Prove that *AO* bisects $\angle XOY$.

3. *XY* and *XZ* are tangents from *X* to a circle, centre *P*. Prove that *XP* is the $\perp$ bisector of the chord of contact, *YZ*.

4. *PA* and *PB* are tangents from *P* to a circle, centre *P*. Prove that $\angle PAB = \angle PBA$.

5. *TP* and *TQ* are tangents from *T* to a circle, centre *O*. If *TO* cuts the circle at *X*, prove that $PX = QX$.

6. *P* is 3·75 cm from *O*, the centre of a circle whose radius is 2·25 cm. Find the length of a tangent from *P* to the circle.

7. *AB*, a tangent to a circle at *B*, is 27·6 cm. If *A* is 29·9 cm from *O*, the centre of the circle, find *OB*.

8. The radius of a circle, centre *O*, is 2 cm. *XY* is a tangent to the circle at *Y* and subtends an angle of 63° 30′ at the centre. Find the length of *XY*. If *XZ* is a second tangent from *X*, find the length of the chord of contact.

9. If two tangents to a circle are joined by a diameter, prove that the tangents are parallel.

10. Two concentric circles are drawn with centre *O*. A tangent to the smaller circle at *P* cuts the larger circle at *A* and *B*. Prove that $AP = BP$.

11. *ABCD* is a parallelogram whose sides are tangents to an inscribed circle. Prove that *ABCD* is a rhombus.

12. *AB* and *CD* are parallel tangents to a circle, centre *O*. If a third tangent cuts *AB* and *CD* at *X* and *Y* respectively, prove that *XY* subtends a rt. $\angle$ at *O*.

13. Two circles touch internally at *A*. The centre of the larger circle is *O* and the centre of the smaller is *P*. The line *APO* is produced to cut the larger circle again at *B*. Tangents from *B* to the smaller circle cut the larger circle at *X* and *Y* respectively. Prove: (i) $BX = BY$; (ii) the common tangent at *A* is parallel to *XY*.

14. Two circles, centres *O* and *P*, each of radius 1 cm, touch externally at *X*. With centres *A* and *B* on the common tangent through *X*, two circles are drawn, each of radius 0·5 cm, to touch each of the larger circles externally. Prove that centres *A*, *O*, *B*, *P* form the vertices of a rhombus. Calculate the length of *AB*.

15. *XYZ* is an equilateral triangle inscribed in a circle, centre *O*. If the three tangents at *X*, *Y* and *Z* meet at *A*, *B* and *C*, prove that $\triangle ABC$ is also equilateral.

THEOREM 38

If a straight line touches a circle, and from the point of contact a chord is drawn, the angles which this chord makes with the tangent are equal to the angles in the alternate segments.

Given: Straight line APB touches a circle at P. PQ is a chord such that PXQ is a major arc and PYQ is a minor arc.

To prove: (i) $\angle BPQ = \angle PXQ$ ($\angle$ in alt. seg. PXQ)
     (ii) $\angle APQ = \angle PYQ$ ($\angle$ in alt. seg. PYQ)

Construction: Draw the diameter PR and join RQ.

Proof: (i) From fig. C : 136.

In $\triangle PQR$, $\angle PQR = 90°$ ($\angle$ in semicircle)
     $\therefore \angle \alpha_1 + \angle \theta = 90°$ (sum of $\angle$s in $\triangle = 180°$)
But $\angle \beta + \angle \theta = 90°$ (tang. $AB \perp$ rad. OP)
     $\therefore \angle \alpha_1 + \angle \theta = \angle \beta + \angle \theta$
          $\therefore \angle \alpha_1 = \angle \beta$
But $\angle \alpha_1 = \angle \alpha_2$ ($\angle$s in same seg.)
     $\therefore \angle \beta = \angle \alpha_2$
     $\therefore \angle BPQ = \angle PXQ$

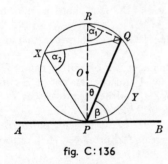

fig. C:136

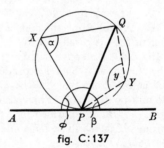

fig. C:137

(ii) From fig. C : 137.

$PXQY$ is a cyclic quadrilateral.

$$\therefore \angle \alpha + \angle y = 180°$$

APB is a straight line.

$$\therefore \angle \phi + \angle \beta = 180°$$
$$\therefore \angle \phi + \angle \beta = \angle \alpha + \angle y$$

But $\angle \beta = \angle \alpha$ (proved above)
     $\therefore \angle \phi = \angle y$
     $\therefore \angle APQ = \angle PYQ$ Q.E.D.

Exercise 33

In fig. C : 138, *AB* is a tangent to the circle at *P*.

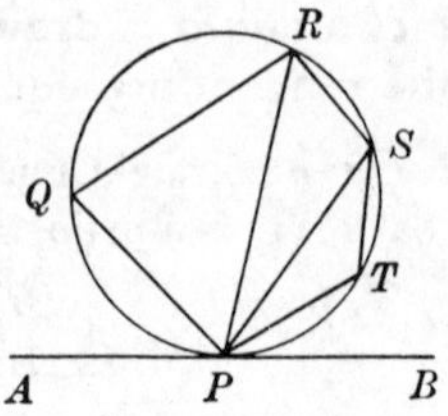

1. $\angle APS = 120°$, find $\angle PRS$.

2. $\angle PQS = 70°$, find $\angle APS$.

3. $\angle PQR = 90°$, find $\angle RPB$.

4. $\angle QRS = 112°$, find $\angle QPS$.

5. *RP* is a diameter and $\angle APS = 140°$, find $\angle PRS$.

6. *RP* is a diameter and $\angle PQS = 65°$, find $\angle RPS$.

7. $\angle APS = 130°$ and $PT = ST$, find $\angle SPT$.

fig. C:138

8. $\angle QPB = 150°$, find $\angle QTP$.

9. $\angle QPB = 140°$ and $QR = QP$, find $\angle BPR$.

10. $\angle RSP = 100°$ and $SR = SP$, find $\angle SPB$.

11. *AX* and *AY* are tangents to a circle at *X* and *Y*. If $\angle XAY = 60°$, find the value of the angle subtended by *XY* in the minor segment of the circle.

12. $\triangle ABC$ is inscribed in a circle and contains angles of 40°, 60°, 80°. If the tangents to the circle at *A*, *B* and *C* are produced to meet at *X*, *Y* and *Z*, find the $\angle$s of $\triangle XYZ$.

13. *XAY* is a tangent to a circle at *A*. *BC* is a chord parallel to *XAY*. Prove $AB = AC$.

14. $\triangle ABC$ is inscribed in a circle. If $AB = AC$ prove that the tangent at *A* is parallel to *BC*.

15. $\triangle XYZ$ is constructed on the diameter *XY* of a circle, centre *O*. The tangent at *X* cuts *YZ* produced at *A* and the tangent at *Z* cuts *AX* at *B*. Prove that $\angle BAZ = \angle BZA$, and hence prove that *B* is the mid-point of *AX*. Prove also that $\angle BXZ = \angle OZY$.

16. $\triangle ABC$ is inscribed in a circle. *BD* is the bisector of $\angle ABC$ cutting *AC* at *D*. The tangent at *B* meets *CA* produced at *T*. Prove that $\triangle BTD$ is isosceles.

17. *PQRS* is a cyclic quadrilateral in which *PR* bisects $\angle SRQ$. Why is $PS = PQ$?

18. *PQRS* is a cyclic quadrilateral in which *PR* bisects $\angle SRQ$. Prove that *QS* is parallel to the tangent at *P*.

19. *PQRS* is a cyclic quadrilateral in which $PQ = PS$. Prove that *QS* is parallel to the tangent at *P*.

20. *PQRS* is a cyclic quadrilateral in which *QS* is parallel to the tangent at *P*. *RS* is produced to cut the tangent at *P* at point *T*. Prove $\angle PTS = \angle QPR$.

TANGENT CONSTRUCTIONS
CONSTRUCTION 16

To construct a tangent at a given point on the circumference of a circle.

Given: P is a point on the circumference of a circle, centre O.

To construct: The tangent to the circle at P.

Construction:

    (i) Join OP and produce the line to Q.

    (ii) At P construct $TS \perp OQ$. Then TPS is the required tangent to the circle at P.

fig. C:139

Proof: A straight line perpendicular to the end of a radius of a circle is a tangent to the circle. (TH. 35, COR. 1.)

    Since OP is a radius and $TS \perp OP$ at P

            TPS is a tangent at P.   Q.E.F.

NOTE. If the position of O, the centre of the circle, is not given, it must first be found by means of CONST. 14.

CONSTRUCTION 17

To construct a tangent from a given point outside a circle.

Given: P is a point outside a circle, centre O.

To construct: A tangent to the circle from P.

Construction:

    (i) Join OP.

    (ii) Bisect OP at Q.

    (iii) With centre Q and OP as a diameter, construct a semicircle to cut the given circle at T.

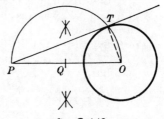

fig. C:140

    (iv) Join PT. Then PT is the required tangent to the given circle.

Proof: Join OT. OT is a radius of the given circle.

        $\angle PTO = 90°$ ($\angle$ in a semicircle)

        $\therefore PT \perp$ radius OT

        $\therefore PT$ is a tangent to the given circle.   Q.E.F.

CONSTRUCTION 18

To construct an exterior common tangent to two circles of unequal radius.

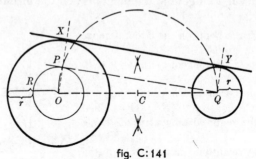

fig. C: 141

Given: Circle centre O, of radius R, and circle centre Q, of radius r.
To construct: An exterior common tangent to the two circles.
Construction:

(i) With centre O, draw a circle of radius $(R - r)$.
(ii) Construct a tangent to this circle from Q.

    (*a*) Describe a semicircle, centre C and diameter OCQ.
    (*b*) Let the semicircle cut circle radius $(R - r)$ at P.
    (*c*) PQ is a tangent to this circle.

(iii) Produce radius OP to cut circle radius R at X.
(iv) Construct radius $QY\|OX$.
(v) Join XY, which is an exterior common tangent to the circles, centres O and Q.

Proof: By construction $PX = QY$ and $PX\|QY$.
    ∴ $PQYX$ is a parallelogram.
But   ∠$OPQ = 90°$ (∠ in a semicircle)
    ∴ ∠$XPQ = 90°$ (adj. ∠s on a str. line)
∴ $PQYX$ is a rectangle.
    ∴ ∠$PXY = $ ∠$XYQ = 90°$
But OX and QY are radii.
∴ XY is a common tangent at X and Y.   Q.E.F.

NOTE

(i) A second tangent may be constructed *below OCQ*.

(ii) An exterior common tangent may also be described as a **direct common tangent**.

(iii) If the two given circles are of equal radius, XO and YQ will each be ⊥ to OQ and XY will be ∥ to OQ.

CONSTRUCTION 19

To construct an interior common tangent to two non-intersecting circles.

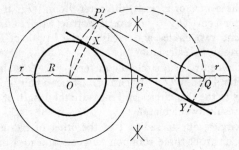

fig. C:142

Given: Circle centre O, of radius R, and circle centre Q, of radius r. The circles do not intersect.

To construct: An interior common tangent to the two circles.

Construction:

   (i) With centre O, draw a circle of radius $(R + r)$.

   (ii) Construct a tangent to this circle from Q.

      (*a*) Describe a semicircle, centre C and diameter OCQ.

      (*b*) Let the semicircle cut circle radius $(R + r)$ at P.

      (*c*) PQ is a tangent to this circle.

   (iii) Construct radius OP to cut circle radius R at X.

   (iv) Construct radius $QY \| OX$

   (v) Join XY which is an interior common tangent to the circles, centres O and Q.

Proof: By construction $PX = QY$ and $PX \| QY$.

   $\therefore$ $PQYX$ is a parallelogram.

But $\angle OPQ = 90°$ ($\angle$ in a semicircle)

     $\therefore$ $\angle OXY = 90°$ (Corr. $\angle$ with $\angle OPQ$)

And $\angle QYX = 90°$ (Opp. $\angle$ of $\|^{gram.}$)

     $\therefore$ $\angle OXY = \angle QYX = 90°$

But OX and QY are radii.

   $\therefore$ XY is a common tangent at X and Y.   Q.E.F.

NOTE

   (i) An interior common tangent may also be described as a **transverse** common tangent.

   (ii) If the two given circles are of equal radius, the method of construction is unaffected.

EXERCISE 34

1. Draw a circle of radius 3 cm. From the centre O, produce a radius to P so that OP is 5 cm. From P construct a tangent to meet the circle at T. Measure PT and check the result by calculation.

2. O is the centre of a circle, diameter 9 cm. Q is the centre of a circle, diameter 4 cm. $OQ = 6·5$ cm. Construct a pair of direct common tangents to the two circles and measure the length of one of them. Check your result by calculation.

3. O is the centre of a circle, diameter 8 cm. Q is the centre of a circle, diameter 4 cm. $OQ = 10$ cm. Construct a pair of interior common tangents to the two circles and measure the length of one of them. Check your result by calculation.

4. Two circles, one of radius 1 in the other of radius $\frac{1}{2}$ in, touch externally. Construct three common tangents and measure the length of each. Check your results by calculation.

5. The sides of $\triangle ABC$ are 2 in, 3 in, 4 in. Three circles are constructed, one at each vertex of the triangle, to touch each other externally. Calculate the radius of each circle.

6. Three circles are constructed, as described in Question 5, at the vertices of a triangle whose sides measure 4 cm, 5 cm, 6 cm. Calculate the radius of each circle.

7. Three circles whose respective radii are 2 cm, 4 cm, 6 cm are constructed to touch each other externally. Make the construction, drawing only sufficient arc of each circle for the purpose of touching.

8. Construct the incircle of a triangle whose sides are 10 cm, 8 cm and 6 cm. Measure the radius of the incircle and check your result by calculation.

9. The tangents to a circle, centre O, from a point P, touch the circle at X and Y respectively and PO is produced to cut the circle at Z. If $\angle XOP = 58°$, calculate $\angle XPY$ and $\angle XZY$.

10. A circle, centre O and radius 2 cm, is inscribed in $\triangle ABC$. If $\angle BAC = 70°$ and $\angle OBC = 30°$, calculate the length of BC.

11. PQ is a straight line 2 in long. A circle with radius $1\frac{1}{2}$ in is drawn with centre Q. Construct a second circle of radius 1 in to touch the first circle and also to pass through P.

12. PQ and PR are tangents to a circle, centre O, at Q and R. QS and RS are equal chords. If $\angle QPR = 40°$, calculate $\angle QRS$.

13. P is a fixed point and O is the fixed centre of a circle whose radius is variable. Construct the locus of the point of contact, T, of a tangent from P to the circle. Describe the locus.

14. X and Y are two points 12 cm apart. A circle of radius 8 cm is constructed with X as centre. A second circle is constructed to touch the first circle and also to touch XY at Y. Find the radius of this circle.

CONSTRUCTION 20

To construct the escribed circle of a given triangle.

Given: $\triangle ABC$.

To construct: A circle to touch BC externally and AB, AC each produced.

Construction:

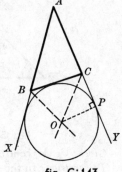

   (i) Produce AB to X and AC to Y.

   (ii) Bisect $\angle CBX$ and $\angle BCY$, let the bisectors intersect at O.

   (iii) Erect the perpendicular OP from BX, BC or CY.

   (iv) The required circle has centre O and radius OP.

Proof: BO is the locus of a point equidistant from BX and BC.

       CO is the locus of a point equidistant from CB and CY.

fig. C:143

   $\therefore$ O is equidistant from BX, BC and CY.   Q.E.F.

CONSTRUCTION 21

To construct, on a given straight line, a segment of a circle containing an angle equal to a given angle.

Given: A straight line AB and $\angle\alpha$.

To construct: On AB a segment of a circle containing an angle equal to $\angle\alpha$.

Construction:

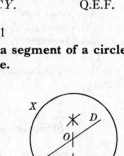

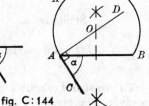

   (i) At A, construct $\angle BAC = \angle\alpha$. (*Use* CONST. 3.)

   (ii) At A, erect $AD \perp AC$.

   (iii) Construct the perpendicular bisector of AB to meet AD at O.

fig. C:144

   (iv) With centre O and radius AO or BO, describe the segment AXB. Then AXB is the required segment of a circle containing $\angle\alpha$.

Proof: The $\perp$ bisector of AB is the locus of points equidistant from A and B.

   $\therefore$ Circle, centre O, passes through A and B.

      AC is a tangent to the circle at A.   (constr.)

      AB is a chord of the circle.

            $\therefore$ $\angle BAC = \angle AXB$ ($\angle$ in alt. segment)

But $\angle BAC = \angle\alpha$ (constr.)

            $\therefore$ $\angle AXB = \angle\alpha$ Q.E.F.

CONSTRUCTION 22

To construct, in a given circle, a triangle equiangular to a given triangle.

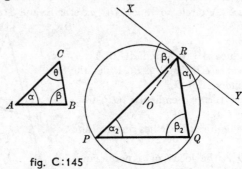

fig. C:145

Given: Circle, centre O, and $\triangle ABC$ containing $\angle \alpha$, $\angle \beta$, $\angle \theta$.
To construct: In the circle a triangle equiangular to $\triangle ABC$.
Construction:

(i) At any point R on the given circle, construct the tangent XRY.
(ii) Construct $\angle YRQ = \angle \alpha$, letting RQ cut the circle at Q.
(iii) Construct $\angle XRP = \angle \beta$, letting RP cut the circle at P.
(iv) Join PQ. Then $\triangle PQR$ is the required triangle constructed in the given circle and equiangular to $\triangle ABC$.

Proof:

$$\angle \alpha_1 = \angle \alpha \quad \text{(constr.)}$$

But

$$\angle \alpha_2 = \angle \alpha_1 \quad (\angle \text{ in alt. seg.})$$

$$\therefore \angle \alpha_2 = \angle \alpha$$

Similarly

$$\angle \beta_2 = \angle \beta$$

$$\therefore \angle PRQ = \angle \theta \quad \text{(remaining } \angle \text{s of } \triangle \text{s)}$$

$$\therefore \triangle PQR \text{ and } \triangle ABC \text{ are equiangular.} \qquad \text{Q.E.F.}$$

CONSTRUCTION 23

To construct, about a given circle, a triangle equiangular to a given triangle.

Given: Circle, centre O, and $\triangle ABC$ containing $\angle \alpha$, $\angle \beta$, $\angle \theta$.
To construct: About the circle a triangle equiangular to $\triangle ABC$.
Construction:

(i) Produce AB to D, forming ext. $\angle \delta$. Produce BA to E, forming ext. $\angle \phi$.

(ii) At O, construct $\angle ROP = \phi$ and $\angle ROQ = \delta$; letting OP, OQ, OR cut the given circle at P, Q, R respectively.

(iii) At P, Q, R, construct tangents to the circle letting them intersect at X, Y and Z.

(iv) Then $\triangle XYZ$ is the required triangle constructed about the given circle and equiangular to $\triangle ABC$.

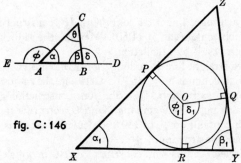

fig. C:146

Proof: In quadrilateral $XPOR$,

$$\angle P + \angle R = 180°$$ (tang. $\perp$ radii)

$\therefore$ $XPOR$ is a cycl. quad.   (opp. $\angle$s supp.)

$\therefore$ $\angle \alpha_1 + \angle \phi_1 = 180°$ (opp. $\angle$s of cyc. quad.)

But $\angle \phi_1 = \angle \phi$ (constr.)

$\therefore$ $\angle \alpha_1 = \angle \alpha$

Similarly $\angle \beta_1 = \angle \beta$

$\therefore$ $\angle XZY = \angle \theta$ (remaining $\angle$s of $\triangle$s)

$\therefore$ $\triangle XYZ$ and $\triangle ABC$ are equiangular.   Q.E.F.

CONSTRUCTION 24

To construct a regular polygon in or about a given circle.

Given: A circle, centre O.

To construct:

  1. An inscribed regular polygon of n sides.

  2. A circumscribed regular polygon of n sides.

Construction:

  1. If an inscribed regular polygon has n sides, each side will subtend an angle of $\left(\dfrac{360}{n}\right)$ degrees at the centre. Therefore the angle at O must be divided into n equal angles and $\angle \alpha = \left(\dfrac{360}{n}\right)$ degrees.

Let the arms of the angles cut the circle at A, B, C, D, E, . . . Then $ABCDE$ . . . is the required regular polygon inscribed in the given circle.

  2. To obtain a circumscribed regular polygon of n sides, construct tangents to the circle at the vertices of the inscribed polygon.

Let the tangents intersect at P, Q, R, S, T, . . . Then $PQRST$ . . . is the required regular polygon circumscribed about the given circle.

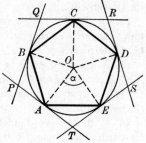

fig. C:147

EXERCISE 35

1. *AB* is a straight line 3 cm long. On *AB*, construct the segment of a circle which contains an $\angle$ of 60°. Measure the radius of the circle and check your result by calculation.

2. *XY* is a straight line 4 cm long. On *XY*, construct the segment of a circle which contains an $\angle$ of 30°. Calculate the radius of the circle and check your construction. Describe your construction and prove it.

3. *PQ* is a straight line 6 cm long. On *PQ*, construct the segment of a circle which contains an $\angle$ of 130°. Calculate the radius of the circle and check your construction.

4. In a circle of radius 2 in, construct $\triangle ABC$ in which $\angle A = 60°$, $\angle B = 50°$, $\angle C = 70°$. Measure *AB*, *BC*, *AC* and check your results by calculation.

5. About a circle of radius 1 in, construct a circumscribed triangle containing angles of 50°, 60°, 70°. Measure the sides of the triangle and check your results by calculation. (*Great care needed.*)

6. Using circles of radius 4 cm construct the following inscribed and circumscribed regular polygons:

(i) 3 sides	(ii) 6 sides	(iii) 4 sides
(iv) 8 sides	(v) 5 sides	(vi) 10 sides

7. *XY* is a straight line $1\frac{1}{2}$ in long. With *XY* as a chord, construct a circle of radius 1 in.

8. Using the results of Question 7, construct a circle of radius $1\frac{1}{2}$ in with *XY* as a chord. Measure the distance between the centres of the smaller and larger circles. (*Two answers.*)

9. *P* is a point on a straight line *XPY* and *Q* is a point outside the line. Construct a circle to touch the line at *P* and also to pass through *Q*.

10. *P* is a fixed point in a straight line *AB*. Plot the locus of the centres of circles drawn to touch *AB* at *P*. Hence state the locus of the centres of circles which are drawn to touch a given circle at a given point.

11. In fig. C : 148, construct a circle touching *AB* and the circle, centre *O*, at *P*. (Externally.)

12. In fig. C : 148, construct a circle touching *AB* and the circle, centre *O*, at *Q*. (Internally.)

13. *X* is a fixed point in a straight line *AB*. A circle, centre *O*, is situated a short distance from *AB*, as in fig. C : 148. Construct a circle to touch *AB* at *X* and the circle, centre *O*. (Externally.)

fig. C:148

14. With *AB* as diameter and radius 5 cm, draw a semicircle. Construct two circles, centres *P* and *Q*, each of radius 2 cm, to touch *AB* and the given arc. Measure *PQ*. Check your result by calculation.

PROPORTIONAL DIVISION

THEOREM 17 (p. 233) showed that if a straight line is drawn parallel to one side of a triangle, it divides the other two sides proportionally. It was convenient to assume that AR, BR could each be divided into a whole number of units (of the same kind) in order that $\dfrac{AR}{BR}$ should be represented by $\dfrac{x}{y}$, where x and y are integers.

Under such conditions AR and BR are said to be **commensurable**. The proposition is still correct for incommensurable values, but the proof is more difficult. The following alternative proof employs the formula for the area of a triangle.

If four quantities a, b, c, d are such that $\dfrac{a}{b} = \dfrac{c}{d}$, then a, b, c, d are said to be in **proportion**.

THEOREM 39 (*Alt. to Th.* 17.)

If a straight line is drawn parallel to one side of a triangle, the other two sides are divided proportionally.

fig. C:149

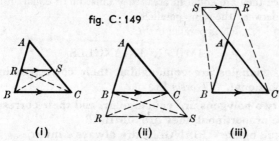

(i)   (ii)   (iii)

Given: ABC is a triangle. RS is a straight line parallel to BC cutting AB and AC (produced if necessary) at R and S respectively. [In figs. (ii) and (iii) AB, AC are divided **externally**.]

To prove: $\dfrac{AR}{BR} = \dfrac{AS}{CS}$

Construction: Join BS, CR.

Proof: If two triangles have equal altitudes, the ratio of their areas will be the same as the ratio of their bases. Using the $\perp$ from S to AB and from R to AC: $\quad \dfrac{\triangle RSA}{\triangle RSB} = \dfrac{AR}{BR}$ and $\dfrac{\triangle SRA}{\triangle SRC} = \dfrac{AS}{CS}$

But $\triangle RSB$ and $\triangle SRC$ are equal in area because they share the same base RS and are between the same parallels RS, BC.

$\therefore \dfrac{\triangle RSA}{\triangle RSB} = \dfrac{\triangle SRA}{\triangle SRC}$ and $\dfrac{AR}{BR} = \dfrac{AS}{CS}$ Q.E.D.

COROLLARY (i) $\dfrac{AR}{AB} = \dfrac{AS}{AC}$ (ii) $\dfrac{BR}{AB} = \dfrac{CS}{AC}$

CONVERSE

If a straight line divides two sides of a triangle proportionally, it is parallel to the third side.

Given: ABC is a triangle. RS is a straight line cutting AB and AC (produced if necessary) at R and S respectively, such that $\dfrac{AR}{BR} = \dfrac{AS}{CS}$.

To prove: RS is parallel to BC.

Construction: Join BS, CR.

Proof: If two triangles have equal altitudes, the ratio of their areas will be the same as the ratio of their bases. Using the $\perp$ from S to AB and from R to AC:

$$\frac{\triangle RSA}{\triangle RSB} = \frac{AR}{BR} \quad \text{and} \quad \frac{\triangle SRA}{\triangle SRC} = \frac{AS}{CS}$$

$$\text{But} \quad \frac{AR}{BR} = \frac{AS}{CS} \text{ (given)} \quad \therefore \quad \frac{\triangle RSA}{\triangle RSB} = \frac{\triangle SRA}{\triangle SRC}$$

$$\therefore \; \triangle RSB = \triangle SRC \; (\triangle RSA = \triangle SRA)$$

But these triangles share the same base RS and are on the same side of it; since they are equal in area, they must be of equal altitude and must lie between the same parallels.

$$\therefore \; RS \text{ is parallel to } BC. \qquad \text{Q.E.D.}$$

SIMILAR TRIANGLES

If two triangles are equiangular, their corresponding sides are proportional. (*Theorem* 18.)

(i) If two polygons are equiangular, and their corresponding sides are proportional, they are SIMILAR.

(ii) Equiangular TRIANGLES are always similar.

THEOREM 40 (*Conv. of Th.* 18.)

If the three sides of one triangle are proportional to the three sides of another, then the triangles are equiangular.

Given: In $\triangle$s ABC, XYZ,

$$\frac{AB}{XY} = \frac{BC}{YZ} = \frac{CA}{ZX}$$

To prove: $\triangle$s ABC, XYZ are equiangular (i.e., $\angle A = \angle X$, $\angle B = \angle Y$, $\angle C = \angle Z$).

Construction: From XY cut off XP equal to AB, construct $PQ \| YZ$ cutting XZ at Q.

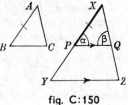

fig. C:150

Proof: In $\triangle$s XPQ, XYZ

$$\angle \alpha = \angle Y \qquad \text{(corres. } \angle \text{s, } PQ \| YZ)$$
$$\angle \beta = \angle Z \qquad (\qquad \text{ditto} \qquad)$$
$$\angle X \text{ is common}$$

$\therefore$ $\triangle$s XPQ, XYZ are equiangular (i.e., similar)

In particular: $\dfrac{XP}{XY} = \dfrac{PQ}{YZ} = \dfrac{QX}{ZX}$

But $XP = AB$ (constr.)

$\therefore$ $\dfrac{AB}{XY} = \dfrac{PQ}{YZ} = \dfrac{QX}{ZX}$

But $\dfrac{AB}{XY} = \dfrac{BC}{YZ} = \dfrac{CA}{ZX}$ (given)

$\therefore$ $PQ = BC$ and $QX = CA$

Also $XP = AB$ (constr.)

$\therefore$ $\triangle XPQ \equiv \triangle ABC$ (S.S.S.)

$\therefore$ $\triangle$s XPQ, ABC are equiangular

But $\triangle$s XPQ, XYZ are equiangular (proved)

$\therefore$ $\triangle$s ABC, XYZ are equiangular. Q.E.D.

THEOREM 41

If two triangles have an angle of the one equal to an angle of the other and the sides about these equal angles are proportional, the triangles are similar.

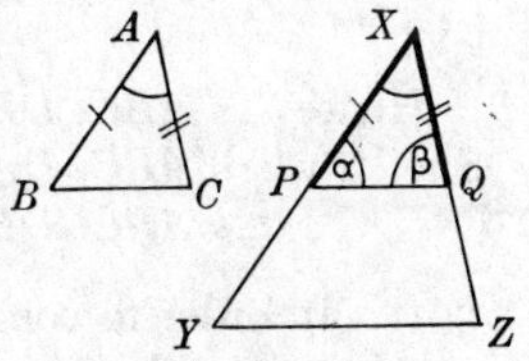

fig. C:151

Given: In $\triangle$s ABC, XYZ,

$$\angle A = \angle X \text{ and } \frac{AB}{XY} = \frac{AC}{XZ}$$

To prove: $\triangle$s ABC, XYZ are similar.

Construction: From XY cut off XP equal to AB, from XZ cut off XQ equal to AC. Join PQ.

Proof: In $\triangle$s XPQ, ABC

$XP = AB$ (const.)

$XQ = AC$ (const.)

$\angle X = \angle A$ (given)

$\therefore$ $\triangle XPQ \equiv \triangle ABC$ (S.A.S.)

In particular: $\angle \alpha = \angle B$ and $\angle \beta = \angle C$

Since $\dfrac{AB}{XY} = \dfrac{AC}{XZ}$ (given)

Then $\dfrac{XP}{XY} = \dfrac{XQ}{XZ}$ ($XP = AB, XQ = AC$)

$\therefore$ $PQ \parallel YZ$ (XY, XZ div. proporty.)

$\therefore$ $\angle \alpha = \angle Y$ and $\angle \beta = \angle Z$ (corresp. $\angle$s)

But $\angle \alpha = \angle B$ and $\angle \beta = \angle C$ (proved)

$\therefore$ $\angle B = \angle Y$ and $\angle C = \angle Z$

Then $\triangle$s ABC, XYZ are equiangular and therefore similar.

Q.E.D.

THEOREM 42

If a perpendicular is drawn from the right-angle of a right-angled triangle to the hypotenuse, the triangles on each side of the perpendicular are similar to the whole triangle and to one another.

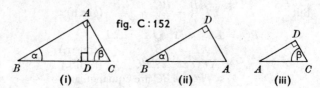

fig. C : 152

(i)   (ii)   (iii)

Given: $\triangle ABC$ in which $\angle BAC = 90°$ and $AD \perp BC$.

To prove: $\triangle$s ABC, DBA, DAC are similar.

Proof: In $\triangle$s ABC, DBA [figs. (i) and (ii)]

$$\angle BAC = \angle BDA = 90°$$
$$\angle \alpha \text{ is common}$$
$$\therefore \ \angle \beta = \angle BAD \qquad (\text{3rd } \angle \text{ of } \triangle)$$

Hence $\triangle$s ABC, DBA are equiangular and $\therefore$ similar.

Similarly $\triangle$s ABC, DAC are equiangular and $\therefore$ similar.

$\therefore \ \triangle$s ABC, DBA, DAC are similar.   Q.E.D.

NOTE. In order to compare the ratio properties of complex figures, it may be advisable to construct separate diagrams of the components as shown in fig. C : 152.

EXERCISE 36

From fig. C : 153.

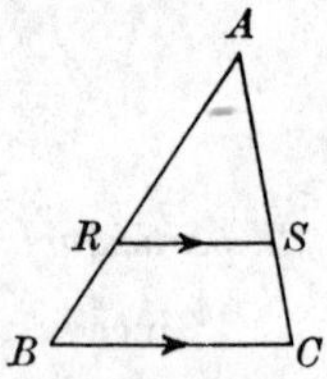

1. $AR = 3$ cm, $BR = 1$ cm, $AS = 3$ cm. Find CS and AC.

2. $AC = 5$ cm, $CS = 2$ cm, $AR = 3$ cm. Find BR and AB.

3. $AB = 4$ cm, $AR = 2$ cm, $CS = 2\cdot5$ cm. Find AC.

4. $AR = 1$ cm, $BR = 1\cdot5$ cm, $AS = 2$ cm. Find AC.

5. $AB = 6$ in, $AR = 1$ in, $CS = 3$ in. Find AS.

6. $AB = 5$ in, $AR = 3$ in, $RS = 1\cdot8$ in. Find BC.

7. $RS = 4$ cm, $BC = 5$ cm, $AC = 8$ cm. Find CS.

8. $RS = 7$ cm, $BC = 10$ cm, $BR = 6\cdot3$ cm. Find AB.

9. $RS = 6$ in, $BC = 7$ in, $CS = 5\cdot4$ in. Find AC.

10. $RS = 8\cdot2$ cm, $BC = 12\cdot3$ cm, $BR = 5\cdot6$ cm. Find AR.

fig. C:153

11. XY is a diameter of a circle, chord $AB \| XY$. AY and BX intersect at P. If $AB = 5$ cm, $AP = 3$ cm, $PY = 6$ cm, find the radius of the circle.

12. AB and XY are two chords of a circle intersecting at P and BY is a diameter. If $XP = 3$ cm, $BP = 5$ cm and the radius is 7 cm, find the length of AX.

13. In $\triangle ABC$, $BY \perp AC$ and $CX \perp AB$. If $AB = 9$ in, $AC = 8$ in, $CX = 6$ in, find the length of BY.

14. $\triangle ABC$ is an isosceles triangle in which $AB = AC$, D is a point on AC such that $BD = BC$. If $BC = 3$ in, $CD = 1$ in, find AD.

15. $\triangle ABC$ is right-angled at B. D is a point on AC such that $BD \perp AC$. If $BC = 10$ cm, $CD = 8$ cm, find AB and AD.

16. Equilateral triangle ABC of side 2 in is inscribed in a circle. The bisector of $\angle A$ cuts the circle again at D. Find the length of CD and the radius of the circle.

17. If two triangles are of equal altitude, prove that the ratio of their areas is equal to the ratio of their bases.

18. ABC is a triangle in which CA is produced to X and BA is produced to Y so that $XY \| BC$. Prove that $\dfrac{AX}{CX} = \dfrac{AY}{BY}$.

19. $\triangle$s FGH, PQR are such that $\dfrac{FG}{PQ} = \dfrac{GH}{QR} = \dfrac{HF}{RP}$. Prove that the triangles are equiangular.

20. X, Y, Z are the mid-points respectively of the sides AB, BC, CA of $\triangle ABC$. Prove that $\triangle XYZ$ and $\triangle ABC$ are similar.

21. $ABCD$ is a cyclic quadrilateral in which the diagonals intersect at Q. Prove that $\triangle AQB$ and $\triangle DQC$ are similar.

22. $PQRS$ is a trapezium in which $PQ \| RS$. If the diagonals intersect at X, prove that $\dfrac{PX}{RX} = \dfrac{QX}{SX}$.

23. P is any point on side AC of $\triangle ABC$. AX, PY are the perpendiculars respectively from A and P to base BC. Prove that $\dfrac{\text{area of } \triangle PBC}{\text{area of } \triangle ABC} = \dfrac{PY}{AX}$, hence prove that $\dfrac{\triangle PBC}{\triangle ABC} = \dfrac{PC}{AC}$.

24. $\triangle XYZ$ is inscribed in a circle, centre O. XO, produced if necessary, cuts YZ at P. Prove that $\dfrac{\triangle XYO}{\triangle XZO} = \dfrac{YP}{ZP}$.

25. $ABCD$ is a rectangle. A semicircle is described on AD and chord DE is drawn parallel to diagonal AC. Prove that $\dfrac{DE}{AD} = \dfrac{AD}{AC}$.

THEOREM 43

The bisector of an internal (or external) angle of a triangle divides the opposite side internally (or externally) in the ratio of the sides containing the bisected angle.

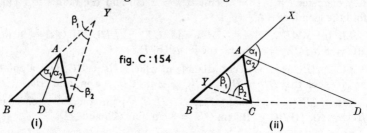

fig. C:154

(i)   (ii)

Given: In $\triangle ABC$, AD bisects $\angle BAC$ internally in (i) and externally in (ii), cutting BC or BC produced at D.

To prove: $\dfrac{DB}{DC} = \dfrac{AB}{AC}$

Construction: Through C draw $CY \parallel AD$, cutting BA or BA produced at Y.

Proof: Since

$$AD \parallel CY \quad \text{(const.)}$$
$$\angle \alpha_1 = \angle \beta_1 \quad \text{(corresp. } \angle s)$$
$$\angle \alpha_2 = \angle \beta_2 \quad \text{(alt. } \angle s)$$

But $\quad \angle \alpha_1 = \angle \alpha_2 \quad \text{(given)}$

$\therefore \quad \angle \beta_1 = \angle \beta_2$

$\therefore \quad AC = AY \quad (\triangle ACY \text{ is isos.})$

But $\quad \dfrac{DB}{DC} = \dfrac{AB}{AY} \quad \left(\begin{array}{c} AD \parallel CY, \text{ sides } BC, BY \\ \text{div. proportionally} \end{array} \right)$

$\therefore \quad \dfrac{DB}{DC} = \dfrac{AB}{AC} \qquad\qquad\qquad \text{Q.E.D.}$

CONVERSE

If the base BC, of a triangle ABC, is divided internally (or externally) at D in the ratio of the sides $AB : AC$, then AD bisects the angle at A internally (or externally).

RATIO CONSTRUCTIONS
CONSTRUCTION 25 (*Ext. of Const.* 11.)

To divide a given straight line (i) internally, (ii) externally, in a given ratio.

Given: AB is a straight line.

To construct: 1. A point E on AB such that $\dfrac{EA}{EB} = \dfrac{x}{y}$.

2. A point F on AB produced such that $\dfrac{FA}{FB} = \dfrac{x}{y}$.

Construction 1. [fig. C : 155 (i)]

   (i) Draw AQ at a convenient angle to AB.

   (ii) Along AQ cut off AD equal to x and DC equal to y. (Then $DA : DC = x : y$.)

   (iii) Join BC and construct $DE\|BC$ cutting AB at E.

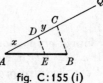

fig. C: 155 (i)

Then E divides AB internally in the ratio $EA : EB = x : y$.

Construction 2. [fig. C : 155 (ii)]

   (i) Draw AQ at a convenient angle to AB.

   (ii) Along AQ cut off AD equal to x and from AD cut off DC equal to y. (Then $DA : DC = x : y$.)

   (iii) Join BC and construct $DF\|BC$ cutting AB produced at F.

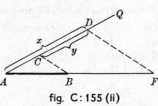

fig. C: 155 (ii)

Then F divides AB externally in the ratio $FA : FB = x : y$.

Proof: Proportional division. (Theorem 39.)   Q.E.F.

CONSTRUCTION 26 (*Alt. to Constr.* 25.)

To divide a given straight line (i) internally, (ii) externally, in a given ratio.

Given: AB is a straight line.

To construct: 1. A point E on AB such that $\dfrac{EA}{EB} = \dfrac{x}{y}$.

            2. A point F on AB produced such that $\dfrac{FA}{FB} = \dfrac{x}{y}$.

Construction 1. [fig. C : 156 (i)]

   (i) With A as centre, describe an arc, radius equal to x. With B as centre, describe an arc, radius equal to y.

   (ii) Let the arcs intersect at C. Join AC, BC.

   (iii) Let the bisector of interior $\angle C$ cut AB at E.

fig. C: 156 (i)

Then E divides AB internally in the ratio $EA : EB = x : y$.

Construction 2. [fig. C : 156 (ii)]

   (i) Construct AC, BC as for Const. 1.

   (ii) Produce AC and let the bisector of the exterior $\angle C$ cut AB produced at F.

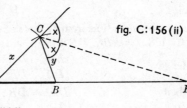

fig. C: 156 (ii)

Then F divides AB externally in the ratio $FA : FB = x : y$.

Proof: (Theorem 43.)   Q.E.F.

FOURTH PROPORTIONAL

If $a : b = c : d$, then a, b, c, d are in proportion and d is said to be the **fourth proportional** to a, b and c.

If $a : b = b : c$, then a, b, c are said to be in **continued proportion** and c is the **third proportional** to a and b.

CONSTRUCTION 27

To construct a fourth proportional to three given straight lines.

Given: a, b, c are three straight lines.

To construct: A fourth proportional d to a, b, c such that $a : b = c : d$.

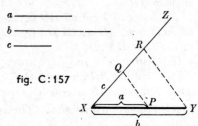

fig. C:157

Construction:

(i) Draw straight line XY equal to b.

(ii) From XY cut off XP equal to a. (Then $XP : XY = a : b$.)

(iii) At X, construct a convenient angle ZXY.

(iv) From XZ cut off XQ equal to c. Join PQ.

(v) Through Y construct $RY \| PQ$, letting YR cut XZ at R.

Then the required fourth proportional d is given by XR.

Proof: Since $PQ \| RY$, $\dfrac{XP}{XY} = \dfrac{XQ}{XR}$

Then $\dfrac{a}{b} = \dfrac{c}{XR}$

And XR is the fourth proportional to a, b, c Q.E.F.

MEAN PROPORTIONAL

In such a case as $a : b = b : c$, b is said to be the **mean proportional** between a and c.

If the proportion is expressed as $\dfrac{a}{b} = \dfrac{b}{c}$ we may cross-multiply and obtain: $b^2 = a \cdot c$

Let us consider Theorem 42 once again. From fig. C:158, $\triangle$s ABC, DBA, DAC are similar.

Hence $\dfrac{DC}{DA} = \dfrac{DA}{DB}$

fig. C:158

Then $DA^2 = DB \cdot DC$ ———————— (i)

DB and DC are called the **segments** of the line BC and the result (i) may be expressed as a corollary to Theorem 42.

COROLLARY 1 (THEOR. 42)

If a perpendicular is drawn from the right-angle of a right-angled triangle to the hypotenuse, the square on the perpendicular is equal to the rectangle contained by the segments of the hypotenuse (i.e., the product of the segments).

Also from fig. C : 158 we obtain the following:

$$\frac{BC}{BA} = \frac{BA}{BD} \quad \text{and} \quad \frac{CD}{CA} = \frac{CA}{CB}$$

Hence
$$BA^2 = BC \cdot BD \quad \text{——————} \quad \text{(ii)}$$
$$CA^2 = CB \cdot CD \quad \text{——————} \quad \text{(iii)}$$

These conclusions may be expressed in a second corollary.

COROLLARY 2 (THEOR. 42)

If a perpendicular is drawn from the right-angle of a right-angled triangle to the hypotenuse, the square on either of the other sides is equal to the rectangle contained by the hypotenuse and the segment adjacent to that side.

Bearing in mind that $a : b = b : c$ reduces to $b^2 = a \cdot c$ and b is the mean proportional between a and c, we may employ either of the preceding corollaries to enable us to construct a mean proportional to two given lines.

CONSTRUCTION 28

To construct a mean proportional to two given straight lines.

Given: a and c are two straight lines.

To construct: A mean proportional b such that $a : b = b : c$.

Construction:

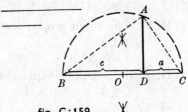

fig. C:159

    (i) Draw a straight line BD equal to c and produce to C so that DC is equal to a.

    (ii) Bisect BC at O.

    (iii) With O as centre and BC as diameter, describe a semicircle.

    (iv) Erect a perpendicular at D to cut the semicircle at A. Then the required mean proportional b is given by AD.

Proof:

$$\angle BAC = 90° \qquad (\angle \text{ in semicircle})$$
$$\therefore \triangle ABC \text{ is right-angled at } A$$
$$\therefore AD^2 = DC \cdot DB \quad (\text{Th. 42, Cor. 1.})$$
$$\therefore AD^2 = a \cdot c$$

And
$$\frac{a}{AD} = \frac{AD}{c}$$

$\therefore AD$ is the mean proportional between a and c.   Q.E.F.

EXERCISE 37

1. In $\triangle XYZ$, YX is produced to A. XB, the bisector of $\angle AXZ$, meets YZ produced at B. Prove that XB divides YZ externally in the ratio $XY : XZ$.

2. X is a point on side PQ of $\triangle PQR$ such that $XP = XR$. XY is parallel to PR meeting QR at Y. Prove that $YQ : YR = XQ : XR$.

3. In a right-angled triangle ABC, $AC = 6$ cm, $BC = 8$ cm. If AX, the bisector of $\angle A$, meets BC at X, calculate BX and CX.

4. In a right-angled triangle PQR, $PQ = 3 \cdot 25$ in, $PR = 3$ in. If PX, the bisector of $\angle P$, meets QR at X, calculate QX and RX. Test your calculations by a careful drawing.

5. $ABCD$ is a cyclic quadrilateral in which $AB = BC$. If the diagonals AC and BD intersect at P, prove that $PA : PC = DA : DC$.

6. $PQRS$ is a cyclic quadrilateral in which $\angle P = \angle Q$. PR and QS intersect at T and O is the centre of the circle. If OT is produced to meet SP produced at X, prove that $TP : TS = XP : XS$.

7. AB is a straight line 4 cm long. Using a straight-edge and compasses only, divide AB externally at F in the ratio $FA : FB = 5 : 3$. Measure FB and check your result by calculation.

8. AB is a straight line 5 cm long. Divide AB internally at E so that $EB : EA = 3 : 7$ and externally at F so that $FB : FA = 3 : 7$. Measure EF and check your result by calculation.

9. If $a : b = c : d$, prove:

(i) $d = \dfrac{bc}{a}$ (ii) $c = \dfrac{ad}{b}$

(iii) $d(a + b) = b(c + d)$ (iv) $ad(1 + b) = b(c + ad)$

10. Find the fourth proportional to:

(i) 3, 5, 9   (ii) $2 \cdot 5$, 6, 10

(iii) a^2b, bc^2, a^2 (iv) ay, bx, a^3y

11. Find the third proportional to:

(i) 2, 3   (ii) 4, 5   (iii) $\frac{1}{3}$, $\frac{1}{4}$ (iv) x^2, bx

12. Find the mean proportional between:

(i) 4, 16   (ii) 12, $16\frac{1}{3}$ (iii) 8, 25   (iv) x^3yz^2, xyz^2

13. Construct a fourth proportional to three straight lines given $a = 4$ cm, $b = 5$ cm, $c = 6$ cm. Describe your construction and check the result by calculation.

14. Construct a fourth proportional to three straight lines given $a = 5$ cm, $b = 6$ cm, $c = 8$ cm. Prove your construction and check the result by calculation.

15. Construct a third proportional to two straight lines given $a = 4$ cm, $b = 6$ cm. Check your result by calculation.

16. Construct a mean proportional to two straight lines given

$a = 2$ cm, $c = 8$ cm. Describe your construction and check your result by calculation.

17. Construct a mean proportional to two straight lines given $a = 5$ cm, $c = 9\cdot8$ cm. Prove your construction and check your result by calculation.

18. Triangle ABC is right-angled at A. AD is the perpendicular from A to BC. Prove: (i) $DA^2 = DB \cdot DC$; (ii) $CA^2 = CB \cdot CD$.

19. $ABCD$ is a rectangle and Y is a point on CD such that $\angle AYB$ is a right-angle. Prove that $YD \cdot YC = AD \cdot BC$.

20. $ABCD$ is a rectangle and Y is a point on AB such that $\angle DYC$ is a right-angle. Prove that AD is a mean proportional between AY and BY.

21. YZ is a tangent to a circle at Z and XZ is a diameter. Chord XP is produced to meet YZ at Y. Prove that PZ is a mean proportional between PX and PY.

22. $ABCD$ is a trapezium in which $AB\|CD$. The diagonal $AC \perp AD$ and $\angle B = 90°$. Prove that AC is a mean proportional between AB and CD.

SEGMENTS OF CHORDS
THEOREM 44

If two chords of a circle intersect either inside or outside the circle, the rectangle contained by the segments of the one is equal to the rectangle contained by the segments of the other.

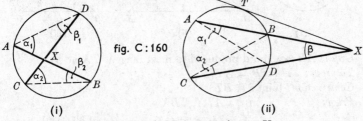

fig. C : 160

(i)   (ii)

Given: Two chords AB and CD intersecting at X:
(i) inside the circle; (ii) outside the circle.

To prove: In each case $XA \cdot XB = XC \cdot XD$.

Construction: Join AD, BC.

Proof: In $\triangle$s AXD, CXB
$\angle \alpha_1 = \angle \alpha_2$ ($\angle$s in same segm.)
In (i) $\angle \beta_1 = \angle \beta_2$ (ditto   )
In (ii) $\angle \beta$ is common
In each case the third angles must be equal.

$\therefore$ $\triangle$s AXD, CXB are equiangular and similar.

$\therefore \dfrac{XA}{XC} = \dfrac{XD}{XB}$ $\therefore XA \cdot XB = XC \cdot XD$

Q.E.D.

288 PART C: GEOMETRY

CONVERSE

If two straight lines AB and CD intersect at X such that $XA . XB = XC . XD$, then points A, B, C, D are concyclic.

NOTE

(i) These propositions are probably best remembered by referring to the point of intersection first. I.e., $XA . XB = XC . XD$

(ii) [Fig. (ii).] As the chord AB becomes shorter, A and B will eventually coincide and ABX will become the tangent TX. In this case XA and XB will each be equal to TX and $XA . XB = XT^2$

$$\therefore XT^2 = XC . XD \text{ (see Theor. 45)}$$

THEOREM 45

If, from any point outside a circle, a secant and a tangent are drawn, the rectangle contained by the whole secant and the part of it outside the circle is equal to the square on the tangent.

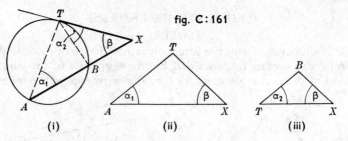

fig. C:161

(i) (ii) (iii)

Given: AB is a chord produced to meet tangent TX at X.
To prove: $XA . XB = XT^2$.
Construction: Join AT, BT.
Proof: In $\triangle$s ATX, TBX
$$\angle \alpha_1 = \angle \alpha_2 \qquad (\angle \text{ in alt. segm.})$$
$$\angle \beta \text{ is common}$$
$$\therefore \angle ATX = \angle TBX \quad (3rd \angle \text{ of } \triangle)$$
$\therefore \triangle$s ATX, TBX are equiangular and similar.
$$\therefore \frac{XT}{XB} = \frac{XA}{XT}$$
$$\therefore XA . XB = XT^2 \qquad\qquad \text{Q.E.D.}$$

CONVERSE

If two straight lines XBA and XT intersect at X such that $XA . XB = XT^2$, then points A, B, T are concyclic and XT is a tangent to the circle.

CONSTRUCTION 29

To construct a square equal in area to a given rectangle.

Given: Rectangle *ABCD*.

To construct: A square equal in area to *ABCD*.

Construction:

(i) Produce *AB* to *E*, making *BE = BC*.

(ii) With centre *O* and *AE* as diameter, describe a semicircle.

(iii) Produce *CB* to cut the semicircle at *X*.

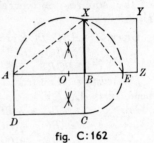

fig. C:162

(iv) Construct a square *BXYZ*. Then *BXYZ* is the required square.

Proof: △*AXE* is right-angled at *X*.   (∠ in semicircle)

$$BX \perp AE \quad (ABCD \text{ a rectangle})$$
$$\therefore BX^2 = BA \cdot BE \quad (\text{Theor. 42})$$
$$\text{But} \quad BE = BC \quad (\text{const.})$$
$$\therefore BX^2 = BA \cdot BC$$
$$\therefore \text{Area } BXYZ = \text{area } ABCD \quad \text{Q.E.F.}$$

CONSTRUCTION 30

To construct a square equal in area to a given polygon.

Given: ABCDE is a polygon.

To construct: A square equal in area to *ABCDE*.

Construction:

Stage 1. Reduce *ABCDE* to △*XBY*, equivalent in area. (Const. 10.)

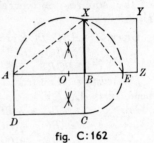

fig. C:163(i)

Stage 2. Area of △ = ½ Base × Height.

(i) Erect *BP* ⊥ *XY*

(ii) Bisect base *XY* at *Z*.

(iii) Area of △*BXY* = *XZ . BP*

fig. C:163(ii)

Stage 3.

(i) Construct a rectangle of area *XZ . BP*

(ii) By Const. 29 produce *KLMN* = *XZ . BP*

(iii) Then area *KLMN* = area *ABCDE*
     Q.E.F.

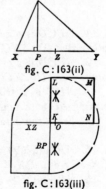

fig. C:163(iii)

THEOREM 46

The ratio of the areas of similar triangles is equal to the ratio of the squares on corresponding sides.

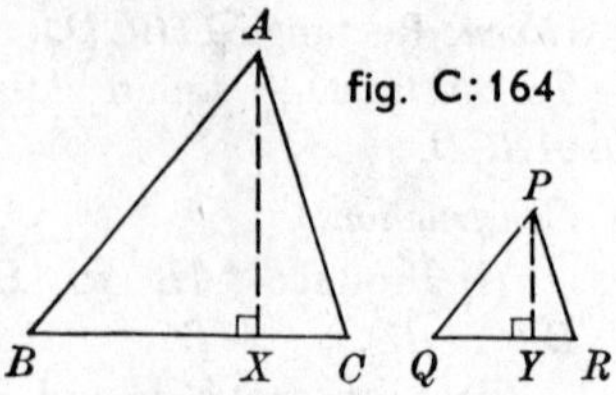

Given: $\triangle$s ABC, PQR are similar.

To prove: $\dfrac{\triangle ABC}{\triangle PQR} = \dfrac{BC^2}{QR^2}$

fig. C:164

Construction: Construct the perpendiculars AX, PY.

Proof: $\triangle ABC = \frac{1}{2}(BC \cdot AX)$ and $\triangle PQR = \frac{1}{2}(QR \cdot PY)$

$$\frac{\triangle ABC}{\triangle PQR} = \frac{\frac{1}{2}(BC \cdot AX)}{\frac{1}{2}(QR \cdot PY)} = \frac{BC \cdot AX}{QR \cdot PY}$$

In $\triangle$s AXB, PYQ

$$\begin{aligned}
\angle B &= \angle Q & \text{(similar } \triangle\text{s)} \\
\angle AXB &= \angle PYQ & \text{(rt. } \angle\text{s, const.)} \\
\therefore \ \angle BAX &= \angle QPY & \text{(3rd } \angle \text{ of } \triangle\text{)}
\end{aligned}$$

$\therefore$ $\triangle$s AXB, PYQ are equiangular and similar.

$$\therefore \ \frac{AX}{PY} = \frac{AB}{PQ}$$

But $\dfrac{AB}{PQ} = \dfrac{BC}{QR}$ ($\triangle$s ABC, PQR similar)

$$\therefore \ \frac{AX}{PY} = \frac{BC}{QR}$$

Then $\dfrac{\triangle ABC}{\triangle PQR} = \dfrac{BC \cdot AX}{QR \cdot PY} = \dfrac{BC \cdot BC}{QR \cdot QR}$

$$\therefore \ \frac{\triangle ABC}{\triangle PQR} = \frac{BC^2}{QR^2}$$

                                           Q.E.D.

NOTE

Linear ratio $= \dfrac{BC}{QR}$ Area ratio $= \left(\dfrac{BC}{QR}\right)^2 = $ (Linear ratio)^2

EXERCISE 38

1. $ABCD$ is a cyclic quadrilateral. If the diagonals intersect at X, prove that $XA \cdot XC = XB \cdot XD$.

2. $ABCD$ is a cyclic quadrilateral. If the diagonals intersect at X, prove that $AB \cdot XC = CD \cdot XB$ and also that $AB \cdot XD = CD \cdot XA$.

3. ABC is a triangle. A circle is drawn with AB as diameter cutting AC at X and BC at Y. Prove that $CA \cdot CX = CB \cdot CY$.

4. From the data given in Question 3, prove that
$$CA \cdot BX = CB \cdot AY.$$

5. AB and CD are two chords of a circle intersecting at X. If $CX = 5$ cm, $DX = 3$ cm, $BX = 4$ cm, calculate AX.

6. AB and CD are two chords of a circle intersecting at X. If

$AX = 8$ cm, $BX = 3$ cm, $CD = 10$ cm, calculate CX and DX.

7. AB and CD are two chords of a circle intersecting at X. If $AX = 8$ cm, $BX = 4$ cm, $CD = 11\cdot4$ cm, calculate CX and DX.

8. ABC is a triangle in which $AH \perp BC$ and $BK \perp AC$. If AH and BK intersect at X, prove that $AH \cdot AX = AC \cdot AK$.

9. From the data of Question 8, prove that $BK \cdot BX = BC \cdot BH$.

10. Two circles intersect at A and B. BA is produced to P from which PXY cuts one circle at X, Y and PQR cuts the other circle at Q, R. Prove that $PY \cdot PX = PR \cdot PQ$.

11. Two circles intersect at X and Y. If XY is produced to P and a tangent to each circle is drawn from P, prove the tangents are equal.

12. From the data of Question 10, prove that points Y, X, Q, R are concyclic.

13. From the data of Questions 10 and 12, construct the circle passing through Y, X, Q, R.

14. From the data of Question 11, prove that the two tangents are tangents to a common circle. Hence construct the common circle.

15. Two circles intersect at P and Q. If XY is a common tangent to the circles at X and Y, prove that PQ produced bisects XY.

16. PQ and PR are tangents from P to a circle, centre O. If PO cuts QR at X, prove that $OQ \cdot OR = OP \cdot OX$.

17. AB is a diameter of a circle and chord $PXQ \perp AB$ at X. Prove that $XP \cdot XQ = XA \cdot XB$ and hence that $XP^2 = XA \cdot XB$.

18. $ABCDE$ is a regular polygon on AE as base and of side 2 in. By producing AE both ways, to X and Y, construct $\triangle XCY$ equal in area to the polygon. Calculate the area of the polygon. Measure XY and the altitude CH from C to XY. Check the area of the triangle by calculation.

19. Construct a square equal in area to a rectangle 4 cm by 3 cm. Measure the side of the square and check your result by calculation.

20. Construct a square equal in area to a rectangle 9 cm by $4\cdot7$ cm. Describe your construction and measure the side of the square.

21. Construct a square equal in area to a rectangle $3\cdot2$ in by $1\cdot5$ in. Prove your construction and measure the side of the square.

22. $ABCD$ is an irregular cyclic quadrilateral inscribed in a circle of radius 2 in. Construct a square equal in area to $ABCD$.

23. ABC is an equilateral triangle of side 4 cm. Construct a square equal in area to $\triangle ABC$. Prove your construction and measure the side of the square.

24. AB and CD are two parallel chords. If AD and BC intersect at X and $AB = \frac{1}{2}CD$, prove that $\triangle AXB = \frac{1}{4}\triangle CXD$.

25. A tangent to a circle at T meets chord AB produced at X. If the tangent $XT = 3$ in, and chord $AB = 2\frac{1}{2}$ in, prove that $\dfrac{\triangle BXT}{\triangle TXA} = \dfrac{4}{9}$.

THEOREM 47 (*Ext. of Pyth. Theor.*)

In an obtuse-angled triangle, the square on the side opposite the *obtuse* angle is equal to the sum of the squares on the other two sides *plus* twice the product of one of these sides and the projection on it of the other.

Given: $\triangle ABC$, in which $\angle A$ is obtuse, CD is $\perp$ to BA produced and AD is the projection of AC on BA.

To prove:

$BC^2 = AC^2 + AB^2 + 2AB \cdot AD$.

Proof: Let BC, AC, AB, AD, CD be a, b, c, x, h units respectively.

It is required to prove $a^2 = b^2 + c^2 + 2cx$

In $\triangle BDC$, $\angle D = 90°$

$\therefore a^2 = (c + x)^2 + h^2$ (Pythagoras' Th.)

$\therefore a^2 = c^2 + 2cx + \underline{x^2 + h^2}$

In $\triangle ADC$, $\angle D = 90°$

$\therefore b^2 = \underline{x^2 + h^2}$ (Pythagoras' Th.)

$\therefore a^2 = c^2 + 2cx + b^2$

or $a^2 = b^2 + c^2 + 2cx$

i.e. $BC^2 = AC^2 + AB^2 + 2AB \cdot AD$

                          Q.E.D.

fig. C:165

NOTE

(i) Since $x = b \cos \theta$, then $x = -b \cos \alpha$.

$\therefore a^2 = b^2 + c^2 - 2bc \cdot \cos A$

(See Trigonometry: Cosine Formula, p. 322.)

(ii) In $\triangle ABC$, when $BC^2 > AC^2 + AB^2$, then $\angle A > 90°$.

THEOREM 48 (*Ext. of Pyth. Theor.*)

In any triangle, the square on the side opposite an *acute* angle is equal to the sum of the squares on the other two sides *minus* twice the product of one of these sides and the projection on it of the other.

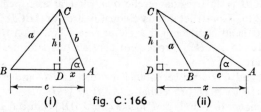

(i)   fig. C:166   (ii)

Given: $\triangle ABC$, in which $\angle A$ is acute. CD is $\perp$ to AB or AB produced and AD is the projection of AC on AB.

To prove: $BC^2 = AC^2 + AB^2 - 2AB \cdot AD$.

Proof: Let BC, AC, AB, AD, CD be a, b, c, x, h units respectively.

It is required to prove $a^2 = b^2 + c^2 - 2cx$

In $\triangle BDC$, $\angle D = 90°$

$\therefore$ In fig. (i), $a^2 = (c - x)^2 + h^2$ $\left.\begin{array}{l}\\\\\end{array}\right\}$ {Pythagoras' Th.}
In fig. (ii), $a^2 = (x - c)^2 + h^2$

In each case $a^2 = c^2 - 2cx + \underline{x^2 + h^2}$

In $\triangle ADC$, $\angle D = 90°$

$\qquad \therefore b^2 = \underline{x^2 + h^2}$ (Pythagoras' Th.)

$\qquad \therefore a^2 = c^2 - 2cx + b^2$

or $\qquad a^2 = b^2 + c^2 - 2cx$

i.e. $\quad BC^2 = AC^2 + AB^2 - 2AB \cdot AD$ Q.E.D.

NOTE

(i) Since $x = b \cos \alpha$, $a^2 = b^2 + c^2 - 2bc \cdot \cos A$
(See Trigonometry: Cosine Formula, p. 322.)

(ii) In $\triangle ABC$, when $BC^2 < AC^2 + AB^2$, then $\angle A < 90°$.

THEOREM 49 (*Apollonius' Theor.*)

In any triangle, the sum of the squares on any two sides is equal to twice the square on half the third side *plus* twice the square on the median which bisects the third side.

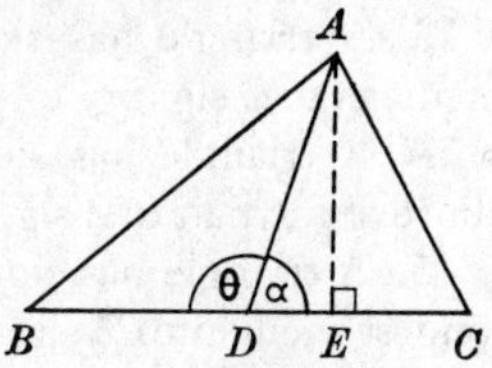

Given: $\triangle ABC$ in which D is the mid-point of BC and AD is a median.

To prove: $AB^2 + AC^2 = 2BD^2 + 2AD^2$.

Construction: Construct $AE \perp BC$.

Proof: Let $\angle \theta$ be obtuse and $\angle \alpha$ be acute.

fig. C:167

In $\triangle ABD$, since $\angle \theta$ is obtuse:

From TH. 47: $\qquad AB^2 = AD^2 + BD^2 + 2BD \cdot DE$ ——— ①

In $\triangle ADC$, since $\angle \alpha$ is acute:

From TH. 48: $\qquad AC^2 = AD^2 + CD^2 - 2CD \cdot DE$ ——— ②

But $\qquad BD = CD$

$\qquad \therefore BD^2 = CD^2$ and $BD \cdot DE = CD \cdot DE$

By adding ① and ②:

$\qquad AB^2 + AC^2 = 2BD^2 + 2AD^2$ Q.E.D.

EXERCISE 39

1. Find by calculation whether the following triangles are obtuse-, acute-, or right-angled:

(i) 3, 4, 5   (ii) 5, 6, 7   (iii) 6, 10, 15   (iv) 15, 36, 39
(v) 5, 7, 11   (vi) 7, 11, 13   (vii) 7, 12, 17   (viii) 5, 15, 16

2. *ABC* is an acute-angled triangle. If *AB* = 8 in, *BC* = 16 in, what is the maximum length of *AC* if it is to be an integral number of inches?

3. *ABC* is an acute-angled triangle. If *AB* = 15 cm, *BC* = 20 cm, prove that *AC* $\angle$ 25 cm.

4. In fig. C : 165, *AB* = 10 cm, *BC* = 13 cm, *AC* = 7 cm. Calculate *AD* and *CD* to 2 sig. fig.

5. In fig. C : 165, *AB* = 7 cm, *BC* = 13 cm, *AC* = 10 cm. Calculate *AD* and *CD* to 2 sig. fig.

6. In fig. C : 166, *AB* = 12 cm., *BC* = 10 cm., *AC* = 8 cm. Calculate *AD* and *CD* to 2 sig. fig.

7. In fig. C : 166, *AB* = 12 cm, *BC* = 8 cm, *AC* = 10 cm. Calculate *AD* and *CD* to 2 sig. fig.

8. In fig. C : 166, *AB* = 16 cm, *BC* = 18 cm, *AC* = 30 cm. Calculate *AD*, *BD* and *CD* to 2 sig. fig.

9. In $\triangle ABC$, *AB* = 5 cm, *AC* = 12 cm, *BC* = 16 cm. Calculate the area of the triangle to 3 sig. fig.

10. In $\triangle ABC$, *AB* = 10 cm, *AC* = 8 cm, *BC* = 9 cm. Calculate the area of the triangle to 3 sig. fig.

11. Calculate the length of the projection of the shortest side on the longest side in a triangle whose sides are 8, 12, 14 in.

12. Calculate the length of the projection of the longest side on the shortest side in a triangle whose sides are 24, 10, 15 in.

13. A triangle has sides 4, 6, 8 cm. Calculate the lengths of the medians to 2 sig. fig.

14. A triangle has sides, 6, 8, 10 cm. Calculate the length of the shortest median to 2 sig. fig.

15. A triangle has sides of 5, 12, 13 cm. Calculate the length of the longest median to 2 sig. fig.

16. Two sides of a triangle are 4 in, 6 in and the median to the third side is 5 in. Find the length of the third side. From your result construct the triangle. What conclusion can be reached?

17. Two sides of a triangle are 6 in, 12 in and the median to the third side is 8 in. Find the length of the third side to 2 sig. fig.

18. A parallelogram has sides 8 in, 10 in and one diagonal is 6 in. Find the length of the other diagonal to 2 sig. fig.

19. A parallelogram has one side of 10 cm and the diagonals are 10 cm and 12 cm. Find the length of the other side to 2 sig. fig.

20. The medians of a triangle are 3 cm, 4 cm, 5 cm. Find the lengths of the sides to 2 sig. fig. in each case.

21. In $\triangle ABC$, *BC* = *AC* and *CD* $\perp$ *AB*. Prove $AB^2 = 2AB \cdot AD$.

22. In $\triangle ABC$, *BE* $\perp$ *AC* and *CD* $\perp$ *AB*. Prove $AB \cdot AD = AC \cdot AE$.

23. In equilateral $\triangle ABC$, *AC* is produced to *D* so that *AC* = *CD*. Prove that $BD^2 = 3BC^2$.

24. BC is the diameter of a circle and A is a point outside the circle. AB and AC cut the circle at D and E respectively. Prove that $AC \cdot AE = AB \cdot AD$.

25. In $\triangle ABC$, $AB = AC$ and D is a point on BC. Prove $AB^2 = AD^2 + BD \cdot CD$.

26. ABC is an obtuse-angled triangle in which $AB = AC$ and AD is the projection of AB on CA produced. If $AC^2 = \frac{1}{3}BC^2$, prove that $AD = \frac{1}{2}AC$.

27. In $\triangle ABC$, $\angle A > 90°$, $AF \perp BC$ and $CE \perp BA$ produced. Prove that $AC^2 = BC \cdot CF - AB \cdot AE$.

28. In $\triangle ABC$, $AB = AC$ and BC is produced to D so that $CD = BC$. Prove that $AD^2 = 2BC^2 + AC^2$.

29. Prove that twice the sum of the squares on two adjacent sides of a parallelogram is equal to the sum of the squares on the two diagonals.

30. AF, BD, CE are the medians of $\triangle ABC$. Prove that

$$AF^2 + BD^2 + CE^2 = \tfrac{3}{4}(AB^2 + AC^2 + BC^2).$$

31. In $\triangle ABC$, AD is a median and $\angle C = 90°$. Prove that $\frac{1}{2}BC^2 = BD^2 + AD^2 - AC^2$.

32. $ABCD$ is a quadrilateral in which $AC = CD$, AE and DF are each perpendicular to BC such that $BF < BE$. Prove that $AB^2 = BD^2 + 2BC \cdot EF$.

PART D: TRIGONOMETRY

The Greek alphabet is given to assist you (small letters only), the characters in bold type are generally the most common in science and mathematics.

α	alpha	ι	iota	ρ	**rho**
β	**beta**	κ	kappa	σ	**sigma**
γ	**gamma**	λ	**lambda**	τ	**tau**
δ	**delta**	μ	**mu**	υ	upsilon
ε	epsilon	ν	nu	φ	**phi**
ζ	zeta	ξ	xi	χ	chi
η	eta	ο	omicron	ψ	psi
θ	**theta**	π	**pi**	ω	omega

THE TANGENT RATIO

The **tangent ratio** is always given by $\dfrac{\textbf{Opposite}}{\textbf{Adjacent}}$ and it is most important that the sides are **named with regard to the angle under consideration.**

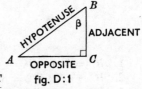

fig. D:1

Consider fig. D : 1. TAN $\beta = \dfrac{\text{OPP.}}{\text{ADJ.}} = \dfrac{AC}{BC}$

EXERCISE 1

Find the tangent ratio of these angles:

1. 15°	**2.** 29°	**3.** 34°	**4.** 39°	**5.** 44°
6. 57°	**7.** 65°	**8.** 75°	**9.** 80°	**10.** 17° 12′
11. 23° 24′	**12.** 37° 42′	**13.** 48° 54′	**14.** 59° 18′	**15.** 63° 24′
16. 63° 30′	**17.** 71° 30′	**18.** 71° 36′	**19.** 78° 36′	**20.** 78° 42′
21. 20° 40′	**22.** 32° 32′	**23.** 41° 41′	**24.** 49° 3′	**25.** 52° 47′
26. 63° 5′	**27.** 67° 49′	**28.** 70° 45′	**29.** 1° 5′	**30.** 0° 8′

Find the angles whose tangent ratios are:

31. 0·0000	**32.** 0·3739	**33.** 0·9965	**34.** 1·1960	**35.** 1·4659
36. 1·8190	**37.** 2·0872	**38.** 3·6554	**39.** 0·1718	**40.** 0·2749
41. 0·4831	**42.** 0·6780	**43.** 0·8205	**44.** 0·9890	**45.** 1·7651
46. 2·6559	**47.** 0·2842	**48.** 0·6069	**49.** 1·7873	**50.** 5·0737

USING THE TANGENT RATIO: PROBLEMS

Elevation is the **raising** of a line of sight, depression is the **lowering** of a line of sight; each taken from the **horizontal.**

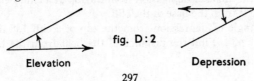

fig. D:2

Elevation   Depression

297

EXAMPLE. *The top of a ladder touches a wall at a point 30 dm above the ground. If the ladder is inclined at an angle of 59° 2′ to the ground, find the distance from the foot of the ladder to the wall.*

$$\text{TAN } \alpha = \frac{\text{OPP.}}{\text{ADJ.}}$$

$$\therefore \text{TAN } 59° 2′ = \frac{30}{YZ}$$

$$\therefore 1\cdot6666 \triangle YZ = 30$$

$$\therefore YZ = \frac{30}{1\cdot6666} *$$

∴ Distance = 18 dm (*corr. to* 1 *dm.*)

fig. **D:3**

The disadvantage with this method is the calculation*. This can be avoided by using the angle at X, i.e., $90° - 59° 2′$.

EXERCISE 2

Questions 1–10 *refer to fig.* D : 4.

1. $AC = 5$ cm, $BC = 4$ cm, find $\angle \alpha$ and $\angle \beta$.

2. $AC = 6$ ft, $BC = 8$ ft, find $\angle \alpha$ and $\angle \beta$.

3. $AC = 9$ m, $BC = 4$ m, find $\angle \alpha$ and $\angle \beta$.

4. $AC = 45$ ft, $BC = 16$ ft, find $\angle \alpha$ and $\angle \beta$.

5. $BC = 18$m, $\angle \beta = 30°$, find AC to three figures.

6. $BC = 30$ ft, $\angle \beta = 45°$, find AC to three figures.

fig. **D:4**

7. $BC = 28$ m, $\angle \beta = 50° 16′$, find AC to three figures.

8. $BC = 16\cdot25$ ft, $\angle \beta = 38° 46′$, find AC to three figures.

9. $BC = 32\cdot5$ m, $\angle \alpha = 16° 27′$, find AC to three figures.

10. $BC = 386$ ft, $\angle \alpha = 54° 32′$, find AC to three figures.

11. The angle of elevation of the sun is $35° 50′$. Find the height of a tree whose shadow is 63 dm.

12. If an 80-ft building casts a 47-ft shadow, find the angle of elevation of the sun.

13. The sun's elevation is $72° 16′$. Find the length of shadow cast by a building 76 ft high.

14. From the top of a cliff the angle of depression to a boat, 1000 m from the cliff, is $4° 29′$. Find the height of the cliff.

15. Find the angle of depression from the top of a cliff, height 167 ft, to a boat 876 ft from the base of the cliff.

16. The angle of depression from the top of a cliff 138 dm high to a boat is $17° 43′$. Find the distance of the boat from the base of the cliff.

USING THE TANGENT RATIO: PROBLEMS 299

17. A ladder, whose foot is 8 ft from a wall, makes an angle of 63° 26′ with the ground. If the top of the ladder slides 1 ft down the wall the angle with the ground is now 56° 58′. Find the distance of the foot of the ladder from the wall.

18. The top of a ladder rests against a wall 18·5 dm above the ground, the ladder making an angle of 67° 56′ with the ground. If the foot of the ladder is pulled a further 1·5 dm from the wall the angle decreases by 4° 44′. How far does the ladder slide down the wall?

19. From a horizontal distance of 68 ft the angle of elevation to the top of a tree is 21° 39′. The observer moves back to a position from which the angle of elevation is decreased by 11° 20′. What distance did the observer move?

20. A tree is 32 dm high. From a certain position the angle of elevation to the top of the tree is 50°. By moving a further 5·15 dm from the tree a new angle of elevation is obtained. What is the change in the angle of elevation?

21. The angle of elevation of the top of a tower from a point 50 dm away on level ground is 50°. Find the height of the tower. (To three figures.)

22. The angle of depression at the top of a vertical cliff to a boat 3000 ft from the base of the cliff is 10° 12′. Find, correct to three figures, the height of the cliff.

23. Two trees, on opposite banks of a river, are in line forming an angle of 90° with the river. At a distance of 75 m from the tree on one bank, the angle of sight between the bank and the second tree is 50° 12′. Find, correct to three figures, the width of the river.

24. An isosceles triangle of base 7 cm has base angles of 36° 30′. Find, correct to three figures, the perpendicular height of the triangle. Hence, find the area of the triangle, correct to three figures.

25. The angle between the diagonal and the longer side of a rectangular field is 35° 4′. If the longer side measures 50 yd, find the area of the field in square yards. (Correct to three figures.)

26. An aircraft, attempting to land, is estimated to be 400 ft above the approach to the runway. Calculate the angle of elevation of the aircraft as seen from a runway control vehicle 3000 ft away at the other end of the runway.

27. The foot of a ladder is placed 6 ft from the base of a vertical wall. If the ladder touches the wall at a point 25 ft above the ground, find the angle of elevation of the ladder.

28. If the ladder in Question 27 is pulled an extra foot away from the wall at the base the angle of elevation becomes 74° 12′. Find how far the top of the ladder slides down the wall. (Correct to two figures.)

THE COTANGENT

It may frequently be found advantageous to employ the **reciprocal of the tangent ratio,** this is known as the **cotangent** and is given by:

$$\text{COTANGENT } \theta \ (\text{COT } \theta) = \frac{\text{ADJ.}}{\text{OPP.}}$$

EXAMPLE. *At what distance from the base of a tower* 50 *ft. high will the angle of elevation of the top be* 31° 33′?

$$\text{TAN } 31° 33′ = \frac{50}{AC}$$

$$\therefore \ \text{COT } 31° 33′ = \frac{AC}{50}$$

Cross multiplying: $AC = 1 \cdot 6287 \times 50$ *

$$\therefore \ AC = 81 \cdot 435$$

$\therefore \ \underline{\text{Distance} = 81 \cdot 4 \text{ ft}}$ *(Corr. to 3 fig.)*

fig. **D:5**

NOTE

(i) The stage marked * shows clearly that the cotangent produces an easier calculation.

(ii) You are advised to begin a solution by setting down the **tangent** ratio. If the unknown appears as the denominator of the fraction on the R.H.S., rewrite the data using the **cotangent.**

N.B. (iii) The figures in the difference columns of the cotangent tables are **subtracted.** The reason is explained below.

MORE ABOUT TAN AND COT

(i) Since the **tangent** ratios **increase** as β increases, we must **add** the figures in the difference columns. (Fig. D : 4.)

(ii) Since the **cotangent** ratios **decrease** as β increases, we must **subtract** the figures in the difference columns.

If we consider angle α we shall see that α and β are complementary (their sum is 90°); therefore as β increases, α decreases. Thus, tan α will be the same as cot β.

Verify from your tables that:

$$\text{COT } 31° 33′ = \text{TAN } 58° 27′$$

i.e., **co**tangent of a given angle = tangent of the **co**mplement

EXAMPLE. *Find angle* θ, *given that cot* θ = 0·4133.

Required cot value = 0·4133

Nearest value *(not higher)* = 0·4122 for 67° 36′

Difference = 0·0011 for 0° 3′ $\left(\begin{array}{c}\text{Diff. of } 10 \\ \text{is nearest}\end{array}\right)$

$$\therefore \ \underline{\text{Angle } \theta = 67° 33′}$$

NOTE. The angle obtained from the difference columns is **subtracted.**

EXERCISE 3

Use the most convenient method and give answers correct to three figures:

From tables give the cotangents of the following angles: (4 *figs.*)

1. 30° 30'   **2.** 51° 52'   **3.** 70° 9'   **4.** 79° 41'   **5.** 83° 33'

Find the angles whose cotangents are:

6. 0·2068   **7.** 0·4529   **8.** 0·7799   **9.** 0·9201   **10.** 1·5840

11. At what distance from the base of a tower 100 ft high will the angle of elevation be 20°?

12. How far is a boat from the base of a cliff 50 m high if the angle of depression of the boat from the top of the cliff is 11° 18'?

13. The base angle of an isosceles triangle is 40° 15' and the perpendicular height is 10 cm. Find the area of the triangle in dm².

14. A man walks from A due North to point B, a distance of 3 miles. At B he turns due East and walks to C, a distance of 1¾ miles. What is the bearing of C from A?

15. From a ship a lighthouse can be seen due South at a distance of 2½ km. A second ship is due West of the lighthouse on a bearing S. 53° 20' W. from the first ship. Find the distance between the lighthouse and the second ship.

16. A ramp is constructed so that there is a vertical rise of 0·48 m for every 15 m along the horizontal. What is the angle of elevation of the ramp?

17. Two buildings are separated by a road 65 ft wide. From the roof of one building, height 60 ft, the angle of elevation of the top of the second building is 23° 52'. Find the height of the second building.

18. If the sun's elevation is 38° 52', find the length of shadow cast by a building 32 m high.

19. ABC is an equilateral triangle of side 2 in inscribed in a circle centre O. (A, B and C lie on the circumference.) Find the distance from O to AB.

20. The angle of elevation from a boat to the top of a vertical cliff 238 ft high is 14° 49'. Find the distance from the boat to the base of the cliff.

21. X is the base of a vertical tower XY, 68 ft high. A and B are positions on opposite sides of XY, AXB forming a straight line.

From A the angle of elevation of Y is 73° 18' and from B, 17° 44'. Find the distance AB.

22. The base diameter of a metal cone is 14·75 cm. If the vertical angle is 127°, find the vertical height.

23. P is the base of a vertical column PQ. L and M are positions on opposite sides of PQ, LPM forming a straight line. L is 128 m from P and the angle of elevation of Q from L is 32° 25'. The angle of elevation of Q from M is 50° 45'. Find the distance from P to M.

THE SINE AND COSINE RATIOS

The **sine ratio** (abbr. sin), is the ratio of the **side opposite** to a given angle and the **hypotenuse.**

The **cosine ratio** (abbr. cos), is the ratio of the **side adjacent** to a given angle and the **hypotenuse.**

Thus:
$$\text{SIN } \theta = \frac{\text{OPP.}}{\text{HYP.}} \qquad \cos \theta = \frac{\text{ADJ.}}{\text{HYP.}}$$

THE BEHAVIOUR OF SIN AND COS

(i) As angle θ increases, sin θ increases but cos θ decreases.

(ii) The figures in the difference columns of the **sine table** are **added**, but in the **cosine table** they are **subtracted.**

(iii) As the names suggest, the sine of a given angle is equal to the cosine of the complement.

$$\text{SIN } \theta = \cos (90° - \theta)$$

Sine of a given angle = **Co**sine of the **co**mplement

Learn: $\sin \theta = \dfrac{\text{Opp.}}{\text{Hyp.}}$, $\cos \theta = \dfrac{\text{Adj.}}{\text{Hyp.}}$, $\tan \theta = \dfrac{\text{Opp.}}{\text{Adj.}}$

[SIN = O/H, COS = A/H, TAN = O/A]

EXERCISE 4

Use tables to find the sines of the following. (*Place a ruler under the required line.*)

1. 5°	**2.** 19°	**3.** 32°	**4.** 53°	**5.** 88°
6. 12° 18′	**7.** 27° 36′	**8.** 41° 48′	**9.** 72° 4′	**10.** 3° 59′
11. 14° 38′	**12.** 34° 27′	**13.** 49° 34′	**14.** 65° 47′	**15.** 84° 22′

Use tables to find the cosines of the following:

16. 7°	**17.** 23°	**18.** 39°	**19.** 56°	**20.** 77°
21. 14° 12′	**22.** 32° 42′	**23.** 47° 24′	**24.** 53° 5′	**25.** 69° 17′
26. 83° 51′	**27.** 4° 52′	**28.** 28° 7′	**29.** 34° 21′	**30.** 46° 46′

Use tables to find the angles whose sines are:

31. 0·2284	**32.** 0·5721	**33.** 0·8131	**34.** 0·9841	**35.** 0·2599
36. 0·5066	**37.** 0·1818	**38.** 0·3312	**39.** 0·4140	**40.** 0·5630

Use tables to find the angles whose cosines are:

41. 0·9641	**42.** 0·8131	**43.** 0·5075	**44.** 0·2602	**45.** 0·1610
46. 0·8203	**47.** 0·7572	**48.** 0·5774	**49.** 0·4900	**50.** 0·3463

From tables verify:

51. sin 30° ⇌ cos 60°   **52.** sin 25° ⇌ cos 65°

53. sin 85° ⇌ cos 5°   **54.** sin 12° 40′ ⇌ cos 77° 20′

55. sin 70° 25′ ⇌ cos 19° 35′   **56.** sin 83° 34′ ⇌ cos 6° 26′

PROBLEMS ON SINE AND COSINE

EXERCISE 5

Give distances correct to three figures.

ABC is a triangle, right-angled at B (fig. D : 6).

Find a and c in Questions 1–10.

1. $b = 10$ in, $\theta = 50°$ 2. $b = 20$ in, $\alpha = 30°$
3. $b = 15$ cm, $\theta = 37°$ 4. $b = 25$ m, $\alpha = 74°$
5. $b = 14$ ft, $\theta = 62° 30'$ 6. $b = 12$ km, $\alpha = 18° 42'$
7. $b = 22$ yd, $\theta = 78° 47'$ 8. $b = 18$ mm, $\alpha = 42° 41'$
9. $b = 32·5$ ft, $\theta = 28° 28'$ 10. $b = 8$ dm 6 cm, $\alpha = 53° 19'$

fig. D:6

Find α and θ in Questions 11–14.

11. $a = 8$ in, $b = 10$ in   12. $b = 2·5$ dm, $c = 18$ cm
13. $b = 34$ ft, $c = 28$ ft   14. $a = 3·8$ km, $b = 4·2$ km.

15. The diagonal of a rectangular field is 700 m long and makes an angle of 25° with the longer side. Find: (a) the length; (b) the breadth; (c) the area, in hectares, of the field.

16. An isosceles triangle has equal sides of 15·5 cm and base angles of 63° 15'. Find: (a) the height; (b) the length of the base; (c) the area, of the triangle.

17. The legs of a pair of dividers, 4 in long, are opened to form an angle of 20° 46'. Find the distance between the points.

18. A kite is flying at a height of 115 dm. If the length of the kite string (presumed taut) is 175 dm, find the angle of elevation of the kite string.

19. A ladder 20 ft long is resting against a vertical wall with the foot of the ladder 10 ft from the base of the wall. Find: (a) the angle of elevation of the ladder; (b) the height up the wall to the point where the ladder rests.

20. AB and AC are tangents to a circle centre O, touching the circle at B and C. If AO is 15·5 cm long and angle BAC is 15° 40', find: (a) the radius of the circle, and (b) the length of AB.

21. The guy-rope of an aerial mast is 65 m long and is inclined to the mast at an angle of 34° 45'. Find: (a) the height of the mast; (b) the distance from the base of the mast at which the rope is anchored to the ground.

22. A road is on an incline of 4° 49' to the horizontal. Find the vertical rise for a kilometre of road. Answer in metres.

23. A man walks $4\frac{1}{2}$ km on a bearing N. 38° 17' E. Find how far he is (a) North, (b) East, of his starting-point.

24. A buoy is due South of a lighthouse at a distance of 5 miles. A ship, due West of the buoy, is 6 miles from the lighthouse. Find the bearing of the ship from the lighthouse.

MISCELLANEOUS PROBLEMS
EXERCISE 6
Give distances correct to three figures.

1. The sloping sides of a bell tent are 28·5 dm long and meet a vertical canvas wall 3 dm high. If the angle at the apex of the tent is 51° 22′, find the height of the tent pole.

2. A man walks 5 km on a bearing N. 30° E. then changes direction to N. 60° E. and walks a further 5 km. Find how far he is now (*a*) North, (*b*) East, of his starting-point, and (*c*) the bearing of his new position from his starting-point.

3. P is 120 ft from A, the base of a vertical mast AB. The angle of elevation of B from P is 21° 48′ and from Q, a position on the opposite side of A, the angle of elevation of B is 53° 30′. If PAQ is a straight line, find AQ.

4. A ship sails from A for 8 km on a bearing N. 21° 6′ W. to point B, then changes course to a bearing N. 70° 3′ W. and sails for 10 km to point C. Find how far C is (*a*) North, (*b*) West, of point A.

5. A rocket is projected vertically. After 1 min of flight it tilts over through an angle of 60° to the vertical and maintains this direction for a further 1 min of flight. If the average velocity of the rocket is 4800 mile/h, find: (*a*) the height reached after the period described above; (*b*) the horizontal distance from the launching-pad to a tracking station directly below the rocket after the 2 min of flight.

6. Town A is on a bearing S. 15° 40′ E. from town A. If A is 7 miles East of B, find the distance in a straight line from A to B.

7. An aircraft flies from A to B, a distance of 50 km, on a course N. 32° 33′ E. From B the aircraft flies to C, a distance of 60 km on a course N. 68° 8′ E. Find (*a*) the distance North, (*b*) the distance East, of C from A.

8. B is 225 dm from the base Q of a vertical pylon PQ, and the angle of elevation of P from B is 64° 3′. A is in line with B and Q on the same side of the pylon as B, and the angle of elevation of P from A is 31° 4′. Find the distance AB.

9. C is due East of A at a distance of 4 km. B is due North of C on a bearing N. 38° 10′ E. from A. D is due East of B and due South of X, a distance of $2\frac{3}{4}$ km. X is on a bearing N. 63° 34′ E. from B. Find how far X is (*a*) North, (*b*) East, of A, and (*c*) the bearing of X from A.

10. F is the base of a vertical mast EF. G and H are positions on the same side of F so that GHF is a straight line. The distance FH is 17·5 m, the angle of elevation of E from G is 16° 35′ and from H, 63° 26′. Find the distance GH.

11. A boat sails 8 miles on a bearing S. 40° 10′ W. Find the distance (*a*) South, (*b*) West, of the starting-point.

THE RECIPROCALS OF SINE AND COSINE

$$\text{COTANGENT } \theta = \frac{1}{\text{TANGENT } \theta} \quad \text{COSECANT } \theta = \frac{1}{\text{SINE } \theta} \quad \text{SECANT } \theta = \frac{1}{\text{COSINE } \theta}$$

Ordinary Ratio: *Reciprocal Ratio:*

$$\text{TAN } \theta = \frac{\text{OPP.}}{\text{ADJ.}} \qquad \text{COTAN } \theta = \frac{\text{ADJ.}}{\text{OPP.}}$$

$$\text{SIN } \theta = \frac{\text{OPP.}}{\text{HYP.}} \qquad \text{COSEC } \theta = \frac{\text{HYP.}}{\text{OPP.}}$$

$$\text{COS } \theta = \frac{\text{ADJ.}}{\text{HYP.}} \qquad \text{SEC } \theta = \frac{\text{HYP.}}{\text{ADJ.}}$$

NOTE. When using tables of ratios employing the prefix **CO**, the figures in the difference columns are *subtracted*.

In earlier problems involving the use of the tangent ratio, we have seen that when the "unknown" appears as the denominator of the ratio fraction the calculation is very much simplified by employing the reciprocal ratio.

The same is true for the sine and cosine ratios, but all trigonometrical calculations can be made even easier by the use of logarithms.

EXAMPLE. *A ship is due East of a lighthouse. A buoy, 2·38 km due North of the lighthouse, is observed from the ship on a bearing N. 27° 17′ W. Find, correct to 3 figures, the distance from the ship to the buoy.*

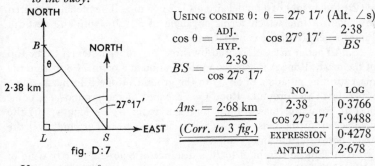

USING COSINE θ: $\theta = 27° 17′$ (Alt. $\angle$s)

$$\cos \theta = \frac{\text{ADJ.}}{\text{HYP.}} \qquad \cos 27° 17′ = \frac{2·38}{BS}$$

$$BS = \frac{2·38}{\cos 27° 17′}$$

Ans. = 2·68 km
(Corr. to 3 fig.)

	NO.	LOG
	2·38	0·3766
	cos 27° 17′	1̄·9488
	EXPRESSION	0·4278
	ANTILOG	2·678

fig. D:7

USING SECANT θ:

$$\theta = 27° 17′ \text{ (Alt. } \angle \text{s)}$$

$$\text{SEC } \theta = \frac{\text{HYP.}}{\text{ADJ.}}$$

$$\text{SEC } 27° 17′ = \frac{BS}{2·38}$$

$$BS = 2·38 \times \text{SEC } 27° 17′$$

	NO.	LOG
	2·38	0·3766
	sec 27° 17′	0·0512
	EXPRESSION	0·4278
	ANTILOG	2·678

Ans. = 2·68 km *(Corr. to 3 fig.)*

EXERCISE 7

Give all lengths and distances correct to three significant figures.
From fig. D : 8:

fig. **D:8**

1. $a = 16\cdot48$ in, $\angle\ \alpha = 28°\ 10'$. Find c.
2. $a = 35\cdot54$ m, $\angle\ \alpha = 75°\ 25'$. Find c.
3. $a = 8\frac{5}{8}$ km, $\angle\ \alpha = 54°\ 16'$. Find c.
4. $a = 2145$ yd, $\angle\ \alpha = 82°\ 32'$. Find c.
5. $b = 25\cdot22$ yd, $\angle\ \beta = 36°\ 40'$. Find c.
6. $b = 3\cdot05$ km, $\angle\ \beta = 78°\ 35'$. Find c in metres.
7. $b = 7\cdot374$ chains, $\angle\ \alpha = 65°\ 28'$. Find c.
8. $b = 3964$ cm, $\angle\ \alpha = 24°\ 46'$. Find c in decametres.
9. $a = 1586$ m, $\angle\ \beta = 69°\ 52'$. Find c in kilometres.
10. $a = 4410$ ft $\angle\ \beta = 14°\ 12'$. Find c.
11. $a = 3587$ yd, $c = 4973$ yd. Find $\angle\ \alpha$.
12. $a = 258\cdot2$ m, $b = 0\cdot3457$ km. Find $\angle\ \beta$.
13. $b = 478\cdot3$ in, $c = 14\cdot52$ yd. Find $\angle\ \alpha$.
14. $a = 47\cdot36$ ft, $b = 18\cdot84$ yd. Find c, $\angle\ \alpha$, $\angle\ \beta$.

MISCELLANEOUS EXAMPLES

EXERCISE 8

Give all lengths and distances correct to three significant figures.

1. A kite is flying at a height of 224 ft, the string forming an angle of 67° 23′ with the ground. Find the length of the string.

2. The height of a model aircraft is 28·45 dm. The model is secured to a control line which forms an angle of 32° 16′ with the vertical. Find the length of the control line.

3. An aerial mast is secured by three guy-ropes fastened to the top of the mast. Rope A is secured to the ground at a point 74·8 ft from the mast forming an angle of 54° with the ground. Rope B forms an angle of 42° with the ground. Rope C is secured to the ground at a point 96·4 ft from the mast. Find for each rope: (i) its length; (ii) the distance from the mast at which it is secured to the ground; (iii) the angle between rope and ground.

4. A ship sails from A to B, a distance of 8·56 km, on a bearing of 036° (N. 36° E.). From B it sails to C, a distance of 14·07 km, on a bearing of 110° (S. 70° E.). Find the distance AC and the bearing of C from A.

5. A regular eight-sided polygon is inscribed in a circle. If the length of side of the polygon is 2·05 cm calculate: (i) the radius of the circle; (ii) the area of the polygon.

6. Three buoys A, B and C are positioned such that C is due North of A, B is 042° (N. 42° E.) of A, C is 316° (N. 44° W.) of B. If $AB = 3592$ m and $BC = 2347$ m, find AC in kilometres.

7. From a boat the angle of elevation of the top of a vertical cliff, 138 ft high, is 42° 8′. Find the distance from the boat to the base of the cliff.

8. Fig. D:9 shows the end section of a glass-house. The dimensions are in feet and ∠ α is 62°. Find the distance *AB*.

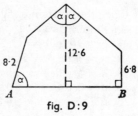

9. A path slopes upwards at an angle of 17° 18′. How far must a person walk along the path to obtain a vertical rise of 35·4 m?

fig. **D:9**

10. The base of a house and the foot of a flagpole are level. From the top of the house, 114·6 dm high, the angle of depression of the foot of the flagpole is 22° 16′, and the angle of elevation of the top of the flagpole is 18° 25′. Find the height of the flagpole.

AREA OF TRIANGLE AND PARALLELOGRAM BY TRIGONOMETRY

Area of Triangle

In fig. D : 10, *BD* is the perpendicular from *B* to *AC*.

In △*ABD* (rt. ∠d at *D*):

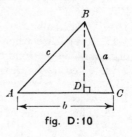

$$\text{SIN } A = \frac{BD}{c} \qquad \therefore \underline{\underline{BD = c \text{ . SIN } A}}$$

$$\text{Area of } \triangle ABC = \frac{\text{Base} \times \text{Height}}{2} = \frac{b \text{ . } BD}{2}$$

$$\text{Area of } \triangle ABC = \frac{bc \text{ . SIN } A}{2} \qquad \text{or } \tfrac{1}{2}bc \text{ SIN } A$$

fig. **D:10**

Area of △*ABC* = ½*bc* SIN *A*

Area of Parallelogram

In fig. D:11 *WXYZ* is a parallelogram having adjacent sides *a* and *b* to include ∠θ.

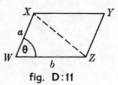

Area of △ *WXZ*	= ½*ab* SIN θ
But area of ‖gram. *WXYZ*	= 2(△ *WXZ*)
∴ Area of ‖gram. *WXYZ*	= 2(½*ab* SIN θ)

fig. **D:11**

Area of ‖gram. *WXYZ*   = *ab* SIN θ

The "S" Formula

By a more extensive use of theoretical trigonometry it can be proved that the area of a triangle is given by the formula:

$$\textbf{Area of } \triangle = \sqrt{S(S-a)(S-b)(S-c)}$$

where $S = \tfrac{1}{2}(a + b + c)$, i.e., S = half the sum of the sides.

EXERCISE 9

Using the notation of fig. D : 12, find the area of $\triangle ABC$, correct to 3 significant figures.

1. $b = 10$ m, $c = 12\cdot5$ m, $A = 52°$

2. $a = 4$ in, $b = 5$ in, $c = 6$ in

3. $a = 2$ cm, $b = 10$ cm, $C = 70° \, 40'$

4. $a = 5\cdot8$ ft, $c = 6\cdot4$ ft, $B = 65° \, 32'$

5. $a = 7\cdot2$ m, $b = 10\cdot8$ m, $c = 14$ m

6. $a = 2\cdot15$ ft, $b = 3\cdot08$ ft, $c = 4\cdot65$ ft

7. $b = 124\cdot2$ cm, $c = 58\cdot7$ dm, $A = 60°$. (Ans. in m².)

8. $a = 17\cdot45$ in, $b = 14\cdot38$ in, $C = 58° \, 22'$

9. $a = 32\cdot55$ cm, $b = 46\cdot4$ cm, $c = 38\cdot88$ cm. (Ans. in m².)

10 $a = 7\cdot744$ yd, $c = 12\cdot76$ yd, $B = 28° \, 16'$

11. Find the area of ‖gram. *ABCD*, in which $AB = 173\cdot2$ m, $AD = 1500$ m, $\angle A = 62° \, 34'$. (Ans. in hectares.)

12. Find the area of ‖gram. *WXYZ* in which $WX = 35\cdot27$ hm, $XY = 98\cdot1$ dam, $\angle Y = 74° \, 41'$ (Ans. in km².)

13. A regular octagon is inscribed in a circle of radius $3\cdot34$ in. Find the area of the octagon.

14. Calculate the area of $\triangle ABC$ in fig. D : 13. (HINT. *Consider the value of the $\perp$ar from A to BC produced.*)

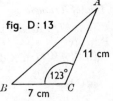

fig. D : 13

15. The perimeter of a regular hexagon is 27 m. Find its area.

PROBLEMS INVOLVING THREE DIMENSIONS

The solution of problems in **solid** geometry requires first the construction of a clear perspective diagram showing the angles and dimensions given. The various aspects of the problem may then be considered in terms of the appropriate **right-angled triangles**, which are *usually* found to be in either the **horizontal or vertical planes.**

Study the following diagrams and vocabulary. (Fig. D : 14.)

VOCABULARY

ABCD is a horizontal plane. *DCXY* is a vertical plane. *ABXY* is a sloping plane.

θ is the angle between the two planes *ABCD* and *ABXY*.

AB is the line of intersection of the two planes *ABCD* and *ABXY*.

YD is a line perpendicular to the plane *ABCD*.

PR is any line in the plane *ABXY* drawn **perpendicular** to the line of intersection *AB*. *PR* is called a line of **greatest slope.**

Compare *PR* with *YB*, which is not at right-angles to *AB*. Angle θ is, clearly, greater than angle α. *Consider tangent ratio of each angle.*

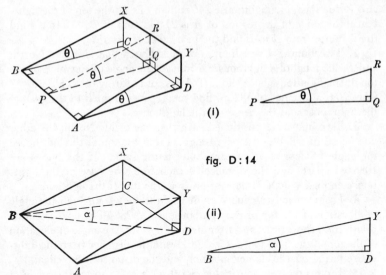

fig. D : 14

EXAMPLE. *Two buoys, A and B, are positioned such that A is due North, and B due East, of a lighthouse 128 ft high. From the top of the lighthouse the angle of depression of A is 32° 17' and of B, 28° 20'. Find the distance from A to B and the bearing of A from B.*

EXPLANATION. The problem has been represented by a series of diagrams showing each stage of the solution. With experience it is possible to appreciate the problem from the perspective diagram alone [fig. (i)]. Fig. (ii) represents a vertical plane, North–South, from

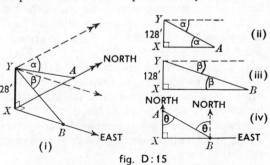

fig. D : 15

which we can obtain AX. From fig. (iii), a vertical plane East–West, we can obtain BX. Fig. (iv) shows a right-angled triangle ABX in the horizontal plane, from this we can see how to obtain AB (Pythagoras' Theorem) and θ.

EXERCISE 10

Give heights and distances correct to three significant figures.

1. Two buoys, *A* and *B*, are positioned such that *A* is due North,

and B due East, of a lighthouse 246 ft high. From the top of the lighthouse the angle of depression of A is $27°$ $32'$ and of B, $21°$ $16'$. Find the distance from A to B and the bearing of A from B.

2. Two buoys, A and B, are positioned such that A is due North, and B due East, of a lighthouse. From B, the angle of elevation of the top of the lighthouse is $18°$ $40'$. If A lies on a bearing of N. $58°$ $16'$ W. from B at a distance of 382 m, find the distance from B to the base of the lighthouse and the height of the lighthouse.

3. Two buoys are positioned such that one is due North, the other due West, of a lighthouse 64 yd high. From the top of the lighthouse the angle of depression of each buoy is the same. If the buoys are 1086 yd apart, find the distance of each from the base of the lighthouse and each angle of depression from the top of the lighthouse.

4. From a boat lying due West of a lighthouse, 294 dm high, the angle of elevation of the top of the lighthouse is $62°$ $28'$. The boat sails due South for a short time, and from its new position the angle of elevation of the top of the lighthouse is $32°$ $28'$. Find the distance from the lighthouse to the boat in its new position and the distance it travelled.

5. From point A, 450 ft due South of a radio mast, the angle of elevation of the top of the mast is $29°$ $45'$. From point B, due East of A, the angle of elevation of the top of the mast is $22°$. Find the distance of B from A and the bearing of B from the mast.

6. A pyramid has a square base. Each slanting edge is 4·3 cm long and forms an angle of $67°$ with a diagonal of the base. Find: (i) the height of the pyramid, and (ii) the area of the base.

7. A pyramid has a square base and each *face* of the pyramid forms an angle of $58°$ with the base. A line of greatest slope is 6·4 dm long from the apex of the pyramid to the base. Find: (i) the height of the pyramid, and (ii) the length of a slant edge.

8. Two paths, diverging from the same point, lead up a hill. Path x lies along a line of greatest slope and forms an angle of $24°$ with the horizontal. Path y forms an angle $\theta°$ with x. A third path z lies horizontally, meeting x at right-angles and cutting y 324 m away. If z is 45 m vertically above their starting-point, find θ, the angle between the paths of x and y.

9. $ABCD$ is the horizontal base of a rectangular prism. $WXYZ$ is the top surface and the edges AW, BX, CY and DZ are vertical. AB is 8 cm, AD is 12 cm and AW is 6 cm. Find: (i) the length of the line joining the mid-points of AB and WZ, and (ii) the angle formed by this line and the horizontal base.

10. A vertical mast PX stands at the corner of a square horizontal parade-ground, $PQRS$, of area 3136 m². The angle of elevation of X from S is $26°$. Find: (i) the height of the mast PX; (ii) the angle of elevation of X from R, and (iii) angle QXS.

TRIGONOMETRICAL RATIOS OF SPECIAL ANGLES

An Angle of 45°

ABC is a right-angled, isosceles triangle in which
$$AC = BC = 1 \text{ unit of length.}$$
$$\therefore \ \angle A = \angle B = 45°$$

fig. D : 16

By Pythagoras' Theorem:
$$AB^2 = AC^2 + BC^2$$
$$\therefore \ AB^2 = \ 1 \ + \ 1 \ . \ \therefore \ AB^2 = 2 \text{ and } \underline{AB = \sqrt{2}}$$

Since $\text{SIN } \beta = \dfrac{\text{OPP.}}{\text{HYP.}}$ Then $\text{SIN } 45° = \dfrac{1}{\sqrt{2}}$

Since $\text{COS } \beta = \dfrac{\text{ADJ.}}{\text{HYP.}}$ Then $\text{COS } 45° = \dfrac{1}{\sqrt{2}}$

Since $\text{TAN } \beta = \dfrac{\text{OPP.}}{\text{ADJ.}}$ Then $\text{TAN } 45° = 1$

Angles of 30° and 60°

XYZ is an equilateral triangle in which
$$XY = YZ = ZX = 2 \text{ units of length.}$$
$$\therefore \ \angle X = \angle Y = \angle Z = 60°$$
XP is $\perp$ to YZ and $\therefore$ bisects YZ and $\angle X$.

fig. D : 17

In $\triangle XYP$, by Pythagoras' Theorem:
$$XY^2 = XP^2 + YP^2 \qquad \therefore \ XP^2 = XY^2 - YP^2$$
$$\therefore \ XP^2 = 4 - 1 \qquad \text{and } \underline{XP = \sqrt{3}}$$

Then $\text{SIN } 30° = \dfrac{1}{2}$ and $\text{SIN } 60° = \dfrac{\sqrt{3}}{2}$

        $\text{COS } 30° = \dfrac{\sqrt{3}}{2}$ $\text{COS } 60° = \dfrac{1}{2}$

        $\text{TAN } 30° = \dfrac{1}{\sqrt{3}}$ $\text{TAN } 60° = \dfrac{\sqrt{3}}{1}$

SURDS

Numbers of the form $\sqrt{2}$ are called **irrational numbers.**

We can give the fraction $\dfrac{1}{\sqrt{2}}$ a **rational denominator** by multiplying the numerator and denominator by $\sqrt{2}$. Thus:

$$\text{SIN } 45° = \frac{1}{\sqrt{2}} = \frac{\sqrt{2}}{2} = \frac{1 \cdot 414}{2} = 0 \cdot 707$$

By employing rational denominators, use the ratios from figs. D : 16 and D : 17 to find:

(i) $\cos 45°$ (ii) $\cos 30°$ (iii) $\tan 30°$ (iv) $\sin 60°$ (v) $\tan 60°$

EXERCISE 11

Evaluate without using tables. *Express fractions with rational denominators where necessary.*

1. $\dfrac{\sin 45°}{\cos 45°}$ 2. $\sin 45° \times \cos 45°$ 3. $\dfrac{\sin 30°}{\cos 60°}$

4. $\dfrac{\tan 30°}{\tan 60°}$ 5. $\dfrac{\tan 60°}{\tan 30°}$ 6. $\tan 30° \times \tan 60°$

7. $\dfrac{\cos 30°}{\sin 60°}$ 8. $\cos 60° \times \sin 30°$ 9. $\dfrac{\sin 60°}{\cos 30°}$

10. $\sin 60° \times \cos 30°$ 11. $\dfrac{\sin 60°}{\sin 30°}$ 12. $\dfrac{\cos 60°}{\cos 30°}$

13. $\sin 30° \times \cos 30°$ 14. $\dfrac{\sin 60°}{\cos 60°}$ 15. $\dfrac{\sin 30°}{\sin 60°}$

16. $\dfrac{\sin 30°}{\cos 30°}$ 17. $\dfrac{\cos 30°}{\cos 60°}$ 18. $\sin 60° \times \cos 60°$

19. A ladder rests against a vertical wall at an angle of 60° to the ground. If the foot of the ladder is 18 dm from the wall, find the length of the ladder.

20. A road slopes upwards at an angle of 30° to the horizontal. Find the vertical rise for a distance of 54 m along the road.

21. A regular hexagon is inscribed in a circle of radius 2 in. Express the area of the hexagon as a surd in its simplest form.

22. The radius of a bell-tent is 12 dm and the slant height of the tent is 24 dm. Find the angle formed by the side of the tent and the ground.

TRIGONOMETRICAL EQUATIONS

fig. D:18

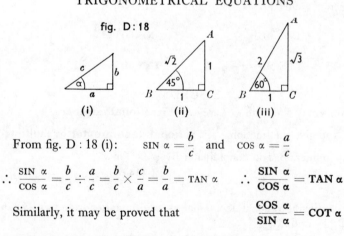

(i)   (ii)   (iii)

From fig. D : 18 (i): $\text{SIN } \alpha = \dfrac{b}{c}$ and $\text{COS } \alpha = \dfrac{a}{c}$

$\therefore \dfrac{\text{SIN } \alpha}{\text{COS } \alpha} = \dfrac{b}{c} \div \dfrac{a}{c} = \dfrac{b}{c} \times \dfrac{c}{a} = \dfrac{b}{a} = \text{TAN } \alpha$ $\therefore \dfrac{\text{SIN } \alpha}{\text{COS } \alpha} = \text{TAN } \alpha$

Similarly, it may be proved that $\dfrac{\text{COS } \alpha}{\text{SIN } \alpha} = \text{COT } \alpha$

Such results are sometimes called **identities**.

FURTHER IDENTITIES

1. From fig. D : 18 (i), by Pythagoras' Theorem: $c^2 = a^2 + b^2$

Since $\qquad\qquad \sin^2 \alpha = \dfrac{b^2}{c^2} \qquad$ and $\qquad \cos^2 \alpha = \dfrac{a^2}{c^2}$

Then $\qquad \sin^2 \alpha + \cos^2 \alpha = \dfrac{b^2}{c^2} + \dfrac{a^2}{c^2} = \dfrac{a^2 + b^2}{c^2}$

But $\qquad\qquad a^2 + b^2 = c^2 \quad \therefore \ \sin^2 \alpha + \cos^2 \alpha = \dfrac{c^2}{c^2} = 1$

$$\mathbf{SIN^2\ \alpha + COS^2\ \alpha = 1}$$

Check this identity by substituting for

$\qquad\qquad$ (i) $\alpha = 45°$ $\qquad$ (ii) $\alpha = 30°$ $\qquad$ (iii) $\alpha = 60°$

2. Since $\qquad\qquad\qquad\qquad \sin^2 \alpha + \cos^2 \alpha = 1$

Dividing both sides ⎫ $\qquad \dfrac{\sin^2 \alpha}{\cos^2 \alpha} + 1 = \dfrac{1}{\cos^2 \alpha}$
by $\cos^2 \alpha$: ⎭

$$\therefore \left(\frac{\sin \alpha}{\cos \alpha} \right)^2 + 1 = \left(\frac{1}{\cos \alpha} \right)^2$$

But $\quad \dfrac{\sin \alpha}{\cos \alpha} = \tan \alpha \quad$ and $\quad \dfrac{1}{\cos \alpha} = \sec \alpha$

$$\therefore \ \tan^2 \alpha + 1 = \sec^2 \alpha$$

or $\qquad \mathbf{1 + TAN^2\ \alpha = SEC^2\ \alpha}$

3. Similarly, by dividing $\sin^2 \alpha + \cos^2 \alpha = 1$ by $\sin^2 \alpha$ it may be proved that

$$\mathbf{1 + COT^2\ \alpha = COSEC^2\ \alpha}$$

EXERCISE 12

Express as simply as possible:

1. $\dfrac{\sin \alpha}{\cos \alpha}$ $\qquad\qquad$ **2.** $\tan \alpha \cos \alpha$ $\qquad\qquad$ **3.** $\dfrac{\tan \alpha}{\sin \alpha}$

4. $\dfrac{\cos \alpha}{\sin \alpha}$ $\qquad\qquad$ **5.** $\cot \alpha \sin \alpha$ $\qquad\qquad$ **6.** $\dfrac{\cot \alpha}{\cos \alpha}$

7. $\cos \alpha \sec \alpha$ $\qquad\qquad$ **8.** $\sin \alpha \csc \alpha$ $\qquad\qquad$ **9.** $\tan \alpha \cot \alpha$

10. $\cos \alpha \csc \alpha$ $\qquad\qquad$ **11.** $\sin \alpha \sec \alpha$ $\qquad\qquad$ **12.** $\csc \alpha \tan \alpha$

13. $\sin \alpha \cot \alpha$ $\qquad\qquad$ **14.** $\dfrac{\tan \alpha}{\sec \alpha}$ $\qquad\qquad$ **15.** $\dfrac{\sec \alpha}{\csc \alpha}$

16. Prove that $(1 + \cos \alpha)(1 - \cos \alpha) = \sin^2 \alpha$

17. Prove that $1 + \dfrac{\sin^2 \alpha}{\cos^2 \alpha} = \sec^2 \alpha$

18. Prove that $\dfrac{\tan \alpha}{1 + \tan^2 \alpha} = \sin \alpha \cos \alpha$

19. Prove that $\dfrac{1}{\sin^2 \alpha} - 1 = \cot^2 \alpha$

20. Prove that $\sin^4 \alpha - \cos^4 \alpha = 1 - 2 \cos^2 \alpha$

HARDER PROBLEMS ON TRIGONOMETRY
EXERCISE 13

1. F is the base of a vertical mast EF. G and H are positions on the same side of F so that GHF is a straight line. The distance GH is 100 m, the angle of elevation of E from G is 16° 42' and from H, 63° 26'. Find the height of EF.

2. An aircraft is attempting to land on the flight-deck of an aircraft carrier. From the stern end of the deck the angle of elevation of the aircraft is 11° 19' and from a position 450 dm farther forward the angle of elevation is 5° 43'. If the deck is 180 dm above the water, find the height of the aircraft above the water.

3. From the bow of a ship the angle of elevation of the "crow's nest" is 58°, from the ship's stern the angle of elevation is 28° 4'. If the ship is 200 ft long, find the height of the "crow's nest" above the deck.

4. R and S are two points, 80 m apart, both in line with Q, the base of a vertical tower PQ. The angle of elevation of P from R is 35° and from S, 70°. Find the distance QS.

5. C and D are two points, 60 ft apart, both in line with B, the base of a vertical tower AB. The angle of elevation of A from D is 20° 33' and from C, 36° 52'. Find the distance BC.

6. Two boats R and S are both in line with Q, a point at the base of a vertical cliff, such that QR is 100 ft and RS is 150 ft. From P, a point vertically above Q, the angles of depression of S and R are $\alpha°$ and $\beta°$ respectively. If $\tan \beta° - \tan \alpha° = 0.6$, find α.

7. Two ropes are attached to T, the top of a mast TG, and secured to the ground at points X and Y on the same side of G forming a straight line XYG. The angles of elevation of T from X and Y are $\alpha°$ and $\beta°$ respectively, XT is 120 m and YT is 80 m. If $\sin \beta - \sin \alpha = 0.25$, find α and the height of the mast.

8. Two ropes are attached to A, the top of a mast AB, and secured to the ground at points C and D on opposite sides of B, forming a straight line CBD. AC is 100 ft long and forms an angle $\beta°$ with the mast, whilst AD is 57·75 ft long, forming an angle $\alpha°$ with the mast. If $\cos \alpha + \cos \beta = 1.366$, find α and the height of the mast.

9. A and B are two buoys 1500 ft apart; A is due South and B due East of a lighthouse SL. The angle of elevation of L, the top of the lighthouse, from A is 5° 43' and $BS = \frac{3}{4}AS$. Find the height of the lighthouse and the angle of elevation of L from B.

10. A, B and C are three points on the ground, in the same horizontal plane, forming a straight line from East to West with B mid-way between A and C. X lies due North of A on a bearing N. 26° 34' W. from B, and Y lies due North of C on a bearing N. 55° E. from B. If AX is 10 km greater than CY, find the distance CY.

DEGREES AND RADIANS

Fig. D : 19 shows a circle, centre O, in which the arc AB is equal in length to the radius (r). Angle θ is the angle subtended by AB.

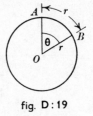

The ratio of θ to the complete circle $= \dfrac{\theta°}{360°}$

The ratio of AB to the complete circle $= \dfrac{r}{2\pi r}$

and $\quad \dfrac{\theta}{360°} = \dfrac{r}{2\pi r} \quad \therefore \theta = \left(\dfrac{360r}{2\pi r}\right)°$

$\therefore \theta = \left(\dfrac{180}{\pi}\right)°$ or 1 radian

fig. D : 19

NOTE. The value of the **radian** does NOT depend upon the value of the radius.

DEFINITION

A **radian** is the angle subtended at the centre of a circle by an arc of the circle equal in length to the radius.

$$1 \text{ radian} = \frac{180}{\pi} \text{ degrees}$$

From this formula we obtain

$$\pi \text{ radians} = 180 \text{ degrees}$$

and $\quad$ **1 degree** $= \dfrac{\pi}{180}$ **radians**

EXAMPLE. *Change 30° to radians.*

$$30° = \frac{\pi}{180} \times \frac{30}{1} \text{ radians} = \frac{\pi}{6} \text{ radians}$$

$$= \frac{3\cdot142}{6} = 0\cdot5236 \text{ radians}$$

Ans. $= \underline{0\cdot524 \text{ radians } (Corr. \text{ to } 3 \text{ fig.})}$

EXERCISE 14

Change the following to radians, *correct to three significant figures*:

1. 45°	**2.** 60°	**3.** 90°	**4.** 180°	**5.** 270°
6. 360°	**7.** 20°	**8.** 40°	**9.** 120°	**10.** 150°
11. 32° 15′	**12.** 56° 30′	**13.** 65° 45′	**14.** 74° 12′	**15.** 88° 50′

Change the following angles in radians to degrees and minutes:

16. 1	**17.** 0·2	**18.** 3	**19.** 0·4	**20.** 1·5
21. 0·5	**22.** 2·5	**23.** 3·25	**24.** 4·15	**25.** 3·142
26. π	**27.** 2π	**28.** $\dfrac{\pi}{2}$	**29.** $\dfrac{\pi}{3}$	**30.** $\dfrac{2\pi}{3}$

TRIGONOMETRICAL RATIOS OF OBTUSE ANGLES

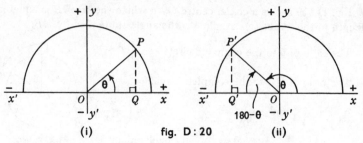

(i)   **fig. D : 20**   (ii)

Fig. D : 20 shows a line OP, of unit length, rotating in an anti-clockwise direction about point O (the intersection of the two axes xOx' and yOy').

As OP rotates it forms an angle θ. PQ is the perpendicular from P to xOx' and OQ is the projection of OP on xOx'.

When θ is between $0°$ and $90°$ [fig. D : 20 (i)]

Remember:
OP is unit length.
$$\text{SIN } \theta = \frac{PQ}{OP} = +\frac{PQ}{1}$$

$$\cos \theta = \frac{OQ}{OP} = +\frac{OQ}{1} \qquad \text{TAN } \theta = \frac{PQ}{OQ} = +\frac{PQ}{OQ}$$

When θ is between $90°$ and $180°$ [fig. D : 20 (ii)]

OP has been rotated to form a second quadrant angle θ, between $90°$ and $180°$, such that $P'Q' = PQ$. It follows that $OQ' = OQ$ (Congruent $\triangle$s: R.H.S.)

By considering θ in terms of its supplement we obtain the following:

$$\text{SIN } (180° - \theta) = \frac{P'Q'}{OP'} = +\frac{PQ}{1}$$

$$\cos (180° - \theta) = \frac{OQ'}{OP'} = -\frac{OQ}{1} \quad \begin{cases} \text{Because } OQ' \\ \text{is negative.} \end{cases}$$

$$\text{TAN } (180° - \theta) = \frac{P'Q'}{OQ'} = -\frac{PQ}{OQ} \quad \begin{cases} \text{Because } OQ' \\ \text{is negative.} \end{cases}$$

CONCLUSIONS

$$\text{SIN } (180° - \theta) = \text{SIN } \theta \qquad \text{or} \quad \textbf{SIN } \boldsymbol\theta = \textbf{SIN } (\textbf{180°} - \boldsymbol\theta)$$
$$\cos (180° - \theta) = -\cos \theta \qquad\qquad \textbf{COS } \boldsymbol\theta = -\textbf{COS } (\textbf{180°} - \boldsymbol\theta)$$
$$\text{TAN } (180° - \theta) = -\text{TAN } \theta \qquad\qquad \textbf{TAN } \boldsymbol\theta = -\textbf{TAN } (\textbf{180°} - \boldsymbol\theta)$$

NUMERICALLY

$$\text{SIN } 120° = \quad \text{SIN } (180° - 120°) = \quad \text{SIN } 60° = 0\cdot8660$$
$$\cos 120° = -\cos (180° - 120°) = -\cos 60° = -0\cdot5$$
$$\text{TAN } 120° = -\text{TAN } (180° - 120°) = -\text{TAN } 60° = -1\cdot7321$$

RATIOS OF THIRD AND FOURTH QUADRANT ANGLES

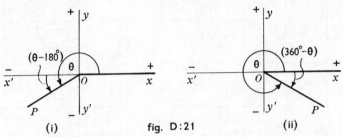

(i)   fig. D:21   (ii)

By rotating OP still further about point O, it can be shown that:

When θ is between 180° *and* 270° [fig. D : 21 (i)]

$$\text{SIN } \theta = -\text{SIN } (\theta - 180°) \quad \text{i.e.,} \quad \text{SIN } 240° = -\text{SIN } 60°$$
$$\cos \theta = -\cos (\theta - 180°) \qquad \cos 240° = -\cos 60°$$
$$\text{TAN } \theta = \text{TAN } (\theta - 180°) \qquad \text{TAN } 240° = \text{TAN } 60°$$

When θ is between 270° *and* 360° [fig. D : 21 (ii)]

$$\text{SIN } \theta = -\text{SIN } (360° - \theta) \quad \text{i.e.,} \quad \text{SIN } 300° = -\text{SIN } 60°$$
$$\cos \theta = \cos (360° - \theta) \qquad \cos 300° = \cos 60°$$
$$\text{TAN } \theta = -\text{TAN } (360° - \theta) \qquad \text{TAN } 300° = -\text{TAN } 60°$$

GENERAL CONCLUSIONS

(i) Given an angle θ, for numerical purposes we consider the angle formed between the rotating line and the xOx' axis.

(ii) The sign to be attached to the ratio depends upon the ratio required and the quadrant in which the rotating arm rests. (See fig. D : 22.)

(iii) It is helpful to represent the data on a simple sketch. See the following example.

SIN+VE	ALL+VE
TAN+VE	COS+VE

fig. D:22

EXAMPLE. *Evaluate the three trig. ratios for* 220°.

$$\sin 220° = -\sin 40° = -0.6428$$
$$\cos 220° = -\cos 40° = -0.7660$$
$$\tan 220° = \quad \tan 40° = 0.8391$$

220°

40°

TAN+VE

fig. D:23

EXERCISE 15

Evaluate the following:

1. sin 100°	**2.** cos 150°	**3.** tan 145°	**4.** sin 110°
5. cos 125°	**6.** tan 160°	**7.** sin 200°	**8.** cos 260°
9. tan 230°	**10.** sin 195°	**11.** cos 225°	**12.** tan 250°
13. sin 290°	**14.** cos 330°	**15.** tan 280°	**16.** sin 315°
17. cos 320°	**18.** tan 345°	**19.** sin 150° 30′	**20.** tan 342° 15′

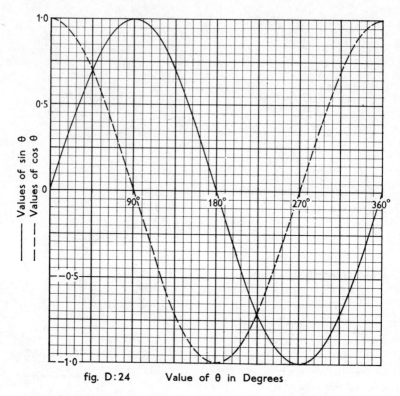

fig. D:24   Value of θ in Degrees

GRAPHS OF TRIGONOMETRICAL FUNCTIONS

The value of sin θ may be shown graphically for values of θ between 0° and 360°. (See fig. D : 24.)

The graph of cos θ is represented by the broken line.

Fig. D : 25 shows an alternative construction for the graph of sin θ. OB is a radius of unit length; as it rotates anti-clockwise through θ°, BC varies in length and sign. The horizontal axis of the graph represents angle θ (in degrees) and the vertical axis represents the value of BC. For convenience angle θ is measured by intervals of 9°.

The graph of tan θ may be similarly represented. Note the breaks in the graph line as it "disappears" to +∞ and −∞. (See fig. D : 26.)

The shapes of these graphs should lead you to conclude that values of the sin, cos and tan ratios would repeat themselves for angles greater than 360°. I.e., if the line OP in fig. D : 21 were to continue to rotate.

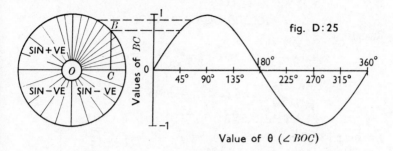

fig. D:25

Value of θ (∠ BOC)

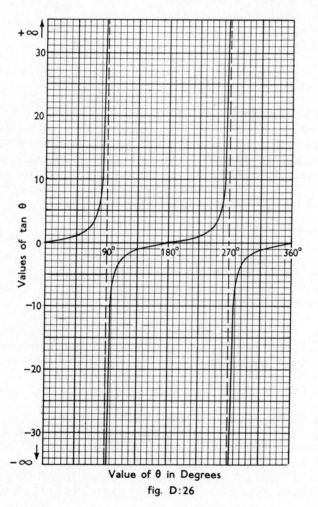

Value of θ in Degrees

fig. D:26

RECIPROCAL RATIOS

REMEMBER

$$\text{COT } \theta = \frac{1}{\text{TAN } \theta}; \quad \text{COSEC } \theta = \frac{1}{\text{SIN } \theta}; \quad \text{SEC } \theta = \frac{1}{\text{COS } \theta}$$

$$\text{e.g., } \text{cosec } 225° = \frac{1}{\sin 225°} = \frac{1}{-\sin 45°} = -\text{cosec } 45°$$

FURTHER EVALUATION OF RATIOS

EXAMPLE. *Find the values of* θ *between* $0°$ *and* $360°$ *for which sin* $\theta = -0·342$.

From tables, a sine ratio of $0·342$ gives an angle of $20°$. A negative sine means that θ cannot be in the 1st or 2nd quadrant.

$$\therefore \quad \theta = 200° \text{ or } 340°$$

fig. D : 27

EXERCISE 16

Evaluate the following:

1. sin 150°	**2.** cos 164°	**3.** tan 175°	**4.** sin 268°
5. cos 230°	**6.** tan 220°	**7.** sin 342°	**8.** cos 335°
9. tan 352°	**10.** sec 78°	**11.** cosec 62°	**12.** cot 38°
13. cosec 115°	**14.** sec 156°	**15.** cot 128°	**16.** cot 268°
17. sec 224°	**18.** cosec 256°	**19.** cosec 324°	**20.** sec 344°
21. cot 336°	**22.** sin 460°	**23.** cos 520°	**24.** cot 680°

Find the values of θ between $0°$ and $360°$ if:

25. $\sin \theta = 0·342$ **26.** $\sin \theta = -0·866$ **27.** $\cos \theta = 0·891$

28. $\cos \theta = -0·891$ **29.** $\tan \theta = 0·404$ **30.** $\tan \theta = -1·804$

31. $\sin \theta = -0·309$ and $\tan \theta$ is positive.

32. $\sin \theta = 0·809$ and $\cos \theta$ is negative.

33. $\cos \theta = 0·788$ and $\sin \theta$ is negative.

34. $\cos \theta = -0·682$ and $\sin \theta$ is negative.

35. $\tan \theta = -1·327$ and $\cos \theta$ is negative.

36. Construct the graphs of sin θ, cos θ, tan θ, each on separate axes, for values of θ from $0°$ to $720°$. From your graphs verify the values of the ratios given in Exercise 15.

37. From fig. D : 24, find values of θ between $0°$ and $360°$ for which sin θ = cos θ.

38. On the same axes, construct the graphs of cos θ and tan θ for values of θ between $0°$ and $180°$. From your graphs find values of θ, between the given limits, for which cos θ = tan θ.

39. Using intervals of $20°$ and on the same pair of axes, plot the graphs of 3 sin θ and 2 cos θ for values of θ between $0°$ and $180°$. From your graphs determine the value of θ for which 3 sin θ = 2 cos θ.

40. Plot the graph of $3 \sin \theta + 2 \cos \theta$ for values of θ between $0°$ and $180°$. Use your graph to solve within these limits: (i) $3 \sin \theta + 2 \cos \theta = 0$; (ii) $3 \sin \theta = 3 - 2 \cos \theta$.

THE SINE FORMULA

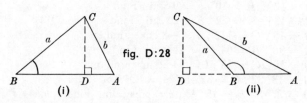

fig. D:28

In an acute-angled triangle: [Fig. D : 28 (i)]

In $\triangle ACD$: $\sin A = \dfrac{CD}{b}$ $\therefore$ $\underline{CD = b \sin A}$

In $\triangle BCD$: $\sin B = \dfrac{CD}{a}$ $\therefore$ $\underline{CD = a \sin B}$

In an obtuse-angled triangle: [Fig. D : 28 (ii)]

In $\triangle ACD$: $\sin A = \dfrac{CD}{b}$ $\therefore$ $\underline{CD = b \sin A}$

In $\triangle BCD$: $\sin (180° - B) = \dfrac{CD}{a}$ and $CD = a \sin (180° - B)$

But $\sin (180° - B) = \sin B$ $\therefore$ $\underline{CD = a \sin B}$

In each case, $a \sin B = b \sin A$ $\therefore$ $\dfrac{a}{\sin A} = \dfrac{b}{\sin B}$

Similarly, with $AE \perp BC$, it may be
proved: $\dfrac{b}{\sin B} = \dfrac{c}{\sin C}$

CONCLUSION $\dfrac{a}{\sin A} = \dfrac{b}{\sin B} = \dfrac{c}{\sin C}$

EXAMPLE. *In $\triangle ABC$, $A = 116°$, $B = 38°$, $a = 12$ in. Find b.*

$\dfrac{a}{\sin A} = \dfrac{b}{\sin B}$ $\therefore$ $\dfrac{12}{\sin 116°} = \dfrac{b}{\sin 38°}$

But $\sin 116° = \sin 64°$
Cross-multiply:

$b = \dfrac{12 \sin 38°}{\sin 64°}$

Ans. $\underline{\underline{b = 8 \cdot 22 \text{ in}}}$

(Corr. to 3 fig.)

NO.	LOG
12	1·0792
SIN 38°	$\bar{1}$·7893
NUMER.	0·8685
SIN 64°	$\bar{1}$·9537
8·219	0·9148

EXERCISE 17

Give lengths correct to 3 sig. fig.

1. $A = 30°$, $B = 40°$, $a = 10$ in. Find b.
2. $B = 30°$, $C = 70°$, $c = 8$ cm. Find b.
3. $A = 50°$, $C = 20°$, $c = 6\cdot4$ m. Find a.
4. $A = 120°$, $B = 30°$, $b = 9\cdot5$ in. Find a.
5. $B = 135°$, $C = 22°$, $b = 22$ dm. Find c.
6. $A = 15°$, $C = 112°$, $a = 4\cdot8$ cm. Find c.
7. $A = 38°$, $B = 64°$, $c = 16\cdot3$ m. Find a, b.
8. $B = 25°$, $C = 33°$, $b = 14\cdot25$ in. Find a, c.
9. $A = 18°\ 32'$, $C = 32°\ 14'$, $b = 12\cdot7$ cm. Find a, c.
10. $A = 27°\ 8'$, $B = 45°\ 43'$, $c = 36\cdot42$ dm. Find a, b.

THE COSINE FORMULA

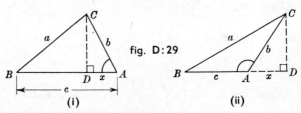

fig. D:29

(i)   (ii)

In an acute-angled triangle: [Fig. D : 29 (i)]

In $\triangle ACD$: $\cos A = \dfrac{x}{b}$ $\therefore \underline{x = b \cos A}$ (i)

In an obtuse-angled triangle: [Fig. D : 29 (ii)]

In $\triangle ACD$: $\cos(180° - A) = \dfrac{x}{b}$ and $x = b \cos(180° - A)$

But $\cos(180° - A) = -\cos A$ $\therefore \underline{x = -b \cos A}$ (ii)

In each case: $a^2 = BD^2 + CD^2$

(i) $a^2 = (c - x)^2 + CD^2$ (ii) $a^2 = (c + x)^2 + CD^2$

(i) $a^2 = c^2 - 2cx + x^2 + CD^2$ (ii) $a^2 = c^2 + 2cx + x^2 + CD^2$

In each case: $b^2 = x^2 + CD^2$

(i) $a^2 = b^2 + c^2 - 2cx$ (ii) $a^2 = b^2 + c^2 + 2cx$

(i) $x = b \cos A$ (ii) $x = -b \cos A$

In each case: $a^2 = b^2 + c^2 - 2bc \, . \, \cos A$

Similarly: $b^2 = a^2 + c^2 - 2ac \, . \, \cos B$
                $c^2 = a^2 + b^2 - 2ab \, . \, \cos C$

CONCLUSION $\mathbf{a^2 = b^2 + c^2 - 2bc \, . \, \cos A}$

EXAMPLE 1. *In* $\triangle ABC$, $b = 4 \cdot 8$ *in*, $c = 6 \cdot 4$ *in*, $A = 128°$. *Find a.*

$a^2 = b^2 + c^2 - 2bc \cdot \cos A \qquad \cos 128° = -\cos 52°$

		NO.	LOG
$a^2 = (4 \cdot 8)^2 + (6 \cdot 4)^2 - 2(4 \cdot 8)(6 \cdot 4)(-\cos 52°)$		2	0·3010
$a^2 = 23 \cdot 04 + 40 \cdot 96 + 2(4 \cdot 8)(6 \cdot 4)(\cos 52°)$		4·8	0·6812
$a^2 = 23 \cdot 04 + 40 \cdot 96 + 37 \cdot 82$		6·4	0·8062
$a^2 = 101 \cdot 82$		$\cos 52°$	$\bar{1}$·7893
$a = 10 \cdot 09$ *(From 4-fig. tables.)*		37·82	1·5777

Ans. $a = 10 \cdot 1$ in (Corr. to 3 fig.)

EXAMPLE 2. *In* $\triangle ABC$, $a = 5 \cdot 6$ *cm*, $b = 6 \cdot 2$ *cm*, $c = 9 \cdot 5$ *cm*. *Find C.*

$$c^2 = a^2 + b^2 - 2ab \cdot \cos C \qquad \therefore \cos C = \frac{a^2 + b^2 - c^2}{2ab}$$

	NO.	LOG
$\cos C = \dfrac{(5 \cdot 6)^2 + (6 \cdot 2)^2 - (9 \cdot 5)^2}{2(5 \cdot 6)(6 \cdot 2)}$	20·45	1·3107
	2	0·3010
$\cos C = \dfrac{31 \cdot 36 + 38 \cdot 44 - 90 \cdot 25}{2(5 \cdot 6)(6 \cdot 2)}$	5·6	0·7482
	6·2	0·7924
$\cos C = -\dfrac{20 \cdot 45}{2(5 \cdot 6)(6 \cdot 2)}$		1·8416
	72° 52′	$\bar{1}$·4691

The negative cosine indicates a 2nd quadrant angle.

$\therefore C = 180° - 72° 52′$

Ans. $\underline{\underline{C = 107° 8′}}$

NOTE. **Subtract** figures in difference columns when using cosine or log-cosine tables.

EXERCISE 18

Give lengths correct to 3 sig. fig.

1. $b = 4$ in, $c = 5$ in, $A = 72°$. Find a.
2. $a = 5$ in, $b = 6$ in, $c = 7$ in. Find A.
3. $a = 5 \cdot 2$ cm, $b = 7 \cdot 4$ cm, $C = 58°$. Find c.
4. $a = 3 \cdot 8$ cm, $b = 4 \cdot 3$ cm, $c = 5 \cdot 5$ cm. Find C.
5. $a = 2 \cdot 7$ in, $c = 14 \cdot 2$ in, $B = 28°$. Find b.
6. $b = 6 \cdot 4$ cm, $c = 7 \cdot 2$ cm, $A = 126°$. Find a.
7. $a = 4 \cdot 8$ cm, $b = 9 \cdot 5$ cm, $c = 6 \cdot 2$ in. Find B.
8. $a = 5 \cdot 6$ in, $c = 6 \cdot 8$ in, $B = 132°$. Find b.
9. $a = 13 \cdot 4$ cm, $b = 15 \cdot 2$ cm, $c = 22 \cdot 7$ cm. Find C.

Using first the Cosine Formula, then the Sine Formula:

10. $b = 3 \cdot 5$ in, $c = 4 \cdot 6$ in, $A = 68°$. Find B.
11. $a = 4 \cdot 25$ cm, $b = 18 \cdot 2$ cm, $C = 18°$. Find B.
12. $a = 4 \cdot 75$ cm, $b = 4 \cdot 18$ cm, $c = 8 \cdot 34$ in. Find A, B, C.

NOTE

(i) The examples given in Exercises 17, 18 have been based on data conforming to S.A.S., A.A.S. or S.S.S.

(ii) If data refers to two sides and a "not-included angle" the result will be an example of **the ambiguous case** and two interpretations may be possible. In fig. D : 30, there are two possible values for AC and two for $\angle B$.

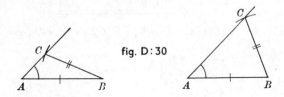

fig. D : 30

AIR NAVIGATION—VOCABULARY

COURSE (Co). The direction in which an aircraft (A/c) is **pointing**; measured by magnetic compass in A/c.

TRUE AIR SPEED (T.A.S.). The speed of an A/c through the air in the direction of the Course.

TRUE COURSE [Co(T)]. After correction of Co as given by compass, it is given as a bearing from *true* north instead of *magnetic* north, measured clockwise, e.g., 220° T = S. 40° W.

VELOCITY. Speed in a given direction.

WIND VELOCITY (W/V). A combination of Wind *Direction* (W/D) and Wind *Speed* (W/S), e.g., W/V 048°/30 mile/h indicates a wind **from** 048° T at 30 mile/h.

TRACK (Tr.). The direction in which an A/c **moves** over the ground below it. A resultant of True Course and Wind Direction. [Track Made Good (T.M.G.).]

DRIFT. The angle between Co(T) and Tr. To PORT (*left*) or STARBOARD (*right*).

GROUND SPEED (G/S). The speed with which an A/c moves over the ground (i.e., on the Track course). A resultant of True Air Speed and Wind Speed.

RECIPROCAL BEARING. Fig. D : 31 shows that the bearing of A from B is 240°. The bearing of B from A is 60°, this is a reciprocal bearing and is obtained quickly thus: 240° − 180° = 60°, alternatively 60° + 180° = 240°.

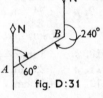

fig. D : 31

NOTE

(i) Although the **Course** of an A/c may be in a given direction, the Track may be quite different (see fig. D : 32). Similarly, the G/S may be quite different from the T.A.S. due to the influence of W/S.

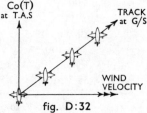

fig. D:32

(ii) It should be clear that ships (and swimmers) may be similarly affected by the influences of wind and current.

(iii) Attention is drawn to the notation: Course →-, Wind →-→-→-, Track or resultant →-→-.

TRIANGLE OF VELOCITIES

A velocity may be represented, in magnitude and direction, by a vector, i.e., a line drawn to a suitable scale to show speed and in an appropriate direction for the bearing.

A number of the quantities mentioned recently can be dealt with in this way:

The true velocity of an A/c: [T.A.S.; Co(T)]
Wind velocity: [W/S; W/D]
Resultant velocity: [G/S; Tr]

These may be combined to form a **triangle of velocities**, see fig. D : 33.

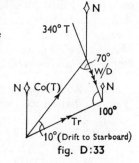

fig. D:33

EXAMPLE 1. *An A/c is flying on a course 050° T at T.A.S. 200 mile/h. The direction of the wind is 340° T and track is 060° T. Find the W/S and G/S.*

NOTE

(i) In order not to overload the diagram, fig. D : 33 (not accurately drawn) deals only with the *directions* of the three velocities.

(ii) The reader should verify the values of the angles of the triangle, i.e., 10°, 70°, 100° with special emphasis on the 70°.

(iii) The arrows of the "component velocities" follow each other round the triangle, the arrows of the resultant (track) are in the opposite direction.

CALCULATION USING SINE FORMULA:

$$\frac{200}{\sin 100°} = \frac{W/S}{\sin 10°} = \frac{G/S}{\sin 70°}$$

$$W/S = \frac{200 \sin 10°}{\sin 100°}; \quad G/S = \frac{200 \sin 70°}{\sin 100°}$$

$$\sin 100° = \sin 80°$$

$$\underline{W/S = 35\cdot3 \text{ mile/h}; \ G/S = 191 \text{ mile/h}}$$

(Each corr. to 3 fig.)

NO.	LOG
200	2·3010
SIN 10°	1̄·2397
NUMER.	1·5407
SIN 80°	1̄·9934
35·27	1·5473
200	2·3010
SIN 70°	1̄·9730
NUMER.	2·2740
SIN 80°	1̄·9934
190·8	2·2806

SOLUTION BY SCALE DRAWING:

A rough sketch from the data should enable you to determine a number of useful angles. (Fig. D : 33.)

Stage (i). Construct an angle of 10° with the arms Co(T) and Tr fairly long. [Fig. D : 34 (i).]

Stage (ii). Along Co(T), using a convenient scale (e.g., 1 in = 50 mile/h), measure a distance AB to represent 200 mile/h.

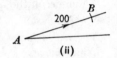

Stage (iii). At B construct ∠ABC = 70°. Measure BC, AC and by using the same scale employed for AB, evaluate W/S and G/S.

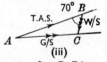

fig. D:34

NOTE. An alternative scale might have been 2 cm = 50 km/h for a velocity in km/h.

EXAMPLE 2. *Find Co(T) and T.A.S. of an A/c if W/V is 030°T/60 km/h, T.M.G. is 090°T and G/S is 320 km/h.*

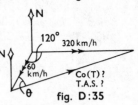

fig. D:35

CALCULATION USING COSINE AND SINE FORMULAE:

$$\cos 120° = -\cos 60°$$
$$(T.A.S.)^2 = (60)^2 + (320)^2 - 2(60)(320)(\cos 120°)$$
$$(T.A.S.)^2 = 3600 + 102400 + 38400 (\cos 60°)$$
$$(T.A.S.)^2 = 3600 + 102400 + 19200 = 125200$$
$$\underline{T.A.S. = 353\cdot9 \text{ km/h}}$$

$$\frac{353\cdot9}{\sin 120°} = \frac{320}{\sin \theta}$$

$$\sin \theta = \frac{320 \sin 60°}{353\cdot9}$$

$$\therefore \text{Co(T)} = 30° + 51° 32' \quad (Why?)$$

$$\sin 120° = \sin 60°$$

NO.	LOG
320	2·5051
SIN 60°	$\overline{1}$·9375
NUMER.	2·4426
353·9	2·5489
51° 32'	$\overline{1}$·8937

Ans. Co(T) = 81° 32'; T.A.S. = 354 km/h. (*Corr. to 3 fig.*)

CHECK. By constructing a suitable scale drawing, verify these results.

NOTE. The scale should be stated clearly.

EXERCISE 19

(*Give angles to nearest degree; speeds, etc., to 3 sig. fig.*)

Find, by scale drawing and calculation, the magnitude and direction of the resultant of the given A/c and wind velocities. (REMEMBER: *wind blows* **from** *the direction given.*)

1. 000°/100 mile/h; 090°/10 mile/h.
2. 000°/100 mile/h; 270°/10 mile/h.
3. 090°/100 mile/h; 180°/22 mile/h.
4. 180°/120 km/h; 270°/10 km/h.
5. 220°/250 km/h; 330°/30 km/h.
6. 220°/250 km/h; 150°/30 km/h.
7. 300°/300 mile/h; 030°/60 mile/h.
8. 120°/300 mile/h; 030°/60 mile/h.
9. 200°/420 km/h; 340°/58 km/h.
10. 140°/512 km/h; 212°/45 km/h.

Find, by scale drawing and calculation, the Co(T) and T.A.S., given W/V; Tr/G.S.:

11. 180°/20 mile/h; 250°/120 mile/h.
12. 180°/30 mile/h; 060°/150 mile/h.
13. 270°/50 mile/h; 180°/120 mile/h.
14. 325°/40 km/h; 220°/280 km/h.
15. 075°/45 km/h; 338°/360 km/h.
16. 020°/62 km/h; 082°/412 km/h.

17. Point B is 1200 yd due East, on the opposite side of a river, from A. A launch, whose speed in still water is 12 mile/h, leaves A steering due East but reaches the opposite bank at a point C, 500 yd South of B, due to the current flowing from North to South. Find: (i) the drift; (ii) speed of the current; (iii) speed from A to C; (iv) time taken from A to C; (v) the necessary Co(T) from A in order that T.M.G. is due East from A to B.

18. An A/c is on course 320° T at T.A.S. 316 km/h. If W/D is 220° T and Tr is 325° T, find W/S and G/S.

19. An A/c is flying Co(T) 200°, W/V is 145° T/58 km/h and Tr is 220° T. Find T.A.S. and G/S.

20. A ship is steaming Co(T) 048° and a current is flowing from 150° T. If the T.M.G. is 032° at 28 knots, find the true speeds of the ship and the current.

21. Find Co(T) and T.A.S. of an A/c if W/V is 140° T/58 mile/h, T.M.G. is 050° T and G/S is 278 mile/h.

22. Find Tr and G/S of an A/c if Co(T) is 194°, T.A.S. is 364 km/h and W/V is 318° T/32 km/h.

23. Find W/V if an A/c is flying Co 028° T at T.A.S. 264 km/h and Tr is 020° T at G/S 242 km/h.

24. A lighthouse B, lies to the North and on a line 600 yd to the West of a jetty A, the distance AB being 2000 yd. A launch, whose speed in still water is 15 knots, wishes to travel directly from A to B. Find the Co(T) of the launch and its speed from A to B if a current of 4 knots is running from East to West.

25. With the basic conditions of Question 24, find the Co(T) of the launch and its speed from A to B if a current is flowing 110° T/5 knots.

26. A swimmer, whose speed in still water is 6 ft/s, wishes to cross a river directly from West to East, a distance of 900 ft. If the river flows from 040° at $2\frac{1}{2}$ ft/s, find the Co(T) which the swimmer must set and the time taken for the crossing.

27. A is 5 miles due North of B on a straight coast line. A ship lies on a bearing 140° from A at a distance of 4 land miles. Calculate the bearing and distance (in land miles) of the ship from B. If the ship is moving due North at 15 knots, and given that 1 knot is 6080 ft/h, find the time taken for the ship to be due East of A. (1 mile = 5280 ft).

28. An A/c leaves A at 12 noon on a course 70° T at T.A.S. 400 km/h. At what time must a second A/c, whose maximum speed is 350 km/h, leave B, 800 km due East of A, in order to intercept the first A/c by the shortest route, if W/V is 190° T/50 km/h.

LATITUDE AND LONGITUDE

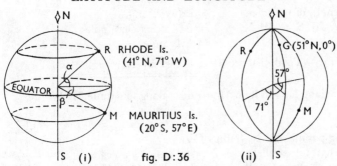

fig. D:36

A parallel of latitude, $\alpha°$ N., is the locus of all points on the earth's surface whose angle of elevation at the centre of the earth is $\alpha°$ from the plane through the Equator. Similarly, $\beta°$ S. is an angle of depression at the centre of the earth. [Fig. D : 36 (i).]

The planes bounded by lines of latitude are **small circles**, perpendicular to the earth's axis; the Equator is an exception, it is a great circle. **Great circles** are planes which pass through the earth's **centre**, they have the same radius as the earth. Great circles which pass through the earth's **axis** (i.e., through the N. and S. poles) are called **meridians, lines of longitude** are semi-meridians. The semi-meridian through Greenwich (51° N., 0°) is used as a reference, in terms of the angles at the earth's centre, from which longitudes can be measured East or West. [Fig. D : 36 (ii).]

The shortest distance between two points on the earth's surface is the arc of the great circle which passes through them, but navigating such a course is extremely difficult because the bearing is constantly changing. "Parallel sailing" consists of following a parallel of latitude until the required line of longitude is reached and then proceeding due north or south, as necessary. In some cases the distance may be shorter by following a longitude until the required parallel of latitude is reached, see the following example.

EXAMPLE. *Assuming the earth to be a sphere of radius 3960 miles and*
$\pi = 3\cdot142$, *calculate the distances from Mauritius Is. (20° S., 57° E.) to Rhode Is. (41° N., 71° W.) by travelling:*
(i) due W. and then due N.; (ii) due N. and then due W.

Along parallel 20° S (radius AB):

$OB = 3960$ miles $AB = (3960 \cos 20°)$ miles
Length of parallel $= 2\pi(3960 \cos 20°)$ miles
Arc 57° E. to 71° W. $= \frac{128}{360} \times 2\pi(3960 \cos 20°)$ miles
                $= 8310$ miles

fig. D:37

Along longitude 71° W.:

Length of meridian $= 2\pi(3960)$ miles
Arc 20° S. to 41° N. $= \frac{61}{360} \times 2\pi(3960)$ miles
                            $= 4216$ miles

Total distance (i) $= \underline{\underline{12\ 500\ \text{miles.}}}$ (*Corr. to 3 fig.*)

Dist. along longitude 57° E. $=$ dist. along longitude 71° W.
                            $= 4216$ miles

Along parallel 41° N. (radius *CD*):
Length of parallel $= 2\pi(3960 \cos 41°)$ miles
Arc 57° E. to 71° W. $= \frac{128}{360} \times 2\pi(3960 \cos 41°)$ miles
                       $= 6674$ miles

Total distance (ii) $= \underline{\underline{10\ 900\ \text{miles}}}$ (*Corr. to 3 fig.*)

fig. D:38

EXERCISE 20

(*Give answers to 3 sig. fig.: earth's radius = 3960 miles or as given.*)

Expressing your answer in terms of π, find the distances between the following points on the earth's surface:

1. (15° N., 0°); (75° S., 0°)
2. (18° N., 20° E.); (27° S., 20° E.) Radius 6360 km.
3. (80° N., 42° W.); (20° N., 42° W.)
4. (72° S., 72° E.); (18° S., 72° E.) Radius 6360 km.

Take π as 3·142 and evaluate the distances between:

5. (40° N., 20° W.); (40° N., 70° E.)
6. (35° N., 22° W.); (35° N., 23° E.) Radius 6370 km.
7. (60° S., 16° W.); (60° S., 24° E.)
8. (73° S., 10° W.); (73° S., 70° W.) Radius 6370 km.
9. (62° N., 30° W.); (62° N., 150° E.)
10. (25° S., 12° E.); (25° S., 132° E.) Radius 6370 km.
11. (45° N., 20° W.); (45° S., 20° E.)
12. (30° N., 30° W.); (30° S., 30° E.) Radius 6370 km.
13. (40° N., 20° W.); (40° S., 30° E.)
14. (20° N., 32° W.); (20° S., 108° W.) Radius 6370 km.
15. Calculate the distances from (20° N., 72° E.) to (30° S., 18° W.) by travelling: (i) W. and S.; (ii) S. and W.
16. Calculate the distances from (42° N., 22° W.) to (18° S., 38° E.) by travelling: (i) E. and S.; (ii) S. and E. Radius 6370 km.
17. Calculate the distances from (21° N., 110° W.) to (87° N., 8° W.) by travelling: (i) E. and N.; (ii) N. and E.

18. Calculate the distances from (78° S., 168° E.) to (16° S., 12° E.) by travelling: (i) W. and N.; (ii) N. and W. Radius 6370 km.

19. Calculate the distances from (40° N., 35° W.) to (10° S., 145° E.) by travelling: (i) E. and S.; (ii) the Great Circle route over the North Pole.

20. A is a point (64° N., 15° E.), B is also on latitude 64° N. but West of A. The meridians through A and B are 6660 km apart at the Equator. Calculate: (i) the longitude of B; (ii) the distance AB along parallel 64° N. Radius 6360 km.

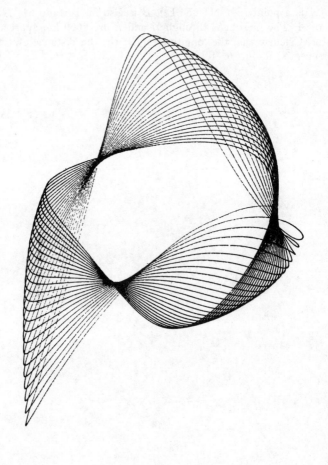

PART E: CALCULUS

GRADIENT

From the graph of a linear function of x (e.g., $y = 3x + 2$) an increase in the value of x can be compared with a corresponding increase in y by considering the tangent of the angle between the graph line and the x-axis. This relationship, $\dfrac{\text{Increase in } y}{\text{Increase in } x}$, is called

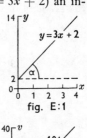

fig. E:1

the **gradient** of the line (i.e., tan α in fig. E : 1).

The gradient of a graph measures the "rate of change" of values on the vertical axis for a given change of values on the horizontal axis. The gradient of a Velocity/Time graph will indicate the **acceleration** of a moving body, i.e., we can find the increase in velocity for a given increase in time. Fig. E : 2 is an example of *constant acceleration.*

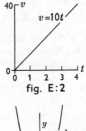

fig. E:2

If a body moves with *variable acceleration* the velocity/time graph would be a curve (see p. 176) and we must now consider the gradient of a function which produces a curve instead of a straight line (e.g., $y = 3x^2 + 2$). Such a curve may be likened to a chain in which very small links form a series of minute straight lines, each with its own gradient and each acting as a tangent to the actual curve, instead of a single rod whose gradient is constant as

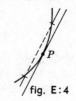

fig. E:3

in fig. E : 1. Since the gradient changes, we must specify the point at which we wish to find the gradient, let the point be P.

GRADIENT BY CONSTRUCTING A TANGENT

The gradient of the selected 'link', P, will be the same as a tangent drawn to the curve at this point. The tangent is probably best constructed by drawing it parallel to the chord through two points on the curve equidistant from P and fairly close to it. (See fig. E : 4.)

fig. E:4

333

EXERCISE 1

Draw the graph of $y = x^2$ from $x = 0$ to $x = 4$. Construct tangents and use them to find the gradient of the curve at the following points:

(i) $x = 1$;   (ii) $x = 2$;   (iii) $x = 3$;   (iv) $x = 2\frac{1}{2}$;   (v) $x = 0.75$

THE GRADIENT OF A CHORD

If PQ is a chord through P the gradient of the chord approaches ($\longrightarrow$) the gradient of the tangent RPS as $Q \longrightarrow P$ (fig. E : 5).

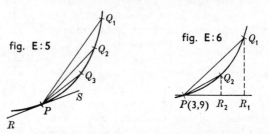

Consider the following table, which shows the gradient of successive chords as Q approaches P $(3, 9)$, a point on the graph $y = x^2$ (fig. E : 6).

COORDINATES		LENGTH		GRADIENT OF PQ $\left(\dfrac{QR}{PR}\right)$
P	Q	PR	QR	
3, 9	Q_1 (4, 16)	1·0	7·0	$7 \div 1 \quad = 7\cdot0$
3, 9	Q_2 (3·5, 12·25)	0·5	3·25	$3\cdot25 \div 0\cdot5 = 6\cdot5$
3, 9	Q_3 (3·1, 9·61)	0·1	0·61	$0\cdot61 \div 0\cdot1 = 6\cdot1$
3, 9	Q_4 (3·01, 9·0601)	0·01	0·0601	6·01
3, 9	Q_5 (3·001, 9·006 001)	0·001	0·006 001	6·001

The results show that as $Q \longrightarrow P$ the gradient of the chord $\longrightarrow 6$ and 6 is said to be the *limit of the gradient of the chord*, therefore the **gradient of the tangent** at P will be 6. Thus, the gradient of the graph $y = x^2$, at the point $x = 3$, is 6.

REMEMBER. The gradient of the tangent is the limit of the gradient of the chord.

THE DERIVED FUNCTION OF $y = x^2$

Consider two points P, Q on the curve $y = x^2$. Let the coordinates of P be (x, y) and of Q be $(x + b, y + h)$ such that b and h are very small increases in the values of x and y respectively (fig. E : 7).

Since $\quad y = x^2$

$y + h = (x + b)^2$ fig. E : 7

$y + h = x^2 + 2bx + b^2$

$\therefore \; h = 2bx + b^2$

$\therefore$ Gradient of chord $PQ = \dfrac{h}{b} = \dfrac{2bx + b^2}{b} = 2x + b$

But as $Q \longrightarrow P$, $b \longrightarrow$ zero and the gradient of $PQ \longrightarrow 2x$.

Since the limit of the gradient of the chord is $2x$, the gradient of the tangent to the graph $y = x^2$ at any point $P(x, y)$ is $2x$.

$2x$ is called the derived function of $y = x^2$

NOTE

(i) The process of finding the *derived function* is called **differentiation.**

(ii) The derived function may also be described as the **differential coefficient of y with respect to x.**

(iii) The **derived function** is indicated by the symbol $\dfrac{dy}{dx}$ and since $y = x^2$ (in this example), our result could be expressed thus:

$$\frac{d}{dx}(x^2) = 2x$$

(iv) $\dfrac{dy}{dx}$ is a symbol complete in itself, it is not a ratio nor a division sum.

USING THE DERIVED FUNCTION

EXAMPLE. *Find the gradient of the tangent to the curve $y = x^2$ at the point* (3, 9).

The derived function gives the gradient at **any** point.

$$\frac{d}{dx}(x^2) = 2x$$

To find the gradient at point (3, 9) substitute for $x = 3$ in the derived function $2x$.

$\therefore$ Gradient at (3, 9) = 6

EXERCISE 2

1. Find the gradient of the tangent to the curve $y = x^2$ at the points:
(i) (1, 1);   (ii) (2, 4);   (iii) $(2\frac{1}{2}, 6\frac{1}{4})$;   (iv) (0·75, 0·5625).

2. Find the gradient of the graph of $y = x^2$ at the points where y has the following values: (i) 16;   (ii) 25;   (iii) 0.

3. Use the method for finding the derived function of $y = x^2$ to differentiate: (i) $2x^2$;   (ii) $3x^2$;   (iii) $4x^2$;   (iv) $\frac{1}{2}x^2$;   (v) $1\frac{1}{2}x^2$.

TO DIFFERENTIATE x^3

Instead of the earlier use of b and h to indicate small increases in the coordinate of x and y, we shall now employ the more formal δx (delta x) to indicate "a small increase in x" and δy (delta y) to indicate "a small increase in y".

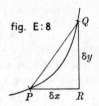

fig. E:8

Consider two points $P(x, y)$ and $Q(x + \delta x, y + \delta y)$ close together on the curve $y = x^3$.

$$\text{Since} \quad y = x^3$$
$$y + \delta y = (x + \delta x)^3$$
$$y + \delta y = x^3 + 3x^2 \cdot \delta x + 3x(\delta x)^2 + (\delta x)^3$$
$$\therefore \; \delta y = 3x^2 \cdot \delta x + 3x(\delta x)^2 + (\delta x)^3$$
$$\therefore \; \text{Gradient of chord } PQ = \frac{\delta y}{\delta x} = \frac{3x^2 \cdot \delta x + 3x(\delta x)^2 + (\delta x)^3}{\delta x}$$
$$\therefore \; \frac{\delta y}{\delta x} = 3x^2 + 3x \cdot \delta x + (\delta x)^2$$

But $\dfrac{\delta y}{\delta x}$ is the gradient of the chord PQ and as $\delta x \longrightarrow 0$, $\dfrac{\delta y}{\delta x} \longrightarrow 3x^2$ so that at the limit the gradient of the tangent at P is $3x^2$.

$$\left(\underset{\delta x \to 0}{\text{Limit}}\right) \frac{\delta y}{\delta x} = \frac{\mathbf{dy}}{\mathbf{dx}} = \mathbf{3x^2}$$

NOTE

(i) The gradient of the *chord* is given by the ratio $\dfrac{\delta y}{\delta x}$.

(ii) The gradient of the *tangent*, i.e., the derived function, is given by $\dfrac{dy}{dx}$. (*Not a ratio.*)

(iii) The method used for finding the derived function of $y = x^2$ and $y = x^3$ is described as **differentiation from first principles.**

Exercise 3

1. Find the gradient of the tangent to the curve $y = x^3$ at the points: (i) $(1, 1)$;   (ii) $(2, 8)$;   (iii) $(4, 64)$;   (iv) $(2\frac{1}{2}, 15\frac{5}{8})$.

2. Find the gradient of the graph of $y = x^3$ at the points where y has the following values: (i) 27;   (ii) 125;   (iii) 0.

3. Use the method for finding the derived function of $y = x^3$ to differentiate: (i) $2x^3$;   (ii) $3x^3$;   (iii) $4x^3$;   (iv) $\frac{1}{2}x^3$;   (v) $1\frac{1}{2}x^3$.

4. Given: $(p + q)^4 = p^4 + 4p^3q + 6p^2q^2 + 4pq^3 + q^4$ differentiate x^4 from first principles.

RESULTS

$$\frac{d}{dx}(x^2) = 2x \qquad \frac{d}{dx}(3x^2) = 6x \qquad \frac{d}{dx}(1\tfrac{1}{2}x^2) = 3x$$

$$\frac{d}{dx}(x^3) = 3x^2 \qquad \frac{d}{dx}(4x^3) = 12x^2 \qquad \frac{d}{dx}(\tfrac{1}{2}x^3) = 1\tfrac{1}{2}x^2$$

$$\frac{d}{dx}(x^4) = 4x^3 \qquad \text{It can be shown } \frac{d}{dx}(4\tfrac{1}{2}x^4) = 18x^3, \text{ etc.}$$

The derived function of $ax^n = n \,.\, ax^{(n-1)}$

TO DIFFERENTIATE A LINEAR FUNCTION

REMEMBER. The *derived function* or *differential coefficient* is the term used for the *gradient* of a function.

1. To find $\frac{d}{dx}(mx)$:

We have seen elsewhere (p. 42) that the gradient of a linear function (e.g., $y = mx$) is given by the coefficient of x.

$$\therefore \frac{d}{dx}(mx) = m$$

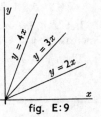

fig. E:9

2. To find $\frac{d}{dx}(c)$:

If a function is in the form of a constant only, then the graph will be a straight line *parallel to the x-axis*, hence the gradient will be zero.

$$\therefore \frac{d}{dx}(c) = 0$$

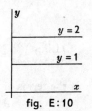

fig. E:10

3. To find $\frac{d}{dx}(mx + c)$:

In such a function as $y = mx + c$, we should ememember (p. 42) that the value of m controls the ɹradient; the value of c governs the intercept on the y-axis.

$$\therefore \frac{d}{dx}(mx + c) = m$$

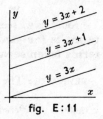

fig. E:11

EXERCISE 4

1. Write down the derived functions of: x, $2x$, $-3x$, $4x$, $\tfrac{1}{2}x$, $0.24x$, $-6.5x$.

2. Write down the derived functions of: 1, 2, 3, -4, $\tfrac{1}{2}$, $-\tfrac{1}{4}$, 7.25.

3. Write down the derived functions of: $x + 3$, $2 - 2x$, $5x - 4$, $8x + \frac{1}{2}$, $12x - \frac{3}{4}$, $6x + 1$, $4 - 5x$.

4. Write down the values of the y-intercepts if graphs were drawn of the functions in Question 3.

5. What would the graphs of the following functions have in common: $5x - 1$, $5x + 2$, $5x - 6$, $5x + 8$?

6. What would the graphs of the following functions have in common: $2x + 3$, $3 - 5x$, $x^2 + 3$, $3 - 2x^2$?

TO DIFFERENTIATE A SUM

EXAMPLE. *Find, from first principles, the derived function of* $3x^2 - 5x + 2$.

Let $y = 3x^2 - 5x + 2$. Suppose an increase in x of δx gives an increase in y of δy.

Then
$$y + \delta y = 3(x + \delta x)^2 - 5(x + \delta x) + 2$$
$$y + \delta y = 3[x^2 + 2x \cdot \delta x + (\delta x)^2] - 5x - 5\delta x + 2$$
$$y + \delta y = 3x^2 + 6x \cdot \delta x + 3(\delta x)^2 - 5x - 5\delta x + 2$$
But $y = 3x^2 - 5x + 2$
$$\therefore \delta y = \qquad 6x \cdot \delta x + 3(\delta x)^2 \qquad - 5\delta x$$
$$\therefore \frac{\delta y}{\delta x} = \frac{6x \cdot \delta x + 3(\delta x)^2 - 5\delta x}{\delta x} = 6x + 3\delta x - 5$$
$$\left(\underset{\delta x \to 0}{\text{Limit}}\right)\frac{\delta y}{\delta x} = 6x - 5 \qquad \therefore \frac{dy}{dx} = 6x - 5$$

If the terms of the sum are taken separately:
$$\frac{d}{dx}(3x^2) = 6x; \quad \frac{d}{dx}(-5x) = -5; \quad \frac{d}{dx}(2) = 0$$

From this we see that: **The derived function of the sum of several terms is the same as the sum of the derived functions of the terms taken separately.**

REMEMBER. The derived function gives the gradient at *any* point.

EXAMPLE. *Differentiate* $2x^3 - 3x - 4$ *and find the gradient at the point* $(2, 6)$.
$$\frac{d}{dx}(2x^3 - 3x - 4) = 6x^2 - 3$$

Gradient at $(2, 6)$: Substitute for $x = 2$ in $6x^2 - 3$
$$\therefore \text{Gradient} = 21$$

EXERCISE 5

Write down the derived functions of:

1. $x^2 + 1$ **2.** $x^3 - 3$ **3.** $x^4 + 2$ **4.** $x^5 - 4$

5. $2x^2 - 3$ **6.** $3x^2 + x$ **7.** $2x^2 - 3x$ **8.** $3x^2 + 2x$

9. $x^3 - x$ **10.** $2x^3 + x^2$ **11.** $x^3 + 2x$ **12.** $2x^3 - 3x^2$

13. $x^3 + x^2 - x$ **14.** $2x^3 - 2x^2 - 2$ **15.** $3x^3 + x^2 + 4x$

16. $x^3 - 4x^2 + 3x$ **17.** $2x^4 + 5x + 1$ **18.** $x^3 - x^2 + x - 1$

19. $\frac{1}{2}x^2 - 3x + 4$ **20.** $\frac{1}{3}x^3 + \frac{1}{4}x^2$ **21.** $\frac{1}{2}x^4 - \frac{2}{3}x^3$

22. $\frac{2}{3}x^3 + \frac{3}{4}x^2$ **23.** $\frac{5}{6}x^3 - 1\frac{1}{2}x^2$ **24.** $\frac{3}{8}x^4 + 2\frac{1}{2}x^2$

Find the gradients of the following, at the given points:

25. $3x + 2$ $(2, 8)$ **26.** $x^2 - x$ $(3, 6)$ **27.** $x^3 + 4$ $(2, 12)$

28. $2x^2 + 3x$ $(1, 5)$ **29.** $2x^3 - 2x$ $(2, 12)$ **30.** $x^4 - 3x^3$ $(3, 0)$

31. $x^3 - 3x^2 + 5x$ $(2, 6)$ **32.** $3x^2 - 5x - 6$ $(3, 6)$

33. $3x^3 - 3x^2 - 3x$ $(3, 45)$ **34.** $\frac{3}{4}x^3 + \frac{3}{4}x^2$ $(4, 60)$

TO DIFFERENTIATE $\dfrac{1}{x}$

Let $y = \dfrac{1}{x}$. Suppose an increase in x of δx gives an increase in y of δy.

Then $y + \delta y = \dfrac{1}{x + \delta x}$ But $y = \dfrac{1}{x}$

By subtraction: $\delta y = \dfrac{1}{x + \delta x} - \dfrac{1}{x}$ $= \dfrac{x - (x + \delta x)}{x(x + \delta x)}$

$$\therefore \frac{\delta y}{\delta x} = \frac{x - x - \delta x}{x^2 + x \cdot \delta x} \div \frac{\delta x}{1} = -\frac{\delta x}{x^2 + x \cdot \delta x} \times \frac{1}{\delta x}$$

$$\therefore \frac{\delta y}{\delta x} = -\frac{1}{x^2 + x \cdot \delta x} \quad \left[\text{As } \delta x \to 0, \frac{\delta y}{\delta x} \to -\frac{1}{x^2} \right]$$

$$\left(\underset{\delta x \to 0}{\text{Limit}} \right) \frac{\delta y}{\delta x} = -\frac{1}{x^2} \quad \therefore \frac{\mathbf{d}}{\mathbf{d}x} \left(\frac{1}{x} \right) = -\frac{1}{x^2}$$

NOTE. The rule $\dfrac{\mathbf{d}}{\mathbf{d}x}(ax^n) = n \cdot ax^{(n-1)}$ still applies.

Consider the problem in this form: $\dfrac{1}{x} = x^{-1}$

In accordance with the rule:

Multiply the coefficient by the power and decrease the power by 1:

$$\text{Then} \quad \frac{\mathbf{d}}{\mathbf{d}x} \left(\frac{1}{x} \right) = -x^{-2} = -\frac{1}{x^2}$$

EXAMPLES

(i) $\dfrac{d}{dx}\left(\dfrac{1}{x^2}\right) = \dfrac{d}{dx}(x^{-2}) \quad = -2x^{-3}$ or $-\dfrac{2}{x^3}$

(ii) $\dfrac{d}{dx}\left(\dfrac{2}{x^3}\right) = \dfrac{d}{dx}(2x^{-3}) \quad = -6x^{-4}$ or $-\dfrac{6}{x^4}$

(iii) $\dfrac{d}{dx}(\sqrt{x}) = \dfrac{d}{dx}(x^{\frac{1}{2}}) \quad = \frac{1}{2}x^{-\frac{1}{2}}$ or $\dfrac{1}{2\sqrt{x}}$

(iv) $\dfrac{d}{dx}\left(2x^2 - 3x + \dfrac{2}{x} - \dfrac{3}{x^2}\right) = 4x - 3 - \dfrac{2}{x^2} + \dfrac{6}{x^3}$

TO DIFFERENTIATE A PRODUCT OR A QUOTIENT

EXAMPLES

(i) $\dfrac{d}{dx}[x(2x^2 + 3x)] = \dfrac{d}{dx}(2x^3 + 3x^2) = 6x^2 + 6x$

(ii) $\dfrac{d}{dx}\left(\dfrac{3x^4 - 2x^3}{x^2}\right) = \dfrac{d}{dx}(3x^2 - 2x) = 6x - 2$

NOTE. There are, in fact, special methods for differentiating *products* and *quotients*, but for the present such functions must be simplified to form a *sum* of several terms.

EXAMPLE. *Find, on the graph of $y = 3x(3 - 2x)$, the coordinates of the point at which the gradient is 3.*

The gradient at *any* point is given by $\dfrac{d}{dx}[3x(3 - 2x)]$

$\dfrac{d}{dx}(9x - 6x^2) = 9 - 12x$. But at one point the gradient = 3.

$\therefore\ 9 - 12x = 3$ and $x = \frac{1}{2}$ at this point.

From the function $y = 9x - 6x^2$, if $x = \frac{1}{2}$, $y = 3$.

$\therefore$ The point has coordinates $(\frac{1}{2}, 3)$

EXERCISE 6

Differentiate the following:

1. $\dfrac{1}{2x}$ 2. $\dfrac{1}{3x}$ 3. $-\dfrac{1}{x}$ 4. $\dfrac{2}{x}$ 5. $\dfrac{x}{2}$

6. $-\dfrac{3}{x}$ 7. $\dfrac{x}{3}$ 8. $\dfrac{2}{3x}$ 9. $\dfrac{3x}{2}$ 10. $-\dfrac{3x}{4x^2}$

11. $\dfrac{2}{x^2}$ 12. $\dfrac{x^2}{2}$ 13. $-\dfrac{1}{x^3}$ 14. $\dfrac{x^3}{3}$ 15. $\dfrac{3}{x^4}$

16. $2x^{\frac{1}{2}}$ 17. $\sqrt[3]{x}$ 18. $\dfrac{\sqrt{x}}{x^2}$ 19. $\dfrac{x^2}{\sqrt{x}}$ 20. $\dfrac{1}{3x^{\frac{1}{3}}}$

21. $x(3x - 2)$ **22.** $2x(2 - 3x)$ **23.** $3x(x^2 + 3x - 2)$

24. $x^2(x + 2)$ **25.** $2x^2(3x - 2)$ **26.** $3x^2(x^2 - 2x + 3)$

27. $(x + 2)(x - 1)$ **28.** $(2x - 1)(x + 3)$ **29.** $(3x - 2)(x - 3)$

30. $2(x + 1)(x - 1)$ **31.** $3(1 - 2x)(2 + x)$ **32.** $4(3x + 2)(2x - 3)$

33. $\dfrac{2x^2 + 3x}{x}$ **34.** $\dfrac{3x^3 - 2x^2}{x^2}$ **35.** $\dfrac{4x^4 - 6x^3}{2x^2}$ **36.** $3x^2 - 2x + \dfrac{3}{x}$

37. $\dfrac{4}{x^2} - \dfrac{5}{x} + 2x$ **38.** $\dfrac{2x^2 + x - 6}{x + 2}$ **39.** $\dfrac{6x^2 - 23x + 20}{2x - 5}$

40. Find the gradient of $y = \dfrac{1}{x}$ at $(3, \tfrac{1}{3})$.

41. Find the gradient of $y = 4 - x^2$ where it cuts (i) the x-axis, (ii) the y-axis.

42. Find the gradient of $y = 4x^2 - 9$ where it cuts (i) the x-axis, (ii) the y-axis.

43. Find the gradient of $y = 2x^2 + 3x$ at its points of intersection with $y = 6x - 1$.

44. Find the gradient of $y = \dfrac{1}{x}$ at its points of intersection with $y = 3x - 2$.

45. Find the gradients of $y = 2x^2$ and $y = \dfrac{16}{x}$ at their point of inter-section.

46. Find the gradients of $y = 3x^2$ and $y = \dfrac{3}{x^2}$ at their points of inter-section.

47. Find, on $y = 2x(3x + 7)$, the coordinates of the point at which the gradient is 2.

48. Find, on $y = 3x(2 - 3x^2)$, the coordinates of the points at which the gradient is -21.

49. Find, on $y = \tfrac{3}{4}\left(\dfrac{x^3 + \tfrac{1}{2}x^2 - 2x}{3x}\right)$, the coordinates of the point at which the gradient is $3\tfrac{1}{8}$.

50. Find the gradients of successive tangents to $y = 2x^2 + 2$ at the points $x = -2, -1, 0, 1, 2$. Sketch the tangents in sequence and use them to observe the shape of the graph. Is it hill-shaped or bowl-shaped?

MAXIMUM AND MINIMUM VALUES

From fig. E : 12:

In the section AB the gradient is positive, i.e., $\dfrac{dy}{dx}$ is +ve.

In the section BC the gradient is negative, i.e., $\dfrac{dy}{dx}$ is −ve.

In the section CD $\dfrac{dy}{dx}$ is again positive.

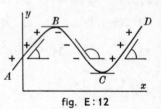

fig. E : 12

Now consider the gradients at the turning-points B and C. If tangents were drawn at these points they would be parallel to the x-axis, and consequently their gradients would be zero. Thus, **at a turning-point $\dfrac{dy}{dx}$ is zero.**

By comparison with adjacent points, B is clearly a **maximum** value of y, though there may be higher values farther away (e.g., beyond D). Similarly, C is described as a **minimum** value of y, though beyond A there may be lower values.

CONCLUSION

y [i.e., $f(x)$] **has a maximum or minimum value when $\dfrac{dy}{dx}$ is zero.**

As x increases (i.e., moving left to right):

y **is a maximum if $\dfrac{dy}{dx}$ changes from +ve to −ve.**

y **is a minimum if $\dfrac{dy}{dx}$ changes from −ve to +ve.**

NOTE. Fig. E : 13 shows a point P on a curve such that the gradient at P is zero, yet the gradient on either side does not change in sign. P is called a **point of inflexion.**

fig. E : 13

EXAMPLE. *Find the point for which there is a maximum or minimum value of $3x^2 - 6x$ and sketch the curve.*

$$\text{Let} \quad y = 3x^2 - 6x \qquad \text{Then} \quad \frac{dy}{dx} = 6x - 6$$

y is a max. or min. when $6x - 6 = 0$ (i.e., $x = 1$) $\therefore \dfrac{dy}{dx} = 0$ when $x = 1$ and $y = -3$.

We now know the turning-point $(1, -3)$ but not whether it is a max. or min. value of y. We will consider the sign of the gradient for values of x on each side of 1 by substituting in $(6x - 6)$.

$(1-)$ "a little less than 1"; $(1+)$ "a little greater than 1".

If $x = 1-$, $\dfrac{dy}{dx}$ is $-$ve (On the left)

If $x = 1+$, $\dfrac{dy}{dx}$ is $+$ve (On the right)

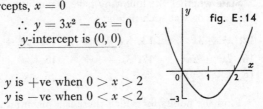

$\therefore$ y is a minimum at $(1, -3)$

To sketch the graph, we also need to know: (i) the points at which the graph cuts the x-axis; (ii) the point at which the graph cuts the y-axis.

(i) For the x-intercepts, $y = 0$ $\therefore$ $3x^2 - 6x = 0$

$\therefore$ $3x(x - 2) = 0$ and $x = 0$ or 2

x-intercepts are $(0, 0)$ and $(2, 0)$

(ii) For the y-intercepts, $x = 0$

$\therefore$ $y = 3x^2 - 6x = 0$

y-intercept is $(0, 0)$

In this example:

y is $+$ve when $0 > x > 2$

y is $-$ve when $0 < x < 2$

fig. E:14

NOTE. Examination of the gradient on each side of a turning-point requires great care, particularly if the derived function contains several terms. The task may be simplified if the derived function can be expressed in factor form.

EXAMPLE. *Find the turning-points on $y = \frac{1}{3}x^3 + x^2 - 3x$ and distinguish between them.*

If $\dfrac{dy}{dx} = x^2 + 2x - 3 = (x + 3)(x - 1) = 0$

$x = -3$ or 1 and $y = 9$ or $-1\frac{2}{3}$

$\therefore$ The turning-points are $(-3, 9)$ and $(1, -1\frac{2}{3})$

To test the gradient left and right of these points, we must consider $x = (-3-)$ and $(-3+)$; $(1-)$ and $(1+)$.

Consider the derived function in the form $(x + 3)(x - 1)$ and examine the sign value of each bracket:

If $x = -3-$; $(-ve)(-ve)$; product is $+ve$⎱
If $x = -3+$; $(+ve)(-ve)$; product is $-ve$⎰

If $x = 1-$; $(+ve)(-ve)$; product is $-ve$⎱
If $x = 1+$; $(+ve)(+ve)$; product is $+ve$⎰

$$\therefore \quad \underline{\underline{(-3, 9) \text{ is a maximum value of } y.}}$$

$$\underline{\underline{(1, -1\tfrac{2}{3}) \text{ is a minimum value of } y.}}$$

NOTE. $(-3\cdot1)$ is "a little less" than (-3) and $(-2\cdot9)$ is "a little greater" than (-3).

EXERCISE 7

Find the turning-points on the following graphs of y:

1. $x^2 + 3$ 2. $2x^2 - 5$ 3. $x^2 + 2x$ 4. $3x^2 - 4$

5. $3x^2 - 2x$ 6. $x^3 - 4$ 7. $x^3 - 3x$ 8. $3x^3 - x$

9. $1\tfrac{1}{3}x^3 - x$ 10. $x^3 - x^2 - x$ 11. $\tfrac{1}{3}x^3 - \tfrac{1}{2}x^2 - 6x$

State whether the following have maximum or minimum values and the value of x at such points:

12. $x^2 + 3x$ 13. $3x - 2x^2$ 14. $3x^2 + 4x$ 15. $2x^3 - 6x$

16. $2x^3 + 2x^2$ 17. $3x - 4x^3$ 18. $\tfrac{1}{3}x^3 + \tfrac{1}{2}x^2 - 2x$

19. $12x - 2\tfrac{1}{2}x^2 - x^3$ 20. $3x - 2x^2 - 1\tfrac{1}{3}x^3$

Find the points for which there is a maximum or minimum value of y and sketch the curve:

21. $y = x^2 - 4x$ 22. $y = 2x^2 - 4$ 23. $y = 9 - x^2$

24. $y = 3x^2 - 12x$ 25. $y = 5 + 3x - 2x^2$ 26. $y = 3x - x^3$

27. Find the point on $y = 4 - x^4$ at which the tangent is parallel to the x-axis. State whether y has a maximum or minimum value at this point.

28. Find the turning-points on $y = 4x + \dfrac{1}{x}$ and distinguish between them.

29. Find the turning-point on $y = 4x^2 + 3x - 10$ and use it to help you to sketch the graph. For what values of x is y: (i) positive; (ii) negative?

30. Find the turning-points on $y = 4 + 9x - 3x^2 - x^3$ and distinguish between them.

PROBLEMS ON MAXIMA AND MINIMA

EXAMPLE. *An open tank having a volume of $13\frac{1}{2}$ ft³ and a square base is to be made from sheet metal. Find the length of side of the base and the height such that the area of metal is a minimum.*

Volume $= x^2h = 13\frac{1}{2}$ $\therefore$ $h = \dfrac{13\frac{1}{2}}{x^2}$

Area of metal $= x^2 + 4(xh)$

$$\therefore A = x^2 + 4x\left(\frac{13\frac{1}{2}}{x^2}\right) = x^2 + \frac{54}{x}$$

fig. E:15

A is a max. or min. when the derived function $= 0$

$$\frac{\mathrm{d}A}{\mathrm{d}x} = 2x - \frac{54}{x^2} = 0 \quad \therefore\ 2x^3 = 54$$

$$\therefore\ x = 3 \quad \text{(A min. by testing gradient.)}$$

$$\therefore\ h = \frac{13\frac{1}{2}}{9} = 1\frac{1}{2}$$

Side of base $= 3$ ft; height $= 1\frac{1}{2}$ ft

NOTE

(i) The problem begins with several "unknowns" (x, h, A). By expressing one in terms of another, these are reduced to two, A and x, A being expressed as a function of x.

(ii) Since $A = f(x)$, the derived function is in the form $\dfrac{\mathrm{d}A}{\mathrm{d}x}$.

EXERCISE 8

1. An open tank having a volume of 32 ft³ and a square base is to be made of sheet metal. Find the length of side of the base and the height such that the area of metal is a minimum.

2. Repeat the requirements of Question 1 for a closed tank having a volume of 8 ft³.

3. An open box having a square base is made from 108 dm² of material. Find the dimensions to give maximum volume.

4. A sheet of metal 8 cm square has squares of side x cm removed from each corner, and the sides are bent up to form an open box x cm deep. Find the value of x for the volume to be a maximum.

5. Find the greatest rectangular area which can be enclosed by 1000 ft of fencing.

6. A rectangular enclosure is to be constructed using a straight wall as one boundary and 48 yd of fencing from which to make the other three. Find the dimensions of the enclosure and the area if this is to be a maximum.

7. With the conditions of Question 6, find the minimum length of fencing required to enclose an area of 50 m².

8. A rectangular pen is to be constructed from 100 ft of fencing wire, but one of the shorter ends is to have two layers of wire. Find the greatest area that can be enclosed.

9. Divide 12 into two parts so that (i) their product is a maximum; (ii) the sum of their squares is a minimum.

10. If $x - y = 4$, find the minimum value of $x^2 - xy + y^2$.

11. A closed cylinder of volume 432π cm³ is to be made from sheet metal. Find the radius so that the area of metal is a minimum.

12. A cylinder, open at one end and having a surface area of 243π cm², is to be made from sheet metal. If the volume is to be a maximum, find the radius and also the volume in terms of π.

13. Regulations require that the dimensions of a parcel be such that the sum of the length and girth does not exceed 6 ft. (Girth = perimeter of cross-section.) Find the greatest volume which complies with the requirements: (i) for a square cross-section; (ii) for a cylinder.

14. If a body is projected vertically upwards, the height after t seconds is $(40t - 16t^2)$ ft. Find the time taken to reach maximum height and the height so reached.

15. Repeat Question 14 for $(160t - 16t^2)$ cm.

VELOCITY AND ACCELERATION

Brief reference to Velocity/Time graphs was made earlier to introduce the idea of gradient of a curved line. Velocity and acceleration are now discussed in greater detail.

If a body is moving such that it travels over equal distances in equal periods of time it is said to be moving with **uniform velocity.** Fig. E : 16 shows a *Travel Graph* or **Distance/Time Graph** (s = Space = Distance), and we see that in the 1st second the body moves 10 ft, in the 1st two seconds it moves 20 ft; during *each* second of time it moves a distance of 10 ft. This is represented by the relationship $s = 10t$

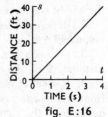

fig. E:16

Velocity, then, is a measure of the distance covered in a given period of time and is given in miles/h, ft/s, km/h or cm/s, **average velocity** $= \dfrac{\text{Total distance}}{\text{Total time}}$.

In fig. E: 16 $\dfrac{\text{Total distance}}{\text{Total time}} = \dfrac{40}{4} = 10$ ft/s. The velocity of the body at some **particular moment in time**, however, is given by the gradient of the Distance/Time Graph at the precise moment in question $\left(\text{i.e., } \dfrac{ds}{dt}\right)$. In fig. E : 16 the velocity at the end of the 3rd second precisely is given by the gradient of the graph at the point where $t = 3$; we see that the velocity is 10 ft/s $\left(\text{i.e., } \dfrac{ds}{dt} = 10\right)$. This is the same as the average velocity, but we are dealing with a special case: *uniform velocity*.

Let us now consider the more probable case in which **velocity is varying**, e.g., a car starts from rest and gradually increases its velocity (i.e., it **accelerates**). Fig. E : 17 illustrates such a situation in which $s = 5t^2$. We see that at the end of the 1st second the body has moved 5 ft; at the end of 2 sec., 20 ft; at the end of 3 sec., 45 ft, etc. Thus, *during* the 2nd second it travelled 15 ft; *during* the 3rd second, 25 ft; *during* the 4th second, 35 ft. Clearly, the body is increasing its speed by an extra 10 ft per sec. *every* second, and we say that its acceleration is 10 ft/s² (i.e., 10 ft per sec. per sec.).*

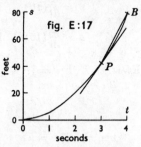

fig. E:17

As in the case of uniform velocity, we can make a comparison between average velocity during a **period of time** and the velocity at some **precise instant in time**:

$$\begin{array}{l}\text{Average velocity} \\ \text{from 3rd sec. to 4th sec.}\end{array} = \frac{\text{Dist. travelled}}{\text{Time taken}} = \frac{35}{1} = \underline{35 \text{ ft/s}}$$

$$\begin{array}{l}\text{Velocity at the end of} \\ \text{the 3rd sec. (precisely)}\end{array} = \text{gradient of } (s = 5t^2) \text{ at } (t = 3)$$

$$\frac{d}{dt}(5t^2) = 10t \qquad \text{After 3 sec. velocity} = \underline{30 \text{ ft/s}}$$

In fig. E : 17 average velocity from the 3rd second to the 4th second is illustrated by the gradient of the chord $PB\left(\dfrac{\delta s}{\delta t} \text{ as it were}\right)$, whereas the velocity after precisely 3 sec. is illustrated by the gradient of the tangent at $P\left(\text{i.e., } \dfrac{ds}{dt}\right)$.

Fig. E : 18 is the **Velocity/Time Graph** obtained by plotting, from the above data, the velocities at the ends of successive second intervals. The gradient of this graph $\left(\dfrac{dv}{dt}\right)$ is a measure of acceleration, i.e., the extent to which velocity is increasing with the passage of time. The gradient shows that $a = 10$ ft/s², which agrees with the

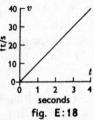

fig. E:18

result obtained by previous calculation.* The linear graph and $\dfrac{dv}{dt}$ indicate that, in this example, acceleration is uniform.

CONCLUSION

When distance (s ft) is expressed in terms of time (t sec.)

$$\textbf{Velocity } (v) = \frac{ds}{dt}; \quad \textbf{Acceleration } (a) = \frac{dv}{dt}$$

EXAMPLE 1. *If $s = 5t^2$, find the velocity and acceleration after 3 sec.*

$$v = \frac{d}{dt}(5t^2) = 10t \quad \therefore \text{ After 3 sec. } v = \underline{30 \text{ ft/s}}$$

$$a = \frac{d}{dt}(10t) = 10 \quad \therefore \text{ After 3 sec. } a = \underline{10 \text{ ft/s}^2}$$

EXAMPLE 2. *If $s = \frac{2}{3}t^3 + \frac{3}{4}t^2 + 4t$, find the velocity and acceleration at the end of 3 sec.*

$$v = \frac{d}{dt}(\tfrac{2}{3}t^3 + \tfrac{3}{4}t^2 + 4t) = 2t^2 + 1\tfrac{1}{2}t + 4$$

When $t = 3$, $v = 18 + 4\tfrac{1}{2} + 4 = \underline{26\tfrac{1}{2} \text{ ft/s}}$

$$a = \frac{d}{dt}(2t^2 + 1\tfrac{1}{2}t + 4) = 4t + 1\tfrac{1}{2}$$

When $t = 3$, $a = 12 + 1\tfrac{1}{2} \qquad = \underline{13\tfrac{1}{2} \text{ ft/s}^2}$

NOTE. Velocity and acceleration may each be +ve or −ve. Graphically, each may be thought of thus: −ve ⟷ +ve. (*Speed* takes no account of direction.) A +ve acceleration would be taken generally to mean an increase in speed and a −ve acceleration a decrease in speed (retardation). However, if the direction of the velocity is also taken into account the results need careful thought, particularly in the case of a −ve acceleration acting on a −ve velocity. (The result is an increase in *speed*!)

EXERCISE 9

(*s = distance; t = time in sec.*)

1. Given that $s = (2t^2 - 3t + 4)$ ft, find expressions for v and a. If $t = 3$, evaluate s, v and a.

2. Given that $s = (3t^2 + 2t - 5)$ cm, find expressions for v and a. If $t = 2$, evaluate s, v and a.

3. A particle moves along a straight line OA, so that after t sec. the distance from O is s ft, where $s = 2 - 4t - 2t^2$. Find expressions for v and a. If $t = 2$, evaluate s, v and a; explain the nature of the answers.

4. A body moves along a straight line so that, after t sec., the distance from a fixed point O is given by $s = (t^3 - 2t^2 + 6)$ ft. Calculate:

(i) the distance from O when motion is about to begin;

(ii) the velocity when motion is about to begin;

(iii) s, v and a after 1 sec.;

(iv) s, v and a after 2 sec.;

(v) the distance travelled *during* the 2nd second;

(vi) the time elapsing for the body to come momentarily to rest again;

(vii) the time elapsing for acceleration to reach 0 ft/s^2;

(viii) the least distance between the body and point O.

5. A particle moves along a straight line OA so that t sec. after passing O the distance from O is s cm., where $s = t^3 - 6t^2 + 9t$. Calculate:

(i) the distance and position from O when $t = 0$; $t = 2$;

(ii) the velocity when $t = 0$; $t = 2$;

(iii) the acceleration when $t = 0$; $t = 2$;

(iv) the times at which the velocity is momentarily zero;

(v) the distances from O at these times.

6. A body moves along a straight line OA so that after t sec. the distance from a fixed point O is given by $s = (6 - 4t + 3t^2)$ cm. Calculate:

(i) the distance from O at the instant from which t is reckoned;

(ii) the velocity at this moment;

(iii) the acceleration at this moment. Does the acceleration vary?

(iv) the time at which the body is momentarily at rest. What changes does the velocity undergo an instant after this time?

(v) the velocity after 2 sec.; after 3 sec.;

(vi) the distance travelled between the 2nd and 3rd seconds (i.e. *during* the 3rd second).

7. A crane lifts a load from the ground. The height above ground after rising for t sec. is given by $h = (6t^2 - t^3)$ ft. Calculate:

(i) the velocity after t sec.;

(ii) the time and the height when the load comes to rest;

(iii) the height at which the acceleration is zero.

8. A body starts from rest at O and moves in a straight line so that after t sec. its distance from O is given by $s = (4\frac{1}{2}t^2 - t^3)$ cm. Calculate:

(i) the time required for the body to begin to return to O and the distance from O at that time;

(ii) the speed and acceleration of the body as it reaches O on the return journey.

9. A particle moves along a straight line so that t sec. after passing point O the distance travelled is given by $s = (t^3 - 2t^2 + 5t)$ ft. (i) Calculate the distance OP if the particle passes point P 5 sec. after passing O. (ii) Show that the particle has travelled $\frac{1}{10}OP$ when the acceleration is 8 ft/s².

10. A shell is fired vertically upwards so that after t sec. its height is given by $h = (800t - 16t^2)$ ft. Calculate:

(i) after what time the shell is at maximum height;
(ii) the maximum height;
(iii) the average velocity for the period during which the shell rises;
(iv) the height and velocity at the middle of this period of time.

OTHER APPLICATIONS OF GRADIENT

11. After t min the volume (dm³) of water in a tank is given by $V = 8 + 9t - t^2$. Calculate:

(i) the rate at which the volume is increasing after t min;
(ii) the rate at which the volume is increasing after 3 min;
(iii) after what time the volume ceases to increase.

12. An aircraft flying from A to B at v miles/min consumes F gal of fuel where $F = 10v^2 + \dfrac{810}{v^2}$. Calculate:

(i) the rate at which F is increasing with respect to v when $v = 5$;
(ii) the least quantity of fuel required by this A/c to fly this particular journey and the most economical speed of the A/c in mile/h.

13. The surface area, A cm², of a toy balloon after t sec. is given by $A = 12t^2 + 12t$. Calculate:

(i) the rate at which the surface area is increasing when $t = 2$ sec.;
(ii) the rate at which the surface area is increasing when the balloon bursts if the maximum area it can attain is 240 cm².

14. Man A starts at a point 4 km due West of O walking due West at 3 km/h. Man B starts at a point 16 km due South of O walking due North at 4 km/h. Find:

(i) an expression of the area (km²) of the triangle AOB after t h;
(ii) the rate, in km²/h at which the area of the triangle is changing after t h; 1 h; 2 h;
(iii) the time at which the area of the triangle stops increasing and begins to decrease;
(iv) the maximum area of the triangle.

15. The load-bearing properties of a rectangular beam of fixed length vary as x^2y, where x is the depth and y the breadth, each in decimetres. Find the dimensions of the cross-section of the strongest beam whose perimeter is 6 dm.

16. The production cost, C pence, per item manufactured varies as n, the number produced, such that $C = \dfrac{1}{2} + \dfrac{4n^2}{10^6} + \dfrac{10^3}{n}$. Find:

    (i) the number of items to be manufactured for minimum production cost;

    (ii) the minimum production cost per item.

17. The production cost, C pence, per item manufactured varies as t, the time (min) taken to make it, such that $C = \frac{1}{2}t + \dfrac{16}{t^2}$. Calculate:

    (i) the rate at which the cost is changing with respect to time when $t = 2$ sec.;

    (ii) the minimum production cost per item.

18. A ship covers a distance of 2304 sea miles at v knots. The cost of fuel for the ship is $£\left(3 + \dfrac{v^3}{1152}\right)$ per hour. Find:

    (i) an expression for C, the cost in £s of fuel on this voyage;

    (ii) the most economical speed for the ship;

    (iii) the cost of fuel for the voyage.

INTEGRATION

Integration is the name given to the process of finding the original function from which a derived function has been obtained; it is the reverse of differentiation, and hence may be described as *anti-differentiation*.

Thus: given $\dfrac{dy}{dx}$, we require to find y in terms of x.

EXAMPLE 1. *If* $\dfrac{dy}{dx} = 3$, *find y in terms of x.*

In differentiating y to arrive at 3, y must have been equal to $3x$. But $y = 3x + 1$, $y = 3x + 2$, $y = 3x + 3$ all give the derived function **3** because the derived function of 1, 2, 3 (constants) is zero in every case.

$$\left[\frac{dy}{dx}(mx + c) = m\right.$$

Without more information, we are able to give only the *general solution*: $\underline{\underline{y = 3x + c}}$

NOTE. c is called an **arbitrary constant,** i.e., its *precise* value is unknown.

EXAMPLE 2. *If* $\dfrac{dy}{dx} = 4x^2$, *find y in terms of x.*

(i) *Index figure*: x^2 has been derived from x^3.

(ii) *Coefficient*: 4 has been derived by multiplying the original coefficient by the original index figure, 3. Thus:

$$4 = a \times 3 \qquad\qquad \therefore \frac{a}{3} = \frac{4}{3} = 1\tfrac{1}{3}$$

(iii) *Original function*: $y = 1\tfrac{1}{3}x^3$

CHECK the function $y = f(x)$ by differentiating.

(iv) *Constant term*: $1\tfrac{1}{3}x^3$; $1\tfrac{1}{3}x^3 + 2$; $1\tfrac{1}{3}x^3 - 2$, etc., each produce the derived function $4x^2$ and without more information we are able to give only the general solution: $\qquad \underline{y = 1\tfrac{1}{3}x^3 + c}$

NOTE. The general solution indicates the nature of a family of curves whose gradient function $\left(\dfrac{dy}{dx}\right)$ is the same (fig. E : 19).

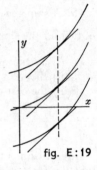

CONCLUSION

If $\dfrac{dy}{dx} = ax^n$, then $y = \dfrac{ax^{(n+1)}}{n+1} + c$

The statement is true for all values of n, **except $n = -1$**; this requires special consideration at a later stage. [If $n = -1$, $(n+1) = 0$.]

fig. E : 19

EXERCISE 10

Integrate the following with respect to x:

1. 3	**2.** 5	**3.** $2x$	**4.** $3x^2$	**5.** $4x^3$
6. $4x$	**7.** $6x$	**8.** $5x$	**9.** $7x$	**10.** $6x^2$
11. $9x^2$	**12.** $2x^2$	**13.** $5x^2$	**14.** $8x^3$	**15.** $12x^3$
16. $3x^3$	**17.** $6x^3$	**18.** x^{-2}	**19.** $-2x^{-2}$	**20.** $\tfrac{1}{2}x^{-3}$

21. $2x + 2$ **22.** $4x - 3$ **23.** $4 - 3x$ **24.** $5 - 5x$

25. $3x^2 + 2x - 4$ **26.** $6x^2 - 3x + 1$ **27.** $x - 2x^2 - 3x^3$

28. $\dfrac{x^2}{3} + \dfrac{x}{4} - \dfrac{1}{2}$ **29.** $\dfrac{x^3}{4} - \dfrac{x^2}{3} + \dfrac{x}{2}$ **30.** $\dfrac{3x^2}{4} - \dfrac{5x^4}{8}$

31. $\dfrac{4}{x^3} - \dfrac{2}{x^2}$ **32.** $\dfrac{3}{x^2} + \dfrac{2}{x^3}$ **33.** $\dfrac{2}{3x^2} - \dfrac{3}{2x^3}$

THE INTEGRAL NOTATION

$$\text{If } \frac{dy}{dx} = 3x^2, \quad \text{then} \quad y = x^3 + c$$

This fact may be stated thus:

If the derived function of y with respect to x is $3x^2$, then the original function was $y = x^3 + c$.

Or we may say:

The integral of $3x^2$ with respect to x is $(x^3 + c)$.

In symbol form we have:

$$\int (3x^2)\, dx = x^3 + c$$

The integral sign, an elongated $S\left(\int \right)$, has been taken from the word SUM, an indication that summation is an important feature of integration; dx tells us that the expression is to be integrated "with respect to x".

EXERCISE 11

Find the following integrals:

1. $\displaystyle\int (3x^2 - 2x + 1)\, dx$
2. $\displaystyle\int (4x^3 + 6x^2 - 2x)\, dx$
3. $\displaystyle\int (4t^2 + 5t + 4)\, dt$
4. $\displaystyle\int (5ap^4 - 3bp^2 + 5)\, dp$
5. $\displaystyle\int (at^3 - 2bt^2 - 3gt)\, dt$
6. $\displaystyle\int (4ax^5 + 3bx^3 - 5tx)\, dx$

EXAMPLE. *A curve passes through the point* (1, 0) *and its gradient at any point* (x, y) *is* 2x + 3. *Find:* (i) *the equation of the curve;* (ii) *the points where the curve cuts the x-axis;* (iii) *the point at which the gradient is zero.*

(i) *Equation of curve:*

If $(2x + 3)$ is the gradient function:

$$y = \int (2x + 3)\, dx = x^2 + 3x + c$$

The curve passes through point (1, 0):

$\therefore$ When $x = 1$, $y = 0$.   Substitute for $x = 1$

$$y = x^2 + 3x + c = 1 + 3 + c = 0 \quad \therefore \quad c = -4$$
$$\therefore \quad \underline{y = x^2 + 3x - 4}$$

(ii) *x-intercepts:*

When curve cuts x-axis, $y = 0$

$\therefore y = x^2 + 3x - 4 = 0$ $\therefore (x + 4)(x - 1) = 0$

$\therefore$ Graph cuts x-axis at $(-4, 0)(1, 0)$

(iii) *Zero gradient:*

Gradient of graph is zero when $\frac{dy}{dx} = 0$

$\therefore 2x + 3 = 0$ When $x = -1\frac{1}{2}, y = -6\frac{1}{4}$

$\therefore$ Gradient is zero at $(-1\frac{1}{2}, -6\frac{1}{4})$

EXERCISE 12

Given first the gradient and second a point (x, y) on the curve, find in the following: (i) the equation of the curve; (ii) the x-intercepts; (iii) the point at which the gradient is zero:

1. $2x$; $(1, -3)$
2. $2x - 1$; $(2, 0)$
3. $2x - 1$; $(1, -6)$
4. $2x - 7$; $(2, 2)$
5. $4x + 5$; $(0, -12)$
6. $24x - 23$; $(0, 10)$

Find: (i) the equation; (ii) the x-intercepts:

7. $6x^2 + 6x - 2$; $(1, 3)$
8. $6x^2 - 6x - 9$; $(1, -10)$
9. $6x^2 - 24x + 16$; $(0, 0)$
10. $18x^2 + 30x - 75$; $(1, -54)$
11. $54x^2 - 6x - 6$; $(1, 9)$
12. $12x^2 - 4$; $(0, 0)$
13. $6x^{-2}$; $(6, 0)$
14. $4x^{-3} - x^{-2}$; $(1, -1)$
15. $6x^{-3} - 6x^{-2}$; $(1, 3)$
16. $2 + 6x^{-2}$; $(2, -3)$

Find the equation:

17. $6x - 4 - 5x^{-2}$; $(2, 6\frac{1}{2})$
18. $12x^{-4} - 6x^{-3}$; $(2, \frac{1}{4})$
19. $\frac{3}{x^4} - \frac{4}{3x^3} - \frac{3}{2x^2}$; $(1, 5\frac{1}{6})$
20. $2x + \frac{3}{x^3}$; $(2, -2\frac{3}{8})$

VELOCITY AND ACCELERATION

We have seen earlier (p. 348) that if distance (s ft) is expressed in terms of time (t sec.), we can derive expressions in terms of t for both velocity and acceleration. Thus: $v = \frac{ds}{dt}$; $a = \frac{dv}{dt}$.

If we are given an expression for v in terms of t, we may obtain an expression for s by integrating v, $\left[\text{i.e., } \int (v)\, dt = s \right]$ and an expression for a by differentiating v, $\left[\text{i.e., } \frac{dv}{dt} = a \right]$; each with respect to t.

Further, if we are given an expression for a in terms of t we may obtain an expression for v by integrating a, and an expression for s by integrating v; each in terms of t. (And each with an arbitrary constant added during integration.)

EXAMPLE. *Starting from rest, a body acquires an acceleration of* $(12 - 2t)$ *ft/s² after t sec. Find, after 4 sec.; (i) acceleration: (ii) velocity; (iii) distance travelled.*

(i) *Acceleration:*

$$a = 12 - 2t. \text{ When } t = 4, \ a = 4 \qquad \underline{a = 4 \text{ ft/s}^2}$$

(ii) *Velocity:*

$$v = \int (12 - 2t) \, dt = 12t - t^2 + c$$

If time is zero, velocity is zero. (*From rest.*)
   Subst. $t = 0$, $v = 0$ in $v = 12t - t^2 + c$: $\therefore \ c = 0$
$\therefore \ v = 12t - t^2$ and after 4 sec.:

$$v = 48 - 16 = 32 \qquad \underline{v = 32 \text{ ft/s}}$$

(iii) *Distance:*

$$s = \int (12t - t^2) \, dt = 6t^2 - \tfrac{1}{3}t^3 + k$$

If time is zero, distance is zero. (*From rest.*)
   Subst. $t = 0$, $s = 0$ in $s = 6t^2 - \tfrac{1}{3}t^3 + k$: $\therefore \ k = 0$
$\therefore \ s = 6t^2 - \tfrac{1}{3}t^3$ and after 4 sec.:

$$s = 96 - 21\tfrac{1}{3} = 74\tfrac{2}{3} \qquad \underline{s = 74\tfrac{2}{3} \text{ ft}}$$

EXERCISE 13

A body starts from rest. Given first the acceleration (ft/s²) after t sec. and second the time to be considered, find: (i) acceleration; (ii) velocity; (iii) distance:

1. $6 + 4t$; 3 sec.   2. $4 + 6t$; 4 sec.
3. $12t - 4$; 2 sec.   4. $18t - 6$; 5 sec.
5. $20 - 6t$; 3 sec.   6. $24 - 9t$; 2 sec.
7. 32; 2 sec.   8. 20; 5 sec.

A body starts from rest. Given first the velocity (cm/s) after t sec. and second the time to be considered, find: (i) acceleration; (ii) velocity; (iii) distance:

9. $t^2 + 4t$; 3 sec.   10. $8t - 6$; 4 sec.
11. $16t - 12t^2$; 2 sec.   12. $8t^2 - 24t$; 3 sec.
13. $4t^3 - 6t$; 2 sec.   14. $12t - t^3$; 4 sec.

15. A body is thrown vertically upwards so that its velocity after t sec. is $(64 - 32t)$ ft/s. Find: (i) acceleration; (ii) time to reach maximum height; (iii) maximum height.

16. A stone falls from the top of a vertical cliff 256 ft high with a constant acceleration of 32 ft/s². Find: (i) the time taken to reach the base of the cliff; (ii) the velocity at this time.

17. A body starts from rest and moves in a straight line from a point O such that after t sec. the velocity is $(18t - 6t^2)$ cm/s. Find: (i) the time taken for the body to come momentarily to rest; (ii) the distance from O at this time; (iii) the speed and acceleration with which the body reaches O on the return journey.

18. A body starts from rest and moves in a straight line from a point O such that after t sec. the acceleration is $(6t - 12)$ cm/s². Find: (i) the time taken for the body to come momentarily to rest; (ii) the distance from O at this time; (iii) the velocity and acceleration with which the body reaches O on the return journey.

19. A body moves in a straight line and passes a point O with a velocity of 60 cm/s and, t sec. after passing O, the acceleration is $(24 - 6t)$ cm/s². Find: (i) the maximum velocity; (ii) the time at which the body is momentarily at rest; (iii) the distance from O at this instant.

20. A body moves in a straight line such that after t sec. the velocity is $(24t - 6t^2)$ ft/s. Find: (i) the distance travelled between the end of the 1st second and the end of the 3rd second (i.e., $t = 1$ and $t = 3$); (ii) the times at which the body is at rest; (iii) the acceleration at these moments.

DEFINITE INTEGRALS

Question 20 (i) of the previous exercise required us to find the distance travelled *during the interval of time* from $t = 1$ to $t = 3$.

Since $s = 12t^2 - 2t^3 + c$, the distance after 3 sec. $= (54 + c)$ ft. and after 1 sec. $= (10 + c)$ ft. The distance travelled during the interval is given by:

$$(54 + c) - (10 + c) = 44 \text{ ft.}$$

We should observe that although the value of the arbitrary constant (c) remained unknown, the result of the calculation was unaffected, since c disappeared in the course of subtraction.

When an integral includes an arbitrary constant it is called an ***indefinite integral***, but when the integral is taken between certain prescribed ***limits***, as above, the integral is called a ***definite integral*** and it is unnecessary to include the constant. In symbol form we have:

$$s = \int_1^3 (24t - 6t^2)\, dt = \left[12t^2 - 2t^3 \right]_1^3 = (54 - 10) = \underline{\underline{44 \text{ ft}}}$$

Exercise 14

Evaluate the following:

1. $\displaystyle\int_0^2 (2x)\, dx$ **2.** $\displaystyle\int_1^3 (2x)\, dx$ **3.** $\displaystyle\int_{-1}^2 (2x)\, dx$

4. $\displaystyle\int_0^1 (3x^2)\, dx$ **5.** $\displaystyle\int_2^4 (3x^2)\, dx$ **6.** $\displaystyle\int_{-1}^2 (4x^3)\, dx$

7. $\displaystyle\int_2^3 (2x + 2)\, dx$ **8.** $\displaystyle\int_{-2}^{-1} (3x^2 - 2x)\, dx$ **9.** $\displaystyle\int_3^4 (4x^3 + 3x^2)\, dx$

10. $\displaystyle\int_2^3 \left(\frac{1}{x^2}\right) dx$ **11.** $\displaystyle\int_{-2}^{-1} \left(\frac{2}{x^3}\right) dx$ **12.** $\displaystyle\int_1^2 \left(\frac{2}{3x^2}\right) dx$

13. $\displaystyle\int_1^3 \left(3x^2 - \frac{1}{x^2}\right) dx$ **14.** $\displaystyle\int_2^4 \left(2x^3 - \frac{3}{x^4}\right) dx$ **15.** $\displaystyle\int_{-2}^2 (2x - 3)^2\, dx$

16. A body moves in a straight line so that t sec. after passing point O, the velocity is $(18t - 3t^2)$ cm/s. Find the distance travelled *during* the 3rd second.

17. A body moves in a straight line so that t sec. after passing point O the velocity is $(12t - 6t^2)$ cm/s. Find the distance travelled *during* the 2nd second.

18. A body moves in a straight line so that t sec. after passing point O the acceleration is $(12t - 6)$ ft/s^2. If velocity at O is 8 ft/s, find the distance travelled *during* the 4th second.

19. A body moves in a straight line so that t sec. after passing point O the acceleration is $(2t - 5)$ cm/s^2. If velocity at O is 2 cm/s, find the distance travelled *during* the 3rd second.

20. A non-stop train travels from A to B so that t hours after leaving A the velocity is $(48t - 12t^2)$ km/h. Find: (i) the maximum velocity for the journey; (ii) the distance from A to B; (iii) the average velocity for the journey.

AREAS BY INTEGRATION

In Book I (pp. 287–288) we have seen methods for assessing the area of curvilinear figures: (i) by counting squares, and (ii) by using mid-ordinates.

If the curved boundary of a figure represents the graph of an equation we may determine the area of the figure by integral calculus.

EXAMPLE. *Find the area bounded by the curve $y = 3x^2 + 4$, the x-axis and the ordinates at $x = 1$, $x = 6$.*

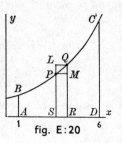

fig. E:20

From fig. E : 20 we require to find the area *ABCD*.

Let P be any point (x, y) on the curve and PS the ordinate through that point. If the area of *ABPS* is A units, then, as PS moves to left or right, A will decrease or increase as x decreases or increases; hence **A is a function of x.**

From fig. E : 20, if $P(x, y)$ moves a small distance to $Q(x + \delta x; y + \delta y)$:

    x increases by δx
    y increases by δy
    A increases by δA

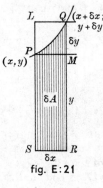

fig. E:21

From fig. E : 21, if QR is the ordinate through Q the area of *PQRS* is equal to δA, and this area is somewhere between the area of *PMRS* and the area of *LQRS*.

$\therefore$ δA lies between $y \cdot \delta x$ and $(y + \delta y)\,\delta x$

$\therefore$ $\dfrac{\delta A}{\delta x}$ lies between y and $(y + \delta y)$

But as $Q \longrightarrow P$, $\delta x \longrightarrow 0$ and $\delta y \longrightarrow 0$; $\therefore$ $(y + \delta y) \longrightarrow y$

$$\therefore \left(\underset{\delta x \to 0}{\text{Limit}}\right)\frac{\delta A}{\delta x} = \frac{\mathrm{d}A}{\mathrm{d}x} = y \qquad \text{But } y = 3x^2 + 4$$

$$\therefore \frac{\mathrm{d}A}{\mathrm{d}x} = 3x^2 + 4 \qquad \text{And } A = \int (3x^2 + 4)\,\mathrm{d}x$$

$$\therefore \underline{\underline{A = x^3 + 4x + c}}$$

CALCULATION

Since we are integrating between the definite integrals $x = 1$ and $x = 6$, the arbitrary constant $c(*)$ will disappear.

$$\text{Hence: } A = \int_1^6 (3x^2 + 4)\,\mathrm{d}x = \left[x^3 + 4x\right]_1^6 = 240 - 5$$

$$\underline{\underline{\text{Area} = 235 \text{ units}^2}}$$

NOTE

(i) *In fig. E : 20, if the area is so reduced that *CD* falls along *BA* the area would be zero when $x = 1$.

$$\therefore 0 = 1 + 4 + c \quad \text{and} \quad c = -5$$

(ii) The result $A = \int (y) \, \mathbf{dx}$ is true for any curve of the form $y = f(x)$.

EXAMPLE. *Find the area enclosed by the curve $y = 10x - 2x^2$ and the line $y = x + 4$.*

The graphs of $y = 10x - 2x^2$ and $y = x + 4$ will intersect when:

$$10x - 2x^2 = x + 4$$
$$\therefore\ 2x^2 - 9x + 4 = 0$$
$$(2x - 1)(x - 4) = 0$$
$$x = \tfrac{1}{2} \text{ or } 4;\ y = 4\tfrac{1}{2} \text{ or } 8$$

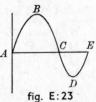

fig. E : 22

From fig. E : 22 we require to find the area $ABPCD$ (i.e., under $y = 10x - 2x^2$ between $x = \tfrac{1}{2}$ and $x = 4$), and from this we must subtract the area of the trapezium $ABCD$.

(i) *Area under curve:*

$$A = \int_{\tfrac{1}{2}}^{4} (10x - 2x^2)\, dx = \left[5x^2 - \tfrac{2}{3}x^3 \right]_{\tfrac{1}{2}}^{4}$$
$$= (80 - 42\tfrac{2}{3}) - (1\tfrac{1}{4} - \tfrac{1}{12}) \qquad\qquad = 36\tfrac{1}{6}$$

(ii) *Area of trapezium:*

$$A = \text{Aver. Width} \times \text{Height} = \frac{4\tfrac{1}{2} + 8}{2} \times 3\tfrac{1}{2} = 21\tfrac{7}{8}$$

$$\therefore\ \text{Area required} = 36\tfrac{1}{6} - 21\tfrac{7}{8}$$

$$\textit{Ans. } 14\tfrac{7}{24} \text{ units}^2$$

NOTE

(i) In fig. E : 23 part of the area enclosed by the curve is below the x-axis; such an area is given a negative sign. However, if the *total* area is required (i.e., $ABC + CDE$) the negative sign of the lower portion is disregarded.

(ii) In circumstances requiring the total area of such a figure it is necessary to calculate separately the areas ABC and CDE, then add (discounting the $-$ve sign of CDE). *A sketch of the curve will indicate when this procedure must be followed.*

fig. E : 23

EXERCISE 15

1. Find the area enclosed by the curve $y = 12 - 3x^2$, the x-axis, and the following pairs of ordinates. *Sketch the curve.*

(i) $x = 0$, $x = 2$;   (ii) $x = -1$, $x = 1$;   (iii) $x = -2$, $x = 2$

2. Find the area enclosed by the curve $y = 18 + 3x - x^2$, the x-axis, and the following pairs of ordinates. *Sketch the curve.*

(i) $x = -3$, $x = 0$; (ii) $x = 0$, $x = 1\frac{1}{2}$; (iii) $x = 1\frac{1}{2}$, $x = 6$

Hence give the area between $x = -3$ and $x = 6$. Check your answer by integrating between these limits.

Find the area enclosed by the given curve, the x-axis and the given ordinates. *Sketch the curve.*

3. $y = x^2 + 2$; $x = 0$, $x = 3$ **4.** $y = x^3 + 6$; $x = 2$, $x = 6$
5. $y = 3x^2 + 4x$; $x = 1$, $x = 4$
6. $y = 10 + 3x - x^2$; $x = -1$, $x = 4$
7. $y = 8 - 8x^3$; $x = -1$, $x = 1$. (*Sketch needs care.*) What name is given to the point $(0, 8)$?

8. $y = \dfrac{1}{x^2}$; $x = 1$, $x = 4$ **9.** $y = \dfrac{2}{x^3} - \dfrac{3}{x^2}$; $x = 2$, $x = 3$
10. $y = x^2 + x - 12$; $x = -3$, $x = 2$

Find the area enclosed by the given curve and the x-axis. *Sketch the curve.*

11. $y = 4 - x^2$ **12.** $y = 3x^2 - 12$
13. $y = 15 + 12x - 3x^2$ **14.** $y = 6x^2 + 18x - 24$
15. $y = 18x - 3x^2$ **16.** $y = 2x^2 + 7x$
17. $y = x^3 - 4x$ (*Care!*) **18.** $y = 18x - 2x^3$ (*Care!*)
19. $y = x^3 - 6x^2 - 4x + 24$ (*Care!*)
20. $y = 8 + 8x - 2x^2 - 2x^3$ (*Care!*)

Find the area enclosed by the given curve and straight line. *Sketch the curve and line.*

21. $y = 9 - x^2$; $y = 5$ **22.** $y = 12 + 4x - x^2$; $y = 12$
23. $y = x^2 + 2x - 15$; $y = -7$
24. $y = x^2 - 8x - 20$; $y = 13$ (*Care!*)
25. $y = 25 - x^2$; $y = x + 13$
26. $y = 16 + 6x - x^2$; $y = 16 - x$
27. $y = x^2 - 2x + 4$; $y = x + 14$
28. $y = x^2 + 8x + 22$; $y = 46 - 2x$
29. Find the area enclosed by the curves $y = 6x - x^2$ and $y = \frac{1}{2}x^2 - 3x$.
30. Find the area enclosed by the curves $y = 40 + 6x - x^2$ and $y = x^2 - 14x + 40$.

SOLIDS OF REVOLUTION

From fig. E : 24, if $\triangle OCD$ is rotated about the x-axis, it produces a **solid of revolution.** The volume of such a solid may be found by integration.

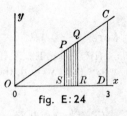

EXAMPLE. *Find the volume of the circular cone formed by rotating about the x-axis the area under* $y = 2x$ *between the ordinates* $x = 0$ *and* $x = 3$.

fig. E : 24

When the line $y = 2x$ is rotated about the x-axis it will generate a cone.

Let V be the volume of the cone cut off between O and PS, the ordinate through $P(x, y)$. From fig. E : 24, as PS moves left or right, V decreases or increases; hence **V is a function of x.**

From fig. E : 24, if $P(x, y)$ moves a small distance to Q $(x + \delta x; y + \delta y)$:

x increases by δx
y increases by δy
V increases by δV

From fig. E : 25, δV lies between the volume generated by the rotation of $PMRS$ and the volume generated by the rotation of $LQRS$.

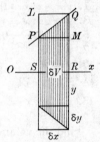

Since each of these solids is a cylinder ($V = \pi r^2 h$):

Vol. generated by $PMRS = \pi y^2 \cdot \delta x$
Vol generated by $LQRS = \pi(y + \delta y)^2 \delta x$
$\therefore\ \delta V$ lies between $\pi y^2 \cdot \delta x$ and $\pi(y + \delta y)^2 \delta x$

fig. E : 25

$\therefore\ \dfrac{\delta V}{\delta x}$ lies between πy^2 and $\pi(y + \delta y)^2$

But as $Q \longrightarrow P$, $\delta x \longrightarrow 0$ and $\delta y \longrightarrow 0$, $\therefore\ (y + \delta y)^2 \longrightarrow y^2$

$$\therefore\ \left(\underset{\delta x \to 0}{\text{Limit}}\right)\frac{\delta V}{\delta x} = \frac{dV}{dx} = \pi y^2 \qquad \text{But } y = 2x$$

$$\therefore\ \frac{dV}{dx} = \pi(2x)^2 \qquad = \pi 4x^2$$

$$\therefore\ V = \int (4\pi x^2)\, dx = 4\pi(\tfrac{1}{3}x^3) + c$$

$$\therefore\ V = 1\tfrac{1}{3}\pi x^3 + c$$

CALCULATION

Since we are integrating between the definite integrals $x = 0$ and $x = 3$, the arbitrary constant $c(*)$ will disappear.

Hence: $V = \int_0^3 (4\pi x^2)\, dx = \left[1\tfrac{1}{3}\pi x^3 \right]_0^3$

$$\underline{\underline{\text{Volume} = 36\pi \text{ units}^3}}$$

NOTE

(i)* In fig. E : 24, if the volume is so reduced that CD falls along the y-axis (i.e., $x = 0$), the volume would be zero when $x = 0$.

$$\therefore\ 0 = 1\tfrac{1}{3}\pi(0)^3 + c \quad \text{and} \quad c = 0$$

(ii) The result $V = \int (\pi y^2)\, dx$ is true for the rotation of any curve of the form $y = f(x)$.

Exercise 16

(*Give volume answers in terms of* π.)

Find the volume of the solid generated by rotating about the x-axis the area under the given curve between the given ordinates.

1. $y = 4x; x = 0, x = 3$
2. $y = 6x; x = 1, x = 3$
3. $y = 3x; x = -2, x = 2$
4. $y = x; x = -1, x = 3$
5. $y = x^2; x = 0, x = 2$
6. $y = 3x^2; x = 1, x = 3$
7. $y = x^2 - 1; x = -1, x = 1$
8. $y = 2x^2 + 1; x = 1, x = 2$
9. $y = x^2 - 2x; x = -1, x = 2$
10. $y = \tfrac{1}{2}x^2 + 2x; x = 0, x = 3$
11. $y = \sqrt{x}; x = 0, x = 6$
12. $y = \sqrt{(x^2 - 2x)}; x = -1, x = 3$
13. $y = x^3; x = -2, x = 2$
14. $y = \dfrac{1}{x}; x = 1, x = 3$

Find the volume of the solid generated by rotating about the x-axis the area under the given curve between the points at which the curve cuts the x-axis.

15. $y = x^2 - x$
16. $y = x^2 - 4x$
17. $y = x^2 - 1$
18. $y^2 = 3 + 2x - x^2$
19. $y = x^3 - 4x$
20. $y = 5x^2 + 10x$

21. The volume of the solid generated by rotating about the x-axis the area under $y^2 = 4x - 3$ between $x = a$ and $x = 4$ is 18π units of volume. Calculate the value of a.

22. The volume of the solid generated by rotating about the x-axis the area under $y^2 = 2x + 5$ between $x = -1$ and $x = a$ is 28π units of volume. Calculate the value of a.

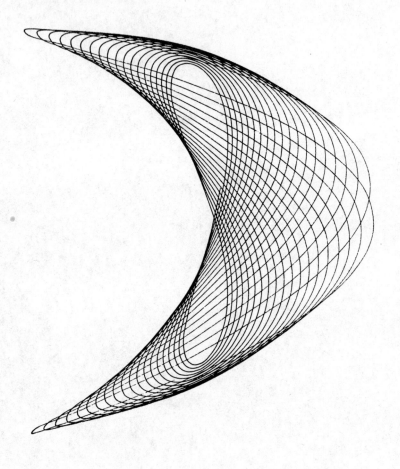

TEST PAPERS

TEST A.1

1. Solve the equation $2x^2 - 3x - 6 = 0$, giving your answers correct to 3 sig. figs.

2. If a, x, y, b are in arithmetic progression, prove that $x = \frac{1}{3}(2a + b)$ and find a similar expression for y in terms of a and b.

3. If $x = \dfrac{a^2b}{a+b}$, calculate: (i) the value of b when $x = 6$ and $a = -3$, and (ii) the values of a, correct to one decimal place, when $x = 2$ and $b = 3$.

4. If $x^3 = ky^2$, where k is a constant, and if $x = 4$ when $y = 6$, find the values of y when $x = 2$.

5. Calculate, as accurately as your tables permit, the sum of the first 15 terms of the geometrical progression $3, 4\cdot5, 6\cdot75, \ldots$

6. Draw the graph of $y = x^2 + \dfrac{20}{x}$ for values of x between $x = 1$ and $x = 5$. Use your graph to find the smallest value of y in this interval.　　　　　　　　　　　　　　　　　　　　　　(L.C.C.)

TEST A.2

1. Simplify: (i) $(5xy^3)^3 \div 5(x^3y^2)^2$

(ii) $\dfrac{2a^2 + 4ab + 2b^2}{2(a^2 - b^2)}$

2. Solve each of the following equations for x:

　(i) $x^2 - 2x - 1 = 0$;　　　(ii) $10{,}000 = 10^{2x}$.

3. Make three separate sketches (*not on graph paper*) of the following relations: (i) $y = -2x$; (ii) $y = -2x^2$; (iii) $y = -\dfrac{1}{2x}$.

4. Solve for x, correct to two decimal places,

$$1 - \frac{3}{x} - \frac{6}{x^2} = 0$$

5. The effective resistance, R ohms, of two wires connected in parallel in an electrical circuit is given by $R = \dfrac{r_1 r_2}{r_1 + r_2}$, where r_1 and r_2 are the resistances in ohms of the individual wires. Express this as a formula for r_2 in terms of R and r_1, and find the value of r_2 when $R = 8$ and $r_1 = 32$.

6. A square sheet of tin, of 12-in sides, is to be used to make an open-top box by cutting equal squares from each corner and bending up the sides. If x in is the length of the side of the squares cut out of the corners then the volume, V in³, of the box is given by $V = 4x(36 - 12x + x^2)$. Calculate the values of V corresponding to the following values of x: 0, 1, 2, 3, 4, 5, 6. Hence draw the graph of V against x and from it find: (i) the maximum value of V; (ii) the value of x for which V is maximum. (M.)

TEST A.3

1. Solve the equation $x^2 + 2x - 2 = 0$, giving your answers correct to two decimal places.

2. If $2x = \dfrac{ay}{3a - 2y}$, express a in terms of x and y, and find the value of a when $x = 2$ and $y = -3$.

3. Given that $p = (x - 2)(x - 3)$, $q = (x - 3)(x - 1)$, and $r = (x - 1)(x - 2)$, find the value of $\dfrac{1}{p} - \dfrac{2}{q} + \dfrac{1}{r}$, and prove that $\dfrac{1}{p^2} - \dfrac{2}{q^2} + \dfrac{1}{r^2} = \dfrac{2}{pqr}$.

4. Solve the simultaneous equations:
$$\frac{3}{x} - \frac{2}{y} = \frac{1}{12}, \quad \frac{4}{x} + \frac{3}{y} = 2$$

5. (i) If $a + b = 1$ and $a^2 + b^2 = 2$, prove that $2ab = -1$, and find the value of $a - b$.

(ii) The sides of a right-angled triangle are of length $(2x - 3)$ cm, $(2x + 1)$ cm, $(2x + 5)$ cm. Find the value of x.

6. The number written abc, where a, b, c are figures, means $100a + 10b + c$. Write down the meanings of the numbers $abcd$ and $dcba$.

Show that the difference between any number with four figures and the number with those figures reversed is always divisible by 9, and that, if the two middle figures are the same, it is also divisible by 37. (O.C.)

TEST A.4

1. (i) Solve the simultaneous equations:
$$3x - y = 1, \quad x + 2y = 0$$

(ii) A rectangle is x cm wide and 3 cm longer than it is wide. Write down its perimeter in cm and its area in cm². Find

x if the number of square centimetres in its area is equal to the number of centimetres in its perimeter.

2. (i) If $4x^2 + ax + 9$ is a perfect square, what are the possible values of a?

(ii) What are the values of a and b if $x^3 + x^2 + ax + b$ is exactly divisible by $x - 1$ and leaves a reminder -8 when divided by $x + 1$?

3. (i) Given that log $1·04 = 0·017\ 033$, find, as accurately as your tables permit, the 21st term of the geometric progression 1, $1·04$, $(1·04)^2$, $(1·04)^3$, . . .

(ii) Find values of x and y so that $6·5$, x, y, 2 are in arithmetic progression.

4. (i) It is given that the roots of the equation $x^2 + px - 3 = 0$ differ by 4. Find p.

(ii) If $\log_{10} y = \frac{2}{3} \log_{10} x - 2$, express y in terms of x.

5. (i) If $g = \dfrac{1}{3 - \sqrt{5}} - \dfrac{1}{\sqrt{5} - 1}$, find g in its simplest form with a rational denominator.

(ii) Using tables, find the value of $(3·162)^{5·791}$.   (N.S.W.)

TEST A.5

1. (i) If $\dfrac{1}{a} = \dfrac{2}{b} + \dfrac{1}{c}$, express b in terms of a and c.

(ii) Solve the simultaneous equations:
$$7x + 4y = 28, \qquad 5x - 2y = 3$$

2. A man cycles to a point x miles away and returns by the same route. On the outward journey his uniform speed is $(u + v)$ m.p.h. and on the return journey it is $(u - v)$ m.p.h. Prove that the time for his whole journey is $\dfrac{2ux}{u^2 - v^2}$. If $x = 7\frac{1}{2}$, $v = 2$ and the total time of the journey is 2 hours, calculate u.

3. (i) Solve the equation $7x^2 + 13x - 5 = 0$, giving the answers correct to two places of decimals.

(ii) Two numbers differ by 15 and their squares differ by 705; find the numbers.

4. Draw on the same diagram the graphs of $y = 2 + \dfrac{4}{x^2}$ and $y = x(4 - x)$ for values of x between 1 and $3\frac{1}{2}$, taking half-unit intervals. Use your graphs to find two of the roots of the equation:
$$x^4 - 4x^3 + 2x^2 + 4 = 0.$$
Give the reason why the readings you make are roots of the equation.

5. The table gives some values of T and p:

p	5·012	6·310	7·943	8·913
T	448·8	634	895·6	1064

If $y = \log T$, $x = \log p$, show that the graph of y against x is a straight line so that x and y satisfy a relation $y = mx + c$. Find the constants m and c. Verify that $c = \log 40$ and deduce the relation between T and p.

(S.U.)

TEST A.6

1. (i) Evaluate: $8^{\frac{2}{3}}$; $(0\cdot5)^0$; 6^{-2}; $(\frac{9}{25})^{-\frac{3}{2}}$

    (ii) Simplify: $x^3 y^{\frac{2}{3}} \times x^{-7} y^{\frac{1}{3}} \div (x^2)^4$, giving your result without any negative indices.

    (iii) Find the value of $\sqrt[5]{\{(3\cdot27)^2 - (3\cdot18)^2\}}$, correct to 3 sig. figs.

2. y is equal to the sum of two parts, one proportional to x and the other proportional to the square of x. When $x = 1$, $y = 5$ and when $x = 2$, $y = 14$. Find: (i) the value of y when $x = 3$; (ii) the values of x, correct to two decimal places, when $y = 8$.

3. The distance between two towns by the most direct route is 200 km. By taking a route advised by one of the motoring associations a car travels 10 km more but is able to maintain an average speed which is 5 km per hour greater than that possible on the direct route, and so shorten the time for the journey by 40 minutes. Find the average speed for the direct route.

4. (i) If the third term of an arithmetical progression is 24 and the fifteenth term is 18, find the sum of the first 16 terms.

    (ii) The second term of a geometrical progression is $\frac{1}{2}$ and the sum of the third and fourth terms is 6. Find the common ratio and the first term.

5. A rectangular area is bounded on one side by a river and on the other three sides with fencing. The total length of fencing required is 100 m. If the two sides of the rectangle which are at right-angles to the river are of length x m, show that the area enclosed is $(100x - 2x^2)$ m².

Draw a graph of $100x - 2x^2$ from $x = 10$ to $x = 50$, taking 2 cm to represent 10 m on one axis and 200 m² on the other axis. From your graph, find the greatest area which can be enclosed.

If the enclosure has an area of 1000 m², find from your graph its possible dimensions.

(D.)

TEST B.1

1. (i) A man notes that for £1 he could obtain 11·2 German marks or 13·72 French francs. How many francs could he obtain for 1960 marks?

(ii) A rectangular field is 220 yards in length and $93\frac{1}{2}$ yards in width. Find its area in acres. (1 acre = 4840 yd².)

(iii) An inkwell holds 9·75 cm³ of ink. How many inkwells can be completely filled from a bottle containing 1 litre of ink?

2. (i) A car cost £875 when new, and its value depreciated each year by 15% of its value at the beginning of that year. Find the value of the car at the end of the two years.

(ii) Calculate what sum of money invested at $2\frac{1}{2}\%$ per annum simple interest would amount to £4200 in 8 years.

3. The rateable value of a city is £5,289,362. Find the amount collected from a rate of $92\frac{1}{2}$p in the £.

If the rateable value of the city is raised to £7,385,276, find, to the nearest penny, the rate in the £ which will be required to collect a sum of £5,924,375.

4. An electrical cable consists of three copper wires embedded in insulating material. The cable has a circular cross-section of diameter 0·22 in, and each copper wire has a circular cross-section of diameter 0·029 in.

If the insulating material weighs 0·042 lb/in³ and copper weighs 0·32 lb/in³, calculate the weight of a 50-yd length of the cable. (Take $\pi = 3\cdot142$.)   (S.)

TEST B.2

1. (i) Taking 1 kilometre to be 0·6214 miles, express 17·5 km in miles, correct to 3 sig. figs.

(ii) Express 140 lb as a decimal of 5 tons.

2. A workman has a rise of 10% in his wage-rate per hour, but there is a drop of $6\frac{2}{3}\%$ in the number of hours worked per week. If his original weekly wage was £18 for 45 hours work, find: (i) his new wage-rate per hour; (ii) the percentage increase in the total weekly wage.

3. A piece of cardboard is in the form of a quadrant of a circle of radius 12 cm, bounded by two perpendicular radii OA, OB and the arc AB. The mid-points of OA, OB are X, Y, and the triangular part OXY is cut off. Find: (i) the area of the piece remaining, correct to the nearest cm²; (ii) its perimeter, correct to the nearest cm. (Take π to be 3·14.)

4. A man starts a business with a capital of £6000, and employs an assistant. From the yearly profits he keeps an amount equal to $4\frac{1}{2}\%$ of his capital and pays his assistant 35% of the remainder of the profits. Find how much the assistant receives in a year in which the profits are £2000.

The following year the assistant receives £770. Calculate the total profits for that year.

5. A rectangular petrol tank weighs 60 lb when empty. It has internal measurements: height $31\frac{1}{2}$ in, base 20 in by 24 in. If a gallon of petrol weighs 7 lb and occupies $\frac{4}{25}$ of a cubic foot, find, correct to the nearest pound, the total weight of the tank and contents when the tank is full.

A second tank with the same capacity as the first is cylindrical in shape, and has an internal height of 21 in. Find its internal radius, correct to the nearest inch. (Take π to be $\frac{22}{7}$.) (C.)

TEST B.3

1. (i) A car travels $122\frac{1}{2}$ miles in 2 h 55 min. Find its speed in miles per hour.

(ii) Three people share £19·53 in the proportion 1:3:5. Find the largest share.

2. (i) A shopkeeper mixes 3 oz of tobacco costing 30p per oz with 7 oz of another tobacco costing 25p per oz. Calculate the price per oz of the mixture.

(ii) The value of a second-hand car is £510. This is 15% less than its value when new. Calculate the value of the car when new.

3. A military map, drawn to a scale of 5 cm to the km, shows a rectangular minefield which measures 3·5 cm by 2 cm on the map. Find the actual dimensions of the minefield in metres and its area in hectares.

4. The rateable value of a house in 1955 was £32, and the annual rate was $117\frac{1}{2}$p in the £. In 1956 the rateable value was increased to £44, and the annual rate was reduced to $87\frac{1}{2}$p in the £. Calculate the change in the total rates paid by the householder, and state whether it was an increase or a decrease.

5. A dealer made 20% profit on his outlay when he sold an article to a shopkeeper. The shopkeeper marked the article at a price which would give him $33\frac{1}{3}\%$ profit on his outlay. When a customer paid cash the shopkeeper allowed a discount of 5% of the marked price.

Calculate: (i) the dealer's outlay on an article for which a customer paid £8·36 cash; (ii) the percentage by which the cash price paid by a customer for any article exceeded the dealer's outlay. (J.)

TEST B.4

1. (i) Without the use of tables find the square root of 15 876.

(ii) The length of a rectangle is increased by 5% and its width decreased by 7%. Calculate the percentage increase or decrease in its area.

2. An aircraft having an average speed of 250 mile/h leaves an airport A and flies direct to another airport B. On the navigator's map, the scale of which is 1 : 500 000, the distance AB is 11·1 in. Find, to the nearest minute, the time taken for the journey.

3. Calculate the difference between the simple and compound interest on £255 for 2 years at $3\frac{1}{2}$% per annum, giving your answer correct to the nearest penny.

4. Given that 1 cm³ of water weighs 1 g, 1 ft³ of water weighs 62·3 lb and 2·2 lb = 1 kg, calculate the number of cubic inches equivalent to 1000 cm³.

5. A room is 18 ft long, $13\frac{1}{2}$ ft wide and $8\frac{1}{2}$ ft high. To enlarge the room a bay in the shape of a trapezium is constructed, its height being 8 ft, the length of wall cut out 10 ft and the length of the new outside parallel wall 8 ft. The non-parallel sides of the bay are each 4 ft long.

Calculate: (i) the increase in floor space correct to the nearest tenth of a square foot; (ii) the percentage increase in the volume of the room.

6. (i) A retailer makes a profit of 30% on the cost price by selling an article for £71·50. If the cost price falls by £5 and he reduces the selling price to £67·50, calculate the new profit on each article expressed as a percentage of the new cost price.

(ii) A man has £1700 of $4\frac{1}{2}$% stock. He sells this stock at 95 and invests the proceeds in a $4\frac{1}{8}$% stock at 85. Find the change in his income if no allowance is made for stamp charges and commission.

(L.)

TEST B.5

1. (i) If 1 cm³ of water weighs 1 g, find the weight in kg of 1·75 litres of water.

(ii) If 1 kg = 2·2 lb express: (a) 0·25 kg in oz; (b) 1 oz in grammes correct to 3 sig. fig.

2. (i) Assuming 1 cm to be 0·394 in, calculate, correct to the nearest inch, the difference between 280 cm and 9 ft.

(ii) The concrete foundation for a wall is to be $3\frac{1}{2}$ ft wide and $1\frac{1}{2}$ ft deep. Calculate the weight in tons, correct to the nearest ton, of the concrete which will be required for the foundation of a wall 52 ft long. (1 ft³ of concrete weighs 125 lb, 1 ton = 2240 lb.)

3. (i) By selling fabric at 51p a yard a tradesman makes a profit of $27\frac{1}{2}\%$ on the cost price. Calculate his percentage profit if he reduces the selling price to 47p a yard.

(ii) Find, correct to the nearest penny, the compound interest on £2165 for three years at 4% per annum.

4. A man bought 1200 shares of nominal value 25p at $77\frac{1}{2}$p per share including all charges. At the end of the year he received a dividend of 15% on his shares. If tax was deducted from this dividend at $42\frac{1}{2}$p in the £, find: (i) his net income from the shares; (ii) the percentage return on capital invested.

5. A cylindrical tank is required to contain 300 gal of water when full. If the height of the tank is to be 6 ft, calculate, correct to 3 sig. fig., the diameter of the circular base. Take 1 ft³ as equivalent to $6\frac{1}{4}$ gal and π as $3\frac{1}{7}$.

If a gallon of water weighs 10 lb, find, correct to 3 sig. fig., the weight of water supported by each square inch of the base of the tank when the depth of the water in it is 3 ft. (A.B.)

TEST B.6

1. A man invested £4560 in 4% stock at 114. Calculate his annual income and express this as a percentage of his original investment. (Give your answer correct to one decimal place.) After selling the stock when its value falls to 108, he invests the money in £1 shares at 90p If a dividend of $3\frac{2}{3}\%$ is declared on these shares, what amount does he receive?

2. A right circular cone of base diameter 11·7 cm is made from metal weighing 12·3 g/cm³. Find the height of the cone if the weight is 3·5 kg. (log π = 0·4971.)

3. (i) Find the compound interest on £840 for $1\frac{1}{2}$ years at 4% per annum if the interest is added half-yearly.

(ii) To enable him to buy goods valued at £840, a man borrowed £840 at $8\frac{1}{2}\%$ per annum for 5 months. At the end of the 5-month period he has sold the goods at a profit of 13%, and has paid back the loan and the interest due on it. What is his gain?

4. (i) Express $3\frac{1}{2}$ metres² as a percentage of 46 ft², given that 1 in = 2·54 cm, 12 in = 1 ft. (Give your answer to the nearest per cent.)

(ii) A lead tube 1 metre long with an internal diameter of 26·2 cm and a thickness of 3·3 cm, is melted down and moulded into a sphere. If there is a 2% loss of material in the process, find the radius of the sphere.

5. Water flows through a pipe, whose internal diameter is 1·76 in, at the rate of 4·3 ft/s. Assuming that the pipe is always full, find, correct to the nearest 10 gallons, the amount of water that the pipe delivers in an hour. (log $\pi = 0.4971$, 12 in = 1 ft, 1 ft$^3 = 6\frac{1}{4}$ gal). (A.)

TEST C.1

1. Draw a circle of radius 2 in with centre O. *Using ruler and compasses only*, construct a chord AB of the circle so that angle $AOB = 60°$. Also construct a triangle APB such that area APB = area AOB and angle $APB = 30°$. Measure angle PAB.

2. State (*but do not prove*) a theorem connected with each of the following:

  (i) Equal chords of a circle.
  (ii) A right-angled triangle.
  (iii) Opposite angles of a cyclic quadrilateral.

3. Prove that the tangents from an external point to a circle are equal. The tangents to a circle centre O from an external point P touch the circle at A and B and OP cuts the circle at R. If angle $APB = 38°$, calculate: (i) angle AOB; (ii) angle ARB; (iii) angle BAR.

4. (i) Prove that the angle which an arc of a circle subtends at the centre is twice the angle which the arc subtends at any point on the remaining part of the circumference.

  (ii) AB is a diameter of a circle centre O, and P is any point on the circumference of the circle. AP meets the tangent at B at the point T. If angle $ATB = x$, find, in terms of x, the angles PBT, OPB, AOP.

5. (i) ABC is an equilateral triangle. P is the mid-point of BC and Q is the mid-point of CP. If $AC = 4x$, prove that $AQ^2 = 13x^2$ and calculate angle PAQ.

  (ii) P is a point outside a circle. The tangent from P touches the circle at T, and a line through P meets the circle at A and B. Prove that $PT^2 = PA \cdot PB$. If $PT = 6$ cm, $AB = 5$ cm, calculate the length of PA. (L.C.C.)

TEST C.2

1. (i) How many sides has a regular convex polygon whose interior angles each contain $144°$?

  (ii) AB is a diameter of a circle. C is a point on the circumference such that angle CAB is $47°$, and AB is produced to a point D. Find the size of angle CBD.

  (iii) $ABCD$ is a trapezium in which AB is parallel to DC. E and F are the mid-points of AB and CD respectively. If the area of the trapezium $ABCD$ is 10 cm², prove that the area of the figure $AEFD$ is 5 cm².

(iv) What is the locus of a point P which moves so that it is always equidistant from two points A and B?

2. The tangents at points A and B of a circle meet at P, and C is a point on the circumference of the circle such that $AC = AP$. If PC passes through the centre of the circle, and PA is produced to a point X, prove that $\angle BPA = \angle CAX$.

3. $OABC$ is a triangular pyramid with ABC as base. X, Y, Z are the mid-points of the sides BC, CA, AB respectively. The edges OA, OB, OC are bisected at D, E, F respectively. Prove that: (i) DE is equal to XY; (ii) the triangles DEF and XYZ are congruent.

4. (i) Draw two circles, one of radius 2 cm and the other of radius 3 cm, with their centres 6 cm apart. Construct a direct common tangent to the two circles and measure its length. (*Leave in all construction lines.*)

(ii) $ABCD$ is a cyclic quadrilateral in which BC is parallel to AD. The diagonals meet at X, angle $CBD = 62°$ and angle $BAC = 15°$. Calculate the size of angles AXD and ACD, showing the steps and reasons in your calculation.   (M.)

TEST C.3

1. Prove that the areas of similar triangles are proportional to the squares on corresponding sides.

2. The base BC of a triangle ABC is 10 cm long and the altitude is 8 cm. A line drawn parallel to BC, at a distance x cm from it, cuts AB, AC at K, L and KM and LM are drawn perpendicular to BC. Prove that the area of the rectangle $KLMN$ is $10x - \dfrac{5x^2}{4}$. Find the values of x when the ratio of the areas of $KLMN$ and ABC is $7 : 32$. Also find the area common to the two rectangles which could thus be drawn.

3. If the internal bisector of the angle A of any triangle ABC meets BC at D, prove that $\dfrac{BD}{DC} = \dfrac{BA}{AC}$. In the triangle ABC, $BC = 3.5$ cm, $AC = 6$ cm, and $AB = 8$ cm; and the internal and external bisectors of the angle A meet BC at D and E. Calculate the length of DE.

4. ABC is a triangle having $AB = AC = 7$ cm and $BC = 4$ cm. From A a line AP is drawn parallel to BC and of length x cm; find the values of PB^2 and PC^2 in terms of x. Calculate the value of x if $PB^2 - PC^2 = 56$ cm².

Find also the possible values of x if the relation between PB^2 and PC^2 is $2PB^2 = 3PC^2$.

5. Prove that, if two chords AB, CD of a circle cut at a point P outside the circle, then $PA \cdot PB = PC \cdot PD$.

O is the centre of the circle, and the radius is 5 cm. If $PA = 9$ cm, $PB = 16$ cm, calculate the length of OP.   (S.U.)

TEST C.4

1. The sides AB, DC of a cyclic quadrilateral $ABCD$ are produced to meet at X. The sides AD, BC are produced to meet at Y. The circumcircles of the triangles BCX, DCY meet again at Z. Prove that XYZ is a straight line.

2. An isosceles triangle ABC has AB equal to AC. Another isosceles triangle DAC, with its angles equal to those of triangle ABC, is described on AC as base, with DA equal to DC and with D on the side of AC opposite to B; the lines AC, BD cut at E. State the reasons why: (i) AD is parallel to BC; (ii) the triangles AEB, DEC are equal in area; (iii) $\dfrac{BE}{ED} = \dfrac{BC}{CD}$.

3. In the triangle ABC, D is the mid-point of AB and E is the mid-point of CD; AE is produced to meet BC at F. Prove that the triangles BDF, AFD, AFC are equal in area.

4. $ABCD$ is a quadrilateral, and AX, drawn parallel to BD, meets CD produced in X. Prove that the triangle BXC is equal in area to $ABCD$.

State a construction necessary to draw a triangle CYD which is equal in area to $ABCD$. (*You are advised to draw a fresh diagram.*)

5. Prove that parallelograms on the same base and between the same parallels are equal in area, and that the area of a triangle is one-half of the product of its height and base.

$ABCD$ is a trapezium of height h cm. The parallel sides AB, CD are of lengths a, b cm. DA and CB produced meet at O. Find an expression for the area of the trapezium in terms of a, b and h and prove that the area of the triangle OAB is $\dfrac{a^2 h}{2(b-a)}$.   (O.C.)

TEST C.5

1. (i) In the parallelogram $ABCD$ the angles A and C are obtuse. Points X and Y are taken on the diagonal BD such that the angles XAD and YCB are right-angles. Prove that $XA = YC$.

(ii) In the acute-angled triangle PQR, points D and E are taken on the sides QR, PR respectively such that the angles PDR, QER are right-angles. If *PD* and *QE* intersect at *H*, prove that the triangles *PHE*, *QHD* are similar.

2. (i) Prove that the internal bisector of an angle of a triangle divides the opposite side internally in the ratio of the sides containing the angle.

(ii) Prove that the ratio of the areas of similar triangles is equal to the ratio of the squares on corresponding sides.

3. The sides *PQ*, *PR* of a triangle *PQR* are produced to *Y*, *Z* respectively, so that $PQ \cdot PY = PR \cdot PZ$. Prove that the triangles *PQR* and *PZY* are similar.

If, in addition, $YZ = 2QR$, prove that the area of the quadrilateral *YZRQ* is three times the area of the triangle *PQR*.

4. In the acute-angled triangle *PQR*, the sides *PQ* and *QR* are equal and *D* is the mid-point of *QR*. If $PD = PR$, prove that:

(i) $PQ^2 = PD^2 + 2QD^2$;   (ii) $PD^2 = 2QD^2$;   (iii) $PL^2 = 7DL^2$,

where *L* is the foot of the perpendicular from *P* to *QR*.   (C.)

TEST D.1

1. (i) *Q* is 500 m from *P* on a bearing S. 50° W., and *R* is 200 m from *P* on a bearing N. 40° W. Calculate the distance and bearing of *R* from *Q*.

(ii) A vertical flagstaff stands on level ground, and a man observes it through a theodolite placed 16·5 dm above the ground some distance away. The angle of elevation of the top is 25°, and the angle of depression of the bottom is 3°. Calculate the height of the flagstaff.

2. *AB*, *BC* and *CD* are three rods fastened together at *B* and *C*, and fastened at *A* and *D* to two points at the same horizontal level. *AB* is 8 dm long and is inclined at 50° to the horizontal; *BC* is 6 dm long and is inclined at 25° to the horizontal; *CD* is 10 dm long. Calculate the depth of *C* below *AD*, the inclination of *CD* to the horizontal, and the distance *AD*.

3. *ABCD* is a field with straight sides. *AB* is 250 m, *AD* is 100 m, and the angle *A* is 110°. The sides *BC* and *CD* are equal in length, and the angle *C* is a right-angle. Calculate the length *DB* to the nearest metre and the area of the field in hectares.   (W.)

TEST D.2

1. (i) The sine of an acute angle is $\frac{20}{29}$. Calculate the cosine of the angle.

(ii) Find the size, in radians and in degrees, of the angle subtended by an arc of length 9·42 in at the centre of a circle of radius 6 in. (Take π as 3·14.)

2. Considering the earth to be a sphere of radius 3960 miles, calculate:

(i) the distance between the 55° parallel of latitude and the equator measured along a meridian of longitude;

(ii) the length of the 55° parallel of latitude. (Take π as 3·14.)

3. From a post, *H*, 520 dm above sea-level two yachts *L* and *M* are observed. The angles of depression of *L* and *M* as viewed from *H* are 23° and 28° and the bearings of *L* and *M* from *H* are 127° (S. 53° E.) and 325° (N. 35° W.) respectively. Find the distance *LM*.

4. Two ships *A* and *B* leave the same port at 9 a.m. steaming with average speeds of 12 and 16 knots respectively. The course of *A* is S. 25° 30′ W. (205° 30′) and that of *B* due S. At 10 a.m. *B* changes course to S. 14° 30′ E. (165° 30′) continuing to steam at 16 knots. Calculate:

(i) the bearing of *A* from *B* at 10 a.m.;

(ii) the distance of *A* from *B* at 10 a.m.;

(iii) the bearing of *A* from *B* at 11 a.m. (A.B.)

TEST D.3

1. (i) The angle *A* is acute and cos $A = \frac{5}{13}$. Calculate, *without using tables*, tan *A* and sin (180° − *A*).

(ii) *Without using tables* calculate the value of $\frac{1 - \cos A}{1 + \sin A}$ when *A* is an acute angle and tan $A = \frac{3}{4}$.

(iii) The angle *A* is acute and tan $A = \frac{5}{12}$. *Without using tables*, find the value of sin *A* + 5 cos *A*.

2. (i) A television mast 500 ft high is sited on level ground. Calculate the angle of elevation of the top of the mast from a point on the ground 1320 yd away.

(ii) A chord of a circle is 2·7 cm long and subtends an angle of 33° at a point on the circumference of the circle. Calculate the radius of the circle.

3. At noon two ships *A* and *B* are sighted from a port *X*. Ship *A* is 30 nautical miles from *X* on a bearing 020°. She is steaming at 20 knots on a course 040°. Ship *B* is 15 nautical miles from *X* on a bearing 080°. She is steaming at 25 knots on a course 140°. Calculate: (i) the distance and bearing of *A* from *X* at 1 p.m.; (ii) the distance by which *A* is north of *B* at 1 p.m.

4. Two towns X and Y have the same latitude 50° N. and their longitudes differ by 22°. Calculate:

(i) the radius of the circle of latitude through X and Y;

(ii) the length of the straight line XY;

(iii) the angle subtended at the centre of the earth by the shorter arc XY of the great circle through X and Y. (*Assume the earth to be a sphere of radius* 3960 *miles.*) (J.)

TEST D.4

1. (i) In the parallelogram $ABCD$ the diagonal AC, which is 12 cm long makes angles of 73° and 49° with the sides AD and CD. Calculate the lengths of the sides of the parallelogram.

(ii) $ABCD$ is a quadrilateral with the diagonal AC 10 cm long and the side AD 7 cm long. The angle $BAC = 31°$, angle $BCA = 42°$, and the angle $DAC = 53°$. Calculate: (*a*) the perimeter of the quadrilateral; (*b*) its area.

2. A and B are two points on a horizontal straight line passing through the foot of a tower and on the same side of the tower. $AB = 225$ dm. The elevations of the top of the tower from A and B are 19° 30' and 31°. Calculate the height of the tower.

3. A and B are two points 120 m apart on a long straight road on a horizontal plane. N, the foot of a vertical tower PN, stands on this plane so that angles NAB and NBA are 23° and 117° respectively. The elevation of P from A is 9° 30'. Calculate: (i) the length of AN; (ii) the height of PN; (iii) the elevation of P from the point on the road at which this elevation is greatest.

4. Three points A, B, C are selected as the corners of a triangular course for an aeroplane race. The course starts from A to B, 75 km due North of A, then direct from B to C and from C to A. C is 95 km due East of a point on AB 30 km from A. Calculate: (i) the total distance round the course; (ii) the bearing of C from B.

One aeroplane on the leg BC is 40 km from B. In an emergency it has to fly back direct to A. Calculate how much shorter this is than completing the course.

5. An observation post P, on the edge of a cliff, is 425 dm above sea-level. From P two buoys A and B are observed in the sea. A is on a bearing of 31° and B on a bearing of 67°. The angle of depression of both buoys is 13° 42'. Calculate: (i) the horizontal distances of A and B from the foot of the cliff vertically below P; (ii) the distance AB; (iii) the angle APB. (O.)

TEST D.5

1. Assuming the earth to be a sphere of radius 3960 miles, calculate the distances from Prestwick ($55\frac{1}{2}°$ N., 5° W.) to Cairo (30° N., 31° E.) by travelling:

(i) due East and then due South;

(ii) due South and then due East.

(Take π to be 3·142.)

2. A is a port on a straight coast 4 km due North of a second port B. A ship is 2·8 km from A on a bearing 132° (S. 48° E.) from A. Calculate the distance and bearing of the ship from B.

If the ship is steaming North at 11 km/h, calculate, to the nearest minute, the time that will elapse before the ship is due East from A.

3. The course of a river flows from NE. to SW. A boat, whose speed relative to the water is $2\frac{1}{2}$ miles per hour, is steered due North from A, a point on the more southerly bank. It reaches a point on the opposite bank whose bearing from A is 330° (N. 30° W.). By drawing or by calculation, find the speed at which the river is flowing.

Find also the direction in which the boat must be steered from A if it is to travel due West.

4. A man, A, in a boat sees 3 landmarks B, C and D on a straight road running up a hillside inclined at 20° to the horizontal. A, B, C and D are all in the same vertical plane. B is at sea-level and the angles of elevation of C and D observed at A are 12° and 17° respectively. Calculate the distance AB, given that the distance from C to D is 124 m.   (L.)

TEST E.1

1. (i) Differentiate $\left(2x + \dfrac{1}{x}\right)(x^2 + 4)$ with respect to x.

(ii) Find the coordinates of points on the curve whose equation is $y = 2x^3 - 6x^2 - 18x + 15$ at which there are maximum and minimum values, and distinguish between them.

By using your results, and making any other calculations you wish, give a rough *sketch* of this curve.

2. A particle moves in a straight line, starting from rest with an acceleration of $(4t - 2)$ cm/s². Calculate:

(i) the number of seconds after the start when the particle returns to its starting-point;

(ii) the distance of the particle from the starting-point after 3s;

(iii) the number of seconds after the start when the particle's velocity is 40 cm/s.

3. Find the area enclosed between the curve and the three straight lines whose equations are $y = 9x^2$, $y = 0$, $x = 2$ and $x = 3$ respectively.

Also, find the volume of the solid obtained by the rotation about the x-axis of the above area. (*Express the number of cubic units in your final answer in terms of* π.)

4. A particle moves in a straight line passing through a point O, and t s after passing O its velocity is v ft/s, v and t being connected by the formula $v = 7 + 5t - 2t^2$.

(i) What is the particle's velocity as it passes O?

(ii) How many seconds after leaving O does it come momentarily to rest?

(iii) How far is the particle from O after 2 s?

(iv) How many seconds after leaving O is its acceleration zero?

(D.)

TEST E.2

1. A body starts from rest and moves in a straight line. Its velocity v cm/s after t s is given by $v = 24t - 3t^2$. Prove that the body returns to the starting-point after 12 s.

Calculate: (i) the greatest velocity of the body on the outward journey; (ii) the greatest distance of the body from the starting-point during the 12 s.

Sketch a graph showing the distance of the body from the starting-point during the 12 s.

2. Sketch the curve $y = \dfrac{x^2}{4}$ and the line $y = 3$. Show that, whatever the value of t, each of the points $A(2t, t^2)$ and $B(-2t, t^2)$ is a point on the curve.

The point A is between the x-axis and the line $y = 3$ and through A a line is drawn parallel to the y-axis to meet the line $y = 3$ at P. Through B a line is drawn parallel to the y-axis to meet the line $y = 3$ at Q. Show that the area of the rectangle $APQB$ is $4t(3 - t^2)$. Calculate the maximum area of the rectangle.

3. (i) Find the maximum and minimum values of $x^3 - 3x - 2$.

(ii) Calculate the volume of the solid formed when the area bounded by the curve $y = x^3 - 3x$, the x-axis and the ordinate at $x = 1$ makes a complete revolution about the x-axis. Give your answer as a multiple of π.

4. Sketch the curve $y = \dfrac{x^3}{4}$ and calculate its gradient at the point $P(2, 2)$.

The tangent to the curve at P cuts the x-axis at T, and the perpendicular from P to the x-axis cuts it at N. Calculate the length of OT where O is the origin.

Calculate the volume of the solid of revolution formed when the area bounded by the arc OP of the curve, the straight line PN and the straight line ON is rotated about the x-axis. Give your answer as a multiple of π. (J.)

TEST E.3

1. (i) Differentiate $3x + 2\sqrt[3]{x} - \dfrac{2}{x^2}$ with respect to x.

(ii) Integrate $4x^4 - \frac{1}{2}\sqrt{x} + \dfrac{6}{x^3}$ with respect to x.

(iii) The cost, £C, per mile of an electric cable is given by $C = \dfrac{120}{x} + 1080x$, where x in^2 is the cross-section of the cable. Find, by differentiation or otherwise, the cross-sectional area for which the cost is least and also the least cost per mile.

2. (i) The distance, s cm, travelled by a particle in t seconds is given by $s = t^3 - 9t^2 + 27t$. Calculate: (a) how far the particle is from its starting-point after 3 seconds; (b) the velocity of the particle after 2 seconds; (c) the time when its acceleration is zero.

(ii) Calculate the values of x for which $x^3 - 3x^2 - 9x + 15$ has a maximum or minimum value. Find these values and state which is the maximum value and which is the minimum value.

3. (i) Differentiate $7x^3 - \dfrac{3}{x^2} - 4\sqrt[4]{x}$ with respect to x.

(ii) Integrate $10x^4 - \dfrac{6}{x^2} - \dfrac{3}{\sqrt[3]{x}}$ with respect to x.

(iii) Plot the graph of $y = x^3 - x^2 - 8x$ from $x = -3$ to $x = +4$. Calculate the area of the region bounded by the curve from $x = 0$ to $x = +3$, the line $x = +3$ and the x-axis from $x = +3$ to $x = 0$.

4. A farmer has 440 m of fencing with which to enclose an area, adjacent to a straight wall, in the shape of a rectangle and semicircle. If the radius of the semicircular boundary is x m, show that

(i) the lengths of the two parallel sides are $\left(220 - \dfrac{\pi x}{2}\right)$ m

(ii) the area of the enclosure is $\left(440x - \dfrac{\pi \cdot x^2}{2}\right)$ m^2

By differentiation, or otherwise, find the value of x for which the area is a maximum and calculate the maximum area. (Take π as $\frac{22}{7}$.)

(A.B.)

TEST E.4

1. If the curve $y = ax^2 - x^3$ has a gradient 3 at the point where $x = 1$, calculate the value of a. Hence determine the points where the curve meets the x-axis and the coordinates of the turning-points. Sketch the curve, indicating the units on the axes.

2. (i) Find $\int (x^2 + 3)\, dx$ (ii) Evaluate $\int_0^3 (2x + x^2)\, dx$

(iii) Evaluate $\int_1^2 (3x^2 - x)\, dx$

3. (i) The curve $y^2 = 3x + 4$ is revolved about the axis of x. Calculate the volume generated by the part of the curve between $x = 0$ and $x = 4$. (*Leave π as a factor in your answer.*)

(ii) Calculate the gradient of the curve $y = 2x^3 - \frac{2}{3}x + 7$ at the point where $x = \frac{2}{3}$. Calculate also the values of x for which y has a maximum or minimum value.

4. A particle moves in a straight line and t seconds after passing a fixed point O its acceleration is $-3t$ cm/s^2. If the particle comes to rest 2 seconds after passing O, calculate:

(i) the velocity with which it passes O;
(ii) the distance from O where it comes to rest;
(iii) the total time taken to reach O again;
(iv) the velocity with which it then passes through O.

5. (i) Sketch the curve $y = (1 + x)(3 - x)$ and calculate the co-ordinates of the points where it is cut by the line $y = 3$. Calculate the area between the curve and the line.

(ii) The curve $y = 3\left(x - \dfrac{1}{x}\right)$ crosses the x-axis at the point $(1, 0)$. The area between the curve and the axis of x from this point to the line $x = 3$ is rotated about the axis of x. Calculate the volume of the solid thus generated. (*Leave π as a factor in your answer.*)   (L.)

ANSWERS

PART A: ALGEBRA

Ex. 1. P. 1.

1. 4   **2.** 18   **3.** 0   **4.** 60   **5.** 2   **6.** 15   **7.** 0
8. 24   **9.** 1   **10.** $\frac{1}{4}$ **11.** $\frac{1}{2}$ **12.** 0   **13.** 3   **14.** 2
15. $\frac{1}{2}$ **16.** 3   **17.** 7   **18.** 0   **19.** 4   **20.** $\frac{1}{3}$ **21.** 1

Ex. 2. P. 1.

1. $x + 3$ **2.** $p - 6$ **3.** $4b$ **4.** $\frac{m}{2}$
5. $6S$ **6.** $2b + a$ **7.** $a - b = 9$ **8.** $bc + xy$
9. $6 - x = 2$ **10.** $x = 2y$ **11.** $\frac{p}{2} = 6$ **12.** $3 - a = a - 5$

Ex. 3. P. 1.

1. $3a$ **2.** $4x$ **3.** $5p$ **4.** 0
5. $7c$ **6.** f **7.** $4x$ **8.** y
9. $2t$ **10.** $2b$ **11.** $7b - 6d$ **12.** $n + 2p$
13. $2r - 7s$ **14.** n.s.f.   **15.** n.s.f.   **16.** $6r$
17. $s + 2t$ **18.** $4k - l + m - 2$ **19.** n.s.f.   **20.** 0

Ex. 4. P. 2.

1. 36   **2.** $G + B$ **3.** $x + y$ miles
4. $38 - t$ miles   **5.** $14 - x$ yr.   **6.** $z - 3$ yr.
7. (i) 100; (ii) 600; (iii) $100S$; (iv) $600S$; (v) $50S$; (vi) $250S$
8. (i) 2; (ii) 6; (iii) $2x$; (iv) $6x$
9. $\frac{x}{2}$; $\frac{3x}{2}$ **10.** $1\frac{1}{2}$; (i) $\frac{x}{y}$; (ii) $\frac{3x}{2y}$; (iii) $\frac{3x}{2y}$

Ex. 5. P. 3.

1. $6a + 9b + 7c$ **2.** $5x + 7y + 6z$ **3.** $5m + 2n + 6p$
4. $3r + t$ **5.** $3p$ **6.** $5x + 3y$ **7.** 0
8. $5ab + 6cz + 9lm$ **9.** $8xy + 5yz + 2$ **10.** $2ax + 2cz$

Ex. 6. P. 3.

1. 0   **2.** x^2y^2 **3.** x^2y^2 **4.** $2x^2y^2$
5. $2x^2y^2$ **6.** $2x^2y^2$ **7.** $6abc$ **8.** $x^2y^3z^4$
9. $24xy^2z^3$ **10.** $6a^2bc$ **11.** $6x^2y^2z$ **12.** $9x^2y^2z^2$

Ex. 7. P. 4.

1. 3   **2.** 9   **3.** 8   **4.** 2
5. 12   **6.** x **7.** y **8.** 4
9. $12b$ **10.** a **11.** $4a$ **12.** $2x$
13. $2a$ **14.** 1   **15.** $2x$ **16.** 0

Ex. 8. P. 5.

1. 3   **2.** a **3.** $2l$ **4.** 2   **5.** 2   **6.** 3
7. 4   **8.** 3   **9.** 2   **10.** $4y$ **11.** y **12.** a
13. $2x^2$ **14.** 6   **15.** 2   **16.** 2   **17.** $4ax$ **18.** $4x$

Ex. 9. P. 5.

1. $2a^2 - 2a$ **2.** $3x^2 + 3x$ **3.** $x^2 + 2x + 5$
4. $4b^2 + 4b + 3$ **5.** $y^2 + 2y + 4$ **6.** $l^3 - 3l^2 + l$
7. $3c^3 + c^2 + c$ **8.** $2a^3b + 2a^2b^2 + 2ab^3$ **9.** $z^2 + z + 5$
10. $2p^2 - p + 2$ **11.** $t^3 - 2t^2 + 4t$ **12.** $7s^3 - 2s^2 + 3s$
13. $4 - a + 3a^2 - 3a^3$ **14.** $2 - y - y^2$ **15.** $3b - b^2 - 2b^3$
16. $1 + x + 3x^2$ **17.** $3 + e + e^3$ **18.** $2g + 2g^2 - g^3 - g^4$
19. $2ab + a^2b + a^3b$ **20.** $3 + ab^3 + a^2b^2 + a^3b$ **21.** $4 + ab^2 - a^2b$
22. $2ab - 4a^2 + 3a^3$ **23.** $3l^2m^2 + 2l^3 - 2m$ **24.** $-3l^2 + l^3 + 3m - 2m^2$

Ex. 10. P. 6.

1. $\dfrac{p+q}{p-q}$ **2.** $a^3 - \sqrt{m}$ **3.** $\dfrac{xyz}{x+y+z}$ **4.** $2a + b$

5. $\dfrac{x}{y} + \dfrac{y}{z}$ **6.** $5t^2$ **7.** $b^2 + \sqrt[3]{a}$ **8.** $\dfrac{p-q}{p+q}$

9. $x^3 - x^2 + x$ **10.** $\sqrt[4]{a} + \sqrt[3]{b} - \sqrt{c}$

Ex. 11. P. 7.

1. $8a^5$ **2.** $6de^2$ **3.** $\frac{2}{3}e$ **4.** $2d$ **5.** $6a^3y^3$
6. $3p^2q$ **7.** z **8.** $3x^3y$ **9.** $3e^2f^2g^2$ **10.** $2p^2q^3$
11. $9a^8$ **12.** $2c^2d^4$ **13.** k^5 **14.** a **15.** b
16. x^4 **17.** h^7 **18.** $3kl^2mn^2$ **19.** $1\frac{1}{2}f$ **20.** x^2
21. x **22.** $2a^2$ **23.** $8a^3$ **24.** 1   **25.** a^2cz^3
26. $2x^2$ **27.** 1   **28.** $\frac{1}{2}$ **29.** $\frac{2}{3}x^3y^3$ **30.** a^{12}
31. $6x^4y^4$ **32.** $\dfrac{x}{a}$ **33.** 1   **34.** 4   **35.** $6x^6$
36. a^2

Ex. 12. P. 7.

1. $2p - q$ **2.** $\dfrac{y}{2}$; y even   **3.** 3p pence; 4p pence; ptx pence

4. $\dfrac{x}{2}$ **5.** $3\frac{1}{2}$ m **6.** (i) 12; (ii) $12F$ **7.** (i) 3; (ii) $3Y$

8. $36Y$ **9.** 22; $\dfrac{Y}{22}$ **10.** 1760; $\dfrac{Y}{1760}$ **11.** lb

12. (i) 216; (ii) $8\frac{3}{4}$ **13.** $2l + 2b$ **14.** 26

Ex. 13. P. 9.

1. $+2$ **2.** -2 **3.** -2 **4.** -3 **5.** -3 **6.** $+4$
7. $+1$ **8.** 0   **9.** -5 **10.** -7 **11.** $+9$ **12.** 0
13. $+5$ **14.** $+11$ **15.** $+11$ **16.** $+3$ **17.** $+11$ **18.** -3
19. $+4$ **20.** $+4$ **21.** $+14$ **22.** -4 **23.** $+2$ **24.** -2
25. 0   **26.** 0   **27.** 0   **28.** 0   **29.** -8 **30.** -8
31. -1 **32.** -1 **33.** 0   **34.** 0   **35.** $+2$ **36.** $+2$
37. $+2x$ **38.** 0   **39.** $-2x$ **40.** $+x$ **41.** $+x$ **42.** $-x$
43. $+2y$ **44.** 0   **45.** $+3b$ **46.** $+3t$ **47.** $9 - 3x$ **48.** $3x - 9$

Ex. 14. P. 11.

1. -4 **2.** 0 **3.** $+16$ **4.** $+12$ **5.** -8 **6.** 0
7. -28 **8.** $+28$ **9.** -45 **10.** $+45$ **11.** -45 **12.** 0
13. $+x^2$ **14.** $-x^2$ **15.** $+x^2$ **16.** $+6x^2$ **17.** $-6x^2$ **18.** $+6x^2$
19. $-48y^2$ **20.** $-16a^2$ **21.** 0 **22.** 0 **23.** $-27x$ **24.** $+27x$

Ex. 15. P. 12.

1. $+2$ **2.** $+\frac{1}{2}$ **3.** -2 **4.** $-\frac{1}{2}$ **5.** $+4$ **6.** $+\frac{1}{4}$
7. 0 **8.** -1 **9.** $-1\frac{1}{2}$ **10.** $-a$ **11.** $-x$ **12.** $+1$
13. -1 **14.** $-x$ **15.** $+\frac{1}{2}x$ **16.** $2a$ **17.** $\frac{1}{2}$ **18.** $3ab$
19. $-2a^2$ **20.** $-\frac{1}{2}b$ **21.** $3ab$ **22.** $-3ac$ **23.** $-xyz$

Ex. 16. P. 12.

1. 4 **2.** -12 **3.** -3 **4.** 8 **5.** 12 **6.** 2
7. a **8.** $4a^2$ **9.** $-a$ **10.** 6 **11.** 3 **12.** $-2a^2$
13. $x^2 - x$ **14.** $-a^2$ **15.** 81 **16.** $2x$ **17.** $2x$ **18.** $4x^3 + 2x^2$
19. $6ab$ **20.** -1 **21.** $-a$ **22.** $2x^2$ **23.** $\frac{1}{2}$ **24.** $8x^2$
25. -2 **26.** 1 **27.** $\frac{1}{2}$ **28.** $\dfrac{b}{a}$ **29.** $\dfrac{a^3}{b^3}$ **30.** $\frac{1}{8}$

Ex. 17. P. 13.

1. $+2$ num.   **2.** -3 num.   **3.** $+6$ num.   **4.** -4 num.
5. $+4$ num.   **6.** $+4$ num.   **7.** -9 num.   **8.** -5 num.
9. $+3$ num.   **10.** $+3$ num.   **11.** -16 num.   **12.** -1 num.
13. $+a$ lit.   **14.** $-b$ lit.   **15.** $+1$ num.   **16.** $+b$ lit.
17. -1 num.   **18.** $+1$ num.   **19.** $-x$ lit.   **20.** $+xy$ lit.
21. $+x$ lit.   **22.** $-ab$ lit.   **23.** $-b$ lit.   **24.** -1 num.
25. $+2$ num. $+a$ lit.   **26.** -2 num.   **27.** -2 num. $+x$ lit.
28. $+16$ num. $+x$ lit.   **29.** -16 num.   **30.** -16 num. $+x$ lit.

Ex. 18. P. 14.

1. $3a + b$ **2.** $a + b$ **3.** $-a - 2b$ **4.** $4a - 2b$ **5.** $2b$
6. $4x + 5y$ **7.** $2x + 5y$ **8.** $3x + 2y$ **9.** $x - 5y$ **10.** $4a - 2b$
11. $5x - 2y$ **12.** $a + b - 3c$ **13.** a **14.** 6

Ex. 19. P. 15.

1. $m - 2n - 4p$ **2.** $2r + s - 3t$ **3.** $-4p + 2q + 6r$
4. $4x + 4y - 4$ **5.** $u + v$ **6.** $4pq$
7. $-ab + 3cz$ **8.** $-2a - b - c - d + e$ **9.** $a^2 + ab - 4b$
10. -4 **11.** $a^3 - a^2 + a - 4$ **12.** $a^2 - 2a + 4$
13. $x - 2y + 6$ **14.** $-2a - 2y$ **15.** 0
16. $a^3 - 4a^2 + 3a - 4$ **17.** $-4x^2 + 4xy + 2y^2$ **18.** $4x^2 - 4xy - 2y^2$
19. $2x^2 + 2xy$ **20.** $-2x^2 - 2xy$

Ex. 20. P. 16.

1. $2a + 2b$ **2.** $4x + 8$ **3.** $-2x - 2y$
4. $-4a + 4b$ **5.** $3a + ab$ **6.** $-3a + ab$
7. $4a + 8b + 12c$ **8.** $-6x + 9y + 12z$ **9.** $2a^4 - 2a^3 + 2a^2$
10. $-6x^4 + 12x^3 + 3x^2$ **11.** $6a^3 + 2a^3b + 2a^2$
12. $-3x^4 - 6x^3y - 3x^2y^2$ **13.** $a^2 + 2ab + b^2$
14. $a^2 - b^2$ **15.** $a^2 - 2ab + b^2$ **16.** $a^2 + 5a + 6$
17. $a^2 + a - 6$ **18.** $a^2 - 4b^2$ **19.** $4a^2 - 4ab + b^2$
20. $2a^2 - 5ab + 2b^2$ **21.** $ax + ay + bx + by$ **22.** $2a^2 - 3a - 2$

Ex. 20. P. 16.

23. $ax - ay - bx + by$ **24.** $a^4 + 2a^3 + a^2$ **25.** $6a^4 + 5a^3 - 6a^2$
26. $a^2c^3 - a^3c^3 + bc^2 - abc^2$ **27.** $x^2 - 2xy + y^2$ **28.** $4a^2 + 4a + 1$
29. $a^3 + 2a^2 + 2a + 1$ **30.** $a^3 - 1$ **31.** $a^3 - 2a - 1$
32. $a^3 - 2a^2 + 1$ **33.** $a^3 + 6a^2 + 2a - 12$ **34.** $a^3 - 9a^2 + 23a - 15$
35. $4a^3 + 4a^2 + 5a + 12$ **36.** $9a^3 - 34a + 24$
37. $x^3 + x^2 + x + x^2y + xy + y$ **38.** $4x^3 - 4x^2y - 4x + 4xy^2 - 4y^3 + 4y$
39. $x^4 + 2x^3 + 4x^2 + 3x$ **40.** $x^2y^2 - 2x^2y + 3x^2 - xy^2 + 2xy - 3x$
41. $x^4 - 3x - 4$ **42.** $6x^4 - 2x^3 - 6x^2 + 4x - 2$
43. $2xy + 2y - y^2$ **44.** $x^2 + 4y - 4y^2 - 1$
45. $2x^2 + 2x + xy - 17y - 6y^2 - 12$ **46.** $a^2 - 2bc - b^2 - c^2$

Ex. 21. P. 17.

1. $1 + b + c$ **2.** $c - d - e$ **3.** $2 - 3y + 4z$ **4.** $-2 - 3y + 4z$
5. $1 - 2ax + 3a^2x^2$ **6.** $-a + x + 2x^2$ **7.** $2x - 3y + 4$
8. $2ac + 3bc^2 - 6$ **9.** $\dfrac{a^2 + b^2 + c^2}{d^2}$ **10.** $1 + 2pqr - 3p^2q^2$

Ex. 22. P. 18.

1. $x + 3$ **2.** $x + 4$ **3.** $x + 5$ **4.** $x + 5$ **5.** $3x + 2$
6. $5x + 4$ **7.** $7x - 10$ **8.** $7a + 4$ **9.** $4l + 5$ **10.** $5p - 6$
11. $4r - 1$ **12.** $4y - 10$ **13.** $2x^2 - x - 1$
14. $a^2 + ab + b^2$ **15.** $2x^3 + 4x^2 - 2x$ **16.** $2x - 2$
17. $x^2 - x + 1$ **18.** $3x - 1$ **19.** $2x - 3$
20. $a^4 - a^3 + a^2 - a$ **21.** $a^2 - ab + b^2$ **22.** $x + y$
23. $a^2 + ab + b^2$ **24.** $a^2 - ab - b^2$ **25.** $4a^3 - 12a^2 + 54a - 162$
26. $x^3 - x^2y + xy^2 - y^3$ **27.** $a^3 + a^2b + ab^2 + b^3$
28. $3a^4b - 6a^3b^2 + 12a^2b^3 - 24ab^4$ **29.** $a^3b + a^2b^2 + ab^3$
30. $x^5 - x^4y + x^3y^2 - x^2y^3 + xy^4 - y^5$

Ex. 23. P. 19.

1. $x^2 + 2x + 1$ r. -2 (add 2)   **2.** $x^2 - x$ r. $2x$ (add $-2x$)
3. $2x^2 + x - 1$ r. 2 (add -2)   **4.** $x^2 + 5x - 3$ r. -6 (add 6)
5. $x^3 + 2x^2 - x + 1$ r. -9 (add 9) **6.** $2a + b$ r. $-2b^2$ (add $2b^2$)
7. $a + b - c$ r. c^2 (add $-c^2$)
8. $x^2 - x - 1$ r. $(xy + 5y)$ add $(-xy - 5y)$
9. $a^3 - a^2 + a - 1$ r. 1 (add -1)   **10.** $x + 1$ r. -2 (add 2)

Ex. 24. P. 19.

1. (i) $a + b$; (ii) $2(a + b)$; (iii) $2(a + b) + (a + b)$
2. (i) $(2 + x)$ lb; (ii) $6(2 + x)$ lb   **3.** $(G - g)$ gal
4. $n(x - 2)$ ft   **5.** (i) $p - q$; (ii) $b(p - q)$
6. (i) $x + 6$; (ii) $(50x + 60)$ pence (iii) $£\dfrac{50x + 60}{100}$ **7.** (i) $59f$ ft; (ii) $\dfrac{60}{f} + 1$
8. (i) r$(5H + 4)$ pence; (ii) $£\dfrac{r(5H + 4)}{100}$ **9.** $(x^2 + 3x + 2)$ in^2
10. (i) $(6x^2 + 10x + 4)$ in^2; (ii) 4 in^2; (iii) $(6x^2 + 10x)$ in^2

Ex. 25. P. 21.

1. x^4y^4 **2.** $8a^3b^3$ **3.** $16a^4$ **4.** $9a^4$ **5.** x^4
6. $-27x^3$ **7.** $9a^2b^2$ **8.** $4a^4b^2$ **9.** a^6b^3 **10.** $4a^2b^6$
11. $8a^6b^9$ **12.** $9a^6b^4$ **13.** a^9b^9 **14.** $81a^8b^8$ **15.** $64a^6x^9y^{12}$
16. ab **17.** xy **18.** $2ay$ **19.** $2a$ **20.** $3x$

Ex. 25. P. 21.

21. $2x$ **22.** $3a^2$ **23.** $2x^2$ **24.** $3a^2$ **25.** $4a^2x$
26. $4ab^2$ **27.** $9a^4$ **28.** $2a^2bc^4$ **29.** $5a^2x^3$ **30.** ab^2c^3
31. $a^3b^3c^3$ **32.** $a^2b^2c^2$ **33.** a^6 **34.** $\frac{1}{64}$ **35.** $16a^2$
36. $\frac{a^6}{b^3c^3}$ **37.** $\frac{y^3}{a^3}$ **38.** $\frac{9}{49}$ **39.** $\frac{k^2m^2}{4l^2}$ **40.** $\frac{1}{16}$ **41.** $\frac{z^3}{x^3}$
42. 27 **43.** $\frac{c}{b}$ **44.** $\frac{3ab}{4xy}$ **45.** 1 **46.** abc^3 **47.** $\frac{x^2y^3}{a^3b^2}$
48. $3a$ **49.** $\frac{2ab}{3xy^2}$ **50.** $\frac{3y}{2a}$ **51.** $\frac{ab^2}{y}$

Ex. 26. P. 22.

1. a **2.** $6x + 2y$ **3.** $9b - 4a$ **4.** 0 **5.** $p - 1$
6. $7y - 4$ **7.** $2x - y + z$ **8.** $d + 2e$ **9.** $3k + 2l$ **10.** $q - p$
11. $2a^3 + a^2$ **12.** $6 + 5a - a^2$ **13.** $2x - 1$ **14.** $2x^2 + 3x^4 - 2x^5$
15. $2a^2x - 2a^2x^2 - 3ax + 2ax^2 - x$ **16.** $a^3 - 2ay$
17. $3e - 5d$ **18.** 4 **19.** $2n - m^2$ **20.** $-5r$
21. $3x^3 - 6x^2 + 18x$ **22.** $3a + b - c$ **23.** $2x + 9y - 9z$
24. a **25.** $a^4x - a^3x + a^3x^2 + a^2x - a^2x^2 - ax^3$

Ex. 27. P. 23.

1. 2 **2.** 4 **3.** 5 **4.** 4 **5.** 3 **6.** 6 **7.** 5
8. 1 **9.** 0 **10.** 5 **11.** 5 **12.** 4 **13.** 3 **14.** 2
15. 4 **16.** 2 **17.** 3 **18.** 3 **19.** 1 **20.** 2 **21.** 1
22. 1 **23.** -1 **24.** -3 **25.** $4\frac{1}{2}$ **26.** 5 **27.** $3\frac{1}{5}$ **28.** -2
29. $-2\frac{1}{2}$ **30.** $-\frac{1}{2}$ **31.** $\frac{3}{4}$ **32.** $-1\frac{1}{6}$ **33.** $-2\frac{1}{2}$ **34.** -1 **35.** $1\frac{1}{3}$
36. $2\frac{3}{8}$ **37.** 2 **38.** 11 **39.** $1\frac{4}{5}$ **40.** $-\frac{3}{16}$

Ex. 28. P. 25.

1. 4 **2.** 2 **3.** $1\frac{1}{2}$ **4.** $\frac{1}{2}$ **5.** 1 **6.** 5
7. 2 **8.** 1 **9.** 2 **10.** 2 **11.** 1 **12.** 5
13. 1 **14.** 2 **15.** 3 **16.** 1 **17.** 9 **18.** 5
19. -2 **20.** -1 **21.** -2 **22.** $-\frac{1}{2}$ **23.** $\frac{1}{2}$ **24.** 6
25. 3 **26.** $\frac{2}{3}$ **27.** $-4\frac{1}{3}$ **28.** 0

Ex. 29. P. 26.

1. 6 **2.** 2 **3.** $\frac{1}{3}$ **4.** 4 **5.** $3\frac{1}{2}$
6. 1 **7.** 8 **8.** -4 **9.** $1\frac{1}{2}$ **10.** $10\frac{1}{2}$
11. 0 **12.** $\frac{7}{12}$ **13.** 3 **14.** 2 **15.** 2
16. 5 **17.** 2 **18.** -4 **19.** 1 **20.** $\frac{3}{4}$

Ex. 30. P. 26.

1. 1 **2.** 1 **3.** $-\frac{1}{7}$ **4.** 1 **5.** $-\frac{5}{8}$ **6.** $2\frac{3}{4}$
7. 1 **8.** 1 **9.** $1\frac{1}{2}$ **10.** $-1\frac{1}{2}$ **11.** $3\frac{1}{6}$ **12.** 2

Ex. 31. P. 27.

1. $2x^2 + 6x + 4$ **2.** $4x^2 + 4x - 24$ **3.** $6x^2 - 24$
4. $3x^2 - 3$ **5.** $5x^2 + 5x - 60$ **6.** $8x^2 + 40x + 48$
7. $6x^2 - 9x - 6$ **8.** $15x^2 - 25x + 10$ **9.** $24x^2 - 16x - 8$
10. $32x^2 - 2$ **11.** $x^3 - x$ **12.** $2x^3 + 3x^2 + x$
13. $6x^3 + 13x^2 + 6x$ **14.** $4x^3 + 15x^2 - 4x$ **15.** $2x^3 - 8x$
16. $8x^3 - 2x$ **17.** $9x^3 + 15x^2 + 6x$ **18.** $3a^3x - 3ax^3$
19. $2a^3x^2 - 2a^3x + 2a^2x^2 - 2a^2x$ **20.** $a^4x^2 + 2a^3x^3 + a^2x^4$

Ex. 32. P. 27.

1. -14 **2.** 7   **3.** -3 **4.** $1\frac{1}{4}$ **5.** $-\frac{7}{8}$
6. 2   **7.** 3   **8.** 1   **9.** -3 **10.** -1

Ex. 33. P. 28.

1. (i) (i) $x + y$; (ii) $x - y$; (iii) xy **2.** $16 - x$ **3.** $y - 12$
4. $3a - 2b$ **5.** $4x + 3y$ **6.** xy **7.** $\dfrac{x}{y}$
8. (i) $3x$; (ii) x^3 **9.** 7; 8   **10.** 5; 4; 3   **11.** $a + 1$; $a + 2$
12. $x - 1$; $x - 2$; $x - 3$ **13.** $x + 1$; $x - 1$ **14.** $n + 2$; $n + 4$
15. $b - 2$; $b - 4$ **16.** 12   **17.** 36   **18.** 8   **19.** 9
20. 5   **21.** 9   **22.** 4

Ex. 34. P. 29.

1. 6   **2.** 2   **3.** 13; 14   **4.** 21; 22   **5.** 3
6. 4; 11   **7.** 12;16   **8.** 12; 24   **9.** 12; 35   **10.** 11; 12; 13
11. 18; 19; 20   **12.** 26; 28   **13.** 17; 19   **14.** 3

Ex. 35. P. 30.

1. A: 17p; B: 11p; C: 8p
2. A: £15; B: £7·50; C: £2·50   **3.** A: 20p; B: 40p; C: 10p
4. A: £1; B: 50p; C: 50p   **5.** A: £2; B: 50p   **6.** 200 in²
7. 25 yr, 50 yr.   **8.** 10 yr.   **9.** A: 14 yr.; B: 4 yr.
10. A: 30 yr.; B: 18 yr.; C: 12 yr.   **11.** A: 26 yr.; B: 14 yr.; C: 20 yr.
12. 30 at 2p, 20 at 10p   **13.** 10 at £5, 50 at £1
14. A: $26\frac{2}{3}$ lb; B: $85\frac{1}{3}$ lb   **15.** 15 at 5p, 10 at 2p, 5 at 1p
16. 90 yd   **17.** 9 yd by 3 yd

Ex. 36. P. 31.

1. $3\frac{1}{6}a$ **2.** $\frac{5}{12}a$ **3.** $-\frac{9}{10}a$ **4.** $4\frac{1}{12}a$ **5.** $\frac{5}{12}a$ **6.** $\frac{7}{12}a$
7. $\dfrac{5}{a}$ **8.** $\dfrac{7}{6a}$ **9.** $\dfrac{31}{12a}$ **10.** $\dfrac{a^2 - ab + b^2}{ab}$
11. $\dfrac{6a^2 - 2ab + 12b^2}{3ab}$ **12.** $\dfrac{a^2c + ab^2 - bc^2}{abc}$ **13.** $\dfrac{a^2 + ab - 3b^2}{2ab}$
14. $\dfrac{37a}{8b}$ **15.** $\dfrac{3a - 4b + 2c}{abc}$ **16.** $\dfrac{9a + 5b + 8c}{6}$
17. $\dfrac{8b - 5a}{12}$ **18.** $\dfrac{5a - b}{12}$ **19.** $-a - 2b$
20. $\dfrac{3a^3 + 2a^2b^2 - 4a^2b + 2ab^2 + 2b^3}{a^2b^2}$ **21.** $\dfrac{3b^6 - 3a^6}{a^3b^3}$
22. $\dfrac{4a^2 + 11ab + 9b^2}{6ab}$ **23.** $a + 4b - 2$

Ex. 37. P. 32.

1. $1\frac{1}{5}$ **2.** 1   **3.** $1\frac{1}{4}$ **4.** 3   **5.** 3
6. 4   **7.** $\frac{5}{6}$ **8.** $15\frac{1}{6}$ **9.** $1\frac{11}{16}$ **10.** 5

Ex. 38. P. 33.

1. 3   **2.** 4   **3.** 2   **4.** $1\frac{1}{2}$ **5.** $3\frac{1}{2}$ **6.** $-5\frac{1}{3}$
7. $\frac{19}{20}$ **8.** $\frac{13}{27}$ **9.** $1\frac{4}{7}$ **10.** $\frac{8}{9}$ **11.** $4\frac{19}{25}$ **12.** -13
12. $\frac{5}{6}$ **14.** -4 **15.** $2\frac{1}{4}$ **16.** 4

Ex. 39. P. 34.

1. $1\frac{1}{2}$ **2.** $2\frac{2}{3}$ **3.** 2 **4.** 5 **5.** $\frac{3}{8}$ **6.** $\frac{3}{8}$

7. $2\frac{2}{3}$ **8.** $\frac{3}{8}$ **9.** 1 **10.** $\frac{2}{3}$ **11.** 5 **12.** 5

13. 1 **14.** $2\frac{2}{5}$ **15.** 13 **16.** $\frac{11}{13}$ **17.** $\frac{11}{13}$ **18.** $1\frac{2}{3}$

19. $\frac{1}{2}$ **20.** $\frac{1}{2}$ **21.** 1 **22.** $\frac{1}{54}$ **23.** $1\frac{5}{6}$ **24.** 0

Ex. 40. P. 35.

1. 3 **2.** 11 **3.** 6 **4.** $\frac{7}{10}$ **5.** 10 miles **6.** 2 miles **7.** 7 miles

8. 12 coloured **9.** 5 plain; 9 fancy **10.** 10 correct

11. 4 correct **12.** 150 mile/h **13.** 3; 5

14. A: 24p per lb; B: 60p per lb **15.** 30 mile/h; 40 mile/h

16. Area: 920 yd²; Perimeter: 378 ft

Ex. 41. P. 37.

1. $\frac{A}{l}$ **2.** $\frac{V}{lb}$ **3.** $\frac{A}{r^2}$ **4.** $\frac{C}{2\pi}$

5. $\frac{V}{h}$ **6.** $\frac{3V}{h}$ **7.** $\frac{W}{I}$ **8.** $\frac{V}{C}$

9. $\frac{W+180}{5}$ **10.** $\frac{S+4}{2}$ **11.** $\frac{P-2l}{2}$ **12.** $2S-b-c$

13. $24 - 2R$ **14.** $A - P$ **15.** $\frac{9C}{5}+32$ **16.** $\frac{2A}{h}$

17. $\frac{A-2bh}{2h}$ **18.** $\frac{2A}{a+b}$ **19.** $\frac{100I}{PT}$ **20.** $\frac{NS}{20}$

21. $\sqrt[3]{\frac{3V}{4\pi}}$ **22.** $\sqrt{\frac{A}{4\pi}}$

Ex. 42. P. 38.

1.
$f(x)$	-5	-4	-3	-2	-1	0	1	2	3	4	5

2.
$f(x)$	-10	-8	-6	-4	-2	0	2	4	6	8	10

3.
$f(x)$	-20	-16	-12	-8	-4	0	4	8	12	16	20

4.
$f(x)$	25	16	9	4	1	0	1	4	9	16	25

5.
$f(x)$	-125	-64	-27	-8	-1	0	1	8	27	64	125

6.
$f(x)$	75	48	27	12	3	0	3	12	27	48	75

7.
$f(x)$	-250	-128	-54	-16	-2	0	2	16	54	128	250

8.
$f(x)$	$-2\frac{1}{2}$	-2	$-1\frac{1}{2}$	-1	$-\frac{1}{2}$	0	$\frac{1}{2}$	1	$1\frac{1}{2}$	2	$2\frac{1}{2}$

9.
$f(x)$	$-1\frac{1}{4}$	-1	$-\frac{3}{4}$	$-\frac{1}{2}$	$-\frac{1}{4}$	0	$\frac{1}{4}$	$\frac{1}{2}$	$\frac{3}{4}$	1	$1\frac{1}{4}$

10.
$f(x)$	-4	-3	-2	-1	0	1	2	3	4	5	6

Ex. 42. P. 38.

11. $f(x)$ | -2 | -1 | 0 | 1 | 2 | 3 | 4 | 5 | 6 | 7 | 8

12. $f(x)$ | 5 | 6 | 7 | 8 | 9 | 10 | 11 | 12 | 13 | 14 | 15

13. $f(x)$ | -8 | -6 | -4 | -2 | 0 | 2 | 4 | 6 | 8 | 10 | 12

14. $f(x)$ | -11 | -8 | -5 | -2 | 1 | 4 | 7 | 10 | 13 | 16 | 19

15. $f(x)$ | -24 | -19 | -14 | -9 | -4 | 1 | 6 | 11 | 16 | 21 | 26

16. $f(x)$ | -21 | -17 | -13 | -9 | -5 | -1 | 3 | 7 | 11 | 15 | 19

17. $f(x)$ | -19 | -16 | -13 | -10 | -7 | -4 | -1 | 2 | 5 | 8 | 11

18. $f(x)$ | -30 | -25 | -20 | -15 | -10 | -5 | 0 | 5 | 10 | 15 | 20

19. $f(x)$ | $-\frac{1}{5}$ | $-\frac{1}{4}$ | $-\frac{1}{3}$ | $-\frac{1}{2}$ | -1 | ∞ | 1 | $\frac{1}{2}$ | $\frac{1}{3}$ | $\frac{1}{4}$ | $\frac{1}{5}$

20. $f(x)$ | $-\frac{1}{10}$ | $-\frac{1}{8}$ | $-\frac{1}{6}$ | $-\frac{1}{4}$ | $-\frac{1}{2}$ | ∞ | $\frac{1}{2}$ | $\frac{1}{4}$ | $\frac{1}{6}$ | $\frac{1}{8}$ | $\frac{1}{10}$

21. $f(x)$ | $-\frac{2}{5}$ | $-\frac{1}{2}$ | $-\frac{2}{3}$ | -1 | -2 | ∞ | 2 | 1 | $\frac{2}{3}$ | $\frac{1}{2}$ | $\frac{2}{5}$

22. $f(x)$ | $-\frac{3}{10}$ | $-\frac{3}{8}$ | $-\frac{1}{2}$ | $-\frac{3}{4}$ | $-1\frac{1}{2}$ | ∞ | $1\frac{1}{2}$ | $\frac{3}{4}$ | $\frac{1}{2}$ | $\frac{3}{8}$ | $\frac{3}{10}$

23. $f(x)$ | $\frac{1}{15}$ | $\frac{1}{12}$ | $\frac{1}{9}$ | $\frac{1}{6}$ | $\frac{1}{3}$ | ∞ | $-\frac{1}{3}$ | $-\frac{1}{6}$ | $-\frac{1}{9}$ | $-\frac{1}{12}$ | $-\frac{1}{15}$

24. $f(x)$ | $-\frac{1}{25}$ | $-\frac{1}{16}$ | $-\frac{1}{9}$ | $-\frac{1}{4}$ | -1 | ∞ | -1 | $-\frac{1}{4}$ | $-\frac{1}{9}$ | $-\frac{1}{16}$ | $-\frac{1}{25}$

Ex. 43. P. 39.

3. *KLMN* is a parallelogram.

4. *RST* is an isosceles triangle.

Ex. 45. P. 41.

1. (a) $y = -4$; (b) $x = 1\frac{1}{2}$

2. (a) $y = 0$; (b) $x = -1\frac{1}{4}$

3. (a) $y = 3$; (b) $x = -\frac{3}{4}$

Ex. 46. P. 42.

The values of y are:

1. 0   **2.** 1   **3.** -1 **4.** -2 **5.** 0   **6.** -1

7. 1   **8.** -2 **9.** 0   **10.** 1   **11.** -1 **12.** 2

13. -3 **14.** 3   **15.** 3   **16.** -3 **17.** 0   **18.** 2

19. $-\frac{1}{2}$ **20.** $1\frac{1}{2}$ **21.** $\frac{1}{2}$ **22.** -2 **23.** 0   **24.** 1

25. 1   **26.** $-\frac{1}{4}$ **27.** 2   **28.** $2\frac{1}{2}$

Ex. 47. P. 44.

All in the form $y =$
1. x 2. $4x$ 3. $4x$ 4. $5x$ 5. $2x$
6. $2x$ 7. $2\frac{1}{2}x$ 8. $\frac{1}{2}x$ 9. $\frac{1}{2}x$ 10. $\frac{1}{4}x$
11. $\frac{2}{3}x$ 12. $\frac{2}{5}x$ 13. $2\frac{1}{2}x$ 14. $3\frac{1}{4}x$ 15. $\frac{1}{3}x$
16. $1\frac{2}{3}x$ 17. $\frac{3}{5}x + 2$ 18. $\frac{2}{5}x + 6$ 19. $2x + 4$ 20. $3x + 2$
21. $5x + 4$ 22. $4x + 5$ 23. $3x - 3$ 24. $2x - 5$ 25. $4x - 1$
26. $5x - 2$ 27. $1\frac{1}{2}x - 4$ 28. $7x - 3$ 29. $\frac{1}{2}x + 3$ 30. $\frac{1}{4}x - 4$
31. $3\frac{1}{4}x + 5$ 32. $2\frac{3}{5}x - 9$

Ex. 49. P. 46.

1. $3 . x$ 2. $5 . a$ 3. $x . y$ 4. $2 . x . y$ 5. $a . x . y$
6. $b . b$ 7. $3 . b . b$ 8. $2 . a . b . b$ 9. $a . a . b . b$ 10. $5 . a . a . b . b$
11. $7 . a . x . x . y . y . y$ 12. $a . a . x . x . y . y$

Ex. 50. P. 46.

1. $a(1 + 2b)$ 2. $2(1 + 2a)$ 3. $4(a + b)$ 4. No factors
5. $a(c - b)$ 6. $a(x + y)$ 7. $2(ac - xy)$ 8. $3cx(a - 2y)$
9. No factors 10. $2(a^2 + b^2)$ 11. $x(a^2 + b^2)$ 12. $2xy(x - y)$
13. $p^3(1 - p)$ 14. $3mn(5m + n)$ 15. $3mn(mn - 5)$ 16. $3(m^2 + 5n^2)$
17. No factors 18. $3mn^2(1 + 5mn)$ 19. $2(p + 2q - 3r)$
20. $p(q + r + s)$ 21. $2p(q - 2r + 3s)$ 22. $2(pq - 2qr - 3rs)$
23. $2x(2 - 4x + 5x^2)$ 24. $4ax(x + 2a - 4)$

Ex. 51. P. 46.

1. $a^2 - ab$ 2. $ax^2 + ax$ 3. $ax^2y^2 + bx^2y^2$
4. $ax - ay + bx - by$ 5. $ab + ay - bx - xy$ 6. $ab + ax + by + xy$
7. $ab - ax - by + xy$ 8. $2ap + aq + 2bp + bq$
9. $aq - 2ap - bq + 2bp$ 10. $2al + 2am - bl - bm$
11. $6al - 6am + 6bl - 6bm$ 12. $p^2y + p^2x + q^2y + q^2x$
13. $ax^2 - ay^2 - bx^2 + by^2$ 14. $abx + acx + by + cy$
15. $ab - ac - bxy + cxy$ 16. $a^2x - abx + aby - b^2y$
17. $ax^2 + axy - bxy - by^2$ 18. $a^2xy - b^2xy$
19. $a^2 + 5a + 6$ 20. $x^2 - y^2$ 21. $c^2 - 1$
22. $a^2 + 2ab + b^2$ 23. $a^2 - 2ab + b^2$

Ex. 52. P. 47.

1. $(a + b)(x + y)$ 2. $(a - b)(x - y)$ 3. $(a + b)(x - y)$
4. No factors 5. No factors 6. $(p + q)(2 - a)$
7. $(l - m)(b + 3)$ 8. $(c + d)(ab - xy)$ 9. $(r^2 - s)(a^2 + b^2)$
10. No factors 11. $(y - z)(3 - t)$ 12. $(x^2 + y^2)(a + b)$
13. $(l + m^2)(x + y)$ 14. No factors 15. $ab(c + d - 1)$
16. $a^2(1 + c - d)$ 17. No factors 18. $(p - q)(a^2 + 1)$
19. No factors 20. $b^3(a^2 + a + b^2 - b)$

Ex. 53. P. 47.

1. $(a + b)(c + d)$ 2. $(c + d)(x + y)$ 3. $(2 + c)(x + y)$
4. $(2 + a)(x + y)$ 5. $(a + 2)(a + x)$ 6. $(a + x)(x + 6)$
7. $(1 + 3a)(2a + 3b)$ 8. $(2a + 3b)(2x + 3y)$ 9. $(a + b)(x - y)$
10. $(a - b)(x + y)$ 11. $(a - b)(a - c)$ 12. $(b - a)(x + y)$
13. $(a + 1)(x - y)$ 14. $(a + b)(x - 1)$ 15. $(x - y)(y - 1)$

Ex. 53. P. 47.

16. $(a - y)(x - y)$ **17.** $(a + x)(a^2 + x^2)$ **18.** $(x + y^2)(x^2 + y)$
19. $(x - y)(x^3 - y^3)$ **20.** $(a^4 + 1)(b - 1)$ **21.** $(a + x)(a + b + c)$
22. $(a + 1)(a + b - c)$

Ex. 54. P. 48.

1. $a^2 + 3a + 2$ **2.** $x^2 + 3x + 2$ **3.** $x^2 + 5x + 6$ **4.** $a^2 + 4a + 3$
5. $x^2 + 4x + 3$ **6.** $x^2 + 4x + 4$ **7.** $a^2 + 5a + 4$ **8.** $x^2 + 6x + 8$
9. $x^2 + 5x + 4$ **10.** $x^2 + 7x + 12$ **11.** $a^2 + 6a + 5$ **12.** $a^2 + 6a + 8$
13. $a^2 + 7a + 10$ **14.** $15 + 8x + x^2$ **15.** $12 + 7a + a^2$ **16.** $a^2 - a - 2$
17. $a^2 + a - 2$ **18.** $x^2 + x - 6$ **19.** $x^2 - 5x + 6$ **20.** $a^2 - 4a + 3$
21. $a^2 + 2a - 3$ **22.** $x^2 - 3x + 2$ **23.** $x^2 - 3x - 4$ **24.** $x^2 + 2x - 8$
25. $x^2 - 5x + 4$ **26.** $x^2 - x - 12$ **27.** $a^2 - 4a - 5$ **28.** $a^2 - 6a + 8$
29. $a^2 - 3a - 10$ **30.** $15 - 2x - x^2$ **31.** $12 - 7a + a^2$ **32.** $a^2 - 1$
33. $6a^2 + 13a + 6$ **34.** $4a^2 - 4$ **35.** $2a^4 - 7a^2 + 3$ **36.** $6a^2 + 5ab - 6b^2$
37. $4a^2 + 6a + 2$ **38.** $4a^2 - 4a + 1$ **39.** $4a^2 - 1$
40. $a^2 + 2ab + b^2$ **41.** $a^2 - 2ab + b^2$ **42.** $a^2 - b^2$
43. $2a^2 + 11a + 12$ **44.** $3x^2 - 14x + 8$ **45.** $6x^2 - 2x - 8$
46. $6x^2 - 5x - 6$ **47.** $9x^2 - 9$ **48.** $2x^2 + 5xy + 2y^2$
49. $4x^2 - y^2$ **50.** $6 + 12a + 6a^2$ **51.** $6 - 13a + 6a^2$
52. $9 - 4a^2$ **53.** $6 + 5a - 6a^2$ **54.** $4a^2 - 9$
55. $x^2 + xy - 6y^2$ **56.** $ax + ay + bx + by$ **57.** $a^4 - 1$
58. $1 - x^6$ **59.** $6a^4 - a^2bx^2 - 2b^2x^4$ **60.** $a^3 + a^2b^2c^3 - ab^3c^4 - b^5c^7$
61. $a^2 + 2a + 1$ **62.** $8a^2 + 2ab - 15b^2$ **63.** $4x^2 - 12x + 9$
64. $16x^2 - 25y^2$ **65.** $a^2 + 2ab + b^2$ **66.** $12y^2 - 25yz + 12z^2$
67. $4x^2 - 4xy + y^2$ **68.** $9a^2 - 25b^2$ **69.** $9a^2 + 18ab + 9b^2$
70. $5a^3 + 25a^2b + ab + 5b^2$ **71.** $a^4 - 2a^2 + 1$ **72.** $9x^4 - 12x^2y^2 + 4y^4$

Ex. 55. P. 49.

1. $a^2 + a$ **2.** $a^2 + ab$ **3.** $a^2 - a$
4. $a^2 - ab$ **5.** $2a^2 + 2a$ **6.** $2a^2 + 2ab$
7. $2a^2 - 2a$ **8.** $2a^2 - 2ab$ **9.** $2a^2 + 2ab + 2ac$
10. $a^2 + 2ab + b^2$ **11.** $2a^2 + 5ab + 2b^2$ **12.** $2a^2 + 4ab + 2b^2$
13. $2x^3 + 7xy + 3y^2$ **14.** $4x^2 + 8xy + 4y^2$ **15.** $a^2 - 2a + 1$
16. $a^2 - 2ab + b^2$ **17.** $4x^2 - 12xy + 9y^2$ **18.** $4a^2 - b^2$

Ex. 56. P. 50.

1. $(x + 1)(x + 1)$ **2.** $(x + 2)(x + 2)$ **3.** $(x + 1)(x + 2)$
4. $(x + 2)(x + 3)$ **5.** $(x + 1)(x + 3)$ **6.** $(x + 1)(x + 4)$
7. $(x + 3)(x + 4)$ **8.** $(x + 2)(x + 4)$ **9.** $(x + 3)(x + 5)$
10. $(x + 2)(x + 5)$ **11.** $(x + 1)(x + 5)$ **12.** $(x + 4)(x + 4)$
13. $(x + 5)(x + 5)$ **14.** $(x + 4)(x + 5)$ **15.** $(x + 2)(x + 8)$
16. $(x - 1)(x - 1)$ **17.** $(x - 2)(x - 2)$ **18.** $(x - 1)(x - 2)$
19. $(x - 1)(x - 3)$ **20.** $(x - 2)(x - 3)$ **21.** $(x - 3)(x - 3)$
22. $(x - 1)(x - 4)$ **23.** $(x - 2)(x - 4)$ **24.** $(x - 3)(x - 4)$
25. $(x - 4)(x - 4)$ **26.** $(x - 5)(x - 5)$ **27.** $(x - 6)(x - 6)$
28. $(x - 1)(x - 6)$ **29.** $(x - 2)(x - 5)$ **30.** $(x - 1)(x - 10)$
31. $(x - 1)(x + 3)$ **32.** $(x - 1)(x + 4)$ **33.** $(x - 1)(x + 5)$
34. $(x - 3)(2x + 1)$ **35.** $(x + 3)(2x + 1)$ **36.** $(x - 1)(2x + 3)$
37. $(x + 1)(2x - 3)$ **38.** No factors   **39.** $(x - 1)(2x + 1)$
40. $(x + 1)(2x - 1)$ **41.** $(x - 1)(2x - 1)$ **42.** $(x + 1)(2x + 1)$
43. No factors   **44.** $(x + 2)(2x + 3)$ **45.** $2(x + 3)(x + 1)$

Ex. 56. P. 50.

46. $(x + 2)(2x - 3)$ **47.** $(x - 2)(2x + 3)$ **48.** $2(x - 3)(x + 1)$
49. No factors **50.** $2(x + 3)(x - 1)$ **51.** $(x + 2)(3x + 1)$
52. $(x + 1)(3x + 2)$ **53.** $(x + 2)(3x - 1)$ **54.** No factors
55. $(x - 1)(3x + 2)$ **56.** $(x + 1)(3x - 2)$ **57.** $(x - 1)(3x - 2)$
58. $(x - 2)(3x + 1)$ **59.** $(x - 2)(3x - 1)$ **60.** No factors
61. $(x + 3)(3x + 2)$ **62.** $3(x + 2)(x + 1)$ **63.** $(x - 3)(3x - 2)$
64. $3(x - 2)(x - 1)$ **65.** $(x - 3)(3x + 2)$ **66.** $(x + 3)(3x - 2)$
67. No factors **68.** $(x - 1)(x + 6)$ **69.** $3(x^2 - 3x - 2)$

Ex. 57. P. 51.

1. $3(a + b)(x - y)$ **2.** $4(2a + 3)(3a - 2)$ **3.** $2(2b - a)(b - a)$
4. No factors **5.** No factors **6.** $(4a + b^2)(4a + b^2)$
7. $8(a + 2b)(2a + b)$ **8.** No factors **9.** $5(x^2 + 5x - 2)$
10. $(3a - 2b)(3x - 2y)$ **11.** $6(3 - b^2)(4 + b^2)$ **12.** $4ac(a - 2ac - c)$
13. $(3a + 4b)(4a + 3b)$ **14.** No factors **15.** $3(3a + b)(3a - 4b)$
16. $6xy(x + 2y)(x - 3y)$ **17.** $(a + 1)(x + y)$ **18.** No factors
19. $(a - 9)(4a + 1)$ **20.** $b(ab + x - y^2 + z)$ **21.** $(2x + 3y)(3x + 4y)$
22. $(x + 12y)(6x + y)$ **23.** $6(x^2 + 9xy + 2y^2)$ **24.** $a(a + 2)(a - 3)$
25. $(a + b)(a - 1)$ **26.** No factors **27.** No factors
28. No factors **29.** $2(2 + x)(3 - 2x)$ **30.** $(3a - 2b)(2x - 3y)$

Ex. 58. P. 52.

1. $x^2 + 2xy + y^2$ **2.** $x^2 - 2xy + y^2$ **3.** $x^2 - y^2$
4. $a^2 + 2a + 1$ **5.** $a^2 - 2a + 1$ **6.** $a^2 - 1$
7. $a^2 + 4a + 4$ **8.** $x^2 - 4$ **9.** $x^2 + 4x + 4$
10. $4x^2 + 4x + 1$ **11.** $4x^2 - 1$ **12.** $4x^2 - 4x + 1$
13. $4a^2 - 4$ **14.** $4x^2 - 4xy + y^2$ **15.** $x^2 + 4xy + 4y^2$
16. $x^2 - 4y^2$ **17.** $x^2 + 6x + 9$ **18.** $9 - 6x + x^2$
19. $9x^2 - 12xy + 4y^2$ **20.** $9 - x^2$ **21.** $1 - x^6$
22. $(x + y)(x - y)$ **23.** $(a + 2)^2$ **24.** $(x - 3)^2$
25. $(2x + y)(2x - y)$ **26.** $(3x + 5)(3x - 5)$ **27.** $(a + 2b)^2$
28. $(3a - b)^2$ **29.** $(3a + b)(3a - b)$ **30.** $(4 + a^3)(4 - a^3)$
31. $(2 + y)^2(2 - y)^2$ **32.** $(y - 4)^2$ **33.** $(a + 10)(a - 10)$
34. $(ax + 1)^2$ **35.** $(3y + 4z)^2$ **36.** $(1 + t^2)(1 + t)(1 - t)$
37. $(2 + t)(2 - t)$ **38.** $(2 - t)^2$ **39.** $(l^2 + p)^2$
40. $(p - l)^2$ **41.** $(2s - 3t)^2$ **42.** $(3r + 2s)(3r - 2s)$
43. $4a^2 + 8ab + 4b^2$ **44.** $4a^2 - 9b^2$ **45.** $9x^4 - 6x^2 + 1$
46. $a^8 - 1$ **47.** $36 + 12ax + a^2x^2$ **48.** $1 - a^2x^2$
49. $25 - 10a^2 + a^4$ **50.** $25 - 30a + 9a^2$ **51.** $a^2x^2 - 100$
52. $25x^2 + 40xy + 16y^2$ **53.** $1 - y^8$ **54.** $a^2x^2 + 2abx^2 + b^2x^2$
55. $a^4 + 2a^2x^3 + x^6$ **56.** $16 - 24t + 9t^2$ **57.** $a^6 - 2a^3b^2 + b^4$
58. $a^6 - b^4$ **59.** $100t^2 - s^4$ **60.** $a^8 + 12a^6 + 36a^4$
61. $9(2t + 1)(2t - 1)$ **62.** No factors **63.** $(a^3 - 3)^2$
64. $(2 + t)^2(2 - t)^2$ **65.** No factors **66.** $(11x + 3y)(11x - 3y)$
67. $(at^2 + 1)^2$ **68.** No factors **69.** $(3 - 2a)^2$
70. $(c + d^2)^2(c - d^2)^2$ **71.** No factors **72.** No factors
73. $(9 - 8a)^2$ **74.** $4(6x^2 + 5y)(6x^2 - 5y)$ **75.** $a^4(3a^2 + 2)^2$

Ex. 59. P. 53.

1. 120 **2.** 160 **3.** 1000 **4.** 2000 **5.** 200
6. 240 **7.** 400 **8.** 280 **9.** 1760 **10.** 2200
11. 7800 **12.** 10200 **13.** 25600 **14.** 32 **15.** 78

Ex. 60. P. 53.

1. $(a - b + c)(a - b - c)$ 2. $x(x + 2y)$ 3. No factors
4. $(p^2 + 1)(p + 1)(p - 1)$ 5. $3(3 + 2a)(3 - 2a)$ 6. No factors
7. $2(a^2 + ab + b^2)$ 8. $(a + c)(a - 2b - c)$ 9. No factors
10. $(x + 4)^2$ 11. $b(b - 2a)$ 12. $9(9 + x^2)$
13. $3(x + y)(x - y)$ 14. No factors   15. $(a + 2b)(3a + 2b)$
16. $a(a + 2b)$ 17. $(\frac{1}{4} + y^2)(\frac{1}{2} + y)(\frac{1}{2} - y)$ 18. $2(2x + 3y)2x - 3y)$
19. $2(a + x)(a + x)$ 20. $-x(8 + x)$ 21. $(x^2y^4 + z^6)(xy^2 + z^3)(xy^2 - z^3)$
22. No factors   23. $9(3x + 1)(3x - 1)$ 24. $(1 + x^2)(1 + x)(1 - x)$
25. $(3 + a + b)(3 - a - b)$ 26. No factors

Ex. 61. P. 54.

1. $9axy(4ay + x)$ 2. $48(a + b)^2$ 3. $a^2b(abc + 1)(abc - 1)$
4. No factors   5. $(a + b)(2a + 3b)$ 6. $7(4a^2 + 1)(2a + 1)(2a - 1)$
7. $x(x + y - 3)(x - y + 3)$ 8. $(5a - b)(5b - a)$ 9. $15(x + 2)(x - 2)$
10. $2(x + y)(x - y)$ 11. No factors   12. $(5ab + 2)^2$
13. $(a - b)(2x - y)$ 14. $(x - y)(x + y + 8)$ 15. $9a(2ax + y)(2ax - y)$
16. $5(2a + 1)(a - 2)$ 17. $x(a + b)(a - b)$ 18. No factors
19. $(7 + 2x)(3 - x)$ 20. $(2 + a)(x + y)$ 21. No factors
22. $5(8x^2 - 5y^2)$ 23. $(\frac{3}{4} + x)(\frac{3}{4} - x)$ 24. $(3a - 2b)^2$

Ex. 62. P. 54.

1. a 2. x 3. ax 4. ay 5. axy
6. a^2xy 7. ax 8. x^2y^2 9. No H.C.F.   10. $3b$
11. $2abc$ 12. $3y^2z$ 13. $4a^4x^4$ 14. $15b^3c^3$ 15. No H.C.F.
16. $14r^2$ 17. $5l^2m^4$ 18. $15a^4x^3z^2$ 19. No H.C.F.   20. $x + y$
21. $2x^2$ 22. $a^2(2x + 3y)$ 23. No H.C.F.   24. $4x^2y^2(a + 1)$
25. $x - y$ 26. $a + b$ 27. $x + 1$ 28. $2(a - x)$
29. $x^2(x + y)$ 30. No H.C.F.   31. $a^3 + a^2 + 1$ 32. $(x + 1)(x - 1)$

Ex. 63. P. 55.

1. a^2b^2 2. abc 3. $12a$ 4. $12abc$
5. $6a^2bc^2$ 6. $6a^3b^3$ 7. $3a^2b^2c^2$ 8. $2abx^3$
9. a^3b^3 10. $a^4x^3y^3$ 11. $12x^2y^3z^3$ 12. $20a^3y^2z^3$
13. $60xyz$ 14. $12y^3z^2$ 15. $18a^2b^3c$ 16. $a^2b^3c^2$
17. $45a^3x^2$ 18. $2a^2b^2$ 19. $p^2q^2r^2s^2t^2$ 20. $60lmn$
21. $l^2m^3n^4p^5$ 22. $12x^3$ 23. $abcx^2y^3$ 24. n^5p^5
25. $3a(a + b)$ 26. $10(a - b)$ 27. $a^3(a + b)$
28. $6x^2(x - y)$ 29. $abxy(a + b)$ 30. $2(a + b)(a - b)$
31. $ab(a + b)(a - b)$ 32. $(2a + b)(a + 2b)$ 33. $ab(x + 2y)(2x + y)$
34. $a^2x(a + 1)(x + 1)$ 35. $a^2x^2(x + 1)(x - 1)$ 36. $a^2x^2(1 + a)(1 + x)$
37. $(a + 1)(a - 1)^2$ 38. $(x + 1)^2(x + 2)$ 39. $x(2 - x)^2$
40. $15(2a + 1)^2$ 41. $(a + 2)(a + 2)(a + 4)$ 42. $(x + 2)(x - 2)(2x - 3)$
43. $(x + 1)(x - 1)(2x + 1)(2x - 1)$ 44. $(x + 1)(x - 1)(2x - 3)$
45. $(a + 1)(x + 1)(x - 1)$ 46. $(2 - a)(3 - a)(3 - x)$
47. $6a^3(a + 1)(a - 1)(a + 2)$ 48. $15x^3(a + 1)(a - 1)(2 - x)$

Ex. 64. P. 56.

1. $\dfrac{2a^2}{3y^2}$ 2. $\frac{5}{6}$ 3. $\dfrac{xz}{4y}$ 4. $\dfrac{3ab}{4xy}$
5. $\dfrac{2p^3}{3r^3}$ 6. $\dfrac{m^3}{l^2n^2}$ 7. r^2t 8. $\dfrac{a^2}{3x}$

Ex. 64. P. 56.

9. $\dfrac{4k}{3lm}$ 10. $\dfrac{a^4d}{b^2c^2}$ 11. $\dfrac{q}{p}$ 12. $\dfrac{b^3x^4}{c^4z^3}$

Ex. 65. P. 56.

1. $\frac{2}{3}$ 2. $\frac{1}{2}$ 3. $\dfrac{a+b}{a-b}$ 4. $\dfrac{ax^2-bx^2}{a^2+bx^2}$

5. $\dfrac{3x}{4y}$ 6. -1 7. $\dfrac{a(b+y)}{b(a+y)}$ 8. $\dfrac{1}{a-b}$

9. $\dfrac{1}{x+1}$ 10. a 11. $\frac{1}{3}$ 12. $\dfrac{3a^2yz^2}{4b^3c^2x}$

13. $\frac{2}{3}$ 14. $\dfrac{a}{x}$ 15. $-\frac{1}{2}$ 16. $(p+1)(p-1)$

17. $4a$ 18. $\frac{1}{2}$ 19. 3 20. -1

Ex. 66. P. 57.

1. $\dfrac{3(a+1)}{(a-1)(a+2)}$ 2. $-\dfrac{a+b}{a-b}$ 3. $\dfrac{2b}{a-b}$ 4. $\dfrac{b^2}{a-b}$

5. $1\frac{5}{6}$ 6. $\dfrac{a-8}{12}$ 7. $\dfrac{6}{a(a+2)(a-1)}$ 8. $-\dfrac{1}{x^2+1}$

9. $-\dfrac{4b}{a+b}$ 10. 0 11. $\dfrac{x+5}{(x+1)(x+2)(x+3)}$

12. $\dfrac{2}{x+y}$ 13. $\dfrac{x+4}{(x-1)(x-4)}$ 14. 0 15. 1

16. $\dfrac{a^2+b^2}{a}$ 17. $\dfrac{2x-1}{9x-2}$ 18. $2(a+1)(a-1)$

Ex. 67. P. 59.

1. $\frac{9}{11}$ 2. $\frac{1}{2}$ 3. $2\frac{1}{46}$ 4. $-1\frac{1}{4}$ 5. $-\frac{1}{2}$ 6. 1

7. $-1\frac{1}{3}$ 8. $-1\frac{1}{8}$ 9. $\frac{2}{7}$ 10. $-\frac{2}{3}$ 11. -2 12. 12

13. $1\frac{1}{4}$ 14. $-\frac{1}{3}$ 15. $-\frac{1}{2}$ 16. 12 17. $\frac{2}{5}$ 18. $5\frac{1}{2}$

Ex. 68. P. 59.

1. $(3x+1)(2x-1)(x-1)$
2. $(x-1)(2x-1)(3x+1)$
3. $(x+1)(x+1)(x-1)$
4. $(x+1)(x-1)(x-1)$
5. $(x-1)(x-1)(x-1)$
6. $(x+1)(x+1)(x+1)$
7. $(x+1)(x+2)(x+3)$
8. $(x-1)(x-3)(x-5)$
9. $(2x+1)(2x+1)(2x-1)$
10. $(x^2+4)^2(x+2)^2(x-2)^2$
11. $(a+b)(a^2-ab+b^2)$
12. $(a-b)(a^2+ab+b^2)$

Ex. 69. P. 60.

1. $(a+b^2)(a^2-ab^2+b^4)$
2. $(x^2-y)(x^4+x^2y+y^2)$
3. $a^2(a+1)$
4. $4(x+2y)(x^2-2xy+4y^2)$
5. $(4+z)(16-4z+z^2)$
6. $(3p-2q)(9p^2+6pq+4q^2)$
7. $(2b-1)(4b^2+2b+1)$
8. $(1+10a)(1-10a+100a^2)$
9. $(a+b)(a-b)$
10. No factors
11. $(5x+6y)(25x^2-30xy+36y^2)$
12. $(1-xy^2)(1+xy^2+x^2y^4)$
13. $2a(a^2+3)$
14. $2b(3a^2+b^2)$

Ex. 70. P. 60.

1. $x^3 + 2x^2 + 2x + 1$ **2.** $x^3 - 2x^2 + 1$ **3.** $1 + 2x - x^3$
4. $x^3 - 1$ **5.** $2x^3 - x^2 + 5x + 3$ **6.** $x^3 - 13x + 12$
7. $6x^3 - 20x^2 + 20x - 8$ **8.** $2x^3 + 7x^2 - 8x - 16$ **9.** 3
10. -4 **11.** 6   **12.** -4 **13.** -28 **14.** -9 **15.** 5
16. -7 **17.** -17 **18.** -1 **19.** 0   **20.** 1

Ex. 71. P. 62.

1. 2, 1   **2.** 1, 2   **3.** 2, 3   **4.** 1, 4   **5.** 5, 2   **6.** 1, 3

Ex. 72. P. 63.

1. 2, -2 **2.** $-1, 3$ **3.** $-2, -3$ **4.** 3, 1   **5.** $2\frac{1}{2}, 3$
6. $2\frac{1}{2}, -3$ **7.** $-2, 4\frac{1}{2}$ **8.** $3\frac{1}{2}, 2\frac{1}{4}$ **9.** $-1\frac{1}{4}, 4\frac{1}{2}$

Ex. 73. P. 63.

1. 2, 1   **2.** 3, 2   **3.** 1, 3   **4.** $-1, 1$ **5.** $-1, -2$
6. 0, 3   **7.** $\frac{1}{2}, 1\frac{1}{2}$ **8.** $5, -\frac{1}{2}$ **9.** $-3, 4$ **10.** $-2\frac{1}{2}, 6$
11. $-\frac{1}{4}, 5$ **12.** $-5, -1$ **13.** $3\frac{1}{2}, -\frac{3}{4}$ **14.** 7, 0   **15.** 9, 3
16. $10, -5$ **17.** $-8, 6$ **18.** $\frac{3}{5}, \frac{2}{5}$

Ex. 74. P. 64.

1. 2, 1   **2.** 1, 3   **3.** 3, 2   **4.** 2, 5   **5.** 4, 1
6. 1, 4   **7.** 2, 2   **8.** 2, 3   **9.** 5, 1   **10.** 4, 7
11. $-1, 4$ **12.** $-3, -2$ **13.** $-4, \frac{1}{2}$ **14.** $2\frac{1}{2}, -5$
15. 3, 4   **16.** $-2, 1\frac{1}{2}$ **17.** $2\frac{1}{3}, \frac{5}{9}$ **18.** $-3\frac{1}{7}, -2\frac{5}{6}$
19. 10, 12   **20.** 8, 6   **21.** $-5, 6$

Ex. 75. P. 65.

1. 1, 2   **2.** 2, 3   **3.** 3, 1   **4.** 2, 2   **5.** $-1, 1$
6. $3, -2$ **7.** 3, 1   **8.** $-4, -3$ **9.** 5, 7   **10.** 12, 10
11. 11, 1   **12.** 6, 2   **13.** $1\frac{1}{2}, -1\frac{1}{3}$ **14.** 5, 6   **15.** $\frac{1}{4}, -\frac{1}{3}$
16. 2, 4   **17.** 16, 24   **18.** $\frac{1}{8}, -\frac{1}{10}$ **19.** 5, 7   **20.** 2, 3

Ex. 76. P. 66.

1. 1, 2, 3   **2.** 3, 2, 1   **3.** $2, -2, 3$ **4.** $4, -1, 1$ **5.** 2, 3, 0
6. $-3, -1, 2$ **7.** $0, -\frac{1}{2}, \frac{3}{4}$ **8.** $3, -2, 4$ **9.** 1, 4, 5

Ex. 77. P. 66.

1. $5, -3$ **2.** 2, 1   **3.** $0\cdot5, -0\cdot4$ **4.** 2, 3, 7
5. $7, -4, 5$ **6.** $\frac{1}{2}, \frac{1}{3}$ **7.** 8, 6   **8.** $-2, 1\frac{1}{2}, -3$
9. $-4, 8$ **10.** $3, -3$ **11.** $-1, 3, -5$

Ex. 78. P. 67.

1. 8, 4   **2.** 36, 20   **3.** 60, 12   **4.** 30, 14   **5.** 25, 15
6. 27, 18   **7.** $\frac{3}{4}$ **8.** $\frac{3}{5}$ **9.** $\frac{5}{9}$ **10.** $\frac{7}{8}$
11. 45   **12.** 39   **13.** 84   **14.** 235   **15.** 742
16. Tea 90p; coffee £1·20   **17.** £35; £25
18. A. 28p; B. 51p.
19. 46 at 10p; 18 at 2p   **20.** 12; 48
21. 45 yr.; 15 yr.   **22.** 50 m; 144 m²
23. 352 tickets   **24.** 8 cows; 12 ducks
25. 360 black, 720 coloured   **26.** Road 250 miles; rail 270 miles
27. 3 h 20 min; 240 miles

Ex. 79. P. 69.

1. $a - b$ 2. $a + b$ 3. ab 4. 1   5. $\dfrac{a}{b}$

6. $\dfrac{b + c}{a}$ 7. 1   8. $a + b$ 9. $\dfrac{b^2}{a}$ 10. $\dfrac{a + b}{a - b}$

11. 1   12. $2(2a + 3b)$ 13. a 14. $-\dfrac{bc}{a}$ 15. $\dfrac{1}{a + b + c}$

16. $\dfrac{2ab}{a + b}$ 17. 1

Ex. 80. P. 70.

1. $a + b, a - b$ 2. $a + b, a + c$ 3. $\dfrac{a + b}{2}, \dfrac{a + b}{3}$ 4. $2a + b, 2a - b$

5. $a + 1, b + 1$ 6. $\dfrac{a + 1}{2}, \dfrac{b + 2}{3}$ 7. $\dfrac{a + b}{3}, \dfrac{a - b}{5}$ 8. $a + 2b, 2a - b$

9. $ab - c, ab + c$ 10. $ab + cd, ac + bd$

Ex. 81. P. 70.

1. 2, 4   2. 6, 3   3. $\frac{1}{2}, \frac{1}{3}$ 4. $1, -\frac{1}{2}$ 5. 3, 2

6. $\frac{1}{3}, -\frac{1}{4}$ 7. $\frac{3}{4}, \frac{5}{8}$ 8. $1\frac{1}{2}, 2\frac{1}{2}$ 9. $1, -1\frac{1}{4}$

Ex. 82. P. 71.

1. $\dfrac{d}{2}$ 2. $\dfrac{C}{2\pi}$ 3. $\sqrt{\left(\dfrac{A}{\pi}\right)}$

4. $\dfrac{A}{2\pi h}$ 5. $\sqrt{\left(\dfrac{V}{\pi h}\right)}$ 6. $\sqrt{\left(\dfrac{3V}{\pi h}\right)}$

7. (i) $\dfrac{EFb}{a}$ (ii) $\dfrac{La}{EF}$ 8. (i) $g\left(\dfrac{T}{2\pi}\right)^2$ (ii) $l\left(\dfrac{2\pi}{T}\right)^2$

9. (i) $\dfrac{4\pi D}{X}$ (ii) $\dfrac{KX}{4\pi}$ 10. $\sqrt{\left(\dfrac{bxz}{a - bz}\right)}$

11. (i) $\dfrac{4\pi IN}{10H}$ (ii) $\dfrac{10Hl}{4\pi N}$ 12. $\dfrac{bc}{\sqrt{(b^2 + c^2)}}$

13. (i) $\dfrac{AKN}{4000\pi C}$ (ii) $\dfrac{4000\pi Cd}{AN}$ 14. $\dfrac{Rr_2 r_3}{r_2 r_3 - Rr_3 - Rr_2}$

15. (i) $\dfrac{FKd^2}{Q_2}$ (ii) $\sqrt{\left(\dfrac{Q_1 Q_2}{FK}\right)}$ 16. $\dfrac{1}{C(2\pi f)^2}$

17. (i) $\dfrac{100I}{RT}$ (ii) $\dfrac{100I}{PR}$ 18. $\dfrac{\sqrt{P(R + b)}}{\sqrt{R}}$

19. (i) $\dfrac{2S}{t^2}$ (ii) $\sqrt{\left(\dfrac{2S}{a}\right)}$ 20. (i) $\dfrac{2s - at^2}{2t}$ (ii) $\dfrac{2(s - ut)}{t^2}$

21. (i) $v - at$ (ii) $\dfrac{v - u}{a}$ 22. (i) $\dfrac{v^2 - u^2}{2s}$ (ii) $\sqrt{(v^2 - 2as)}$

23. $\dfrac{S - 2\pi r^2}{2\pi r}$ 24. (i) $\dfrac{2A}{a + b}$ (ii) $\dfrac{2A - bh}{h}$

25. (i) $\dfrac{2RE}{R - r}$ (ii) $\dfrac{R(L - 2E)}{L}$ 26. $\dfrac{b^2 y}{bx - ay}$

27. $\dfrac{P^2}{(a - K)^2 - P}$ 28. (i) $\dfrac{100(A - P)}{PT}$ (ii) $\dfrac{100A}{100 + RT}$

29. $\sqrt{S^2 - (R - r)^2}$ 30. (i) $\sqrt{2(I^2 - a^2)}$ (ii) $\sqrt{\left(\dfrac{2I^2 - b^2}{2}\right)}$

Ex. 83. P. 72.

1. (i) $1000x$ (ii) $10y$ (iii) $\dfrac{H}{10}$ (iv) $\dfrac{F}{100}$

2. (i) $T(100 - x)$ (ii) $\dfrac{T(100 - x)}{100}$ **3.** (i) $\dfrac{100P}{S}$ (ii) $\dfrac{p}{S}$

4. (i) $\dfrac{D}{V}$ (ii) $\dfrac{60D}{V}$ **5.** $\dfrac{5V}{18}$ **6.** $\dfrac{S(u + 1)}{u(u + 2)}$

7. $\dfrac{S}{100 - l}$ **8.** $\dfrac{PS(P + S)}{P^2 + S^2}$ **9.** $\dfrac{£x(100 - y)}{3}$ **10.** $\dfrac{xy(x + y)}{x^2 + y^2}$

11. $\dfrac{Ms}{s + 5M}$ **12.** $\dfrac{8x^2 + x - 3}{8x}$ **13.** $\dfrac{Mn - HS}{n - H}$ **14.** $4x + 2$

15. $\dfrac{u^2 - 4}{u}$ **16.** $\dfrac{x(10^4 - x^2)}{10^4}$

Ex. 84. P. 73.

1. ± 3 **2.** ± 5 **3.** ± 8 **4.** ± 11 **5.** ± 6
6. ± 7 **7.** ± 10 **8.** ± 12 **9.** $\pm \frac{2}{3}$ **10.** $\pm 2\frac{1}{2}$
11. $\pm 1\frac{1}{2}$ **12.** $\pm \frac{6}{7}$ **13.** ± 2 **14.** ± 3 **15.** ± 4
16. ± 5 **17.** ± 1 **18.** ± 3 **19.** ± 7 **20.** $\pm 2\frac{1}{2}$
21. ± 2 **22.** $\pm 2 \cdot 121$ **23.** $\pm 2 \cdot 828$ **24.** ± 2

Ex. 85. P. 74.

1. $2, 2$ **2.** $1, 3$ **3.** $1, 2$ **4.** $1, -3$
5. $3, -1$ **6.** $-2, -3$ **7.** $3, -2$ **8.** $-3, -3$
9. $1, 4$ **10.** $-2, -2$ **11.** $0, 2$ **12.** $0, 5$
13. $0, -3$ **14.** $0, -7$ **15.** $0, 0, 1$ **16.** $0, 0, -1$
17. $\frac{1}{2}, 2$ **18.** $\frac{1}{3}, -1$ **19.** $\frac{1}{2}, \frac{1}{3}$ **20.** $1\frac{1}{3}, -3$
21. $\frac{2}{3}, -2$ **22.** $\frac{1}{2}, \frac{1}{2}$ **23.** ± 1 **24.** ± 3
25. ± 5 **26.** ± 2 **27.** $\pm \frac{1}{3}$ **28.** $\pm \frac{3}{4}$
29. $2\frac{1}{2}, -3$ **30.** $4, -\frac{2}{3}$ **31.** $5, -3$ **32.** $\frac{1}{2}, -1\frac{1}{3}$
33. $1\frac{1}{2}, \frac{2}{3}$ **34.** $0, 3\frac{1}{2}$

Ex. 86. P. 74.

1. $1, 2$ **2.** $1, -1$ **3.** $-1, -1$ **4.** $2, -1$ **5.** $1, -2$
6. $3, -2$ **7.** $2, -3$ **8.** $3, -1$ **9.** $1, -3$ **10.** $4, 4$
11. $0, 5$ **12.** $0, -2$ **13.** $0, 0, 6$ **14.** $0, 0, -7$ **15.** $0, \frac{1}{2}$
16. $0, 1\frac{1}{3}$ **17.** $0, 2$ **18.** $0, -1\frac{1}{2}$ **19.** $2, \frac{1}{2}$ **20.** $-2, -\frac{1}{2}$
21. $1\frac{1}{2}, \frac{2}{3}$ **22.** $-2\frac{1}{2}, -1\frac{1}{2}$ **23.** $\frac{1}{3}, -1\frac{2}{3}$ **24.** $1\frac{3}{4}, -\frac{5}{6}$

Ex. 87. P. 75.

1. $x^2 - 3x + 2 = 0$ **2.** $x^2 - 5x + 6 = 0$ **3.** $x^2 + x - 6 = 0$
4. $x^2 + 4x + 4 = 0$ **5.** $x^2 - 8x + 16 = 0$ **6.** $x^2 - 1 = 0$
7. $2x^2 - 3x + 1 = 0$ **8.** $4x^2 - 9x + 2 = 0$ **9.** $3x^2 + 2x - 1 = 0$
10. $4x^2 - 4x + 1 = 0$ **11.** $9x^2 - 4 = 0$ **12.** $12x^2 - 17x - 44 = 0$
13. $x^2 - x = 0$ **14.** $x^2 + 2x = 0$ **15.** $4x^2 + x = 0$
16. $2x^3 - x^2 = 0$ **17.** $12x^2 - 23x + 10 = 0$ **18.** $4x^2 - 25 = 0$
19. $4x^2 - 9x = 0$ **20.** $8x^2 - 2x - 15 = 0$

Ex. 88. P. 75.

1. $2, 3$ **2.** $1, -1$ **3.** $1, -2$ **4.** $1, -3$ **5.** $1, 3$
6. $1\frac{1}{3}, -\frac{3}{4}$ **7.** $0, -2$ **8.** $\frac{1}{2}, -\frac{1}{2}$ **9.** $\pm 2 \cdot 236$ **10.** $3, -\frac{1}{3}$

Ex. 88. P. 75.

11. $-3, -9$ **12.** $-\frac{1}{2}, -2$ **13.** $\pm 1 \cdot 732$ **14.** $10, 2\frac{1}{2}$ **15.** ± 2
16. $\frac{1}{7}, -7$ **17.** ± 1 **18.** $1\frac{1}{2}, -\frac{4}{9}$ **19.** ± 1 **20.** $0, \frac{1}{3}$
21. $0, 12$ **22.** $0, \frac{1}{2}$ **23.** $1, 1\frac{2}{5}$ **24.** $2\frac{1}{2}, -1\frac{1}{3}$

Ex. 89. P. 76.

1. 0 or 7 **2.** 3 **3.** $\frac{1}{2}$ **4.** 8 and 10
5. 6 **6.** 16 **7.** 6 and 12 **8.** 11 and 13
9. 9 and 10 **10.** 2 and 5 **11.** $\frac{2}{3}$ **12.** 5 and 7 or 1 and 11
13. 18 in by 6 in **14.** $\frac{5}{8}$ **15.** 32 mile/h **16.** 15 ft by 12 ft
17. $3p$ **18.** 5 h **19.** $1 \cdot 732$ in **20.** 100 yd
21. 30 ft **22.** 6 ft; 10 ft **23.** 20 min; 30 min
24. 5 in; 12 in; 13 in **25.** 16 **26.** 6 in by 4 in
27. 12 min; 18 min **28.** 14 **29.** 75 yd by 40 yd **30.** 4 in

Ex. 90. P. 78.

1. $2\sqrt{3}$ **2.** $3\sqrt{2}$ **3.** $2\sqrt{5}$ **4.** $2\sqrt{6}$ **5.** $3\sqrt{3}$ **6.** $4\sqrt{2}$
7. $3\sqrt{5}$ **8.** $4\sqrt{3}$ **9.** $5\sqrt{2}$ **10.** $3\sqrt{6}$ **11.** $2\sqrt{6}$ **12.** $5\sqrt{2}$
13. $\frac{1}{2}\sqrt{3}$ **14.** $\frac{1}{2}\sqrt{6}$ **15.** $\frac{1}{3}\sqrt{3}$ **16.** $\frac{1}{5}\sqrt{5}$ **17.** $3\sqrt{3}$ **18.** $\frac{1}{2}\sqrt{2}$
19. $3\sqrt{5}$ **20.** $4\sqrt{3}$ **21.** $3\sqrt{2}$ **22.** $7\sqrt{2}$ **23.** $\sqrt{6}$ **24.** $\sqrt{2}$
25. $2\sqrt{2}$ **26.** $7\sqrt{7}$ **27.** $3\sqrt{7}$

Ex. 91. P. 78.

1. $\frac{\sqrt{2}}{4}$ **2.** $\frac{\sqrt{3}}{6}$ **3.** $\frac{\sqrt{5}}{10}$ **4.** $\frac{\sqrt{3}}{3}$ **5.** $\frac{\sqrt{2}}{3}$
6. $\frac{3\sqrt{5}}{5}$ **7.** $\frac{2\sqrt{6}}{3}$ **8.** $\sqrt{2}$ **9.** $\frac{\sqrt{5}}{3}$ **10.** $\sqrt{3}$
11. $\sqrt{2}-1$ **12.** $\sqrt{3}-1$ **13.** $2+\sqrt{3}$ **14.** $\sqrt{5}+2$ **15.** 10
16. $\sqrt{3}-\sqrt{2}$ **17.** 1 **18.** 3 **19.** 12
20. $8+2\sqrt{15}$ **21.** $11-2\sqrt{30}$ **22.** $\sqrt{2}; 1\cdot414$ **23.** $\sqrt{6}; 2\cdot449$
24. $\frac{1}{2}\sqrt{2}; 0\cdot707$ **25.** $2\sqrt{2}; 2\cdot828$ **26.** $\sqrt{6}+\sqrt{3}; 4\cdot181$
27. $\frac{7-2\sqrt{10}}{3}; 0\cdot225$ **28.** $\frac{1}{2}\sqrt{6}; 1\cdot224$

Ex. 92. P. 79.

1. $a^2+2ab+b^2$ **2.** a^2+4a+4 **3.** $4-4a+a^2$
4. $a^2-2ax+x^2$ **5.** $9+6y+y^2$ **6.** $y^2-8y+16$
7. $4x^2-4xy+y^2$ **8.** $x^2-4xy+4y^2$ **9.** $4x^2+12xy+9y^2$
10. $4a^2+4ab+b^2$ **11.** $9a^2-12ab+4b^2$ **12.** $4x^2-12xy+9y^2$
13. $x^2-8xy+16y^2$ **14.** $16x^2-8xy+y^2$ **15.** $16-8y+y^2$
16. $x^2+8x+16$

Ex. 93. P. 79.

1. No **2.** Yes **3.** Yes **4.** No **5.** No
6. Yes **7.** Yes **8.** No **9.** Yes **10.** No
11. No **12.** Yes **13.** Yes **14.** Yes **15.** Yes
16. No **17.** No **18.** No **19.** Yes

Ex. 94. P. 80.

1. 9 **2.** 25 **3.** 64 **4.** 9
5. 25 **6.** 16 **7.** 25 **8.** 4

Ex. 95. P. 80.

1. $9; x + 3; -x - 3$ **2.** $4; x - 2; 2 - x$ **3.** $16; x - 4; 4 - x$
4. $25; x + 5; -x - 5$ **5.** $64; x + 8; -x - 8$ **6.** $36; x - 6; 6 - x$
7. $9; x - 3; 3 - x$ **8.** $4; x + 2; -x - 2$ **9.** $16; x + 4; -x - 4$
10. $25; x - 5; 5 - x$ **11.** $64; x - 8; 8 - x$ **12.** $36; x + 6; -x - 6$
13. $25; 2x - 5; 5 - 2x$
   14. $49; 4x + 7; -4x - 7$
15. $64; 3x - 8; 8 - 3x$
   16. $9; 2x - 3; 3 - 2x$
17. $16; 5x + 4; -5x - 4$
   18. $1; 8x - 1; 1 - 8x$
19. $b^2; a + b; -a - b$
   20. $9y^2; 2x - 3y; 3y - 2x$
21. $2; x + 2; -x - 2$ **22.** $6; x - 3; 3 - x$ **23.** $6; x - 4; 4 - x$
24. $12; x + 3; -3 - x$ **25.** $-4; x - 2; 2 - x$ **26.** $0; x + 1; -x - 1$
27. $2; x - 1; 1 - x$ **28.** $-3; x - 1; 1 - x$ **29.** $\frac{1}{4}; x + \frac{1}{2}; -x - \frac{1}{2}$
30. $-3; 2x - 1; 1 - 2x$ **31.** $6\frac{1}{4}; x - 2\frac{1}{2}; 2\frac{1}{2} - x$ **32.** $0; 3x + 2; -3x - 2$
33. $14; 2x + 3; -2x - 3$
   34. $\frac{9}{16}; x + \frac{3}{4}; -x - \frac{3}{4}$
35. $2y^2; 4x - y; y - 4x$
   36. $-\frac{3}{4}; x - \frac{1}{2}; \frac{1}{2} - x$
37. $1\frac{4}{9}; x - \frac{2}{3}; \frac{2}{3} - x$
   38. $0; x + \frac{1}{4}; -x - \frac{1}{4}$
39. $\frac{1}{9}; 2x - \frac{1}{3}; \frac{1}{3} - 2x$
   40. $-\frac{15}{16}; 3x - \frac{1}{4}; \frac{1}{4} - 3x$

Ex. 96. P. 81.

1. $5, -1$ **2.** $2 \cdot 414, -0 \cdot 414$ **3.** $0 \cdot 732, -2 \cdot 732$
4. $-2, -2$ **5.** $2, -3$ **6.** $0 \cdot 162, -6 \cdot 162$
7. $4 \cdot 732, 1 \cdot 268$ **8.** $1 \cdot 464, -5 \cdot 464$ **9.** $0 \cdot 828, -4 \cdot 828$
10. $3 \cdot 414, 0 \cdot 586$ **11.** $\pm \frac{1}{2}$ **12.** $3 \cdot 236, -1 \cdot 236$
13. $3, -5$ **14.** $3\frac{1}{2}, -3$ **15.** $0 \cdot 472, -8 \cdot 472$
16. $2 \cdot 071, -12 \cdot 071$ **17.** $14 \cdot 071, -0 \cdot 071$ **18.** $12 \cdot 236, 7 \cdot 764$
19. $0 \cdot 5, -1 \cdot 5$ **20.** $2, -1$ **21.** $2 \cdot 5, 0 \cdot 5$ **22.** $0 \cdot 333, -1$
23. $-0 \cdot 5, -2 \cdot 5$ **24.** $-2 \cdot 59, -5 \cdot 41$ **25.** $-1 \cdot 29, -2 \cdot 71$ **26.** $1 \cdot 41, -0 \cdot 0786$
27. $6 \cdot 45, 1 \cdot 55$ **28.** $1 \cdot 39, -0 \cdot 721$ **29.** $6 \cdot 46, -0 \cdot 464$ **30.** $-5 \cdot 59, -8 \cdot 41$
31. $1 \cdot 44, -2 \cdot 44$ **32.** $2 \cdot 23, -1 \cdot 23$ **33.** $2 \cdot 82, 0 \cdot 177$ **34.** $1 \cdot 24, 0 \cdot 0893$
35. $0 \cdot 244, -0 \cdot 911$ **36.** $-5 \cdot 68, -12 \cdot 3$

Ex. 97. P. 82.

1. $4 \cdot 19, -1 \cdot 19$ **2.** $2 \cdot 41, -0 \cdot 41$ **3.** $1 \cdot 28, -0 \cdot 78$ **4.** $2 \cdot 73, -0 \cdot 73$
5. $2 \cdot 62, 0 \cdot 38$ **6.** $0 \cdot 55, -1 \cdot 22$ **7.** $0 \cdot 28, -1 \cdot 78$ **8.** $3 \cdot 73, 0 \cdot 27$
9. $-0 \cdot 70, -4 \cdot 30$ **10.** $1 \cdot 32, -5 \cdot 32$ **11.** $1 \cdot 43, 0 \cdot 23$ **12.** $2 \cdot 47, -0 \cdot 81$
13. $0 \cdot 43, -1 \cdot 18$ **14.** $2 \cdot 71, 1 \cdot 29$ **15.** $0 \cdot 56, -0 \cdot 36$ **16.** $2 \cdot 19, -3 \cdot 19$
17. $1 \cdot 09, -0 \cdot 34$ **18.** $1 \cdot 25, 1 \cdot 25$ **19.** $1 \cdot 20, -0 \cdot 37$ **20.** $-0 \cdot 72, -1 \cdot 95$

Ex. 98. P. 83.

1. $0, 1$ **2.** $1 \cdot 50, -0 \cdot 67$ **3.** $-3 \cdot 27, -6 \cdot 73$ **4.** No roots
5. $0, 1, -1$ **6.** $1, 0 \cdot 67$ **7.** $0 \cdot 14, 3 \cdot 50$ **8.** $0 \cdot 63, -1 \cdot 50$
9. $1 \cdot 04, -1 \cdot 71$ **10.** No roots   **11.** $0, 10$ **12.** $12, -2$
13. $10 \cdot 83, 5 \cdot 17$ **14.** $1 \cdot 62, -1 \cdot 12$ **15.** $1 \cdot 59, 0 \cdot 16$ **16.** $4 \cdot 45, -0 \cdot 45$
17. $3, -2$ **18.** $0, 2, 5$ **19.** $0 \cdot 56, -2 \cdot 16$ **20.** No roots
21. $1 \cdot 54, 0 \cdot 26$ **22.** $\pm 3 \cdot 87$ **23.** $1 \cdot 25, 1 \cdot 25$ **24.** $0 \cdot 83, 0 \cdot 67$

DISCUSSION. P. 85.

  (i) No   (ii) Through $x = -\frac{1}{4}$ (iii) No   (iv) Yes; $-15\frac{1}{8}$
  (v) $2\frac{1}{2} < x < -3$ (vi) $-3 < x < 2\frac{1}{2}$ (vii) $2\frac{1}{2}, -3$
(viii) $3, -3\frac{1}{2}$ (ix) $2, -2\frac{1}{2}$ (x) $1\frac{1}{2}, -2$
  (xi) (a) Draw $y = -5$, (b) Draw $y = -9$, (c) Draw $y = -16$
  (xii) No, no solution

Ex. 99. P. 86.

1. (a) (i) $1\frac{1}{2}$, -2; (ii) 2, $-2\frac{1}{2}$; (iii) $0\cdot78$, $-1\cdot28$
   (b) $x = -\frac{1}{4}$, $y = -6\frac{1}{8}$ (c) $-2 < x < 1\frac{1}{2}$
   (d) (i) $3\frac{1}{2}$, -4; (ii) 0, $-\frac{1}{2}$; (iii) No roots
2. (a) (i) 0, $-2\frac{1}{3}$; (ii) 1, $-3\frac{1}{3}$; (iii) $-1\frac{1}{3}$, -1
   (b) $-1\cdot17$ (c) (i) No; (ii) Yes, $-14\cdot1$ (d) $1 < x < -3\frac{1}{3}$
   (e) (i) $1\cdot36$, $-3\cdot69$; (ii) $0\cdot57$, $-2\cdot91$; (iii) No roots
3. (a) $x = 1\cdot13$ (b) (i) 0, $2\frac{1}{4}$; (ii) $3\cdot87$, $-1\cdot62$; (iii) $3\cdot38$, $-1\cdot12$
   (c) $f(x) < -5\cdot06$
4. (a) Maximum value, $10\cdot13$; $x = 0\cdot75$
   (b) (i) $-1\frac{1}{2} < x < 3$; (ii) $3 < x < -1\frac{1}{2}$
   (c) (i) 3, $-1\frac{1}{2}$; (ii) 4, $-2\frac{1}{2}$; (iii) 2, $-\frac{1}{2}$; (iv) 3, $-1\frac{1}{2}$; (v) $3\frac{1}{2}$, -2;
     (vi) $2\cdot77$, $-1\cdot27$
5. (a) $10\cdot75$ (b) No   (c) $-3\cdot39 < x < 0\cdot39$
   (d) (i) $0\cdot39$, $-3\cdot39$; (ii) $-1\frac{1}{2}$, $-1\frac{1}{2}$; (iii) 0, -3; (iv) $0\cdot6$, $-3\cdot6$
   (e) $4 - 3x - x^2 = 0$ or $x^2 + 3x - 4 = 0$
6. (a) Two: $x = -1\cdot53$, $y = 7\cdot13$; $x = 1\cdot53$, $y = -7\cdot13$
   (b) Neither max. nor min.   (c) $-2\cdot65 < x < 0$ and $x > 2\cdot65$
   (d) $x < -2\cdot65$ and $0 < x < 2\cdot65$
   (e) Three roots: $-2\cdot17$, $-0\cdot78$, $2\cdot95$
   (f) (i) 0, $\pm2\cdot65$; (ii) $-2\cdot95$, $0\cdot78$, $2\cdot17$; (iii) $3\cdot07$; (iv) $-3\cdot07$

VARIATIONS OF QUADRATIC GRAPHS. P. 87.

A. *The graph* $y = 2x^2 + x$
   Constant term in $2x^2 + x$ is zero. $\therefore$ Intercept on y-axis is 0.
   (i) $2\frac{1}{2}$, -3; (ii) 0, $-\frac{1}{2}$; (iii) $1\frac{1}{2}$, -2; (iv) No roots; (v) No roots.
B. *The graph of* $y = 2x^2$
   Constant term in $2x^2$ is zero. $\therefore$ y-intercept is 0.
   $y = 2x^2$ is symmetrical.
   y-intercept of $y = 15 - x$ is 15 because constant $= 15$.
   R: -3; S: $2\frac{1}{2}$.
   (i) Draw $y = 3 - x$; (ii) Draw $y = x + 15$; (iii) Draw $y = x - 15$.
   (i) $x = 1$, $-1\frac{1}{2}$; (ii) $x = 3$, $-2\frac{1}{2}$; (iii) No roots.
C. *The graph of* $y = x^2$
   The roots of $2x^2 + x - 15 = 0$ are -3, $2\frac{1}{2}$.

Ex. 100. P. 89.

1. (a) (i) $\pm3\cdot87$; (ii) $\pm4\cdot8$; (iii) $\pm5\cdot83$; (iv) $\pm6\cdot71$
   (b) (i) 0, 5; (ii) 0, $6\cdot4$; (iii) 0, -4; (iv) 0, $-5\cdot5$
   (c) (i) 3, -2; (ii) 5, -2; (iii) 4, -5; (iv) $4\cdot1$, $-6\cdot1$
   (d) (i) $2x^2 + 3x - 6 = 0$; (ii) $3x^2 + 3x - 2 = 0$; (iii) $x^2 - x = 0$
   (e) (i) $1\cdot14$, $-2\cdot64$; (ii) $0\cdot46$, $-1\cdot46$; (iii) 0, 1
   (f) (i) $2\cdot87$, $-0\cdot87$; (ii) $1\cdot9$, $-4\cdot9$
2. When $x = 0\cdot7$ $y = 5\cdot5$; when $x = 4\cdot3$ $y = 23\cdot5$; $x^2 - 5x + 3 = 0$; $0\cdot7$, $4\cdot3$
3. When $x = 1\cdot18$ $y = 1\cdot98$; when $x = -0\cdot43$ $y = -0\cdot03$; $1\cdot18$, $-0\cdot43$
4. (a) (i) 2, -1; (ii) $2\cdot56$, $-1\cdot56$ (b) $x = \frac{1}{2}$ (c) $3\frac{1}{8}$
5. $x^3 + x^2 + x - 9 = 0$; $x = 1\cdot66$
6. $2\cdot19 < x < -1\cdot52$; $3\cdot08$, $-1\cdot52$
7. (a) $0 < x < 3\cdot8$ (b) $-1\cdot55$, $-0\cdot26$, $3\cdot8$
8. (a) $-2 < x < +2$ (b) (i) $3\cdot37$, $-2\cdot37$; (ii) $3\cdot07$, $-1\cdot74$

Ex. 100. P. 89.

9. (i) 12 in by 15 in (ii) (a) 53·68 in; (b) 13·42 in by 13·42 in
10. (a) (i) $V = 3·375$ in^3, $A = 11·25$ in^2
    (ii) $V = 15·625$ in^3, $A = 31·25$ in^2
    (b) $A = 16·96$ in^2 (c) $V = 10·18$ in^3

Ex. 101. P. 92.

1. $x = 1, y = \frac{1}{4}$ or $x = -1, y = -\frac{1}{4}$
2. $\frac{1}{2}, 2$ or $-\frac{1}{2}, -2$ **3.** $-3, \frac{1}{3}$ or $3, -\frac{1}{3}$ **4.** $\frac{2}{3}, -\frac{1}{2}$ or $-\frac{2}{3}, \frac{1}{2}$
5. $2, 3$ or $-6, -1$ **6.** $-1, -2$ or $3, \frac{2}{3}$ **7.** $1, 2$ or $-1, -2$
8. $2, \frac{1}{4}$ or $-2, -\frac{1}{4}$ **9.** $1, -\frac{2}{3}$ or $-1, \frac{2}{3}$ **10.** $-\frac{1}{2}, \frac{1}{2}$ or $2, -2$
11. $2, 2$ or $-2, -2$ **12.** $1, \frac{1}{2}$ or $-1, -\frac{1}{2}$ **13.** $3, 6$ or $-3, -6$
14. $1, \frac{1}{4}$ or $-1, -\frac{1}{4}$ **15.** $1\frac{1}{2}, -\frac{1}{2}$ or $-1\frac{1}{2}, \frac{1}{2}$ **16.** $-1, 2$ or $1, -2$
17. $-\frac{1}{2}, \frac{2}{3}$ or $\frac{1}{6}, \frac{2}{9}$ **18.** $\frac{1}{4}, \frac{3}{4}$ or $-\frac{1}{4}, -\frac{3}{4}$ **19.** $\frac{1}{2}, -1$ or $2\frac{1}{2}, 3$
20. $1\frac{1}{2}, \frac{2}{3}$ or $1\frac{3}{8}, \frac{1}{2}$ **21.** $2, 3$ or $-1, 0$ **22.** $3, \frac{1}{2}$ or $1\frac{1}{2}, -\frac{1}{4}$
23. $1\frac{1}{2}, \frac{2}{3}$ or $-7, -5$ **24.** $2, -1$ or $-2, -3$ **25.** $2, 1$ or $1\frac{10}{13}, 1\frac{2}{13}$
26. $2\frac{1}{2}, -1\frac{1}{4}$ or $-2\frac{1}{2}, 1\frac{1}{4}$ **27.** $1\frac{1}{2}, \frac{2}{3}$ or $31, -19$ **28.** $1\frac{1}{2}, 2$ or $-2\frac{1}{4}, -\frac{1}{2}$
29. $1, -\frac{1}{2}$ or $-\frac{3}{4}, 1\frac{1}{4}$ **30.** $1·207, 0·207$ or $-0·207, -1·207$
31. $2, -1$ or $-1, 2$ **32.** $2, 1$ (Repeated)   **33.** $2, 1$ or $\frac{1}{8}, -\frac{1}{4}$
34. $-1, -\frac{1}{2}$ or $-\frac{3}{28}, \frac{3}{14}$

Ex. 102. P. 93.

1. (ii), (iii), (vi)   **2.** (i) Nil; (ii) -24; (iii) Nil; (iv) -60; (v) Nil;
    (vi) -120
3. (i), (iii), (iv)   **4.** (ii), (v)   **5.** (i), (iii), (iv)
6. (i) -2; (ii) 16; (iii) 0   **7.** (i) 4; (ii) 0; (iii) -36
8. (i) 0; (ii) 25; (iii) 0

Ex. 103. P. 94

1. $(x - 1)(x^2 + 2)$ **2.** $(x + 1)(x^2 - 3)$ **3.** $(x + 1)(x - 1)(x - 1)$
4. $(x - 1)(x - 1)(x - 1)$ **5.** $x(x + 1)(x + 1)$ **6.** $(x - 1)(x - 1)(x + 2)$
7. $(x + 2)(x + 2)(x + 2)$ **8.** $(x - 2)(x - 2)(x - 2)$
9. $(x - 3)(x + 4)(x - 4)$ **10.** $(x - 4)(x - 4)(x - 4)$
11. $(x - 3)(x^2 + 3x + 9)$ **12.** $(x - 1)(x - 2)(x - 3)(x - 4)$
13. ± 1 **14.** 1   **15.** 1, 2, 3   **16.** $-1, -2, -3$
17. $1, 1\frac{1}{2}, 2$ **18.** $-\frac{1}{2}, \frac{1}{2}, -1$ **19.** $\frac{1}{2}, 1\frac{1}{2}, 2$ **20.** $-\frac{1}{3}, -\frac{2}{3}, -2$

Ex. 104. P. 95.

1. $k = 4, (x + 1)(x + 1)(x + 4)$ **2.** $b = -3, (x - 1)(x - 2)(2x + 3)$
3. $k = -1, (x + 1)(x - 1)(3x - 1)$ **4.** $c = -12(x - 3)(2x + 1)(2x - 1)$
5. $a = 6, x = \frac{1}{3}, \frac{1}{2}, 1$ **6.** $b = -26, x = -\frac{1}{4}, -\frac{1}{3}, -\frac{1}{2}$
7. $b = 6, c = 11; (x - 3)$ **8.** $c = -1, k = -2; (2x + 1)$
9. $a = 3, c = -19; (3x - 1)$
10. $b = -12, k = 15; (x - 1)(2x + 3)(2x + 5)$
11. $a = 24, b = 26; (2x + 1)(3x + 1)(4x + 1)$

Ex. 105. P. 96.

1. $4x + 5$ **2.** $8x - 7$ **3.** $2x^2 - 1$ **4.** $x^3 - 2$
5. $2x^2 - 3$ **6.** $3x^2 + 2$ **7.** $2x^3 - 1$ **8.** $x^2 + x + 1$
9. $x^2 - x + 1$ **10.** $x^2 + 2x - 1$ **11.** $x^2 - 2x + 2$ **12.** $2x^2 + x - 2$
13. $2x^2 - 3x - 1$ **14.** $x^3 - x + 1$ **15.** $x^3 - 2x^2 - 1$ **16.** $2x^3 + x^2 + 3x$
17. $4x^2 + 5x + 3$ **18.** $3x^3 - 2x^2 - 4$ **19.** $2x^2 - xy - y^2$ **20.** $3x^2 - 2xy + 3y^2$

Ex. 106. P. 97.

1. 2	**2.** 3	**3.** 2	**4.** 3	**5.** 2
6. 1	**7.** 7	**8.** 16	**9.** 8	**10.** 27
11. 27	**12.** 64	**13.** 32	**14.** 4	**15.** 125
16. 1	**17.** $\frac{1}{9}$	**18.** $\frac{1}{4}$	**19.** 8	**20.** $\frac{1}{64}$
21. $\frac{1}{25}$	**22.** 32	**23.** $\frac{1}{2}$	**24.** 10	**25.** 1
26. 1	**27.** 1	**28.** 1	**29.** $\frac{1}{4}$	**30.** $\frac{1}{1000}$
31. 8	**32.** $\frac{1}{9}$	**33.** $\frac{1}{8}$	**34.** $\frac{1}{64}$	**35.** $\frac{1}{5}$
36. $\frac{1}{8}$	**37.** 8	**38.** $2\frac{1}{4}$	**39.** $\frac{8}{125}$	**40.** $5\frac{1}{16}$
41. x	**42.** 1	**43.** $\dfrac{1}{x}$	**44.** x^4	**45.** $\dfrac{1}{x}$
46. x^2	**47.** $\dfrac{1}{x^6}$	**48.** $\dfrac{1}{x^2}$	**49.** $\dfrac{1}{4x^4}$	**50.** $\dfrac{1}{27x^6}$

Ex. 107. P. 98.

1. 2	**2.** $\frac{2}{3}$	**3.** $1\frac{1}{2}$	**4.** $\frac{1}{2}$	**5.** 4
6. 2	**7.** $\frac{1}{2}$	**8.** $-\frac{1}{2}$	**9.** -2	**10.** $1\frac{1}{3}$
11. log 9	**12.** log 8	**13.** log 16	**14.** log 81	**15.** log 2
16. log 2	**17.** log 9	**18.** log 625	**19.** $\log\frac{1}{9}$	**20.** $\log\frac{1}{8}$
21. log 6	**22.** log 10	**23.** log 15	**24.** log 4	**25.** $\log 1\frac{1}{4}$
26. $\log 3\frac{3}{8}$	**27.** $\log 6\frac{1}{4}$	**28.** log 54	**29.** log 6	**30.** $\log 1\frac{5}{27}$
31. 0·602	**32.** 1·9084	**33.** 2·097	**34.** 1·1761	**35.** 1·4771
36. 1	**37.** 1·699	**38.** 1·1761	**39.** 1·4313	**40.** 1·204
41. 1·398	**42.** 1	**43.** 1·301	**44.** 1·4771	**45.** 1·699
46. 1·7781	**47.** 1·0791	**48.** 1·2552	**49.** 1·5562	**50.** 1·699
51. 2	**52.** 2	**53.** 2	**54.** 0·4771	**55.** 0·301
56. 0·602	**57.** 1	**58.** 0·699	**59.** 1·7781	**60.** 0·4438
61. 0·1886	**62.** 1·6811	**63.** 1·7323	**64.** 1·903	**65.** 1·8572

Ex. 108. P. 99.

1. $3, -4$	**2.** $-6, -7$	**3.** $9, 8$	**4.** $2, -3$
5. $0, -16$	**6.** $-4, -5$	**7.** $-8, 15$	**8.** $8, 12$
9. $-1, -42$	**10.** $-1\frac{1}{2}, -1$	**11.** $-1\frac{1}{2}, -4\frac{1}{2}$	**12.** $2\frac{1}{2}, -6$
13. $-1\frac{2}{3}, -\frac{2}{3}$	**14.** $-3\frac{2}{3}, 2$	**15.** $-2\frac{2}{3}, -5\frac{1}{3}$	**16.** $0, -\frac{1}{4}$
17. $-2\frac{1}{6}, 1$	**18.** $-2\frac{1}{4}, -\frac{5}{8}$	**19.** $6\frac{4}{5}, 4\frac{4}{5}$	**20.** $\frac{1}{3}, -2\frac{2}{9}$
21. $8, 16$	**22.** $-2; b = -2$	**23.** $-5; b = 8$	**24.** $4; b = -9$
25. $2; b = -7$	**26.** $-2\frac{1}{2}; b = 9$	**27.** $-1\frac{1}{2}; b = 5$	**28.** $-2; c = -6$
29. $1\frac{1}{2}; c = -6$	**30.** $1\frac{2}{3}; c = -20$	**31.** $-2\frac{1}{2}; c = -15$	**32.** $36; 25; 13$
33. $\frac{2}{7}; \frac{1}{4}; 1\frac{1}{6}$	**34.** $26; \frac{1}{25} 1\frac{1}{25}$	**35.** $-\frac{1}{2}; 24; 6$	**36.** $10; -2\frac{1}{20}; \frac{1}{10}$
37. $2\frac{1}{4}$	**38.** $8\frac{1}{3}; 2\frac{1}{4}$	**39.** $1\frac{1}{3}$	**40.** $2x^2 - 13x + 15 = 0$

Ex. 109. P. 101.

1. $y = 1\frac{1}{3}x + 3$; 11; $3\frac{3}{4}$ **2.** $y = 4x + 8$; 24; $9\frac{1}{4}$
3. $y = 3x - 7$; -4; 9   **4.** $v = 30t$; 105 ft/s; 3 s
5. $C = \frac{5}{9}F - 17\frac{7}{9}$; $F = 1\frac{4}{5}C + 32$; $C = F = -40°$
6. $E = 4L + 14$ **7.** $m = 3$; $c = 10$ **8.** $a = \frac{2}{3}$; $b = 5$
9. $u = -4$; $a = \frac{3}{4}$ **10.** $a = -0\cdot8$; $b = 3\frac{1}{2}$ **11.** $k = 5$
12. $k = 4$; $c = 3$ **14.** $k = 72$ **15.** $k = 36$; $c = 6$
16. (a) $y = kx$ (b) $y = \dfrac{k}{x}$ (c) $y = kxz$ (d) $y = \dfrac{kx}{z}$

(e) $y = c + kx$ (f) $y = ax^2 + bx$ (g) $y = ax^2 + \dfrac{b}{x} + c$

(h) $y = ax^3 + bx^2 + \dfrac{c}{x}$

Ex. 109. P. 101.

17. $A = \pi r^2$; a circle. **18.** 8:1

19. $m = a\dfrac{v}{f} + k$; (i) $f = 6$ (ii) $k = -1$ (iii) $m = 1\frac{2}{3}$ (iv) $v = 36$

20. Reduced by 30·6%

Ex. 110. P. 105.

1. 2; 15   **2.** -3; 9   **3.** 7; 54   **4.** $3x$; $22x$
5. -2; -11 **6.** $\frac{1}{2}$; $3\frac{3}{4}$ **7.** $-3x^2$; $-20x^2$ **8.** $-4x$; $-13x$
9. 82   **10.** -16 **11.** $29\frac{1}{2}$ **12.** $75\frac{1}{4}$
13. $5n - 4$ **14.** $12 - 5n$ **15.** $\dfrac{7 + n}{2}$ **16.** $\dfrac{41 - 9n}{4}$
17. 5, 7, 9   **18.** -2, 1, 4   **19.** $4\frac{1}{2}$, 4, $3\frac{1}{2}$ **20.** $\frac{1}{2}$, $\frac{1}{3}$, $\frac{1}{4}$
21. 1; 4   **22.** 4; -4 **23.** $-1\frac{1}{2}$; $30\frac{1}{2}$ **24.** $-3\frac{1}{4}$; $36\frac{1}{4}$
25. 6; 10   **26.** -3; 15   **27.** 22   **28.** 12
29. 45   **30.** 18

Ex. 111. P. 106

1. 110   **2.** 165   **3.** 276   **4.** 630   **5.** 180
6. -225 **7.** 38; 164   **8.** 36; 204   **9.** 82; 882   **10.** $79\frac{1}{2}$; $612\frac{1}{2}$
11. -18; -36 **12.** -216; -2904 **13.** 324   **14.** 975   **15.** 50
16. 165   **17.** 575   **18.** $-142\frac{1}{2}$ **19.** 4   **20.** 20
21. 16   **22.** 16   **23.** 10   **24.** 22   **25.** 390
26. 1152   **27.** 180   **28.** 495   **29.** 705   **30.** 0
31. $2n^2 + 5n$ **32.** $\dfrac{5n^2 - 3n}{2}$ **33.** £950; £11 600
34. 3900   **35.** 12 yr.

Ex. 112. P. 107.

1. 11, 14, 17, 20, 23, 26, 29
2. 4, 10, 16, 22, 28, 34, 40, 46
3. 2, $5\frac{1}{2}$, 9, $12\frac{1}{2}$, 16, $19\frac{1}{2}$, 23, $26\frac{1}{2}$, 30   **4.** 27, 21, 15, 9, 3
5. 35, 26, 17, 8, -1, -10, -19, -28, -37
6. 46, $43\frac{1}{2}$, 41, $38\frac{1}{2}$, 36, $33\frac{1}{2}$, 31
7. a, $\dfrac{3a + b}{4}$, $\dfrac{a + b}{2}$, $\dfrac{a + 3b}{4}$, b
8. $x + y$, $x + \frac{3}{5}y$, $x + \frac{1}{5}y$, $x - \frac{1}{5}y$, $x - \frac{3}{5}y$, $x - y$

Ex. 113. P. 108

1. 2; 96   **2.** $\frac{1}{2}$; $\frac{17}{32}$ **3.** -3; -486 **4.** $-\frac{1}{4}$; $\frac{1}{16}$
5. $\dfrac{x^2}{a^2}$; $\dfrac{x^{11}}{a^4}$ **6.** $-\dfrac{1}{x^2}$; $-\dfrac{1}{x^9}$ **7.** $136\frac{11}{16}$ **8.** $4\frac{67}{125}$
9. 512   **10.** $-56\frac{61}{64}$ **11.** $2^{(n+2)}$ **12.** $(\frac{5}{3})^n$
13. $3^n \cdot (-2)^{(n-1)}$ **14.** $4^{(4-n)} \cdot (-3)^{(n-1)}$ **15.** 2, 4, 8   **16.** 1, 3, 9
17. $\frac{8}{27}$, $\frac{32}{243}$, $\frac{128}{2187}$ **18.** 3, 25, 343   **19.** 3; 2   **20.** $-\frac{1}{2}$; -48
21. $\pm 1\frac{1}{2}$; ± 4 **22.** $-\frac{2}{3}$; $\frac{1}{2}$ **23.** ± 3; $\frac{1}{243}$ **24.** $-\frac{1}{3}$; -81
25. $\frac{4}{5}$; 100   **26.** $-2\frac{1}{2}$; 16   **27.** 14   **28.** 9
29. 7   **30.** 10   **31.** 88·6 MILL. ml.; 1 more fold   **32.** $18\frac{63}{64}°$

Ex. 114. P. 110.

1. 189   **2.** 728   **3.** 1023   **4.** 29,524   **5.** $1593\frac{3}{4}$
6. $221\frac{53}{64}$ **7.** $121\frac{13}{27}$ **8.** $\frac{85}{256}$ **9.** $1\frac{601}{729}$ **10.** $27\frac{1767}{2000}$

Ex. 114. P. 110.

11. 7   **12.** 6   **13.** 7   **14.** 4   **15.** 3,15
16. 13,120   **17.** 648$\frac{3}{8}$ **18.** $\pm\frac{1}{2}$; 126, -42 **19.** £12,600
20. 13 ft.   **21.** ± 6 **22.** ± 8 **23.** ± 12
24. ± 15 **25.** 5, 15, 45, 135   **26.** -14, ± 28, -56, ± 112, -224
27. 1, $-2\frac{1}{2}$, $6\frac{1}{4}$, $-15\frac{5}{8}$ **28.** $\frac{9}{16}$, $\frac{3}{8}$, $\frac{1}{4}$, $\frac{1}{6}$, $\frac{1}{9}$, $\frac{2}{27}$
29. $\pm(x^2 - 2xy + y^2)$; $\pm(x - y)$ **30.** 4, 16

PART B: ARITHMETIC

Ex. 1. P. 113.

1. 20·7   **2.** 0·307   **3.** 4·17   **4.** 30·9   **5.** 5170
6. 80·1   **7.** 0·0613   **8.** 92·140   **9.** 120 100   **10.** 0·000073
11. 4·01   **12.** 5   **13.** 10·1   **14.** 0·001 01   **15.** 0·0001
16. 1001·1   **17.** 1·1   **18.** 0·010 01   **19.** 0   **20.** 0·101
21. 0·101   **22.** 0·01   **23.** 0   **24.** 17·3   **25.** 31·32
26. 72·031   **27.** 121·03   **28.** 121·159   **29.** 1002·102   **30.** 10·101
31. 47·924   **32.** 4·778   **33.** 35·989   **34.** 35·349   **35.** 86·361
36. 27·226   **37.** 351·6   **38.** 35·16   **39.** 3·516   **40.** 0·064
41. 6·4   **42.** 0·64   **43.** 196·8   **44.** 1·968   **45.** 19·68
46. 90·85   **47.** 36·176   **48.** 7·3953   **49.** 93·617   **50.** 150·228
51. 127·556   **52.** 2235·12   **53.** 0·702 99   **54.** 0·409 554   **55.** 399·266
56. 1387·297   **57.** 0·829 17   **58.** 3·7948   **59.** 102·162 06   **60.** 901·8009
61. 2·346   **62.** 124·2   **63.** 0·234   **64.** 1030·301   **65.** 0·001 092 727
66. 0·000 001   **67.** 5·7   **68.** 0·57   **69.** 0·57   **70.** 57
71. 0·000 57   **72.** 57   **73.** 3·8   **74.** 2·4   **75.** 1·06
76. 1·01   **77.** 0·03   **78.** 0·003   **79.** 32·4   **80.** 58·6
81. 73·9   **82.** 45·31   **83.** 3·792   **84.** 5783   **85.** 0·276
86. 10·01   **87.** 0·308   **88.** 0·005 15   **89.** 0·0707   **90.** 0·010 01

Ex. 2. P. 114.

1. 288p; £2·88   **2.** 408p; £4·08   **3.** 368p; £3·68
4. 1056p; £10·56   **5.** 1400p; £14   **6.** 2448p; £24·48
7. 1512p; £15·12   **8.** 2349p; £23·49   **9.** 4275p; £42·75
10. 1416p; £14·16   **11.** 22,320p; £223·20   **12.** 45,024p; £450·24
13. 24p; £0·24   **14.** 32p; £0·32   **15.** 36p; £0·36
16. 45p; £0·45   **17.** 48p; £0·48   **18.** 53p; £0·53
19. 61p; £0·61   **20.** 77p; £0·77   **21.** 89p; £0·89
22. 111p; £1·11   **23.** 805p; £8·05   **24.** 1001p; £10·01

Ex. 3. P. 115.

1. $2^5 \times 3^2$ **2.** $2^4 \times 3^2 \times 5^2$ **3.** $2^2 \times 5^2 \times 7^2$ **4.** $2 \times 3^2 \times 5^2$
5. $2^3 \times 3^3$ **6.** $2^6 \times 3^4$ **7.** $2^2 \times 3$ **8.** 2^4
9. $2^2 \times 3^2$ **10.** $2^4 \times 3$ **11.** 3×5 **12.** 5^2
13. 3×5^2 **14.** $2^3 \times 3^2$ **15.** $2^4 \times 3^2$ **16.** $2^2 \times 3^2 \times 5^2$
17. $2^3 \times 3^2 \times 5$ **18.** $2^3 \times 3^3$ **19.** $3^2 \times 5^2$ **20.** $3^3 \times 7^2$
21. $2^3 \times 3^2 \times 5^2$ **22.** $2^4 \times 3^3$ **23.** $2^5 \times 3^3$ **24.** $2^6 \times 3^4$
25. $2^4 \times 3^3 \times 5^2$ **26.** $3^2 \times 5^2 \times 7^2$

Ex. 4. P. 116.

1. 24	**2.** 12	**3.** 18	**4.** 36	**5.** 15	**6.** 14
7. 90	**8.** 45	**9.** 63	**10.** 27	**11.** 25	**12.** 16
13. 35	**14.** 180	**15.** 225			

Ex. 5. P. 117.

1. 2	**2.** 3	**3.** 4	**4.** 4	**5.** 3	**6.** 6
7. 10	**8.** 16	**9.** 1	**10.** 5	**11.** 7	**12.** 1
13. 12	**14.** 18	**15.** 16	**16.** 48	**17.** 60	**18.** 36
19. 48	**20.** 60	**21.** 210	**22.** 42	**23.** 120	**24.** 36

Ex. 6. P. 117.

	H.C.F.	L.C.M.		H.C.F.	L.C.M.
1.	6	36	**2.**	6	36
3.	7	42	**4.**	6	180
5.	1	216	**6.**	6	216
7.	15	225	**8.**	15	210
9.	6	432	**10.**	1	1296
11.	3	2520	**12.**	3	3150

Ex. 7. P. 118.

1. 3 in square	**2.** 8 in square	**3.** 5	**4.** 7
5. 108 in	**6.** 840 lb	**7.** 1 h later	**8.** 10p
9. 240 miles	**10.** £3·60	**11.** 30	**12.** 60
13. 28 yd	**14.** 33 sea miles	**15.** 30	

Ex. 8. P. 119.

1. 50p   **2.** 10p   **3.** 30 min   **4.** 50p
5. £3   **6.** 40p   **7.** 45°   **8.** 4 h
9. $\frac{1}{4}$ **10.** $\frac{3}{4}$ **11.** $\frac{1}{3}$ **12.** $\frac{1}{10}$ **13.** $\frac{5}{6}$ **14.** $\frac{1}{6}$ **15.** $\frac{2}{3}$
16. $\frac{1}{4}$ **17.** $\frac{2}{3}$ **18.** $\frac{6}{7}$ **19.** $\frac{27}{32}$ **20.** $\frac{7}{12}$ **21.** $\frac{12}{17}$ **22.** $\frac{4}{5}$
23. $\frac{3}{5}$ **24.** $\frac{3}{4}$ **25.** $\frac{3}{4}$ **26.** $\frac{3}{5}$ **27.** $\frac{1}{4}$ **28.** $\frac{3}{4}$ **29.** $\frac{4}{5}$
30. $\frac{7}{12}$ **31.** $\frac{3}{7}$ **32.** $\frac{5}{12}$ **33.** $\frac{2}{5}$ **34.** $\frac{4}{7}$ **35.** $\frac{15}{16}$ **36.** $\frac{8}{9}$
37. $\frac{1}{2}, \frac{3}{5}, \frac{7}{10}$ **38.** $\frac{1}{4}, \frac{3}{10}, \frac{2}{5}$ **39.** $\frac{15}{32}, \frac{1}{2}, \frac{13}{16}$ **40.** $\frac{1}{4}, \frac{3}{8}, \frac{7}{16}$
41. $\frac{11}{15}, \frac{4}{5}, \frac{17}{20}$ **42.** $\frac{3}{7}, \frac{10}{21}, \frac{1}{2}$ **43.** $\frac{2}{9}, \frac{5}{18}, \frac{1}{3}$ **44.** $\frac{7}{9}, \frac{29}{36}, \frac{5}{6}$
45. $\frac{7}{22}, \frac{4}{11}, \frac{13}{33}$ **46.** $\frac{4}{9}, \frac{5}{8}, \frac{3}{4}$ **47.** $\frac{13}{20}, \frac{21}{31}, \frac{7}{10}$ **48.** $\frac{3}{7}, \frac{1}{2}, \frac{5}{9}$
49. $3\frac{1}{5}$ **50.** $3\frac{1}{7}$ **51.** 2   **52.** $2\frac{3}{5}$ **53.** $3\frac{5}{6}$ **54.** $5\frac{1}{4}$
55. $9\frac{1}{2}$ **56.** $3\frac{3}{4}$ **57.** 2   **58.** $4\frac{5}{9}$ **59.** $2\frac{11}{12}$ **60.** $1\frac{5}{6}$
61. $\frac{5}{3}$ **62.** $\frac{13}{5}$ **63.** $\frac{27}{5}$ **64.** $\frac{41}{9}$ **65.** $\frac{61}{11}$ **66.** $\frac{85}{13}$
67. $\frac{113}{15}$ **68.** $\frac{145}{17}$ **69.** $\frac{298}{25}$ **70.** $\frac{537}{32}$ **71.** $\frac{377}{16}$ **72.** $\frac{301}{64}$

Ex. 9. P. 120.

1. $1\frac{1}{12}$ **2.** $5\frac{11}{12}$ **3.** $4\frac{3}{7}$ **4.** $5\frac{15}{16}$ **5.** $6\frac{11}{60}$ **6.** $8\frac{5}{36}$
7. $\frac{13}{18}$ **8.** $1\frac{31}{42}$ **9.** $1\frac{5}{24}$ **10.** $\frac{44}{45}$ **11.** $2\frac{47}{48}$ **12.** $1\frac{25}{33}$
13. $\frac{7}{12}$ **14.** $2\frac{1}{4}$ **15.** $\frac{3}{5}$ **16.** $1\frac{7}{8}$ **17.** $8\frac{3}{4}$ **18.** $1\frac{1}{18}$
19. $\frac{2}{3}$ **20.** $1\frac{3}{8}$ **21.** $\frac{1}{6}$ **22.** $3\frac{3}{4}$ **23.** $\frac{5}{12}$ **24.** $1\frac{11}{18}$
25. $\frac{13}{64}$ **26.** $\frac{1}{90}$ **27.** $\frac{44}{45}$ **28.** 1

Ex. 10. P. 121.

1. $\frac{1}{8}$ 2. $\frac{1}{4}$ 3. $\frac{1}{4}$ 4. $\frac{1}{2}$ 5. $\frac{2}{3}$ 6. $\frac{3}{4}$
7. 1   8. 1   9. 1   10. $3\frac{1}{2}$ 11. 8   12. 4
13. 6   14. 4   15. $4\frac{1}{2}$ 16. 6   17. $1\frac{1}{2}$ 18. $3\frac{3}{5}$
19. $4\frac{1}{2}$ 20. 18   21. $1\frac{1}{3}$ 22. $3\frac{1}{2}$ 23. 0   24. $4\frac{1}{2}$
25. $\frac{1}{4}$ 26. 4   27. $\frac{1}{3}$ 28. 4   29. 4   30. 32
31. $\frac{1}{5}$ 32. $1\frac{1}{8}$ 33. $\frac{1}{2}$ 34. 0   35. 2   36. $\frac{3}{4}$
37. $2\frac{1}{2}$ 38. $1\frac{1}{2}$ 39. $\frac{3}{10}$ 40. $1\frac{1}{2}$ 41. $2\frac{1}{13}$ 42. $1\frac{1}{3}$
43. $\frac{3}{4}$ 44. 3

Ex. 11. P. 122.

1. $\frac{7}{12}$ 2. $\frac{5}{24}$ 3. $\frac{5}{12}$ 4. $\frac{1}{24}$ 5. $2\frac{8}{9}$ 6. $\frac{1}{6}$ 7. 1
8. 1   9. $2\frac{5}{7}$ 10. $\frac{5}{54}$ 11. $5\frac{13}{15}$ 12. $6\frac{11}{12}$ 13. $2\frac{4}{15}$ 14. $3\frac{1}{3}$
15. $1\frac{1}{6}$ 16. $\frac{2}{7}$ 17. $\frac{65}{66}$ 18. $\frac{1}{3}$ 19. $4\frac{1}{6}$ 20. 0

Ex. 12. P. 123.

1. 5   2. 2   3. $\frac{48}{55}$ 4. $2\frac{7}{10}$ 5. $\frac{7}{8}$
6. 15   7. $3\frac{1}{2}$ 8. $3\frac{3}{4}$ 9. $\frac{1}{12}$ 10. $49\frac{1}{21}$

Ex. 13. P. 124.

1. (i) 54 in²; (ii) 95 in²; (iii) $68\frac{1}{4}$ in²; (iv) $29\frac{3}{4}$ ft²; (v) $12\frac{3}{4}$ yd²; (vi) $2\frac{1}{8}$ miles²
2. 1674 ft²   3. 12 100 yd²   4. 32 in; 36 in²
5. 32 in; 31 in²   6. 24 in; 12 in²   7. 26 in; $17\frac{1}{2}$ in²
8. 26 yd; 76 yd   9. 100 yd

Ex. 14. P. 125.

1. (i) 364 ft²; (ii) 367 ft²; (iii) 478 ft²
2. (i) £2·50; (ii) £4·55; (iii) £5·22   3. (i) £79·20; (ii) £59·40
4. (i) £4·75; (ii) £3; (iii) £75; (iv) £82·75
5. (i) £8·40; (ii) £13·96½   6. (i) £31·35; (ii) £1·41; (iii) £32·76
7. No. 2 is cheaper by £5·40   8. £15·20   9. $82\frac{2}{3}$ yd²   10. £5·25

Ex. 15. P. 126.

1. (i) 24 in³; (ii) 315 ft³; (iii) 192 yd²; (iv) 45 ft³; (v) 50 ft³; (vi) 35 ft³
2. (i) 10 ft; (ii) $3\frac{3}{4}$ ft; (iii) 20 ft; (iv) 50 ft; (v) 10 ft; (vi) $2\frac{1}{4}$ ft
3. $2\frac{7}{9}$ ft   4. 25 gal
5. (i) 12 ft; (ii) 24 ft; (iii) 192 ft
6. 30 000 gal   7. 20 min   8. 134 tons

Ex. 16. P. 127.

1. 360 in³   2. 1344 in³   3. 715 in³   4. 685 in³
5. 1407 in³   6. 1416 in³   7. $1232\frac{1}{2}$ in³   8. 1 ft³

Ex. 17. P. 128.

1. 0·1	**2.** 1·3	**3.** 23·5	**4.** 12·8	**5.** 5·1
0·12	1·35	23·46	12·57	5·05
0·123	1·345	23·457	12·567	5·051
6. 7·1	**7.** 9·5	**8.** 6·8	**9.** 0·9	**10.** 2·3
7·11	9·52	6·75	0·90	2·33
7·105	9·515	6·755	0·901	2·335
11. 4·1	**12.** 3·2	**13.** 5·3	**14.** 0·6	**15.** 8·3
4·06	3·15	5·28	0·56	8·33
4·064	3·155	5·275	0·556	8·333
16. 7·5	**17.** 1·0	**18.** 3·1	**19.** 2·1	**20.** 0·0
7·46	1·00	3·10	2·10	0·00
7·456	1·001	3·100	2·103	0·000

21. 1   **22.** 1   **23.** 3   **24.** 3   **25.** 3   **26.** 3   **27.** 1
28. 4   **29.** 4   **30.** 4   **31.** 3   **32.** 3   **33.** 5   **34.** 5
35. 5   **36.** 5   **37.** 4   **38.** 4   **39.** 1   **40.** 1

41. 2000	**42.** 2000	**43.** 2000	**44.** 3000	**45.** 7000	**46.** 6000
1510	2120	2060	3150	7150	6360
47. 7000	**48.** 7000	**49.** 30	**50.** 600	**51.** 0·007	**52.** 0·05
6980	6600	31·1	591	0·00705	0·0470

53. 8·5 **54.** 22·6 **55.** 34 **56.** 40 **57.** 36 **58.** 30 **59.** 64
60. 3·33 **61.** 73·2 **62.** 27·7 **63.** 25·8 **64.** 18·0

Ex. 18. P. 129.

1. £7·05; £7   **2.** £2·72; £3   **3.** £3·51; £4
4. £6·49; £6   **5.** £0·83; £1   **6.** £1·23; £1
7. £0·10; £0·0   **8.** £0·11; £0·0   **9.** £4·62; £5
10. £4·60; £5   **11.** £4·01; £4   **12.** £4·21; £4
13. £2·01; £2   **14.** £2·08; £2   **15.** £2·71; £3
16. £2·77; £3   **17.** £8·15; £8   **18.** £80·54; £81
19. £80·05; £80   **20.** £80·01; £80

Ex. 19. P. 129.

1. $\frac{1}{2}$ **2.** $\frac{1}{4}$ **3.** $\frac{3}{4}$ **4.** $\frac{1}{8}$ **5.** $\frac{3}{8}$ **6.** $\frac{5}{8}$ **7.** $\frac{7}{8}$
8. $\frac{1}{5}$ **9.** $\frac{4}{5}$ **10.** $\frac{6}{25}$ **11.** $\frac{9}{25}$ **12.** $\frac{9}{20}$ **13.** $1\frac{3}{25}$ **14.** $9\frac{1}{40}$
15. $17\frac{21}{200}$ **16.** $5\frac{19}{20}$ **17.** $11\frac{13}{125}$ **18.** $8\frac{31}{100}$ **19.** $4\frac{9}{50}$ **20.** $3\frac{1}{20}$

Ex. 20. P. 129.

1. 0·5   **2.** 0·25   **3.** 0·75   **4.** 0·125   **5.** 0·375
6. 0·625   **7.** 0·875   **8.** 0·667   **9.** 0·8   **10.** 0·45
11. 0·325 **12.** 0·675 **13.** 0·738 **14.** 0·913 **15.** 1·74
16. 0·833 **17.** 1·56   **18.** 3·78   **19.** 5·87   **20.** 9·76

Ex. 21. P. 132.

1. 5432 1   **2.** 46·27 g   **3.** 819·5 m   **4.** 7·4 km
5. 8·06 kg   **6.** 46 1   **7.** 5·004 kg   **8.** 80·2 m
9. 3·5 1   **10.** 7·09 g   **11.** 0·068 m   **12.** 4·03 1
13. 205 cm   **14.** 0·67 g   **15.** 3090 mg   **16.** 50·2 cm
17. 10 000·08 m **18.** 0·5003 kg **19.** 0·007 g **20.** 3 000 003 mg

Ex. 22. P. 132.

1. 12·52 g   **2.** 17·04 km   **3.** 13·5 1   **4.** 15·98 kg
5. 9·12 m   **6.** 17·08 1   **7.** 686 m   **8.** 86 mg
9. 159 1   **10.** 99 cg   **11.** 229 cm   **12.** 409 g

Ex. 23. P. 133.

1. 30·51 fr.   **2.** (i) 8·54 m; (ii) 2·96 m^2
3. (i) 0·0035 km; (ii) 0·035 g   **4.** 13 pieces, *Rem.* 1 cm
5. 2·4 mm   **6.** 61 pkts., *Rem.* NIL   **7.** (i) 15 1; (ii) 0·0015 1
8. (i) 300 000 cm^3; (ii) 0·068 m^2 **9.** 0·454 kg
10. 1·76 pt   **11.** 1180 cm^3 **12.** 64 articles, *Rem.* 7 fr. 20 c.
13. 292 cm   **14.** 105 g   **15.** 0·109 1
16. 75 g   **17.** 103 m^2 **18.** 63 mm
19. 1·61 km   **20.** (i) 2·73 kg; (ii) 6·02 lb

Ex. 24. P. 134.

1. 15 in^2 **2.** 40 in^2 **3.** 130 in^2 **4.** 18 in^2
5. 25·1 in^3 **6.** 26·6 cm^2 **7.** 4·41 m^2 **8.** 0·075 m^2
9. 5 yd^2 **10.** 17·5 ft^2 **11.** 20 in^2 **12.** 25·4 in^2
13. 15·4 cm^2 **14.** 85 in^2 **15.** 3·34 m^2 **16.** 10·1 yd^2
17. 90·3 cm^2 **18.** 17·8 yd^2 **19.** 27 in^2 **20.** 32·5 cm^2
21. 120 in^2 **22.** 98 cm^2 **23.** 162 in^2 **24.** 0·244 m^2
25. 159 in^2 **26.** 5·47 in   **27.** 103 cm   **28.** 10·5 in
29. 4 in   **30.** 20 cm   **31.** 5·24 yd   **32.** 2·33 m
33. 9·6 in   **34.** 10 in   **35.** 1 m

Ex. 25. P. 135.

1. 73 100 gal   **2.** 325 tons   **3.** 59·1 g
4. 3170 ft^3 **5.** 5·33 ft   **6.** 960 ft^2
7. £7·68   **8.** 3 in   **9.** 90 000 tons
10. 22 cm   **11.** 880 cm^3 **12.** (i) 80 ft; (ii) 97 500 gal

Ex. 26. P. 136.

1. 13   **2.** 61   **3.** 348   **4.** 2·59   **5.** 2·152 ft   **6.** £3·14
7. 71 g   **8.** 2·73 km   **9.** $4\frac{7}{10}$ **10.** £1·14

Ex. 27. P. 137.

1. £4·50   **2.** £1·54   **3.** £1·54   **4.** £32·17$\frac{1}{2}$
5. 6$\frac{7}{8}$ lb   **6.** 800 m   **7.** 8 m   **8.** £30
9. 15 days   **10.** 15 days   **11.** 20 days   **12.** 1 h 36 min
13. 26·055 g   **14.** 1 h 36 min   **15.** 42$\frac{7}{9}$ min   **16.** 6 men
17. 7·2 cm   **18.** $\frac{3}{4}$ km^2 **19.** 60 miles   **20.** $\frac{dt}{h}$ miles

21. 3 h   **22.** $\frac{ut}{v}$ h

Ex. 28. P. 138.

1. 4 : 5   **2.** 11 : 9   **3.** A, 3 : 8   **4.** (*a*) 5 : 6; (*b*) 5 : 11
5. 1 : 14   **6.** (*a*) 3 : 8; (*b*) £150   **7.** 4 : 1   **8.** 27 : 1
9. $\dfrac{1}{14\,000\,000}$ **10.** 1·84 : 1

Ex. 29. P. 139.
1. 36　　　　　2. $5\frac{1}{16}$ lb　　　3. £2·80　　　4. $5\frac{1}{3}$ ft
5. $2\frac{1}{4}$ days　　6. 25　　　　　7. $4\frac{11}{16}$ lb　　8. 60p
9. 21 yd　　　10. 7 h 24 min　11. $\frac{1}{2}$　　　　12. $\frac{2}{1}$
13. Decrease 8 : 7　14. £0·80　　　15. £594　　　16. 15 men
17. 24 men　　18. $\dfrac{2x}{y}$ h　　19. 30 oz　　　20. 15 lb
21. 55p　　　22. £44　　　23. 8 men

Ex. 30. P. 140.
1. £288　　　2. £1260　　3. £288　　4. £1080　　5. 3 days
6. $7\frac{1}{2}$ days　7. 15 days　　8. 24 men　9. £16·25　10. £17·50
11. Decrease 40 : 39　　　12. 40 extra men

Ex. 31. P. 141.
1. 3, 6, 9　　　　　2. 3 ft, 9 ft, 15 ft　　　　3. £12, £16, £20
4. 20 lb, 40 lb, 52 lb　5. 21 min, 30 min, 39 min　　6. 6 in, 4 in, 3 in
7. £8, £4, £2, £1　8. 12 : 24 : 7　9. 20 : 15 : 11　10. 3 : 2 : 4
11. A—£250, B—£700, C—£850
12. A—£328·12$\frac{1}{2}$, B—£375, C—£796·87$\frac{1}{2}$
13. A—£285, B—£480, C—£345, D—£390
14. A—£60, B—£20, C—£40　　　　　　15. £3730

Ex. 32. P. 142.
1. (a) (i) 1 kg, (ii) 2·5 kg, (iii) 4·17 kg,; (b) (i) 43·75p, (ii) 17·5p, (iii) 10$\frac{1}{2}$p
2. (a) (i) £7·20, (ii) £12·60, (iii) £18; (b) (i) 15h, (ii) 8·6 h, (iii) 6 h
3. (a) 1·6 km; (b) 3·13 miles　　4. (a) 0·57 l; (b) 6·13 pt
5. (a) 73·3 ft/s; (b) 40·9 mile/h　6. (a) 2·27 kg; (b) 16·5 lb
7. (a) 46·4° F; (b) 71·1° C　8. (a) Approx. 10·5 miles; (b) Approx. 2·5 h
9. (a) Approx. 23 cm; (b) Approx 150 g
10. (a) Approx. 550 cal.; (b) Approx. 7 lb.

Ex. 33. P. 144.
1. 65%　　　2. 40%　　3. 90%　　4. 15%　　5. 7·7%
6. £60　　　7. 22·8 m　8. 0·522 kg　9. £1·40　10. £1·40
11. £0·56　　12. £11·25　13. £4·57　14. 1·39 m
15. 1·29 kg　16. 32 l　　17. 24　　18. 12·1%
19. 0·0875%　20. 485 g

Ex. 34. P. 145.
1. $\frac{53}{50}$　　2. $\frac{11}{10}$　　3. $\frac{5}{4}$　　4. $\frac{59}{50}$　　5. $\frac{9}{8}$　　6. $\frac{13}{10}$
7. $\frac{3}{2}$　　8. 2　　9. $\frac{11}{5}$　　10. $\frac{5}{2}$　　11. $\frac{47}{50}$　　12. $\frac{9}{10}$
13. $\frac{3}{4}$　　14. $\frac{41}{50}$　　15. $\frac{7}{8}$　　16. $\frac{7}{10}$　　17. $\frac{1}{2}$　　18. $\frac{1}{4}$
19. $\frac{99}{100}$　　20. 0　　21. 192　　22. 300　　23. 112　　24. 224
25. 54　　26. 102　　27. 72　　28. 51　　29. 96$\frac{3}{5}$　　30. 121$\frac{1}{2}$
31. 140　　32. £11　　33. £1000　34. £1000　35. 9%　　36. 23·4%
37. Increase of 1·85%　38. 12·25 l　　　　39. Decrease by 32%
40. Suit £20; coat £12·50　41. 0·227%　　42. 0·2 m
43. 100 tonnes　　44. £1·17　　　　　45. £3·04

Ex. 35. P. 146.

1. Prof. 50%   **2.** Loss 40%   **3.** Prof. 20%   **4.** Loss 20%
5. Prof. $16\frac{2}{3}$%   **6.** Loss $33\frac{1}{3}$%   **7.** Loss 4%   **8.** Prof. 20%
9. Prof. $6\frac{2}{3}$%   **10.** Loss 20%   **11.** Prof. $33\frac{1}{3}$%   **12.** Loss 25%
13. Prof. 10%   **14.** Loss 25%   **15.** $\frac{3}{2}$ **16.** $\frac{11}{10}$ **17.** $\frac{53}{50}$
18. $\frac{9}{8}$ **19.** $\frac{97}{100}$ **20.** $\frac{7}{10}$ **21.** $\frac{2}{3}$ **22.** $\frac{1}{3}$
23. $\frac{100}{97}$ **24.** $\frac{40}{41}$ **25.** $\frac{25}{26}$ **26.** $\frac{25}{23}$ **27.** $\frac{20}{17}$
28. $\frac{20}{29}$ **29.** $\frac{40}{39}$ **30.** $\frac{8}{9}$ **31.** £12   **32.** £4
33. £27   **34.** £0·$52\frac{1}{2}$ **35.** £10·20   **36.** £6·25
37. £50   **38.** £25   **39.** £50   **40.** £25

Ex. 36. P. 147.

1. $33\frac{1}{3}$%   **2.** $11\frac{1}{9}$%   **3.** £34·50   **4.** £4·40   **5.** £150
6. £135   **7.** 20%   **8.** $16\frac{2}{3}$%   **9.** £8·75   **10.** £1·09
11. £0·52   **12.** £0·30   **13.** Increase 4%   **14.** Increase $15\frac{1}{3}$%
15. 25%   **16.** 5% profit   **17.** £3·60   **18.** £10·40   **19.** 10%   **20.** $8\frac{4}{7}$%

Ex. 37. P. 148.

1. (i) 0·83; (ii) 83%   **2.** (i) $\frac{7}{25}$; (ii) 28%   **3.** (i) $\frac{11}{40}$; (ii) 0·275   **4.** $69\frac{3}{8}$%
5. 3p   **6.** 165   **7.** 1·6%   **8.** 3·772 m   **9.** 5·048 l
10. Increase 50%   **11.** 72·8%   **13.** 45p   **13.** 80%   **14.** $17\frac{7}{9}$%
15. No change

Ex. 38. P. 149.

1. 88 cm   **2.** 176 cm   **3.** 22 cm   **4.** 44 cm   **5.** 66 cm
6. 3·74 m   **7.** 220 m   **8.** 11 m   **9.** 11 m   **10.** 31·4 cm
11. 62·8 cm   **12.** 6·28 dm   **13.** 31·4 m   **14.** 15·7 m   **15.** 15·7 cm
16. 18·9 mm   **17.** 13·2 dm   **18.** 16·7 m   **19.** 3·5 cm   **20.** 1·75 cm
21. 2·55 cm   **22.** 7 m   **23.** 140 m   **24.** 19·1 dm   **25.** 2·01 cm
26. 4·77 cm   **27.** 13·7 cm   **28.** 200 rev.   **29.** 11 cm   **30.** 2·75 m
31. 11·5 ft   **32.** 250 rev.   **33.** 25·7 cm   **34.** 1310 ft   **35.** 3·5 cm

Ex. 39. P. 150.

1. 154 cm²   **2.** 38·5 m²   **3.** 347 m²   **4.** 1390 m²
5. 24·6 in²   **6.** 312 000 m²   **7.** 28·3 in²   **8.** 113 m²
9. 18·1 ft²   **10.** 283 000 miles   **11.** 25·1 cm²
12. (i) 357 m; (ii) 6960 m²   **13.** 16·9 yd²   **14.** 1410 m²
15. 616 cm²

Ex. 40. P. 151.

1. 226 cm², 251 cm³   **2.** 132 cm², 113 cm³
3. 314 cm², 393 cm³   **4.** 151 cm², 126 cm³
5. 528 cm², 905 cm³   **6.** 352 cm², 503 cm³
7. 408 cm², 628 cm³   **8.** 12·6 ft², 3·14 ft³
9. 134 cm², 118 cm³   **10.** 176 m², 173 m³
11. 39·6 m²   **12.** 619 kg   **13.** 5·25 in.   **14.** 44 cm   **15.** 2 ft

Ex. 41. P. 152.

1. 23   **2.** 32   **3.** 38   **4.** 41   **5.** 55   **6.** 64
7. 76   **8.** 82   **9.** 89   **10.** 97   **11.** 112   **12.** 134
13. 159   **14.** 187   **15.** 222   **16.** 586

Ex. 42. P. 152.

1. 101	**2.** 107	**3.** 109	**4.** 120	**5.** 160
6. 201	**7.** 220	**8.** 204	**9.** 2020	**10.** 3009

Ex. 43. P. 153.

1. 2·5	**2.** 3·4	**3.** 4·8	**4.** 7·2	**5.** 5·06	**6.** 1·08
7. 0·73	**8.** 0·84	**9.** 0·602	**10.** 0·431	**11.** 0·08	**12.** 0·1
13. 0·001	**14.** 0·011	**15.** 0·033	**16.** 0·0102	**17.** 29·9	**18.** 84·1
19. 0·967	**20.** 4·47	**21.** 859	**22.** 65·0	**23.** 1·80	**24.** 8·28
25. 1·33	**26.** 1·67	**27.** 0·8	**28.** 1·2	**29.** 0·833	**30.** 0·875
31. 0·571	**32.** 0·625	**33.** 1·33	**34.** 1·67	**35.** 1·2	**36.** 1·5
37. 1·90	**38.** 2·06	**39.** 2·36	**40.** 2·71		

Ex. 44. P. 153.

1. 2·5 in	**2.** 3·8 cm	**3.** 5·6 cm	**4.** 5·0 ft	**5.** 7·3 yd
6. 34 m	**7.** 22 ft	**8.** 6·2 cm	**9.** 17 ft	**10.** 18 m
11. $a = 5$ in	**12.** $a = 25$ cm	**13.** $a = 20$ cm	**14.** $a = 13$ cm	
15. $a = 13$ ft	**16.** $a = 65$ cm	**17.** $c = 12$ cm	**18.** $b = 36$ cm	
19. $c = 35$ cm	**20.** $b = 33$ cm	**21.** $a = 7·9$ cm	**22.** $c = 4·9$ cm	
23. $b = 2·6$ m	**24.** $a = 8·6$ m	**25.** 71 m	**26.** 17 cm	
27. 100 yd	**28.** 25 m	**29.** 55 km	**30.** 1 mile	

Ex. 45. P. 154.

1. 324·0	**2.** 1156	**3.** 2025	**4.** 26·01	**5.** 3025
6. 3600	**7.** 39·69	**8.** 5184	**9.** 62·41	**10.** 7396
11. 9409	**12.** 104·0	**13.** 36,480	**14.** 800·9	**15.** 13·25
16. 181 500	**17.** 2694	**18.** 0·4186	**19.** 6100	**20.** 692 200
21. 76·56	**22.** 8336	**23.** 0·8336	**24.** 83·36	**25.** 833 600
26. 106·5	**27.** 241·8	**28.** 4·037	**29.** 64 370	**30.** 13·26
31. 2429	**32.** 0·2853	**33.** 375 000	**34.** 54·14	**35.** 7169
36. 0·005 329	**37.** 10·63	**38.** 1·002	**39.** 0·009 801	**40.** 8345

Ex. 46. P. 155.

1. 3·217	**2.** 2·265	**3.** 4·182	**4.** 5·180	**5.** 1·773
6. 6·117	**7.** 2·873	**8.** 7·090	**9.** 1·318	**10.** 7·951
11. 8·218	**12.** 2·599	**13.** 25·99	**14.** 0·8218	**15.** 82·18
16. 10·61	**17.** 4·625	**18.** 19·17	**19.** 20·44	**20.** 25·26
21. 2·526	**22.** 0·2429	**23.** 0·6718	**24.** 99·16	**25.** 0·3135
26. 165·3	**27.** 6·198	**28.** 0·097 31	**29.** 29·55	**30.** 75·59

Ex. 47. P. 156.

1. 11·2 yd	**2.** 12·4 m	**3.** 98·4 yd	**4.** 3·62 km	**5.** 9·91 ft
6. 2·24 miles	**7.** 9·77 in	**8.** 1·37 yd	**9.** 25·5 ft	**10.** 12·2 cm
11. 4·18 km	**12.** 2550 m	**13.** 15·5 ft	**14.** 27·5 cm	**15.** 7·66 ft
16. 159 in²	**17.** (i) 158 miles; (ii) 126 miles	**18.** 22·9 ft		
19. (i) 9·07 cm; (ii) 73·9 cm²		**20.** 70·6 in		

Ex. 48. P. 157.

1. 3	**2.** 3	**3.** 4	**4.** 5	**5.** 3	**6.** 4
7. 2	**8.** 1	**9.** 2	**10.** 3	**11.** 1	**12.** 2
13. 1	**14.** 0	**15.** 3	**16.** 4	**17.** 0	**18.** 2
19. 0	**20.** 1	**21.** 2	**22.** 0	**23.** 4	**24.** 3

Ex. 49. P. 158.

1. 1·1066	**2.** 2·7185	**3.** 0·9717	**4.** 3·6967	**5.** 3·0000
6. 0·7105	**7.** 3·9696	**8.** 2·0004	**9.** 1·0004	**10.** 0·0004
11. 4·6347	**12.** 1·8657	**13.** 0·8657	**14.** 3·3718	**15.** 2·3718
16. 0·0000	**17.** 0·3010	**18.** 1·4771	**19.** 2·6021	**20.** 3·6990
21. 0·9031	**22.** 4·9634	**23.** 4·7301	**24.** 4·8495	**25.** 1·8011
26. 0·4786	**27.** 1·6992	**28.** 3·8457	**29.** 4·9032	**30.** 1·0000

Ex. 50. P. 158.

1. 374·1	**2.** 22·23	**3.** 7·096	**4.** 4·388	**5.** 34·24
6. 12·42	**7.** 124·2	**8.** 1·242	**9.** 540	**10.** 5400
11. 7·573	**12.** 10	**13.** 3765	**14.** 11·91	**15.** 101·2
16. 85 630	**17.** 521·3	**18.** 1127	**19.** 999·5	**20.** 43 630

Ex. 51. P. 159.

1. 56·28	**2.** 25·24	**3.** 124·3	**4.** 4·361	**5.** 1·709
6. 7·240	**7.** 40·61	**8.** 2 320 000	**9.** 5·719	**10.** 3·867
11. 37·21	**12.** 3·673	**13.** 2 732 000	**14.** 204 400	**15.** 3250
16. 9 292 000	**17.** 24·18	**18.** 3022	**19.** 100·8	**20.** 1·777
21. 2·929	**22.** 3·912	**23.** 302 300	**24.** 1·763	**25.** 389 000
26. 99·26	**27.** 1861	**28.** 5·991	**29.** 56·18	**30.** 129·4

Ex. 52. P. 159.

1. 1906	**2.** 2·364	**3.** 15 370	**4.** 4·987	**5.** 3·522
6. 1·114	**7.** 3410	**8.** 8·742	**9.** 154·7	**10.** 7·065
11. 59·38	**12.** 9977	**13.** 8363	**14.** 2·907	**15.** 4·826
16. 25 210	**17.** 34·6	**18.** 1·981	**19.** 3·321	**20.** 1·077

Ex. 53. P. 160.

1. 2	**2.** $\frac{1}{2}$	**3.** $\frac{1}{9}$	**4.** 3	**5.** $\frac{1}{3}$
6. 1	**7.** 1	**8.** 9	**9.** $\frac{1}{2}$	**10.** 8
11. 9	**12.** $\frac{1}{6}$	**13.** 27	**14.** 25	**15.** $\frac{1}{10}$
16. 4	**17.** 3	**18.** $\frac{1}{32}$	**19.** 1	**20.** 1

Ex. 54. P. 160.

1. 2·78	**2.** 6·94	**3.** 4·07	**4.** 4·79	**5.** 12·3
6. 2·30	**7.** 1·81	**8.** 7·34	**9.** 5·00	**10.** 1·57
11. 506	**12.** 197	**13.** 1·72	**14.** 1·32	**15.** 603
16. 916 ha	**17.** 1 350 000		**18.** 4·85 ft	**19.** 431
20. 70·5 in²	**21.** 15·3		**22.** 20·5 in	**23.** 2·35
24. 56·0 ft²	**25.** 5260		**26.** 204 in²	**27.** 8·85 in²
28. 2·33 ft³; 145 lb			**29.** 288 in²	**30.** 19·6 dm²

Ex. 55. P. 162.

1. 0·7001	**2.** $\bar{1}$·5404	**3.** $\bar{1}$·7437	**4.** $\bar{3}$·6721	**5.** 2·1179
6. $\bar{4}$·9294	**7.** $\bar{1}$·8593	**8.** $\bar{2}$·9694	**9.** 1·3735	**10.** $\bar{5}$·6021
11. 0·9031	**12.** $\bar{2}$·3932	**13.** $\bar{4}$·0000	**14.** $\bar{3}$·9777	**15.** 3·4101
16. $\bar{1}$·6433	**17.** $\bar{5}$·7199	**18.** 0·0000	**19.** 1·3617	**20.** $\bar{3}$·7242

Ex. 56. P. 163.

1. 3·431	**2.** 0·3431	**3.** 343·1	**4.** 0·075 01	**5.** 7501
6. 1	**7.** 0·001	**8.** 10 000	**9.** 0·3008	**10.** 0·010 14
11. 149·6	**12.** 0·001 272	**13.** 0·081 34	**14.** 20	**15.** 0·2
16. 0·000 1014	**17.** 0·053 93	**18.** 1585	**19.** 0·009 752	**20.** 0·1659

Ex. 57. P. 164.

1. 0·9764	**2.** 1·0179	**3.** $\bar{2}$·2153	**4.** 2·1203	**5.** 4·0794
6. $\bar{3}$·8936	**7.** $\bar{2}$·8904	**8.** 1·8825	**9.** $\bar{1}$·1270	**10.** 2·2189
11. $\bar{3}$·1812	**12.** 2·9285	**13.** 0·6488	**14.** $\bar{2}$·4354	**15.** $\bar{2}$·5366
16. $\bar{8}$·5492	**17.** 0·7847	**18.** 5·0546		

Ex. 58. P. 164.

1. $\bar{2}$·0713	**2.** $\bar{3}$·1872	**3.** $\bar{8}$·3725	**4.** $\bar{1}$·0867	**5.** $\bar{2}$·1896
6. $\bar{1}$·1947	**7.** $\bar{2}$·0864	**8.** $\bar{1}$·6286	**9.** $\bar{2}$·7493	**10.** $\bar{3}$·6368
11. $\bar{1}$·8097	**12.** $\bar{1}$·4475	**13.** $\bar{1}$·0572	**14.** $\bar{1}$·7972	**15.** $\bar{1}$·3164
16. $\bar{5}$·9504	**17.** $\bar{2}$·1131	**18.** $\bar{1}$·8224	**19.** 0·1773	**20.** 3·7536
21. $\bar{1}$·7043	**22.** $\bar{1}$·7548	**23.** $\bar{1}$·1486	**24.** $\bar{2}$·8124	

Ex. 59. P. 165.

1. 13·48	**2.** 0·000 1553	**3.** 0·1845	**4.** 0·7219
5. 0·5898	**6.** 3·111	**7.** 285·2	**8.** 0·1267
9. 0·2675	**10.** 0·006 754	**11.** 0·021 29	**12.** 48·04
13. 0·001 929	**14.** 0·000 010 11	**15.** 0·000 002 772	**16.** 2·616
17. 0·1762	**18.** 0·030 38	**19.** 1·934	**20.** 0·8802
21. 0·8714	**22.** 0·2861	**23.** 1·513	**24.** 99·77

Ex. 60. P. 165.

1. 188	**2.** 0·293	**3.** 0·0994	**4.** 0·203	**5.** 1·87
6. 0·000 266	**7.** 0·649	**8.** 0·133	**9.** 0·205	**10.** 611
11. 0·0456	**12.** 0·0250	**13.** 50·6	**14.** 5·02	**15.** 5·64
16. 3·02 m²	**17.** 195 dm³	**18.** 5·21 m³	**19.** 6·74 m	
20. 0·369 m³	**21.** 2410 m	**22.** 0·442 m²		
23. 0·138 dm³	**24.** 400 cm²	**25.** 2910 N.F.		

Ex. 61. P. 166.

1. (i) $1\frac{5}{6}$; (ii) 5	**2.** £3·79	**3.** 6; 432	**4.** 37·04
5. $\frac{7}{10}, \frac{21}{31}, \frac{13}{20}$	**6.** £3·15	**7.** £44·66	
8. (i) 56 900; (ii) 0·837; (iii) 0·0243		**9.** 225	**10.** 95 600 gal
11. 12·0 lb	**12.** 364 m	**13.** 18 men	
14. A: £525; B: £600; C: £1275		**15.** 810 kg	**16.** £2
17. 29·5%	**18.** £175	**19.** $12\frac{1}{2}$%	**20.** 32·1 cm
18. £166. 13s. 4d.	**19.** $12\frac{1}{2}$%	**20.** 32·1 cm	

Ex. 62. P. 167.

1. 2·4 × 10⁵	**2.** 1·86 × 10⁵	**3.** 6·0 × 10¹²	**4.** 4·8 × 10¹³
5. 4·35 × 10⁵, 1·8 × 10⁴	**6.** 2·8 × 10⁹	**7.** 2·5 × 10⁴	
8. 6·0 × 10²¹	**9.** 5·8 × 10⁸	**10.** 6·6 × 10⁴	**11.** 3·6 × 10¹²
12. 1·0 × 10⁷	**13.** 3·3 × 10⁴	**14.** 1·0 × 10⁻⁶	**15.** 1·0 × 10⁻³
16. 1·0 × 10⁶	**17.** 8·0 × 10⁵	**18.** 1·5 × 10³	**19.** 1·0 × 10⁻³
20. 6·1 × 10⁻⁵	**21.** 9·144 × 10⁻⁴	**22.** 3·6 × 10¹⁰	**23.** 6·48 × 10⁻⁵
24. 9·84 × 10⁻⁴	**25.** 1·4 × 10⁷		

Ex. 63. P. 168.

1. 0·009 67   **2.** 0·204   **3.** 3·59   **4.** 0·756   **5.** 0·0928
6. 0·000 157   **7.** 3   **8.** 5·03   **9.** 1·41   **10.** 1
11. 2   **12.** 1   **13.** 2   **14.** 2   **15.** 4
16. 15   **17.** 12   **18.** 5   **19.** 1   **20.** 2
21. 0·751   **22.** 9·94 × 10⁶   **23.** 2·74 × 10⁻¹   **24.** 0·676

Ex. 64. P. 169.

1. 8 in³   **2.** 61·3 cm³   **3.** 192 in³   **4.** 864 cm³
5. 274 in³   **6.** 49·3 ft³   **7.** 8·4 in³   **8.** 33·5 in³
9. 139 in³   **10.** 322 cm³   **11.** 289 in²   **12.** 236 cm²
13. 245 in²   **14.** 455 ft²   **15.** 214 cm²   **16.** 310 in²
17. 11·3 in, 12·1 in   **18.** 7·12 in, 10·3 in
19. 253 cm³, 266 cm²   **20.** 192 in³, 226 in²

Ex. 65. P. 170.

1. 131 in³   **2.** 89·6 cm³   **3.** 4·95 in³   **4.** 314 in³
5. 91·6 dm³   **6.** 176 in³   **7.** 654 in³   **8.** 110 cm³
9. 147 in³   **10.** 66 in²   **11.** 151 cm²   **12.** 204 in²
13. 13·5 dm²   **14.** 85 in²   **15.** 48 in²
16. 45·2 in³, 77·4 in²   **17.** 83·5 cm³, 116 cm²

Ex. 66. P. 171.

1. 119 cm³   **2.** 839 in³   **3.** 1·33 in   **4.** 3·14 in
5. 707 cm²   **6.** 1·41 cm   **7.** 3·21 cm   **8.** 48·2 in³
9. 0·372 in³   **10.** 35·6 cm²   **11.** 28·1 in²   **12.** 0·646 in
13. 3·03 cm   **14.** 60·5 cm²   **15.** 3·02 in   **16.** 13·1 cm³
17. 46·4 cm³

Ex. 67. P. 172.

1. 6·28 cm²   **2.** 1·26 cm²   **3.** 25·1 ft²   **4.** 3140 yd²
5. 842 m²   **6.** 2050 cm²   **7.** 1260 in³   **8.** 37·7 dm³
9. 159 cm³   **10.** 3·84 ft³   **11.** 1650 in³   **12.** 8·64 ft³
13. 2·62 ft³   **14.** 37·3 sec   **15.** 21·7 min   **16.** 128 lb
17. 1·09 ft/s   **18.** 2·61 ft/s   **19.** 4·78 ft/s   **20.** 1·15 ft/s
21. 4 in   **22.** 29·6 tons   **23.** 1·04 in   **24.** 0·177 in
25. 2·75 ft   **26.** 2·08 cm   **27.** 2·32 ft   **28.** 109 cm²
29. 9·96 min   **30.** 5·36 in   **31.** 16 500 ft³
32. (i) 1·67 ft³; (ii) 13·6 ft²

Ex. 68. P. 174.

1. 27p   **2.** 35; 86   **3.** 45   **4.** 12p; 15p   **5.** 2 : 7
6. 2·5 kg   **7.** 56   **8.** 26   **9.** £17·50   **10.** £1·01
11. 53·8   **13.** 55 ,   **13.** £1435   **14.** 2% subst.: 18% subst. = 3 : 5
15. 3 : 7   **16.** 10 kg   **17.** Tin 7½%; Nickel 5½%
18. (i) A : B = 1 : 1; (ii) A : B = 5 : 7; 55⅚%

Ex. 69. P. 176.

1. (i) 44 ft/s; (ii) 58⅔ ft/s; (iii) 66 ft/s; (iv) 36⅔ ft/s
2. (i) 15 mile/h; (ii) 90 mile/h; (iii) 12 mile/h; (iv) 50 mile/h
3. 300 m/min   **4.** 54 km/h   **5.** (i) 15 mile/h; (ii) 22 ft/s
6. 37·5 km/h   **7.** 66 700 mile/h   **8.** 10¾ mile/h   **9.** 33·3 mile/h
***

Ex. 69. P. 176.

10. 6·23 ft/s
11. 25·2 mile/h
12. 30 mile/h
13. (i) 400 mile/h; (ii) 283 mile/h
14. 4 min
15. 60 mile/h
16. 180 ft
17. (i) 60 ft, (ii) 135 ft, (iii) 240 ft
18. (i) 93·9 ft, (ii) 147 ft
19. (i) 225 ft; (ii) 22·5 ft/s
20. (i) 1056 ft; (ii) 3 s; 13 s; (iii) 66 ft/s

Ex. 70. P. 178.

1. 30 mile/h; $5\frac{5}{11}$ s
2. $22\frac{1}{2}$ mile/h
3. 36·7 yd; $7\frac{1}{2}$ s
4. 62 m; 26·4 s
5. 100 m
6. 26 m; 3·6 s
7. 36 s; $17\frac{11}{22}$ s
8. $7\frac{1}{2}$ s; $4\frac{1}{2}$ s
9. 39·2 km/h
10. 55·6 km/h
11. $1\frac{2}{3}$ min
12. 69·3 km
13. 45 mile/h ; 110 yd
14. 45 mile/h; 84 yd
15. 15 min

Ex. 71. P. 180.

1. (i) $4\frac{1}{2}$ miles; (ii) 45 min
2. 15 min
3. (i) 15 min; (ii) 60 mile/h
4. (i) 6.20 P.M.; 6.50 P.M.; (ii) 20 km; 5 km;
(iii) From A 6.55 P.M.   From B 7.00 P.M.
5. 60 km/h; 6·50 P.M.; 50 km
6. (i) 25 min; (ii) $13\frac{1}{2}$ miles; (iii) 1.30 P.M.
7. (i) (a) 12.30 P.M., (b) $12.37\frac{1}{2}$ P.M.; (ii) 1·5 km
8. (i) $7\frac{1}{2}$ min; (ii) 12 mile/h; (iii) $26\frac{1}{4}$ miles

Ex. 72. P. 182.

1. 4 : 9
2. 16 : 25
3. 4 : 25
4. 9 : 16
5. 9 : 16
6. 1 : 16
7. 4 : 25
8. 25 : 64
9. 9 : 16
10. (i) 4 : 9; (ii) 8 : 27
11. 23·7 lb
12. 256 cm^3
13. 686 g
14. $6\frac{3}{4}$ pints
15. £3·20
16. 4 lb
17. 1 ft^2
18. 10 cm
19. $4\frac{1}{2}$ in
20. 1·2 in^2

Ex. 73. P. 183.

1. 880
2. 760; 0·912 m^2
3. 8 m
4. 245 lb
5. 64
6. 42 lb
7. 1080
8. 216
9. 19·3 g/cm^3
10. 340; 0·12 m^2
11. 18
12. 2592
13. 972
14. 8
15. 1
16. 144; 3·58 ft^2
17. 57·6 kg
18. 105; 4·37 dm^3

Ex. 74. P. 184.

1. £918
2. Loss: 9·09%
3. £4418
4. 46·3 ft
5. £39·75
6. Loss: 1%
7. £16·50
8. £1·27$\frac{1}{2}$
9. £2
10. 51·2%
11. Decrease: $\frac{1}{4}$%
12. Increase: 2%
13. 20p
14. 20%
15. 35p

Ex. 75. P. 185.

1. £67·50
2. £180
3. £35
4. £81
5. £43·75
6. £94·50
7. £74·25
8. £146·25
9. £86·62$\frac{1}{2}$
10. £238
11. £108·80
12. £19·80
13. £48·40
14. £12·60
15. £73·44
16. £39·24
17. £50·19
18. £40·40
19. £24·40
20. £87·69
21. £4·85
22. £8
23. £7·70

Ex. 76. P. 185.

1. £600	**2.** £300	**3.** £300	**4.** £300
5. $4\frac{1}{6}\%$	**6.** $6\frac{2}{3}\%$	**7.** $6\frac{2}{3}\%$	**8.** $3\frac{1}{3}\%$
9. 2 yr.	**10.** 4 yr.	**11.** 5 yr.	**12.** 1 yr. 8 mth.
13. £600	**14.** £437·50	**15.** £400	**16.** £600

Ex. 77. P. 186.

1. £24·36	**2.** £20·20	**3.** £32·45	**4.** £27·55
5. £15·62	**6.** £25·50	**7.** £18·86	**8.** £6·86
9. £9·21	**10.** £15·36	**11.** £34·33	**12.** £49·28
13. £7·05	**14.** £10·91	**15.** £35·69	**16.** £29·97
17. £66·93	**18.** 26·4 ft	**19.** 3·52 mm³	**20.** 13·3 lb
21. 93·6 mile/h			

Ex. 78. P. 187.

1. 12p	**2.** 50p	**3.** $82\frac{1}{2}$p	**4.** £1·22
5. 19p	**6.** 81p	**7.** $£1·67\frac{1}{2}$	**8.** £12·36
9. £285·80	**10.** £107·27	**11.** £465·65	**12.** 15·6 ft
13. $14·1\%$			

Ex. 79. P. 191.

1. (a) £36 960, (b) £369 600, (c) £2 772 000
2. £132 800 000; £39 840 000   **3.** £1596   **4.** £3·50
5. £2250; £1350   **6.** (a) 20%, (b) £16·20
7. $\frac{9}{28}$; $32\frac{1}{7}\%$ **8.** £496·71   **9.** 80p in the £; £120
10. (a) 60p in the £, (b) £64·80, (c) £120   **11.** £1200
12. £57·05   **13.** £10 050   **14.** £2160; £240   **15.** £2840
16. 5% **17.** £44; 13·2p in the £
18. Decrease 80p   **19.** £200
20. £1 314 000; 58p in the £; £23 652

Ex. 80. P. 193.

1. £125; £5   **2.** £125; £4   **3.** £360; £15
4. £441; £27   **5.** 176 shares; £8·80   **6.** 760 shares; £22·80
7. £600 stock; £27   **8.** £725 stock; £29   **9.** £240
10. £162 **11.** £190 **12.** £445·50   **13.** 80p; £6·30
14. 28p; £3·40   **15.** £112·50; £19·25   **16.** £104·60; £17·55
17. 5% **18.** 4% **19.** $3\frac{5}{7}\%$ **20.** $4\frac{1}{6}\%$ **21.** £158·90; £152·60
22. £117·73; £106·15   **23.** £720; £708·75
24. £2025; £2007   **25.** 1120 shares; 40p; $6\frac{1}{4}\%$
26. £800 stock; £108; $4\frac{1}{6}\%$ **27.** Disc. 20p; $3\frac{3}{4}\%$; $4\frac{11}{16}\%$
28. $5\frac{1}{2}\%$; Incr. £7   **29.** £3·50; $6\frac{1}{4}\%$
30. £125; $4\frac{4}{5}\%$ **31.** Incr. $£4·87\frac{1}{2}$
32. No change.   **33.** 6%; £76·80; $3\frac{1}{3}\%$
34. $5\frac{2}{3}\%$; £11 050; £16 575; £2762·50; 8500

Ex. 81. P. 195.

1. 50p	**2.** 39p	**3.** 43·9 km/h	**5.** 22·80 fr.
6. 23p	**7.** 41·0 ft/s	**8.** 90p	**9.** 6·88 fr.
10. 1·05 kg/cm²			

PART C: GEOMETRY

Ex. 2. P. 201.

1. 120°   **2.** 70°   **3.** $c = 30°$, $a = b = 150°$
4. $x = 60°$, $y = 60°$, $z = 100°$
5. $p = 100°$, $q = 80°$, $r = 100°$, $s = 100°$ $t = 80°$, $u = 80°$, $v = 100°$
6. $a = 100°$, $b = 80°$, $c = 100°$, $d = 100°$, $e = 80°$, $f = 100°$
7. $x = 60°$, $2x = 120°$ **8.** $x = 30°$, $2x = 60°$, $c = 60°$, $d = 120°$
9. $p = 36°$, $2p = 72°$, $3p = 108°$, $4p = 144°$ **10.** $q = 70°$
11. $t = 70°$ **12.** $x = 20°$, $2x = 40°$

Ex. 5. P. 205.

1. (1) 100°; (2) 80°; (3) 80°; (4) 100°; (5) 100°; (6) 80°; (7) 80°; (8) 100°;
(9) 100°; (10) 80°; (11) 80°; (12) 100°; (13) 100°; (14) 80°; (15) 100°
2. (1) 40°; (2) 110°; (3) 70°; (4) 110°; (5) 70°; (6) 110°; (7) 110°; (8) 70°;
(9) 110°; (10) 70°; (11) 110°; (12) 70°; (13) 110°; (14) 70°; (15) 110°
3. (1) 60°; (2) 60°; (3) 60°; (4) 60°; (5) 60°; (6) 120°
4. (1) 150°; (2) 150°; (3) 30°; (4) 80°; (5) 70°; (6) 110°

Ex. 6. P. 207.

1. (i) 720°; (ii) 1080°   **2.** (i) 360°; (ii) 360°   **3.** (i) 140°; (ii) $128\frac{4}{7}°$
4. (i) 30°; (ii) 18°   **5.** (i) 8; (ii) 9; (iii) 10; (iv) 12
6. (i) 3; (ii) 15; (iii) 18   **10.** 120°   **11.** 150°   **12.** 6   **13.** 3

Ex. 7. P. 210.

10. $\angle C$ 120°, BC 3·4 cm, CD 2·8 cm

Ex. 8. P. 213.

1. (i), (iii), (iv)   **8.** $CD = 9$ cm

Ex. 9. P. 214.

1. A.S.A.   **2.** A.A.S.   **4.** A.S.A.   **5.** A.A.S.   **6.** A.A.S.

Ex. 12. P. 218.

3. $BC = 4$ in, $AC = 3·84$ in, $BD = 3·2$ in, $\angle ADC = 127°$,
$\angle BCD = 53°$; trapezium—$BC \parallel AD$: $\angle A + \angle B = 180°$
4. 2·18 in, 3·2 in   **5.** $AC = 1·4$ in   **7.** Hexagon
8. $\angle Z = 60°$ **9.** 36°   **10.** $x = 70°$

Ex. 13. P. 222.

11. 36°, 72°, 72°   **12.** 30°, 30°, 120°; 30°, 60°, 90°

Ex. 15. P. 228.

13. $BE = 12·4$ cm   **14.** $BE = 14·9$ cm   **16.** Area $= 48·5$ cm²

Ex. 18. P. 234.

1. $\frac{1}{4}$ **2.** 4 : 1   **3.** $\frac{1}{2}$ **4.** 9 : 1
5. 4 : 1   **6.** 9 : 4   **7.** 27 : 64; 20·25 cm³
8. 32 cm²   **9.** 216 cm³   **10.** $2\frac{1}{4}$ in

Ex. 20. P. 239.

1. 10   **2.** 26   **3.** 25   **4.** 17   **5.** 20
6. 9   **7.** 16   **8.** 15   **9.** 13 cm   **10.** 24 ft.
11. $BP = 6$ cm, $AQ = 4$ cm, ABC is isosceles
12. $RY = 1\frac{1}{2}$ in, $TY = 3$ in, $XY = 1\frac{1}{3}$ in
13. $AY = 3\cdot75$ m, $AC = 12\cdot75$ m   **14.** 28 cm²   **15.** 50 cm²
16. Dist. 15 dm, wire 39 dm   **17.** $BC = 3$ in, $PQ = 2\cdot4$ in, $QR = 3\cdot6$ in
18. 42·9 cm²   **19.** $AY = 1\cdot25$ m, 9·77 m²   **20.** 80 ft

Ex. 21. P. 240.

1. 20°   **7.** $\angle BOC = 64°$; $\angle OCB = 58°$ **10.** $\frac{1}{7}$ **13.** $CE = 8$ cm
18. 2·83 in   **19.** (i) 156°; (ii) 18 sides
22. $PQ = 2$ in, $QZ = 4$ in, $PZ = 5\frac{1}{3}$ in

Ex. 23. P. 245.

21. $XR = 1\cdot86$ in   **24.** Centre is mid-point of AO; radius = 1 in

Ex. 25. P. 252.

1. $OA = 5$ cm   **2.** $OA = 7\cdot5$ cm   **3.** $AB = 6\cdot4$ cm
4. $AB = 1$ in   **5.** $DC = 0\cdot16$ in   **16.** $AX = 0\cdot669$ in

Ex. 27. P. 257.

1. 50°   **2.** 60°   **3.** 10°   **4.** 65°   **5.** 45°   **6.** 12°
7. 36°   **8.** 62°   **9.** 108°   **10.** 55°   **11.** 80°   **12.** 55°
13. 65°   **14.** 112°   **15.** 106°   **16.** 90°   **17.** 60°   **18.** 40°
19. 66°   **20.** 180°   **21.** 60°   **22.** 80°, 100°   **23.** 160°
24. 10°, 50°, 120°   **25.** 45°   **26.** 65°, 25°, 40°
27. 40°, 50°, 70°   **28.** 80°   **29.** 25°, 115°, 25°   **30.** 60°

Ex. 29. P. 261.

7. $\angle APX = 65°$

Ex. 30. P. 263.

1. 5·5 cm   **2.** 5·5 cm   **3.** 3·14 cm   **4.** 3·14 cm
5. 3·14 cm   **6.** 3·14 cm   **7.** 1·83 cm   **8.** 7·33 cm
9. 6·29 cm   **10.** 5·24 m

Ex. 31. P. 263.

1. 3·14 cm²   **2.** 6·29 cm²   **3.** 3·14 cm²   **4.** 9·43 cm²
5. 1·57 cm²   **6.** 15·7 cm²   **7.** 5·08 cm²   **8.** 8·11 cm²
9. 13·7 cm²   **10.** 0·583 cm²   **11.** (i) 16·1 cm²; (ii) 4·09 cm²
12. (i) 56·6 cm²; (ii) 20·6 cm²

Ex. 32. P. 266.

6. 3 cm   **7.** 11·5 cm   **8.** $XY = 4\cdot01$ cm, $YZ = 3\cdot58$ cm
14. $AB = 2\cdot24$ cm

Ex. 33. P. 268.

1. 60°   **2.** 110°   **3.** 90°   **4.** 68°   **5.** 40°
6. 25°   **7.** 25°   **8.** 30°   **9.** 100°   **10.** 40°
11. 120°   **12.** 20°, 60°, 100°

Ex. 34. P. 272.

1. $PT = 4$ cm   **2.** 6 cm   **3.** 8 cm   **4.** Each is 1·41 in
5. $\frac{1}{2}$ in, $1\frac{1}{2}$ in, $2\frac{1}{2}$ in   **6.** 1·5 cm, 2·5 cm, 3·5 cm   **8.** 2 cm
9. $\angle XPY = 64°$, $\angle XZY = 58°$ **10.** $BC = 7·75$ cm   **12.** 55° or
13. A circle on OP as diameter   **14.** 5 cm   35°

Ex. 35. P. 276.

1. 1·73 cm **2.** 4 cm **3.** 3·92 cm **4.** $AB = 3·76$ in, $BC = 3·46$ in, $AC = 3·06$ in
5. 3·88 in, 3·57 in, 3·16 in   **8.** 1·96 in or 0·638 in   **14.** 4·47 cm

Ex. 36. P. 280.

1. $CS = 1$ cm, $AC = 4$ cm   **2.** $BR = 2$ cm, $AB = 5$ cm
3. $AC = 5$ cm   **4.** $AC = 5$ cm   **5.** $AS = 0·6$ in
6. $BC = 3$ in   **7.** $CS = 1·6$ cm   **8.** $AB = 21$ cm
9. $AC = 37·8$ in   **10.** $AR = 11·2$ cm   **11.** $r = 5$ cm
12. $AX = 8·4$ cm   **13.** $BY = 6·75$ in   **14.** $AD = 8$ in
15. $AB = 7·5$ cm, $AD = 4·5$ cm   **16.** $CD = r = 1·15$ in

Ex. 37. P. 286.

3. $BX = 5$ cm, $CX = 3$ cm   **4.** $QX = 0·6$ in, $RX = 0·65$ in
7. With AB produced $FB = 6$ cm   **8.** $EF = 5·25$ cm
10. (i) 15; (ii) 24; (iii) c^2; (iv) a^2bx **11.** (i) $4\frac{1}{2}$; (ii) $6\frac{1}{4}$; (iii) $\frac{3}{16}$; (iv) b^2
12. (i) 8; (ii) 14; (iii) 14·1; (iv) x^2yz^2 **13.** $d = 7·5$ cm
14. $d = 9·6$ cm   **15.** $c = 9$ cm   **16.** $b = 4$ cm   **17.** $b = 7$ cm

Ex. 38. P. 290.

5. $AX = 3·75$ cm   **6.** $CX = 6$ cm, $DX = 4$ cm
7. $CX = 6·4$ cm, $DX = 5$ cm
18. Area $= 6·88$ in², $XY = 4·47$ in, $CH = 3·08$ in
19. Side $= 3·46$ cm   **20.** Side $= 6·50$ cm
21. Side $= 2·19$ in   **23.** Side $= 2·64$ cm

Ex. 39. P. 293.

1. (i) rt. $\angle$; (ii) ac. $\angle$; (iii) ob. $\angle$; (iv) rt. $\angle$; (v) ob. $\angle$; (vi) ac. $\angle$;
   (vii) ob. $\angle$; (viii) ob. $\angle$ **2.** 17 in
4. $AD = 1$ cm, $CD = 6·9$ cm   **5.** $AD = 1·4$ cm, $CD = 9·9$ cm
6. $AD = 4·5$ cm, $CD = 6·6$ cm   **7.** $AD = 7·5$ cm, $CD = 6·6$ cm
8. $AD = 26$ cm, $BD = 10$ cm, $CD = 15$ cm   **9.** 20·7 cm²
10. 34·2 cm²   **11.** $4\frac{1}{7}$ in   **13.** $22\frac{11}{20}$ in
13. 3·2 cm, 5·6 cm, 6·8 cm   **14.** 5 cm   **15.** 12 cm
16. 2 in; imposs. (sum of 2 sides = 3rd side)   **17.** 10 in.
18. 17 in   **19.** 4·7 cm   **20.** 3·3 cm, 4·8 cm, 5·7 cm

PART D: TRIGONOMETRY

Ex. 1. P. 297.

1. 0·2679	**2.** 0·5543	**3.** 0·6745	**4.** 0·8098	**5.** 0·9657
6. 1·5399	**7.** 2·1445	**8.** 3·7321	**9.** 5·6713	**10.** 0·3096
11. 0·4327	**12.** 0·7729	**13.** 1·1463	**14.** 1·6842	**15.** 1·9970
16. 2·0057	**17.** 2·9887	**18.** 3·0061	**19.** 4·9594	**20.** 5·0045
21. 0·3772	**22.** 0·6379	**23.** 0·8904	**24.** 1·1525	**25.** 1·3166
26. 1·9699	**27.** 2·4524	**28.** 2·8634	**29.** 0·0190	**30.** 0·0023

Ex. 1. P. 297.

31. 0°	**32.** 20° 30'	**33.** 44° 54'	**34.** 50° 6'	**35.** 55° 42'
36. 61° 12'	**37.** 64° 24'	**38.** 74° 42'	**39.** 9° 45'	**40.** 15° 22'
41. 25° 47'	**42.** 34° 8'	**43.** 39° 22'	**44.** 44° 41'	**45.** 60° 28'
46. 69° 22'	**47.** 15° 52'	**48.** 31° 15'	**49.** 60° 46'	**50.** 78° 51'

Ex. 2. P. 298.

1. $\alpha = 38° 40'$, $\beta = 51° 20'$ **2.** $\alpha = 53° 8'$, $\beta = 36° 52'$
3. $\alpha = 23° 58'$ $\beta = 66° 2'$ **4.** $\alpha = 19° 34'$ $\beta = 70° 26'$
5. 10·4 m   **6.** 30 ft   **7.** 33·7 m   **8.** 13·1 ft
9. 110 m   **10.** 275 ft   **11.** 45·5 dm   **12.** 59° 34'
13. 24·3 ft   **14.** 78·4 m   **15.** 10° 47'   **16.** 432 dm
17. 9·75 ft   **18.** 0·683 dm   **19.** 80 ft   **20.** Decrease 5°
21. 59·6 dm   **22.** 540 ft   **23.** 90·0 m   **24.** 2·59 cm, 9·07 cm²
25. 1760 yd²   **26.** 7° 35'   **27.** 76° 30'   **28.** 0·26 ft

Ex. 3. P. 301.

1. 1·6977	**2.** 0·7850	**3.** 0·3610	**4.** 0·1820
5. 0·1130	**6.** 78° 19'	**7.** 65° 38'	**8.** 52° 3'
9. 47° 23'	**10.** 32° 16'	**11.** 275 ft	**12.** 250 m
13. 1·18 dm²	**14.** N. 30° 15' E.	**15.** 3·36 km	**16.** 1° 50'
17. 88·8 ft	**18.** 39·7 m	**19.** 0·577 in	**20.** 900 ft
21. 233 ft	**22.** 3·68 cm	**23.** 66·4 m	

Ex. 4. P. 302.

1. 0·0872	**2.** 0·3256	**3.** 0·5299	**4.** 0·7986	**5.** 0·9994
6. 0·2130	**7.** 0·4633	**8.** 0·6665	**9.** 0·9515	**10.** 0·0695
11. 0·2527	**12.** 0·5657	**13.** 0·7612	**14.** 0·9120	**15.** 0·9952
16. 0·9925	**17.** 0·9205	**18.** 0·7771	**19.** 0·5592	**20.** 0·2250
21. 0·9694	**22.** 0·8415	**23.** 0·6769	**24.** 0·6006	**25.** 0·3537
26. 0·1071	**27.** 0·9964	**28.** 0·8820	**29.** 0·8256	**30.** 0·6850
31. 13° 12'	**32.** 34° 54'	**33.** 54° 24'	**34.** 79° 45'	**35.** 15° 4'
36. 30° 26'	**37.** 10° 29'	**38.** 19° 21'	**39.** 24° 27'	**40.** 34° 16'
41. 15° 24'	**42.** 35° 36'	**43.** 59° 30'	**44.** 74° 55'	**45.** 80° 44'
46. 34° 53'	**47.** 40° 47'	**48.** 54° 44'	**49.** 60° 40'	**50.** 69° 44'

Ex. 5. P. 303.

1. $a = 6·43$ in, $c = 7·66$ in   **2.** $a = 10$ in, $c = 17·3$ in
3. $a = 12·0$ cm, $c = 9·03$ cm   **4.** $a = 24·0$ m, $c = 6·89$ m
5. $a = 6·46$ ft, $c = 12·4$ ft   **6.** $a = 3·85$ km, $c = 11·4$ km
7. $a = 4·28$ yd, $c = 21·6$ yd   **8.** $a = 12·2$ mm, $c = 13·2$ mm
9. $a = 28·6$ ft, $c = 15·5$ ft   **10.** $a = 6·90$ dm, $c = 5·14$ dm
11. $\alpha = 53° 8'$, $\theta = 36° 52'$ **12.** $\alpha = 43° 57'$, $\theta = 46° 3'$
13. $\alpha = 34° 34'$, $\theta = 55° 26'$ **14.** $\alpha = 64° 48'$, $\theta = 25° 12'$
15. 635 m, 296 m, 18·8 ha   **16.** 13·8 cm, 14·0 cm, 96·5/6 cm²
17. 1·44 in   **18.** 41° 4'   **19.** 60°, 17·3 ft
20. 2·11 cm, 15·4 cm   **21.** 43·4 m, 37·1 m
22. 84 m   **23.** 3·53 km, 2·79 km   **24.** S. 33° 34' W.

Ex. 6. P.304.

1. 28·7 dm   **2.** 6·83 km, 6·83 km, N.E.
3. 35·5 ft   **4.** 10·9 km, 12·3 km.

Ex. 6. P. 304.
5. 120 miles, 69·3 miles   **6.** 25·9 miles
7. 64·5 km, 82·6 km   **8.** 543 dm
9. 7·84 km, 9·53 km, N. 50° 34′ E.   **10.** 100 m
11. 6·11 miles, 5·16 miles

Ex. 7. P. 306.
1. 34·9 in   **2.** 36·7 m   **3.** 10·6 km   **4.** 2160 yd
5. 42·2 yd   **6.** 3110 m   **7.** 17·8 ch   **8.** 4·37 dam
9. 4·61 km   **10.** 4550 ft   **11.** 46° 9′   **12.** 53° 15′
13. 23° 48′   **14.** $\alpha = 39° 57′$, $\alpha = 50° 3′$, $c = 73·7$ ft

Ex. 8. P. 306.
1. 243 ft   **2.** 33·6 dm
3. A. (i) 127 ft   B. (i) 154 ft; (ii) 114 ft   C. (i) 141 ft; (iii) 46° 51′
4. 18·4 km, N. 83° 23′ E.   **5.** 2·68 cm, 20·3 cm²   **6.** 4·36 km
7. 153 ft   **8.** 24·8 ft   **9.** 119 m   **10.** 208 dm

Ex. 9. P. 308.
1. 49·3 m²   **2.** 9·92 in²   **3.** 9·44 cm²   **4.** 16·9 ft²
5. 38·3 m²   **6.** 2·73 ft²   **7.** 3·16 m²   **8.** 107 in²
9. 6·24 m²   **10.** 23·4 yd²   **11.** 23 ha   **12.** 3·34 km²
13. 31·6 in²   **14.** 32·3 cm²   **15.** 52·6 m²

Ex. 10. P. 309.
1. 789 ft; N. 53° 15′ W.   **2.** 325 m; 110 m
3. 768 yd; 4° 46′   **4.** 462 dm; 435 dm
5. 450 ft, S.E.   **6.** (i) 3·96 cm; (ii) 5·65 cm²
7. (i) 5·43 dm; (ii) 7·24 dm   **8.** 71° 9′
9. (i) 9·38 cm; (ii) 39° 46′   **10.** (i) 27·3 m; (ii) 19° 2′; (iii) 78° 54′

Ex. 11. P. 312.
1. 1   **2.** $\frac{1}{2}$ **3.** 1   **4.** $\frac{1}{3}$ **5.** 3   **6.** 1
7. 1   **8.** $\frac{1}{4}$ **9.** 1   **10.** $\frac{3}{4}$ **11.** $\sqrt{3}$ **12.** $\frac{\sqrt{3}}{3}$
13. $\frac{\sqrt{3}}{4}$ **14.** $\sqrt{3}$ **15.** $\frac{\sqrt{3}}{3}$ **16.** $\frac{\sqrt{3}}{3}$ **17.** $\sqrt{3}$ **18.** $\frac{\sqrt{3}}{4}$
19. 36 dm   **20.** 27 cm   **21.** $(6\sqrt{3})$ in²   **22.** 60°

Ex. 12. P. 313.
1. tan α   **2.** sin α   **3.** sec α   **4.** cot α   **5.** cos α
6. cosec α   **7.** 1   **8.** 1   **9.** 1   **10.** cot α
11. tan α   **12.** sec α   **13.** cos α   **14.** sin α   **15.** tan α

Ex. 13. P. 314.
1. 35·3 m   **2.** 270 dm   **3.** 80 ft   **4.** 27·4 m
5. 60 ft   **6.** 21° 48′   **7.** 30°; 50 m   **8.** 30°; 50 ft
9. 120 ft; 7° 36′   **10.** 5·39 km

Ex. 14. P. 315.
1. 0·786   **2.** 1·05   **3.** 1·57   **4.** 3·142   **5.** 4·71
6. 6·28   **7.** 0·349   **8.** 0·698   **9.** 2·09   **10.** 2·62
11. 0·563   **12.** 0·986   **13.** 1·15   **14.** 1·30   **15.** 1·55

Ex. 14. P. 315.

16. 57° 19′	**17.** 11° 28′	**18.** 171° 54′	**19.** 22° 55′	**20.** 85° 58′
21. 28° 39′	**22.** 143° 12′	**23.** 186° 12′	**24.** 237° 48′	**25.** 180°
26. 180°	**27.** 360°	**28.** 90°	**29.** 60°	**30.** 120°

Ex. 15. P. 317.

1. 0·9848	**2.** −0·866	**3.** −0·7002	**4.** 0·9397	**5.** −0·5736
6. −0·3640	**7.** −0·3420	**8.** −0·1736	**9.** 1·1918	**10.** −0·2588
11. −0·7071	**12.** 2·7475	**13.** −0·9397	**14.** 0·8660	**15.** −5·6713
16. −0·7071	**17.** 0·7660	**18.** −0·2679	**19.** 0·4924	**20.** −0·3201

Ex. 16. P. 320.

1. 0·5	**2.** −0·9613	**3.** −0·0875	**4.** −0·9994
5. −0·6428	**6.** 0·8391	**7.** −0·3090	**8.** 0·9063
9. −0·1405	**10.** 4·8097	**11.** 1·1326	**12.** 1·2799
13. 1·1034	**14.** −1·0946	**15.** −0·7813	**16.** 0·0349
17. −1·3902	**18.** −1·0306	**19.** −1·7013	**20.** 1·0403
21. −2·2460	**22.** 0·9848	**23.** −0·9397	**24.** −1·1918
25. 20°; 160°	**26.** 240°; 300°	**27.** 27°; 333°	**28.** 153°; 207°
29. 22°; 202°	**30.** 119°; 299°	**31.** 198°	**32.** 126°
33. 322°	**34.** 227°	**35.** 127°	**37.** 45°; 225°
38. 38°; 142°	**39.** 34°	**40.** (i) 146°; (ii) 23°; 90°	

Ex. 17. P. 322.

1. 12·9 in.	**2.** 4·26 cm	**3.** 14·3 m	**4.** 16·5 in
5. 11·7 dm	**6.** 17·2 cm		**7.** 10·3 m; 15·0 m
8. 28·6 in; 18·4 in	**9.** 5·21 cm; 8·74 cm	**10.** 17·4 dm; 27·3 dm	

Ex. 18. P. 323.

1. 5·35 in	**2.** 44° 25′	**3.** 6·41 cm	**4.** 85° 18′	**5.** 11·9 in
6. 12·1 cm	**7.** 118° 54′	**8.** 11·3 in	**9.** 104° 54′	**10.** 44° 38′
11. 156° 42′	**12.** 22° 22′; 19° 34′; 138° 4′			

Ex. 19. P. 327.

1. 354°/101 mile/h	**2.** 006°/101 mile/h	**3.** 078°/102 mile/h
4. 175°/120 km/h	**5.** 214°/262 km/h	**6.** 227°/241 km/h
7. 289°/306 mile/h	**8.** 131°/306 mile/h	**9.** 195°/466 km/h
10. 109°/500 km/h	**11.** 242°/128 mile/h	**12.** 071°/138 mile/h
13. 203°/130 mile/h	**14.** 242°/272 km/h	**15.** 346°/357 km/h
16. 075°/445 km/h	**17.** 23° St.; 5 mile/h; 13 mile/h; 3·41 min; 067°	
18. 28·5 km/h; 322 km/h	**19.** 164 km/h; 139 km/h	
20. 25·3 knots; 7·89 knots	**21.** 062°/284 mile/h	
22. 190°/383 km/h	**23.** 084°/41 km/h	**24.** 357°; 15·7 knots
25. 358°; 17·5 knots	**26.** 071°; 3·68 min	
27. 053°; 3·22 miles; 10·6 min	**28.** 47·4 min past 12.00 noon	

Ex. 20. P. 330.

1. 1980π miles	**2.** 1590π km	**3.** 1320π miles	**4.** 1910π km
5. 4770 miles	**6.** 4100 km	**7.** 1380 miles	**8.** 1950 km
9. 5840 miles	**10.** 12 100 km	**11.** 8180 miles	**12.** 12 400 km
13. 8180 miles	**14.** 12 400 km	**15.** (i) 9300 miles; (ii) 8840 miles	
16. (i) 11 600 km; (ii) 13 000 km	**17.** (i) 11 100 miles; (ii) 4930 miles		
18. (i) 10 500 km; (ii) 23 600 km	**19.** (i) 13 000 miles; (ii) 10 400 miles		
20. (i) 45° W.; (ii) 2920 km			

PART E: CALCULUS

Ex. 1. P. 334.

   (i) 2; (ii) 4; (iii) 6; (iv) 5; (v) 1·5

Ex. 2. P. 335.

  1. (i) 2; (ii) 4; (iii) 5; (iv) 1·5   **2.** (i) 8; (ii) 10; (iii) 0
  3. (i) $4x$; (ii) $6x$; (iii) $8x$; (iv) x; (v) $3x$

Ex. 3. P. 336.

  1. (i) 3; (ii) 12; (iii) 48; (iv) $18\frac{3}{4}$ **2.** (i) 27; (ii) 75; (iii) 0
  3. (i) $6x^2$; (ii) $9x^2$; (iii) $12x^2$; (iv) $1\frac{1}{2}x^2$; (v) $4\frac{1}{2}x^2$ **4.** $4x^3$

Ex. 4. P. 337.

  1. 1, 2, -3, 4, $\frac{1}{2}$, 0·24, $-6·5$ **2.** 0 in each case
  3. 1, -2, 5, 8, 12, 6, -5 **4.** 3, 2, -4, $\frac{1}{2}$, $-\frac{3}{4}$, 1, 4
  5. Parallel: same gradient 5   **6.** Same y-intercept 3

Ex. 5. P. 339.

  1. $2x$ **2.** $3x^2$ **3.** $4x^3$ **4.** $5x^4$
  5. $4x$ **6.** $6x+1$ **7.** $4x-3$ **8.** $6x+2$
  9. $3x^2-1$ **10.** $6x^2+2x$ **11.** $3x^2+2$ **12.** $6x^2-6x$
  13. $3x^2+2x-1$ **14.** $6x^2-4x$ **15.** $9x^2+2x+4$ **16.** $3x^2-8x+3$
  17. $8x^3+5$ **18.** $3x^2-2x+1$ **19.** $x-3$ **20.** $x^2+\frac{1}{2}x$
  21. $2x^3-2x^2$ **22.** $2x^2+1\frac{1}{2}x$ **23.** $2\frac{1}{2}x^2-3x$ **24.** $1\frac{1}{2}x^3+5x$
  25. 3   **26.** 5   **27.** 12   **28.** 7   **29.** 22
  30. 27   **31.** 5   **32.** 13   **33.** 60   **34.** 42

Ex. 6. P. 340.

  1. $-\dfrac{1}{2x^2}$ **2.** $-\dfrac{1}{3x^2}$ **3.** $\dfrac{1}{x^2}$ **4.** $-\dfrac{2}{x^2}$ **5.** $\frac{1}{2}$

  6. $\dfrac{3}{x^2}$ **7.** $\frac{1}{3}$ **8.** $-\dfrac{2}{3x^2}$ **9.** $\frac{3}{2}$ **10.** $\dfrac{3}{4x^2}$

  11. $-\dfrac{4}{x^3}$ **12.** x **13.** $\dfrac{3}{x^4}$ **14.** x^2 **15.** $-\dfrac{12}{x^5}$

  16. $\dfrac{1}{\sqrt{x}}$ **17.** $\dfrac{1}{3\sqrt[3]{x^2}}$ **18.** $-\dfrac{3}{2\sqrt{x^5}}$ **19.** $\dfrac{3\sqrt{x}}{2}$ **20.** $-\dfrac{1}{9\sqrt[3]{x^4}}$

  21. $6x-2$ **22.** $4-12x$ **23.** $9x^2+18x-6$
  24. $3x^2+4x$ **25.** $18x^2-8x$ **26.** $12x^3-18x^2+18x$
  27. $2x+1$ **28.** $4x+5$ **29.** $6x-11$
  30. $4x$ **31.** $-12x-9$ **32.** $48x-20$
  33. 2   **34.** 3   **35.** $4x-3$

  36. $6x-2-\dfrac{3}{x^2}$ **37.** $2+\dfrac{5}{x^2}-\dfrac{8}{x^3}$ **38.** 2

  39. 3   **40.** $-\frac{1}{9}$ **41.** (i) 4, -4; (ii) 0
  42. -12, 12; (ii) 0   **43.** 5; 7   **44.** -9; -1
  45. 8; -4 **46.** 6, -6; -6, 6   **47.** $(-1, -8)$
  48. $(-1, 3)$; $(1, -3)$ **49.** $(6, 9\frac{1}{4})$ **50.** -8, -4, 0, 4, 8; bowl-shaped

Ex. 7. P. 344.

  1. $(0, 3)$ **2.** $(0, -5)$ **3.** $(-1, -1)$ **4.** $(0, -4)$ **5.** $(\frac{1}{3}, -\frac{1}{3})$
  6. $(0, -4)$ **7.** $(-1, 2)$; $(1, -2)$ **8.** $(-\frac{1}{3}, \frac{2}{9})$; $(\frac{1}{3}, -\frac{2}{9})$

Ex. 7. P. 344.

9. $(-\frac{1}{2}, \frac{1}{4})$; $(\frac{1}{2}, -\frac{1}{4})$ **10.** $(-\frac{1}{3}, \frac{5}{27})$; $(1, -1)$ **11.** $(-2, 7\frac{1}{3})$; $(3, -13\frac{1}{2})$
12. Min. at $-1\frac{1}{2}$ **13.** Max. at $\frac{3}{4}$ **14.** Min. at $-\frac{2}{3}$
15. Max. at -1; min. at 1   **16.** Max. at $-\frac{2}{3}$; min. at 0
17. Min. at $-\frac{1}{2}$; max. at $\frac{1}{2}$ **18.** Max. at -2; min. at 1
19. Min. at -3; max. at $1\frac{1}{3}$ **20.** Min. at $-1\frac{1}{2}$; max. at $\frac{1}{2}$
21. T.P. Min. $(2, -4)$; x-inter. $(0, 0)$, $(4, 0)$; y-inter. $(0, 0)$
22. T.P. Min. $(0, -4)$; x-inter. $(-\sqrt{2}, 0)$, $(\sqrt{2}, 0)$; y-inter. $(0, -4)$
23. T.P. Max. $(0, 9)$; x-inter. $(-3, 0)$, $(3, 0)$; y-inter. $(0, 9)$
24. T.P. Min. $(2, -12)$; x-inter. $(0, 0)$, $(4, 0)$; y-inter. $(0, 0)$
25. T.P. Max. $(\frac{3}{4}, 6\frac{1}{8})$; x-inter. $(-1, 0)$; $(2\frac{1}{2}, 0)$; y-inter. $(0, 5)$
26. T.P. Min. $(-1, -2)$, max. $(1, 2)$; x-inter. $(-\sqrt{3}, 0)$, $(\sqrt{3}, 0)$, $(0, 0)$
    y-inter. $(0, 0)$
27. Max. at $(0, 4)$ **28.** Max. $(-\frac{1}{2}, -4)$; min. $(\frac{1}{2}, 4)$
29. Min. at $(-\frac{3}{8}, -10\frac{9}{16})$; x-inter. $(-2, 0)$, $(1\frac{1}{4}, 0)$; y-inter. $(0, -10)$; (i) y is
    $+$ve: $-2 > x > 1\frac{1}{4}$; (ii) y is $-$ve: $-2 < x < 1\frac{1}{4}$
30. Min. $(-3, -23)$; max. $(1, 9)$

Ex. 8. P. 345.

1. Side 4 ft; height 2 ft   **2.** Side 2 ft; height 2 ft
3. Side 6 dm; height 3 dm $1\frac{1}{3}$ cm   **5.** 62 500 ft²
6. 12 yd by 24 yd; 288 yd²   **7.** 20 m   **8.** $416\frac{2}{3}$ ft²
9. (i) 6, 6; (ii) 6, 6   **10.** 12   **11.** 6 cm   **13.** 9 cm; 729π cm³
13. (i) 2 ft³, (ii) 2·55 ft³   **14.** $1\frac{1}{4}$ s; 25 ft   **15.** 5 s; 400 cm

Ex. 9. P. 348.

1. $v = 4t - 3$; $a = 4$; 13 ft; 9 ft/s; 4 ft/s²
2. $v = 6t + 2$; $a = 6$; 11cm; 14 cm/s; 6 cm/s²
3. $v = -4 - 4t$; $a = -4$; -14 ft (on AO produced); -12 ft/s (in the direction AO); -4 ft/s² (speed increasing in direction AO)
4. (i) 6 ft; (ii) 0 ft/s; (iii) 5 ft; -1 ft/s; 2 ft/s²; (iv) 6 ft; 4 ft/s; 8 ft/s²; (v) 1 ft;
    (vi) $1\frac{1}{3}$ s; (vii) $\frac{2}{3}$ s; (viii) $4\frac{22}{27}$ ft
5. (i) 0 cm (at O); 2 cm (on OA); (ii) 9 cm/s; -3 cm/s; (iii) -12 cm/s²;
    0 cm/s²; (iv) 1 s; 3 s; (v) 4 cm; 0 cm
6. (i) 6 cm; (ii) -4 cm/s; (iii) 6 cm/s²; No; (iv) $t = \frac{2}{3}$ s; changes speed and
    direction; (v) 8 cm/s; 14 cm/s; (vi) 11 cm
7. (i) $v = 12t - 3t^2$; (ii) 4 s; 32 ft; (iii) 16 ft
8. (i) 3 s; $13\frac{1}{2}$ cm; (ii) $20\frac{1}{4}$ cm/s; -18 cm/s²
9. (i) 100 ft
10. (i) 25 s; (ii) 10 000 ft; (iii) 400 ft/s; (iv) 7 500 ft; 400 ft/s.
11. (i) $(9 - 2t)$ dm³/min; (ii) 3 dm³/min; (iii) $4\frac{1}{2}$ min
12. (i) $87\frac{1}{25}$ gal/min; (ii) 180 gal; 180 mile/h
13. (i) 60 cm²/s; (ii) 108 cm²/s.
14. (i) $(32 + 16t - 6t^2)$ km²; (ii) $(16 - 12t)$ km²/h; 4 km²/h; -8 km²/h
    (iii) $1\frac{1}{3}$ h; (iv) $42\frac{2}{3}$ km²
15. 2 dm by 1 dm.   **16.** (i) 500; (ii) 42p.   **17.** (i) $3\frac{1}{2}$p/min decrease; (ii) 3p
18. (i) $\pounds\left(\dfrac{6912}{v} + 2v^2\right)$; (ii) 12 knots; (iii) $\pounds 864$

Ex. 10. P. 352.

Each must include the arbitrary constant c.

1. $3x$ **2.** $5x$ **3.** x^2 **4.** x^3 **5.** x^4
6. $2x^2$ **7.** $3x^2$ **8.** $2\frac{1}{2}x^2$ **9.** $3\frac{1}{2}x^2$ **10.** $2x^3$

Ex. 10. P. 352.

11. $3x^3$ **12.** $\frac{2}{3}x^3$ **13.** $1\frac{2}{3}x^3$ **14.** $2x^4$ **15.** $3x^4$

16. $\frac{3}{4}x^4$ **17.** $1\frac{1}{2}x^4$ **18.** $-x^{-1}$ **19.** $2x^{-1}$ **20.** $-\frac{1}{4}x^{-2}$

21. $x^2 + 2x$ **22.** $2x^2 - 3x$ **23.** $4x - 1\frac{1}{2}x^2$ **24.** $5x - 2\frac{1}{2}x^2$

25. $x^3 + x^2 - 4x$ **26.** $2x^3 - 1\frac{1}{2}x^2 + x$ **27.** $\frac{1}{2}x^2 - \frac{2}{3}x^3 - \frac{3}{4}x^4$

28. $\dfrac{x^3}{9} + \dfrac{x^2}{8} - \dfrac{x}{2}$ **29.** $\dfrac{x^4}{16} - \dfrac{x^3}{9} + \dfrac{x^2}{4}$ **30.** $\dfrac{x^3}{4} - \dfrac{x^5}{8}$

31. $-\dfrac{2}{x^2} + \dfrac{2}{x}$ **32.** $-\dfrac{3}{x} - \dfrac{1}{x^2}$ **33.** $-\dfrac{2}{3x} + \dfrac{3}{4x^2}$

Ex. 11. P. 353.

1. $x^3 - x^2 + x + c$ **2.** $x^4 + 2x^3 - x^2 + c$

3. $1\frac{1}{3}t^3 + 2\frac{1}{2}t^2 + 4t + c$ **4.** $ap^5 - bp^3 + 5p + c$

5. $\dfrac{at^4}{4} - \dfrac{2bt^3}{3} - \dfrac{3gt^2}{2} + c$ **6.** $\dfrac{2ax^6}{3} + \dfrac{3bx^4}{4} - \dfrac{5tx^2}{2} + c$

Ex. 12. P. 354.

1. (i) $y = x^2 - 4$; (ii) $(-2, 0)$; $(2, 0)$; (iii) $(0, -4)$

2. (i) $y = x^2 - x - 2$; (ii) $(-1, 0)$; $(2, 0)$; (iii) $(\frac{1}{2}, -2\frac{1}{4})$

3. (i) $y = x^2 - x - 6$; (ii) $(-2, 0)$; $(3, 0)$; (iii) $(\frac{1}{2}, -6\frac{1}{4})$

4. (i) $y = x^2 - 7x + 12$; (ii) $(3, 0)$; $(4, 0)$; (iii) $(3\frac{1}{2}, -\frac{1}{4})$

5. (i) $y = 2x^2 + 5x - 12$; (ii) $(-4, 0)$; $(1\frac{1}{2}, 0)$; (iii) $(-1\frac{1}{4}, -15\frac{1}{8})$

6. (i) $y = 12x^2 - 23x + 10$; (ii) $(\frac{2}{3}, 0)$; $(1\frac{1}{4}, 0)$; (iii) $(\frac{23}{24}, -1\frac{1}{48})$

7. (i) $y = 2x^3 + 3x^2 - 2x$; (ii) $(-2, 0)$; $(0, 0)$; $(\frac{1}{2}, 0)$

8. (i) $y = 2x^3 - 3x^2 - 9x$; (ii) $(-1\frac{1}{2}, 0)$; $(0, 0)$; $(3, 0)$

9. (i) $y = 2x^3 - 12x^2 + 16x$; (ii) $(0, 0)$; $(2, 0)$; $(4, 0)$

10. (i) $y = 6x^3 + 15x^2 - 75x$; (ii) $(-5, 0)$; $(0, 0)$; $(2\frac{1}{2}, 0)$

11. (i) $y = 18x^3 - 3x^2 - 6x$; (ii) $(-\frac{1}{2}, 0)$; $(0, 0)$; $(\frac{2}{3}, 0)$

12. (i) $y = 4x^3 - 4x$; (ii) $(-1, 0)$; $(0, 0)$; $(1, 0)$

13. (i) $y = 1 - \dfrac{6}{x}$; (ii) $(6, 0)$ **14.** (i) $y = \dfrac{1}{x} - \dfrac{2}{x^2}$; (ii) $(2, 0)$

15. (i) $y = \dfrac{6}{x} - \dfrac{3}{x^2}$; (ii) $(\frac{1}{2}, 0)$ **16.** (i) $y = 2x - \dfrac{6}{x} - 4$; (ii) $(-1, 0)$; $(3, 0)$

17. $y = 3x^2 - 4x + \dfrac{5}{x}$ **18.** $y = \dfrac{3}{x^2} - \dfrac{4}{x^3}$

19. $y = \dfrac{3}{2x} + \dfrac{2}{3x^2} - \dfrac{1}{x^3} + 4$ **20.** $y = x^2 - \dfrac{3}{2x^2} - 6$

Ex. 13. P. 355.

1. 18 ft/s²; 36 ft/s; 45 ft   **2.** 28 ft/s²; 64 ft/s; 96 ft

3. 20 ft/s²; 16 ft/s; 8 ft   **4.** 84 ft/s²; 195 ft/s; 300 ft

5. 2 ft/s²; 33 ft/s; 63 ft   **6.** 6 ft/s²; 30 ft/s; 36 ft

7. 32 ft/s²; 64 ft/s; 64 ft   **8.** 20 ft/s²; 100 ft/s; 250 ft

9. 10 cm/s²; 21 cm/s; 27 cm   **10.** 8 cm/s²; 26 cm/s; 40 cm

11. -32 cm/s²; -16 cm/s; 0 cm   **12.** 24 cm/s²; 0 cm/s; -36 cm

13. 42cm/s²; 20 cm/s; 4 cm   **14.** -36 cm/s²; -16 cm/s; 32 cm

15. -32 ft/s²; 2 s; 64 ft   **16.** 4 s; 128 ft/s

17. 3 s; 27 cm; $40\frac{1}{2}$ cm/s; -36 cm/s²

18. 4 s; -32 cm; 36 cm/s; 24 cm/s²

19. 108 cm/s; 10 s; 800 cm

20. 44 ft; 0 s, 4 s; 24 ft/s², -24 ft/s²

Ex. 14. P. 357.

1. 4	**2.** 8	**3.** 3	**4.** 1	**5.** 56
6. 15	**7.** 7	**8.** 10	**9.** 212	**10.** $\frac{1}{6}$
11. $-\frac{3}{4}$	**12.** $\frac{1}{3}$	**13.** $25\frac{1}{2}$	**14.** $119\frac{57}{64}$	**15.** $57\frac{1}{3}$
16. 26 cm	**17.** 4 cm	**18.** 61 ft	**19.** $-4\frac{1}{6}$ cm	

20. 48 km/h; 128 km; 32 km/h

Ex. 15. P. 359. *Square units in each case.*

1. (i) 16; (ii) 22; (iii) 32   **2.** (i) $31\frac{1}{2}$; (ii) $29\frac{1}{4}$; (iii) $60\frac{3}{4}$; $121\frac{1}{2}$
3. 15   **4.** 344   **5.** 93   **6.** $50\frac{5}{6}$ **7.** 16; pt. of inflexion
8. $\frac{3}{4}$ **9.** $-\frac{13}{36}$ **10.** $-50\frac{5}{6}$ **11.** $10\frac{2}{3}$ **12.** -32
13. 108   **14.** -125 **15.** 108   **16.** $-14\frac{7}{24}$ **17.** 8
18. 81   **19.** 128   **20.** $23\frac{2}{3}$ **21.** $10\frac{1}{2}$ **22.** $10\frac{1}{3}$
23. -36 **24.** $457\frac{1}{3}$ **25.** $57\frac{1}{2}$ **26.** $57\frac{1}{6}$ **27.** $57\frac{1}{6}$
28. $457\frac{1}{3}$ **29.** 54   **30.** $333\frac{1}{3}$

Ex. 16. P. 362. *Cubic units in each case.*

1. 144π	**2.** 312π	**3.** 48π	**4.** $9\frac{1}{4}\pi$	**5.** $6\frac{2}{3}\pi$
6. $435\frac{3}{5}\pi$	**7.** $1\frac{1}{15}\pi$	**8.** $35\frac{2}{15}\pi$	**9.** $3\frac{3}{5}\pi$	**10.** $88\frac{13}{20}\pi$
11. 18π	**12.** $1\frac{1}{3}\pi$	**13.** $36\frac{4}{7}\pi$	**14.** $\frac{2}{3}\pi$	**15.** $\frac{1}{30}\pi$
16. $34\frac{2}{15}\pi$	**17.** $1\frac{1}{15}\pi$	**18.** $10\frac{1}{3}\pi$	**19.** $19\frac{53}{105}\pi$	**20.** $26\frac{2}{3}\pi$
21. 2	**22.** 3			

TEST PAPERS

TEST A.1. P. 365.
1. 2·64 or −1·14   **2.** $y = \frac{1}{3}(a + 2b)$ **3.** (i) −6; (ii) 1·8 or −1·1
4. ±2·121   **5.** 2622   **6.** 13·9

TEST A.2. P. 365.
1. (i) $\frac{25y^5}{x^3}$; (ii) $\frac{a + b}{a - b}$ **2.** (i) 2·414 or −0·414; (ii) 2   **4.** 4·37 or −1·37
5. $r_2 = \frac{Rr_1}{r_1 - R}$; $10\frac{2}{3}$ **6.** (i) 128 in³; (ii) 2

TEST A.3. P. 366.
1. 0·73 or −2·73   **2.** $a = \frac{4xy}{6x - y}$; $-1\frac{3}{5}$ **3.** Zero
4. $x = 4$, $y = 3$ **5.** (i) 1·732; (ii) $7\frac{1}{2}$
6. $1000a + 100b + 10c + d$; $1000d + 100c + 10b + a$

TEST A.4. P. 366.
1. (i) $x = \frac{2}{7}$, $y = -\frac{1}{7}$; (ii) $(4x + 6)$ cm; $(x^2 + 3x)$ cm²; 3
2. (i) ±12; (ii) $a = 3$, $b = -5$ **3.** (i) 2·192; (ii) $x = 5$, $y = 3\cdot5$
4. (i) ±2; (ii) $y = \frac{\sqrt[3]{x^2}}{100}$ **5.** (i) $\frac{1}{2}$; (ii) 786·1

TEST A.5. P. 367.

1. (i) $b = \dfrac{2ac}{c-a}$; (ii) $x = 2, y = 3\frac{1}{2}$ **2.** 8

3. (i) 0·33 or −2·18; (ii) 16, 31   **4.** 1·51 or 3·28
5. $m = 1\cdot5, c = 1\cdot6021$; $T = 40p^{1\frac{1}{2}}$

TEST A.6. P. 368.

1. (i) 4; 1; $\frac{1}{36}$; $4\frac{17}{27}$; (ii) $\dfrac{y}{x^{12}}$; (iii) 0·897
2. (i) 27; (ii) 1·39 or −2·89   **3.** 30 km/h
4. (i) 340; (ii) $a = \frac{1}{6}, r = 3$; or $a = -\frac{1}{8}, r = -4$
5. 1250 m²; 36·2 m by 27·6 m, or 13·8 m by 72·4 m

TEST B.1. P. 369.

1. (i) 2401 fr.; (ii) $4\frac{1}{4}\,ac$; (iii) 102   **2.** (i) £632·19; (ii) £3500
3. £4 892 659·85; 80p   **4.** 3·87 lb

TEST B.2. P. 369.

1. (i) 10·9 miles; (ii) 0·0125 tons   **2.** (i) 44p; (ii) $2\frac{2}{3}\%$
3. (i) 95 cm²; (ii) 39 cm   **4.** £605·50; £2470
5. 443 lb; 15 in

TEST B.3. P. 370.

1. (i) 42 mile/h; (ii) £10·85   **2.** (i) $26\frac{1}{2}$p; (ii) £600
3. 700 m by 400 m; 28 ha   **4.** Increase 90p
5. (i) £5·50; (ii) 52%

TEST B.4. P. 371.

1. (i) 126; (ii) Decrease 2·35%   **2.** 21 min
3. 31p   **4.** 61 in³
5. (i) 34·9 ft²; (ii) 13·5%   **6.** (i) 35%; (ii) £1·87$\frac{1}{2}$

TEST B.5. P. 371.

1. (i) 1·75 kg; (ii) (a) 8·8 oz, (b) 28·4 g **2.** (i) 2 in; (ii) 15 tons
3. (i) $17\frac{1}{2}\%$; (ii) £270·33   **4.** (i) £25·87$\frac{1}{2}$; (ii) 2·78%
5. 3·17 ft; 1·31 lb

TEST B.6. P. 372.

1. £160; 3·5%; £176   **2.** 7·94 cm   **3.** (i) £51·42; (ii) £79·45
4. (i) 82%; (ii) 19·3 cm   **5.** 1630 gal

TEST C.1. P. 373.

1. 30°   **3.** (i) 142°; (ii) 109°; (iii) $35\frac{1}{2}$°
4. (ii) $90 - x$; x; $2x$ **5.** (i) 16° 6′; (ii) 4 cm

TEST C.2. P. 373.

1. (i) 10; (ii) 137°; (iv) ⊥ bisector of AB **4.** (i) 5·92 cm; (ii) 56°; 41°

TEST C.3. P. 374.

2. 1 cm or 7 cm; $8\frac{3}{4}$ cm²   **3.** 12 cm
4. $PB^2 = x^2 + 4x + 49$; $PC^2 = x^2 - 4x + 49$; 7; 17·1 or 2·86
5. 13 cm.

Test 3.4. P. 375.

5. Area of trapezium $= h\left(\dfrac{a+b}{2}\right)$ units^2

Test D.1. P. 376.

1. (i) 539 m; 28° 12′; (ii) 163 dm   2. 8·66 dm; 60° 3′; 15·6 dm
3. 299 m; 3·42 ha

Test D.2. P. 376.

1. (i) $\frac{21}{29}$; (ii) 90°; $\dfrac{\pi}{2}$ radians   2. (i) 3799 miles; (ii) 14 270 miles
3. 2176 dm   4. (i) N. 44° 57′ W.; (ii) 7·314 miles; (iii) N. 55° 33′ W.

Test D.3. P. 377.

1. (i) $\frac{12}{5}$; $\frac{12}{13}$; (ii) $\frac{1}{8}$; (iii) 5   2. (i) 20° 44′; (ii) 2·48 cm
3. (i) 49·27 miles; 027° 59′; 60·06 miles
4. (i) 2546 miles; (ii) 971·4 miles; (iii) 14° 6′

Test D.4. P. 378.

1. (i) 10·7 cm, 13·5 cm; (ii) (a) 27·4 cm, (b) 46 cm^2
2. 194 dm   3. (i) 166·3 m; (ii) 27·8 m; (iii) 23° 11′
4. (i) 280 km; (ii) S. 64° 39′ E.; 96·5 km
5. (i) 1740 dm; (ii) 1080 dm; (iii) 34° 57′

Test D.5. P. 379.

1. (i) 3172 miles; (ii) 3918 miles   2. 2·98 km, 044° 21′; 10 min
3. 1·294 mile/h, 291° 29′   4. 30·3 m

Test E.1. P. 379.

1. (i) $6x^2 + 9 - \dfrac{4}{x^2}$; (ii) max. (−1, 25); min. (3, −39)
2. (i) $1\frac{1}{2}$ s; (ii) 9 cm; (iii) 5 s   3. 57 units^2; $3418\frac{1}{5}\pi$ units^3
4. (i) 7 ft/s; (ii) $3\frac{1}{2}$ s; (iii) 18·7 ft; (iv) $1\frac{1}{4}$ s.

Test E.2. P. 380.

1. (i) 48 cm/s; (ii) 256 cm   2. 8 units^2
3. (i) max. 0; min. −4; (ii) $\dfrac{68\pi}{35}$ units^3 4. 3; $1\frac{1}{2}$; $\dfrac{8\pi}{7}$ units^3

Test E.3. P. 381.

1. (i) $3 + \dfrac{2}{3\sqrt[3]{x^2}} + \dfrac{4}{x^3}$; (ii) $\dfrac{4x^5}{5} - \dfrac{\sqrt{x^3}}{3} - \dfrac{3}{x^2} + c$; (iii) $\frac{1}{3}$ in^2; £720
2. (i) (a) 27 cm, (b) 3 cm/s, (c) 3 s
   (ii) max. $x = -1$, $y = 20$; min. $x = 3$, $y = -12$
3. (i) $21x^2 + \dfrac{6}{x^3} - \dfrac{1}{\sqrt[4]{x^3}}$; (ii) $2x^5 + \dfrac{6}{x} - \dfrac{9\sqrt[3]{x^2}}{2} + c$;
   (iii) $24\frac{3}{4}$ units^2 (negative)   4. $x = 140$; 30 800 m^2

Test E.4. P. 382.

1. $a = 3$, curve meets x-axis at (0, 0) and (3, 0) T.P. (0, 0) and (2, 4).
2. (i) $\dfrac{x^3}{3} + 3x + c$; (ii) 18; (iii) $5\frac{1}{2}$ 3. (i) 40π units^3; (ii) 2; $\pm\frac{1}{3}$
4. (i) 6 cm/s; (ii) 8 cm; (iii) $\sqrt{12}$ s; (iv) 12 cm/s
5. (i) (0, 3); (2, 3); $1\frac{1}{3}$ units^2; (ii) 48 π units^3